Essentials of Marine Biotechnology

Se-Kwon Kim

Essentials of Marine Biotechnology

 Springer

Se-Kwon Kim
Department of Marine Life Science
College of Ocean Science and Technology
Korea Maritime and Ocean University
Busan, South Korea

ISBN 978-3-030-20946-9   ISBN 978-3-030-20944-5   (eBook)
https://doi.org/10.1007/978-3-030-20944-5

This Springer imprint is published by the registered company Springer Nature Switzerland AG
The registered company address is: Gewerbestrasse 11, 6330 Cham, Switzerland

Preface

Seventy percent of the earth's surface is covered by sea, and about 80% of all animal species dwells in the sea. The oceans are where life first began; in the three-billion-year history of the evolution of life, it was only in the range of hundreds of millions years ago when life forms first arrived from the sea to land. Those organisms that arrived on terrestrial represented just a small fraction of all marine creatures, and innumerable others continue evolving independently in the sea to this day. It is therefore explicit that marine organisms are living in the extreme condition and produce unique substances that cannot be produced by terrestrial-based organisms, which can perform multitude of functions for human benefits. Furthermore, sea bearing high-pressure environment, averaging about 380 atm, the marine organisms undoubtedly should possess distinctive mechanisms to withstanding this. With average temperatures of 1–4 °C, seawater is also an optimal environment to inhabit psychrophilic (cold-loving) organisms. Interestingly, the sea floor is also home to organisms with enhanced ability to withstand high temperatures, suggesting even further applications. At the bottom of the seas are hydrothermal deposits where water is discharged at temperatures of over 300 °C.

Considering the existence of such extreme environment, it is well understood that marine biomaterials can be exploited for the benefits of human cause in many ways. These marine organisms hold truly great potential for future applications in marine biotechnology. Marine organisms remain integral to human lives either as an alternative source of protein, natural products, or as potential biomaterial source for industrial purpose. In recent years, rapid advancements in scientific application of marine biotechnology have been witnessed, particularly in the area of natural product development, marine bioenergy, seaweed biotechnology, aquaculture, genomics, cosmeceutical, nutraceuticals, pharmaceutical development, transgenic technology, genetic recombination, cell fusion, and ultrafiltration membranes.

For all various aforementioned reasons, hopes for the future of marine biotechnology and their utilization for the product development for human mankind are increasing ever since the last three decades. However, for the complete development of marine biotechnology, it is important to fully utilize modern-day techniques such as satellites (for positioning of vessels and identifying the area of interest of ocean), advanced robotics system, artificial intelligence, to efficiently harvest and maintain marine organisms, along with cutting-edge techniques for deep sea exploration.

The book comprises eleven chapters followed by references and comprehensive list of glossary.

Chapter 1 introduces the present, past, and future of marine biotechnology. In this context, some are now predicting that the twenty-first century will more specifically be called the 'marine bio era.' A major factor in this change is the ever-expanding and understanding of previously less-known marine organisms, along with the rapid development of industry-related biological technology.

In Chap. 2, the basic concepts of genetics have been elaborated to comprehend the various aspects of marine biotechnology. It is very much essential to introduce the basics of various aspects of genetics and other related areas to assist in understanding the marine biotechnology. Furthermore, this chapter will also be beneficial for those students who have not taken biology as one of the subjects in their course curriculum.

The recent development of techniques for analyzing large base sequences has led to efforts to decode the genomes of economically and academically important fish varieties. In the fishing industry, decoding of fish genomes is extremely important in three regards: for understanding biological phenomena, use of genome information in selective breeding, and efficient classification of biological resources.

In this regard, Chap. 3 attempts to bring forth the basic understanding of fish genetics, the knowledge of which can be used to restore marine ecosystems by preserving various endangered marine species. Around the world, recent genomic research has been conducted on fish varieties with high economic or industrial value and applied in their breeding or industry use.

Efforts to produce superior, high value-added varieties, using genetic engineering to maximize the productivity over a short time, are underway worldwide. Much research effort has been focused on fish, which have the highest economic value among marine products. Additionally, the introduction of foreign genes from various farmed fish varieties has recently been observed in laboratories in various advanced economies, with investigations underway on gene expression and transmission to future generations. Hence, in the subsequent Chap. 4, the elementary information of fish breeding and along with the application of marine biotechnology has been elaborated in order to maximize the productivity efforts.

Chapter 5 attempts to give an overview of genetic diversity prevailing in marine bio-system along with DNA markers in fish have been elaborated that can play important role in marine farming research and other marine industries involved in sea products. Like agriculture, fish farming subjects organisms to artificial management. In this chapter, emphasis on future genetic improvements to meet ever-increasing demands of marine biomaterials has been discussed that will revolutionize the marine farming research and industry. Furthermore, in this chapter we explained the use of marker in selective marine organism breeding, along with an overview of its method and the current status of research of the field. An account on achievements already made in other organisms has also been elaborated.

Seaweeds are emerging as a future resource with various potential applications for humankind, including food, production of functional substances, and a source for energy and paper-making. Chapter 6 endeavors to fully address the basic culture techniques for seaweed production and their applications. The field of seaweed biotechnology includes a number of different areas, including tissue culture, callus induction, protoplast production and regeneration, cell fusion, and gene manipulation. At the same time, a number of issues remain to be solved, including the failure to achieve full consistency in terminology. Nevertheless, the technology is very important in opening up new possibilities for the use of seaweed not only as a food resource but also for mass production of useful substances or for bioactive substances (cosmeceutical, nutraceuticals, and pharmaceuticals) and biotechnological resources, and greater development is expected for the field going ahead.

Chapter 7 discusses microalgae, a major biological resource for future, particularly in the area of food, cosmetics, energy production, biofertilizer, bioactive substances, and in wastewater treatment. Microalgae, a primary producer in aquatic environments, encompasses a vast biological resource in terms of both volume and variety, with tens of thousands of species producing more than 20 billion tons of organic matter per year. In this chapter, we discussed culture methods of microalgae for mass production; microalgae typically grow far more quickly than land-based plants and can be easily cultured in both freshwater and seawater, or in any environment with light energy. They possess great potential as a material in bio-industry, allowing for low-cost production of industrially useful high-molecular-weight substances such as proteins, fats, sugars, and pigments as well as substances with specific physiological functions.

The development of new and useful materials from marine organisms is a very important field in both academic and industrial terms. Hence, Chap. 8 attempts to address this issue by particular focusing on various aspects of the development of functional materials using marine organisms. This chapter will examine the potential of biomaterial derived from marine organism derived as antibacterial, anti-inflammatory, and anticancer substances that have already garnered attention of pharmaceutical industries. Production and application of mussel adhesive protein are discussed in this chapter. Also, this chapter focusses on various active research efforts currently underway for deriving marine biomaterials for its effectual use in cosmetics, pharmaceuticals, as well as functional foods. Production and applications of marine polysaccharides (chitin, chitosan, chitooligosaccharides, alginate, etc.) in cosmeceuticals, wound healing, and artificial skin development are presented well. A brief explanation has been given the production and pharmacological applications of fish fatty acids such as eicosapentaenoic acid (EPA) and docosahexaenoic acid (DHA).

With the recent trend of consistently high oil prices and the climate change agreement taking effect, energy has emerged as a serious societal issue. As a solution to this problem, research into developing and commercializing bioenergy-related technology for the production of clean energy and new/renewable energies using biomass is taking place actively around the world. Therefore, in Chap. 9 we attempt to explain the potential use of marine materials for bioenergy production. In particular, macroalgae and microalgae can be the alternative sources of biofuels (ethanol, ethane, methane, and other chemical products). Under stable conditions, energy production efficiency is reportedly around ten times higher as compared to land plants. Furthermore, microalgae can be harvested throughout the year and are well suited to automation due to their ease of farming and harvesting. Today, microalgae are drawing global attention as a final means of producing next-generation biofuels. In addition to this, the important key issues are also discussed for the bioenergy production in industrial scale.

Finally, in Chaps. 10 and 11, a comprehensive account on bioactive marine materials has been explained which can be of potential application to treat various human sufferings. In this chapter, bibliographic studies, collection, and surface morphology of marine organism (seaweed, fish, plankton, microorganisms, etc.) are presented well. Detail explanation has been given on isolation techniques for natural product isolation from the marine organisms including analytical techniques (NMR, mass spectra, UV, etc.), chromatographic techniques, separation and fractionation as well as purification techniques. Marine flora and fauna possess outstanding capabilities in producing high-molecular-weight substances and enzymes. Additionally, some species are used as potential producers of enzymes such as deoxyribonuclease, lipase, alginate lyase, protease, agarase, cellulase, and esterase. Numerous bioactive substances have been already derived from marine microorganisms, including those with antimicrobial, antifungal, antiviral, antitumor, anti-inflammatory, antioxidant, and enzyme inhibitors. Even though, mass production of source materials is an issue to that need to be addressed suitably; nevertheless, the beneficial activities from marine microorganisms are enormous.

Nutritional properties of fish, shellfish, algae, and marine microorganisms are generally well known. However, their functional characteristics have not been fully revealed. It is believed that they contain biologically active compounds, including potential nutraceuticals. For example, marine macroorganisms produce a vast array of secondary metabolites including terpenes, steroids, polyketides, peptides, alkaloids, porphyrins, and polysaccharides. These secondary metabolites serve as several pharmaceutical usages (antitumor, anti-inflammation, anti-allergy, antioxidant, antifungal, anti-HIV, and antihypertensive). However, the development of a new drug requires sufficient amounts of pure compounds that exceed by large quantities, but it is extremely difficult to collect them in higher amounts from a marine resource. Moreover, with the respect to investigation and development of marine bioactive substances for industry applications, many studies have been conducted to develop marine biotechnologies, such as membrane bioreactor, bioconversion, and continuous mass producing process technology. The biotransformation technology consisting of membrane bioreactor-assisted bioconversion and continuous mass production made significant contributions to the commercial development of marine nutraceutical and biomedical substance. Even though several biotechnological method improvements have been achieved for the production of commercial materials using marine bioresources, maintaining the raw materials of marine bioresources is a serious warning for marine biotechnology industries. Several difficulties are still existing for the production of larger quantity of marine biomass.

Above all, the biggest problem is that not all the marine biomass resources are produced in huge quantities while industrialization is commonly required huge amount of raw materials for the production commercial products which also should be consistent and availability throughout the year. In order to solve these problems, advanced sea farming technologies are necessary, but there has not been sufficient research and development investment for sea farming industries due to lack of awareness. In addition to this, it is also unfortunate to secure workers for considering the risk of working at sea. With these difficulties, it will not be easy to cultivate marine biotechnology industries, unless investment to be supported for future production of marine biomass. The solution for these issues, new modern technologies such as artificial intelligence robots, drones, submersibles, and automated raw material harvesting vessels in farming industries should be introduced instead of manpower. By including all the above-mentioned technologies, high value-added marine biotechnology industry will be expected to succeed as a world power industry as a Fourth Industrial Revolution.

I am fully aware of the fact that I cannot completely satisfy the interest of various scientists working in this area, but I hope this book will certainly bring forward new avenues in this ever-growing field. In the future, I intend to fully address any inadequacies which may have left inadvertently, and welcome all suggestions that can be included in coming editions. I am thankful to Dr. Venkatesan J, Research Professor, and Dr. Sandeep Kumar Singh, Postdoctoral fellow, who have extended their helping hands to complete this book.

Busan, South Korea   Prof. Se-Kwon Kim
                                                                                            Distinguished Professor

Acknowledgements

Dr. Yoon Dong-Han, CEO of Kolmar Korea Co., Ltd., established Kolmar Korea in 1990 and introduced a new business model 'ODM Network' to supply cosmetics, pharmaceuticals, and health functional foods.

In recent years, he recognized the value of under-utilized marine resources and felt the need to develop products through researches that can utilize them more effectively. He invited me as Research Adviser and dedicated himself to publish technical books emphasizing the scientific value of marine life as well as related research.

With his help, this book has been published, providing many readers with the knowledge of how scientific values of marine life can enhance human health.

I would like to thank him sincerely for publishing this book.

Busan, South Korea

Prof. Se-Kwon Kim
Distinguished Professor

Contents

About the Author

Prof. Se-Kwon Kim, Ph.D. is presently working as Distinguished Professor in Korea Maritime and Ocean University and Research advisor of Kolmar Korea Company. He worked as Distinguished Professor at Department of Marine Bio Convergence Science and Technology and Director of Marine Bioprocess Research Center (MBPRC) at Pukyong National University, Busan, South Korea.

He received his M.Sc. and Ph.D. degrees from Pukyong National University and conducted his postdoctoral studies at the Laboratory of Biochemical Engineering, University of Illinois, Urbana-Champaign, Illinois, USA. Later, he became a visiting scientist at the Memorial University of Newfoundland and University of British Columbia in Canada.

He served as President of the 'Korean Society of Chitin and Chitosan' in 1986–1990 and the 'Korean Society of Marine Biotechnology' in 2006–2007. To the credit for his research, he won the best paper award from the American Oil Chemists' Society in 2002. He was also Chairman for '7th Asia-Pacific Chitin and Chitosan Symposium,' which was held in South Korea in 2006. He was Chief Editor in the 'Korean Society of Fisheries and Aquatic Science' during 2008–2009. In addition, he is Board Member of the International Society of Marine Biotechnology Associations (IMBA) and International Society of Nutraceuticals and Functional Food (ISNFF).

His major research interests are investigation and development of bioactive substances from marine resources. His immense experience of marine bio-processing and mass production technologies for marine bio-industry is the key asset of holding majorly funded marine bio projects in Korea. Furthermore, he expended his research fields up to the development of bioactive materials from marine organisms for their applications in oriental medicine, cosmeceuticals, and nutraceuticals. To this date, he has authored around 650 research papers, 70 books, and 120 patents.

What Is Marine Biotechnology?

Contents

1.1 Marine Biotechnology: Definition and Scope

We often see terms like "bio" and "biotech" in the newspapers and on TV today. This is obviously an abbreviation for "biotechnology," a compound of *bio*, meaning "life" or "organism," and "technology."

Biotechnology does not necessarily have any set definition, but it is typically understood to mean technology that harnesses or imitates organisms or their functions to produce useful materials. As such, it encompasses techniques for improving organisms themselves—such as animals, plants, and microbes—as well as those for propagating those life forms. In simpler terms, it is technology that *allows organisms' functions to be put to more effective use*. Biotechnology is at work in the microbe-using brewing methods used over the generations in Korea to produce such foods as fermented soybeans, soy sauce, liquor, and vinegar. Recent biotechnology is based on recent knowledge from the field of biology, which underwent rapid development after World War II, as well as experimental technology in general and techniques for genetic manipulation—first used in the 1970s —in particular.

© Springer Nature Switzerland AG 2019
S.-K. Kim, *Essentials of Marine Biotechnology*,
https://doi.org/10.1007/978-3-030-20944-5_1

As a subfield of biotechnology, marine biotechnology has received a great deal of attention in recent years (Kim and Venkatesan 2015). Because its focus is on resources and new materials originating from the sea, its possibilities are arguably more open and unlimited than in other fields where resources require money.

In terms of its definition and general meaning, marine biotechnology refers to a field or industry that studies marine organisms and their components, systems, and biological functions, with the ultimate goal of providing products and services to serve human welfare (Jha and Zi-Rong 2004).

This in turns requires the development of a wide range of basic technologies for preserving and using marine organisms and manipulating their functions. Because of its multidisciplinary nature, biotechnology is rooted in a wide range of traditional marine sciences (including marine biology, marine chemistry, and fishery sciences), while entailing additional biological studies in areas such as molecular biology, immunology, biochemistry, pharmacology, and biotechnology. A number of cutting-edge methods such as genomics, proteomics, metabolomics, and bioinformatics have also been put to use recently in the field of marine biotechnology.

Covering 70.8% of the earth's total surface area, the oceans account for 30% of primary or total production. Yet only a very small fraction of that production is used by humans. By further developing marine bio-resources and biotechnology, we may be able to promote usage of yet unexplored marine resources.

1.2 Biotechnology, Past and Present

A large number of biological and pharmaceutical materials that are highly effective and quite safe to the human body have been developed from natural substances. For the most part, the focus has been on land-based animal and plant life, fungi, or microbes such as bacteria. As a result, the development of new materials and substances using land-based bio-resources has reached a limit in terms of dwindling numbers of subjects. This in turn has led to a recent trend in advanced economies and coastal countries of gradually shifting their bio-material development focus from land-based to marine bio-resources.

Marine organisms inhabit a completely different environment from land-based ones, and differ in many ways in terms of their physiological and metabolic processes. For this reason, it is believed that the metabolites that they produce may yield an abundance of new materials. Marine bio-resources have drawn particular attention as an alternative to land-based ones thanks to their great variety and the knowledge that around 80% of all organisms on Earth—some 300,000 species— dwell in the sea. Because these marine organisms yield only very small quantities of metabolites, it remains difficult to ascertain their biological activities or structure. Recent advancements in structural analysis techniques, however, have led to the development of a wide range of bio-assay methods, resulting in a growing understanding of the functionality of marine organism-derived active components.

Accordingly, marine organisms have progressed beyond use as a mere food resource, allowing for the development of a marine organism industry with potentially enormous value added.

The history of marine biotechnology is not a long one. It was with the discovery of nucleic acid material containing the unusual sugar arabinose in the Caribbean sponge *Tethya crypta* in the 1950s that the buried potential of marine organisms first came to notice (Carroll and Crews 2009). (Ara-A and Ara-C are used clinically even today as anti-cancer and anti-viral treatments.) In 1967, a symposium titled "Drugs from the Sea" was organized for the first time in the United States, calling attention to the development of pharmaceuticals derived from marine organisms and leading to the establishment of marine natural science as a research field (Fusetani 2000).

In addition to this marine natural science research, research in marine biotechnology as a more expanded field took place in the 1980s. The true emerge of the term "marine biotechnology" came with the 1989 organization of the first International Marine Biotechnology Conference (IMBC) in Japan.

A model of success in marine biotechnology research and development came with the U.S. Sea Grant program, which has a history of more than half a century since its beginnings in 1965. It was a research, education, and training network program organized by the National Oceanic and Atmospheric Administration (NOAA) to develop and preserve the coastal ecosystem, with around 32 coastally located universities participating. Today it is seen as having generated a substantial return on its rather small investment. Ultimately, its major research focus was on applying to marine organisms the modern biotechnology methods found in molecular biology and genetic engineering.

Since 2000, South Korea has operated its own Sea Grant program along the same lines. University consortium-centered, network-style Sea Grant project teams were established according to local characteristics, with the goal of healthy and sustainable use of the oceans. A marine biotechnology project attempted by the Ministry of Maritime Affairs and Fisheries in 2004 was the basis for the first real development of South Korean marine biotechnology.

By examining the marine biotechnology research findings from the U.S. Sea Grant project and marine biotechnology strategies suggested in Europe, it may be possible to answer the question of how to conduct efficient research in the field with limited funds and resources going ahead, while providing a reference on how and in what fields marine organisms might be applied in biotechnology under South Korea's current practical conditions (Tables 1.1 and 1.2).

Marine biotechnology can be classified into four types: marine organism source technology, marine food resource development technology, marine new materials development technology, and marine ecosystem and environment preservation technology.

Marine biotechnology is thus a future-oriented, knowledge-based industry in which the world's major corporations are making a priority of investing, as well as an environmentally friendly, energy-conserving industry that is optimal for

Table 1.1 Categories of marine biotechnology (2007 biotechnology white paper, Ministry of Science and Technology)

	Category	Content
Marine biotechnology	Marine organism source technology	Foundational technology for marine organism resource management and use Technology for bio uses of marine organisms Technology for identifying vital phenomena and physiological functions of marine life Technology for discovering and identifying marine genes Technology for omic analysis of marine organisms
	Marine food resource development technology	Technology for breeding and development of new species of marine organisms Technology for controlling and monitoring diseases Technology for advanced farming and mass production Technology for bio-safety assessment
	Marine new materials development technology	Technology for development new industry materials Technology for developing new pharmaceutical materials Technology for developing new functional food materials Development of renewable bio-energy
	Marine ecosystem and environment preservation technology	Technology to ensure biodiversity Technology to monitor and predict environmental changes Technology to control and remove marine pollution

advancement of industry structure. Developing it into a cutting-edge technology field and strategic 21st century industry with high future advancement potential will require efforts to court long-term, focused projects and formulate related policies.

1.3   Marine Bioindustry Today and Its Future Prospects

The rapid progression of climate change and the issue of reducing carbon dioxide have resulted in growing attention to the oceans in recent years and a trend toward seeking to understand their role. In particular, facts are coming to light about the vital role marine microbes and organisms in general play in the Earth's oxygen cycle. Previously, only around one percent of marine microbes could be cultured with existing technology, while the 99% remained outside of our understanding.

Table 1.2 Largest oil spills in history

Spill/tanker	Locations	Date	Tonnes of crude oil (thousands)
Kuwait oil fires	Kuwait	1991	136,000
Kuwaiti oil lakes	Kuwait	1991	3409–6818
Lakeview gusher	United States, Kern County, California	1910	1200
Gulf war oil spill	Kuwait, Iraq, and the Persian Gulf	1991	818–1091
Deepwater horizon	United States, Gulf of Mexico	2010	560–585
Ixtoc i	Mexico, Gulf of Mexico	1980	454–480
Atlantic empress/aegean captain	Trinidad and Tobago	1979	287
Fergana valley	Uzbekistan	1992	285
Nowruz field platform	Iran, Persian Gulf	1983	260
Abt summer	Angola	1991	260
Castillo de bellver	South Africa, Saldanha Bay	1983	252
Amoco cadiz	France, Brittany	1978	223

https://en.wikipedia.org/wiki/Oil_spill

Thanks to rapid advancements in marine microbe genome technology, new attempts are now being made to understand the organisms and apply them in industry (Medini et al. 2005).

Many often speak of the 21st century as the "bio-era." The potential applications of biologically based technology are great indeed, spanning pharmaceutical, materials, the environment, energy, and most of the roles necessary for human survival. So important is the role of marine organisms becoming in every field of biotechnology application that some are now predicting the 21st century will more specifically be called the "marine bio era." A major factor in this change is the expanding understanding of previously unfathomable marine organisms and the rapid development of industry-related biological technology.

The world marine bioindustry market amounts today to around US$3 billion annually, but it has grown by multiples each year. At one-in-6000, the success rate for product development from marine materials is more than twice as high as the one-in-13,000 rate for land-based biological materials, suggesting that research is also high in efficiency. In recent years, advanced economies have invested heavily in the production of chemicals, new materials, and energy products using biological resources. In South Korea, however, both R&D and industry development in the field have failed to occur due to a shortage of biological resources (Lee et al. 2009).

The solution to South Korea's growing resource and energy shortage issue thus lies in the sea. With an area four times greater than land masses, it is a potential course for massive amounts of marine organism resources, and the development of technology for the efficient use of these resources will be a pivotal turning point toward the marine biotech era (Kim and Lee 2015).

One of the main reasons for predictions of a "marine biotech era" is the expectation that new technologies using marine technology will supplant or transform the existing chemical industry. Today's chemical industry uses petroleum to produce countless different kinds of chemical products and transportation fuels. A number of situational changes, including exhaustion of petroleum resources and regulations on CO_2 emissions, are now necessitating innovations in that industry. By combining the globally based chemical industry with the marine biotechnology that is experiencing rapid growth independently in South Korea, it may be possible to create a globally competitive marine bioindustry.

The single largest field in marine bioindustry is the algae industry. Historically, humans have a long tradition of using algae in various ways for food and industry. In Asia, algae have long been used for food alone. In the West, algae have been used to produce valuable chemical materials, which have been the subject of a large amount of research. Polysaccharides, which are important constituents of algae, have been used as materials for food, cosmetics, medicines, and material engineering. Recently, attention has focused on multifunctional oligosaccharide materials possessing biological adjustment functions. The oceans contain a vast variety of algae containing somewhat different bioactive materials from land-based organisms, making them a veritable treasure trove of bioactive substances containing new polysaccharides. For this reason, many researchers are vigorously studying algae in the hopes of obtaining new bioactive substances (Brown et al. 1997; Jiao et al. 2011; Verma et al. 2010; Villa-Carvajal et al. 2014; Wijesekara et al. 2011).

Polysaccharides derived from algae have long been used in daily life by humans, including agar, alginic acid, carrageenan, and fucoidan. The value added from these biopolymers can vary; with agar for food purposes it amounts to US$1.50 per kg, while high-quality agarose for cataphoresis sells for US$100–200 per kg. Around 30,000 ton of alginic acid are produced commercial each year, and while sales prices for ordinary commercial products such as dye fixative, medication additives, and coagulants ranges between US$5 and US$20 per kg, high-purity pharmaceutical-grade alginic acid for use in immune boosters and cell fixation has extreme high value added in the range of US$40,000 per kg. In this way, algae-based polysaccharides are receiving great attention as a resource for generating high value added, producing polymeric materials, and contributing to qualitative improvements in human life (Borowitzka 2013).

Steady advancements in research and new technologies using algal polysaccharides have been made around the world, and the resulting industry market is expected to grow. In South Korea, industry development around algal polysaccharides has been lacking, and a research base and industry infrastructure on part with the advanced economies are sorely required.

Trends in new marine materials industries and research today suggest that the industry possesses high potential for continued development. In addition to marine organism-based polymers, algal materials used for cars, algal fibers, and nano-composite materials, there is a nearly endless range of fields that can be developed through the use of new marine-derived materials. In South Korea, however, research into new marine organism-based materials has been lacking. Fundamentally, a response to the growing market for new marine materials will necessitate the possession of new marine materials and the development of technology. By systematically and scientifically uncovering sources for new material materials, increasing yields, and increasing the distribution of sources for a new marine materials industry through establishment of the source technology for their use, it will be possible both to promote activity in existing areas in the new materials and chemical industries and create new value.

Recently, the depletion of petroleum and coal energy around the world has led to major activity in new and renewable energy development and research. As of 2016, bio-energy accounted for 11% of all new and renewable energy. The first research on bio-energy as a means of reducing CO_2 and providing an environmentally friendly energy course typically focused on land-based resources. As the limits of those resources became apparent, more and more studies and policies were devised around the world to apply the same approaches to the marine industry and obtain fuels from marine resources. In particular, substantial investment has been channels into producing bio-fuels from algae and microalgae. Indeed, marine bio-fuels are seen as the industry with the greatest growth potential in the marine bio-industry sector (Notoya 2010).

Perceptions of the sea have been changing by the day, particularly in the major coastal economies. The accelerating depletion of land-based energy and other resources—most used in petroleum-based industry areas—has only increased the importance of as yet unused resources in general, and marine resources in particular. As of 2016, the world population stood at around seven billion; by 2050, it is predicted to reach 9.4 billion. This and the rapid growth of the world's developing economies suggest that humankind's resource and energy needs are poised to grow exponentially. Prices for crude oil and other land-based resources have risen sharply since the 2000s, suggesting that the predictions are already becoming a reality. Societal and industrial demand for alternative resource development has risen greatly under the circumstances, and continued efforts are underway to find a solution to meet those needs in as yet unexplored marine organism resources (Lal 2006).

The biological characteristics of those resources necessitate new techniques that differ from those used with resources from land-based organisms. It is a time when the role of scientists working in fields such as marine science, biochemistry, genetics, and bioengineering is crucially important for the development of marine biotechnology.

1.4 Marine Life Sciences: The Future Is in Sight

In March 1999, *TIME* predicted that biotechnology fused with information technology would decide the survival of humankind in the 21st century. It is a dream
that is now being sought by over 1000 biotechnology venture businesses in locations such as "Biotech Bay" and "Biotech Beach" near the U.S.'s Silicon Valley
(Salazar et al. 2003). The U.S. has selected three areas as major 21st century
scientific technologies receiving focused investment: information technology,
biotechnology, and microtechnology. Among these areas, several of the world's top
research institutes—including the National Cancer Institute (NCI) and Scripps
Research Institute in the U.S.—are engaged in active research to apply
bio-engineering, and marine biotechnology in particular, to the development of
naturally derived future pharmaceuticals. The NCI has commissioned research
institutions around the world to collect and examine large numbers of marine
organisms, separating out the active substances and determining their structure to
examine the relationship between that structure and the activity. In addition, it has
engaged—on its own and with outside institutions—in a series of studies to develop
anti-cancer agents through activity analysis and pre-clinical and clinical testing, for
which it has committed large amounts of financial investment.

In Japan, the Japanese Society for Marine Biotechnology was established in
1988 under the national government and an industry-academia-research consor-
tium. The following year saw the first International Marine Biotechnology Conference in Tokyo. A first marine biotechnology research paper conference was held
in Japan in 1992, followed by the launch of the *Journal of Marine Biotechnology*.
In this way, the country has not only furnished opportunities for researchers to
present their work, but established itself as a current world leader in the field. One
of the reasons for Japan's interest in the marine biotechnology is field lies in its own
characteristics as a country. An island nation, Japan covers an area amounting to
just 0.25% of the world's land mass, most of which is mountainous terrain
unsuitable for farming. Because of its long coastline and an economic zone
extending for 200 nautical miles, however, it gains an area equivalent to nearly half
the U.S.'s. This explains why Japan is committing more efforts than the other
advanced economies to making maximum use of the seas.

South Korea is mountainous peninsula and covered by water. It has more
mountainous regions than Japan and lacks natural resources of its own, but it is also
surrounded by seas on three sides and possesses an abundance of marine organism
resources. Yet an adequate understanding of the marine field has yet to take shape,
the human resources needed to study them have not been established, and no bold
investments are being made, resulting in its being one of the most underdeveloped
fields of all today. We should therefore consider why it is that the world's advanced
economies are viewing marine biotechnology with such interest.

The oceans carry unlimited potential, but their study cannot be limited solely to examining the waters and the life in them as people generally believe. As much as possible, it is important for research to attempt to use marine organisms in a simple way that generates high amounts of value added. First, it is essential to understand the basic mechanisms and physiological characteristics of marine organisms, acquiring an ample store of information that can then be comprehensively until it reaches the realm of marine biotechnology. In other words, marine biotechnology is field of research toward gaining greater value from the use of marine organisms. When we eat fish caught in the ocean, that is one form of using marine life. But marine biotechnology is not merely about such simple uses. As a natural science, biotechnology offers many ideas for application in daily life. When we raise fish in a single location where it can be caught using a single method, we are hoping for large fish rather than small ones. Research on methods to make them grow quickly, devices to take advantage of the fish's habits to make them easier to catch, and efforts to prevent them from destroying the ocean environment all fall into this category. In other words, it would be more appropriate to regard the many different forms of technology inspired by this idea as falling in the domain of marine biotechnology.

On the whole, marine biotechnology research has a shorter history than studies of land-based organisms, and many things that we cannot yet imagine are believed to exist. We are unlikely to efficiently explore the as yet unknown possibilities simply by grabbing a shovel and diving into the ocean in search of them. We must have a sense that allows us to take fullest advantage of the existing information and be judicious in our searching. What meaning does it hold if we simply overlook a marine organism that could potentially be of great value? The process of finding fascinating organisms, exploring their physiological characteristics, investigating the bioactive substances contained within them, and mapping their genes can be very time-consuming, even tedious work. Because it takes so long to arrive at the actual application stage using the current blind, experimentation-based investigation methods, it is common for researchers to give up midway through. This is why research necessitates a proactive stance of quickly identifying the possibilities of a discovered organism and seeking out a wise approach. In addition to technologies specific to marine biotechnology, importation of technologies from other fields is also essential for efficient marine biotechnology studies. Because marine biotechnology is fundamentally focused on marine organisms, it requires methods from chemistry, physics, and engineering in addition to those of biology. In practical terms, marine biotechnology is rooted in biology—one often sees examples of researchers delving into biological aspects. At the same time, it is also essential to development technology from an engineering perspective to provide support. As an example, it would be beyond even the most committed biologist to launch a satellite to study marine yields. In this case, the assistance of engineers would be required.

While marine biotechnology takes advantage of waters that span the entire globe, it is not a matter of simple technology or a single field. Rather, it is a new, comprehensive area of study that must be developed on a foundation of research in other fields.

Most life forms in the seas float or swim, but some also attach themselves to materials within the water. On the coast, one can see many organisms attaching themselves to large rocks, including shellfish, barnacles, and algae. Farmed oysters and pearl oysters also live a sessile life, but the aforementioned creatures propagate in such a way as to create environments where humans are likely to find themselves stuck. It does not present any problem when these sessile organisms attach themselves to a coastal rock or, like tetrapods, to a concrete security block. But when they attach themselves to hulls and fishing nets, they can create problems of various kinds.

Few of us would see it as strange to find one or two small shellfish on the hull of a boat. But when thousands—even tens of thousands or more—are growing there, the hull becomes completely encircled with them, resulting in strong water resistance and potentially generating increased fuel costs as speeds are reduced. Stripping of paint from the boat's surface is another problem. Similarly, the attachment of sessile organisms to fixed nets or fishing nets placed to trap cultured fish results in obstruction to water flow, and water quality can be greatly diminished by the waste products and CO_2 generated by the organisms themselves. It is therefore essential to devise means of removing sessile organisms that dwell in places where they cause problems for people working in the oceans. What approaches might there be for preventing an organism from attaching to and dwelling on the bottom of a boat? Wouldn't it be possible to make the hull in such a way that organisms could not attach themselves to it? Questions like these led to the application of paints containing a mixture of organotin compounds that prevent organisms from attaching to hulls (Schultz et al. 2011).

Sessile organisms are not uniform; they come in quite a wide variety, including algae, shells, and barnacles. When organic tin compounds were first applied to prevent these different organisms from adhering, large amounts of drugs were used to reduce costs and promote pharmacological effects. It represented one means of removing the organisms—but it also carried a large pitfall of which people had been unaware. Painting the bottoms of boats with organic tin compounds was indeed one way of preventing sessile organisms from living there, but those compounds then dissolved into the seawater, causing serious pollution of the ocean. In other words, because the compounds used to paint the boats left so much residue in the body, high concentrations of them accumulated in fish and algae, which were then passed on to human beings. Organic tin compounds are toxic to all organisms, including humans. Is it reasonable to pollute as important a resource as the ocean by using such compounds? The issues raised by these compounds left humans facing the task of having to find a new substance to use in their place to prevent biofouling. This meant developing something new that would be less expensive and highly effective while leaving no negative impact on organisms.

A number of ideas have been presented in the search for substances to prevent biofouling, and one of them has involved research into the possibilities of natural, organism-derived materials. Marine organisms synthesize such a wide variety of substances within their body, and an active search is now under way to find anti-biofouling agents in the natural materials that they produce. The most widely used sessile organism in studies to verify their efficacy has been the sea mussel Sea mussels are a widely recognized variety of sessile organism that can commonly be found on Korea's coasts. To investigate their use in researching anti-biofouling materials, a circle four centimeters in diameter is first drawn on water-resistant paper measuring around one millimeter in thickness. Sea mussels measuring 20–25 mm are then placed around it and fixed in place according to the following method. First, a five-millimeter piece of rubber is attached to the shell with adhesive, and then glued to the paper. As a result, the sea mussel is immobilized, but its interior remains unaffected. The anti-biofouling substance under investigation is then dissolved in water or ethanol and painted in the four-centimeter circle, which is then dried. The paper is placed in seawater and left in a dark place. From within its shell, the mussel extends its foot and feels for the center of the circle to find a place to attach. Once it finds a suitable location, it produces a few strands of thin protein thread called a byssus and fixes the shell (its body) in place (Fig. 1.1) (Harino et al. 2007).

If the sea mussel does not like the substance painted in the circle, it will not produce a byssus. The substance that prevents the sea mussel's byssus from being extended is thus an anti-biofouling agent. Through investigations using this method, a number of compounds of different structures have been identified as anti-biofouling substances. Among sessile organisms, barnacles cause even greater damage still, and the barnacles themselves must be used for testing to determine effective anti-biofouling substances. Sea mussels can be collected from the ocean for use in experiments, but barnacle different from the mussels in attaching during their larval stage and spending their entire lives in the same location. As such, barnacle larvae must be used in experimentation. While similar in form to shellfish, barnacles are crustaceans, part of the same family as shrimp and crabs. Once incubated, they pass through nauplius and cypris larval stages before reaching adulthood (Murosaki et al. 2009). Barnacles begin attaching during their cypris stage. Incubating a barnacle and raising it to adulthood in the laboratory is a very difficult process that almost always ends in failure. But Japan's Marine Biotechnology Institute has successfully developed a method of raising barnacles to supply larvae for experimentation. From examining the effects of anti-biofouling materials from marine organisms on the larvae, it was determined that a compound known as 2,5,6-tribromo-1-methylgulamine from the bryozoan *Zoobotryon pellucidum* was a strongly active agent. *Z. pellucidum* possesses tentacles of roughly the same thickness as a vermicelli noodle. It typically attaches to rock surfaces, where it can be found dwelling in the waters off of Korea's coast. The discovery meant that 2,5,6-tribromo-1-methylgulamine had been found in an organism that can survive

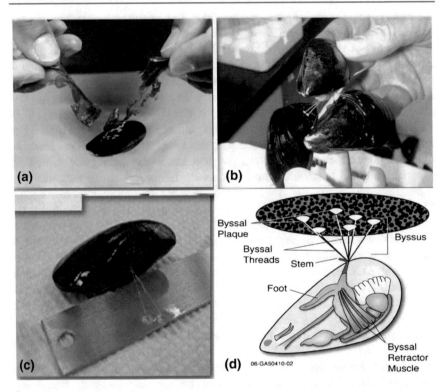

Fig. 1.1 *M. edulis* attachment to **a** seaweed, **b** other mussels, and **c** a stainless steel surface **d** Anatomy of *M. edulis* mussel and byssus structures. Reproduced the figure with permission from (Silverman and Roberto 2007)

more or less anywhere. Further examination showed activity in many similar compounds. From these findings, the process in Japan has already entered the commercialization stage.

Most of the energy used on Earth today comes from fossil fuels, and nearly all of that comes from petroleum. Crude oil refers to petroleum that has not yet been refined; the number of countries that produce crude oil remains very limited. Oil tankers are used to transport crude oil from producing to non-producing countries, and occasionally these thanks will run aground or collide during the transportation process, resulting in large-scale leakage of crude oil. Ocean oil spills also frequently occur at petroleum complexes and oil fields (Harvey et al. 1990). According to the Ministry of Environment, a total of 1798 ocean pollution incidents occurred in South Korea over the six years from 2006 to 2011. The amount of oil spilled during this time totaled 4892 kl; if the 12,437 kl leaked in Taean in the Hebei Spirit oil spill of December 2007 is factored, the total rises to 17,329 kl. In 2011, a total of

369 kl of oil was spilled in 287 ocean pollution incidents. Such spills have declined over the preceding five years after a steady rise between 1994 and 1998, but they continue to occur. In terms of quantities leaked per vessel, miscellaneous ships accounted for the majority at 62% (3046 kl), followed by oil tankers (549.7 kl), cargo ships (354.6 kl), and fishing boats (253 kl) (Kim et al. 2014). Most of these incidents have occurred in the waters off of Ulsan, Yeosu, Busan, or Incheon, which have high volumes of oil passing through. The damage to ecosystems at aquafarms and coastal fisheries near the waters has been immense. To prevent oil from spreading when it has been spilled, oil fences and other means are adopted as physical recovery measures. It is impossible, however, to recover all of the spilled oil using these methods. On average, around one-third of spilled oil can be recovered, one-third will volatilize, and the remaining one-third becomes driftage and a major source of oceanic pollution. Some of the remaining oil in the ocean collects into oil balls, which drift among the waves until they reach a certain weight and sink, adhering to the ocean floor. The grim images of water birds covered in oil should be enough to underline the devastation that these oil spills cause to nature. In December 2007, the 150,000-ton Hong Kong oil tanker Hebei Spirit was at anchor in the waters off of Taean when a Samsung Heavy Industries floating crane collided with it. The accident resulted in 12,547 kl of crude oil spilling out of the tanker and into the ocean, resulting in contamination of a 167-km stretch of coastline from Seosan's Garorim Bay to Taean's Anmyeon Peninsula. Large numbers of fish and shellfish at some 5000 ha of nearby aquafarms were killed. The devastation to fisheries also impacted local livelihood, dealing a blow to the regional economy. An average of 10,000 people per day worked to prevent the disaster from spreading, but some estimate that it may take at least another 50 years for the region's marine ecosystem to fully recover (Fig. 1.2).

Another accident occurred in March 1989 when an Exxon oil tanker ran aground at Prince William Sound in Alaska. Around 40 million l of oil were spilled, covering an area of 8800 km^2 on the coast. Countless aquatic birds and marine organisms were killed, and the coast was left black with oil) While it was possible to remove encrusted oil from the coast using physical washing (the application of forceful, high-pressure jets of warm ocean water), not all of the oil could be removed, and the coast remained black with contamination. To remove the remaining coastal oil that could not be physically washed away, the power of microbes—and their ability to dissolve petroleum components in the sea—had to be employed. Microbes capable of dissolving petroleum components are known to be relatively plentiful in the ocean. This also means that there are many that do not exhibit petroleum dissolution properties under certain environments, but have the potential to do so. Petroleum-dissolving microbes did exist on the Alaskan coast, but they dissolved petroleum components at a very small rate that could not be

Fig. 1.2 Crude oil spill off Taean, South Chungcheong province. http://www.abc.net.au/news/
2013-01-17/an-sth-korea-sets-24694mil-oil-payout/4469898

observed with the human eye. The removal of the coastal oil would take many years
if left to natural self-purification functions alone. After the accident occurred, fertilizer to promote propagation of the microbes was spread along the coast. The
fertilizer contained not only nitrogen, phosphorus, and other essential nutrients for
microbes, but also surfactants, which are capable of dissolving floating oil on the
ocean surface into the surrounding water (Van Hamme et al. 2003).

In 1990, The U.S. Environmental Protection Agency conducted a survey in
which it reported petroleum components to have been dissolved at rates three to five
times faster in ocean trenches where fertilizer was used than in those where it was
not. Even without these numbers to offer proof, mere visual examination of the
coastal regions where fertilizer was spread showed them recovering nearly all of
their original color, prompting a renewed appreciation of the utility of
bio-remediation techniques harnessing the power of microbes.

The Alaskan oil spill is an example of making effective use of
petroleum-dissolving microbes that happened to already be present at the accident
site. Another method under consideration involves artificially introducing
petroleum-dissolving microbes into the sea and allowing them to dissolve and
remove oil spills. Doing so will require the identification of microbes that are
capable of dissolving microbes with great efficiency. The crude oil that is the source
of contamination is a complex mixture of some 10,000 different ingredients. In
particular, they contain many hydrocarbons, such as paraffin compounds, which are
carbon compounds in chain form; aromatic compounds structured around a benzene
nucleus; resins with complex structures; and asphaltene. These components are

known to be relatively difficult to dissolve. The largest oil spills from the tankers are shown in the Table 1.2.

Crude oil components with a low boiling point, such as gasoline, kerosene, and diesel, volatilize when leaked into the ocean and thus present almost no concerns. The problem lies in heavy oil components that do not volatilize readily. Today, the search is on for microbes with outstanding abilities to dissolve these components. In microbiological common sense terms, microbes cannot be viewed as a panacea capable of completely dissolving petroleum components in all their complexity. It is therefore essential to use many different kinds of microbes together to dissolve petroleum mixtures.

To obtain a mixed colony in nature, crude oil is applied to seawater and sand. After about a month, around 20% of the crude oil will have been dissolved by the seawater microbes. A portion of the mixture of these microbes and petroleum must then be extracted and applied to sterilized seawater (with no microbes present) and crude oil. As this process is repeated, it is possible to produce microbes with outstanding crude oil dissolution properties. In Japan, this method was used to separate a natural colony known as SM8 from sand. Among the components of crude oil, SM8 was capable of dissolving 30% of paraffin compounds and 20% of aromatic compounds. As the most any single microbe had been able to dissolve before was 15% of paraffin compounds and 5% of aromatic compounds, SM8 was clearly an outstanding colony.

In some cases, microbes known to have properties absent from natural colonies have been combined to produce artificial colonies. An artificial colony known as M4, which was a mixture of four kinds of microbes, was capable of dissolving 50% of paraffin compounds and 30% of aromatic compounds. A natural colony includes over four kinds of microbes, but it was found that only those four could dissolve crude oil with efficiency equaling a natural colony. This finding will be of some importance going ahead for the establishment of petroleum dissolution techniques. Genetic studies of petroleum-dissolving microbes have also been taking place. In particular, attempts are being made to study the genes of enzymes that are involved in petroleum dissolution and making improvements to increase their effectiveness. For some cases in the development of technology for application in environmental recovery, one may come across situations that could not be predicted from test tube levels alone. Because it takes so long to conduct a single environmental recovery experiment using microbes, it is very difficult to test everything. For this reason, it may be necessary to use techniques that can take the entire situation into account from numerical models that predict activities in seawater affected by an oil spill.

In the past, we have arguably been too casual in our response to the oil spills that have occurred. We have done little more than to apply surfactants or erect oil fences to stop the spreading. In many cases, the surfactants themselves have been of poor quality and failed to function properly; instead, the oil has clumped onto the surfactants and attached to the ocean floor, wreaking havoc on the ecosystem there. Today we stand in the 21st century. Contributing to environmental preservation with the development of new oil pollution prevention technologies suited to the times appears to be an important task that urgently needs to be addressed.

1.5   Developing Industries with New Marine Biotechnology Materials

One of the main reasons that many are predicting the arrival of the "marine biotech era" is the expectation that new materials derived from marine organism resources will take the place of or transform the existing chemical industry. There are countless species of organisms on Earth that could serve as sources to develop new materials. In the past, necessary materials for humankind have been taken directly from the species themselves; the range of species used in developing biological materials has been extremely vast and diverse, encompassing microbes, plants, animals, marine organisms, and insects. The growth potential of the marine-derived new materials industry is expected to be particularly great in view of the many problems related to materials from land-based sources, including the need to use food resources. To date, however, there have been few concrete cases of new materials being developed from marine organisms. The use of new materials obtained from marine organisms that inhabit and have adapted to extreme environments is expected to increase going ahead.

These new marine materials possess high value added and utility, including raw materials for pharmaceuticals to treat and prevent disease; biodegradable polymers; functional polysaccharides; enzyme inhibitors; research reagents; flavoring; functional pigments; and ingredients for cosmetics. Figure 1.3 shows the research stages that will be disease for research and development of pharmaceutical materials in particular (Venkatesan and Kim 2015).

1.5.1   Collecting the Species and Preparing the Extract

The search for biomaterials begins with the collection of species based on human experience. This is a process of choosing bioactive materials to explore and selecting a candidate species based on folk treatments or the literature. When combing the literature, a more effective approach is to consult foreign and domestic materials and related academic journals. This is the stage before bioactivity measurement, with an extraction process using an organic solvent. Methanol or ethanol is typically used as a solvent, while water is used to investigate highly polar bioactive substances. The bioactive substances present in organisms are sensitive to temperature and light and should therefore be extracted in a dark room at a temperature below 40 °C.

1.5.2   Biological and Pharmacological Investigation of the Crude Extract

Once extracted from the organism, the substance is measured for bioactivity. The crude extract contains many compounds, but the quantity of actual bioactive

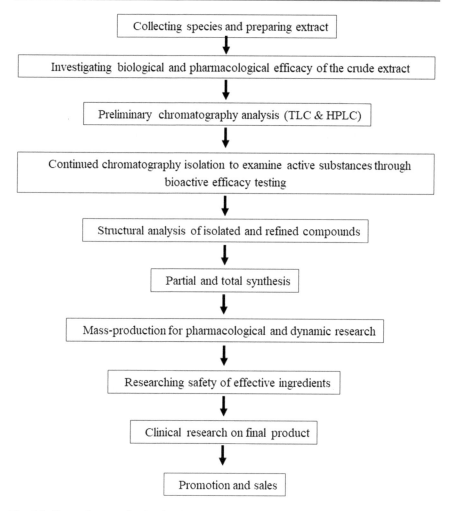

Fig. 1.3 Research stages for the development of marine pharmaceutical materials

materials is very low. The bioactivity measurement must therefore be conducted with great sensitivity. Investigation of bioactive substances in organisms saves time and money, consisting of an initial exploration of the whole biological resource using fast and precise methods and, if development possibilities are found, a second and third exploration. The initial investigation adopts typical experimental methods, with a focus on determining whether the materials are effective or not, while the second investigation adopts a quantitative experimental approach with a focus on the degree of efficacy. The third and final investigation, which involves animal testing, should be done in a way that allows for an accurate determination of development viability.

1.5.3   Isolation and Identification of Active Substances

Isolating and refining active substances from crude extract with confirmed bioactivity involves two kinds of approaches, chemical and instrumental. Prior to the development of instrumental analysis, elemental analysis was principally used. Today, isolation and analysis of substances is performed through innovative chromatography techniques. No matter how advanced the instrumental analyses may be, however, chemical methods should not be neglected; a complementary approach is required.

Among the instrumental approaches used today for isolation analysis and structural identification of bioactive materials are high performance liquid chromatography (HPLC), gas chromatography (GC), UV spectroscopy, mass spectrometry (MS), infrared spectroscopy, and nuclear magnetic resonance (NMR).

1.5.4   Synthesis of Biomaterials

Organisms contain small amounts of useful bioactive substances. Moreover, because these organisms are collected for extraction, there is a risk of ecosystem destruction. The growth of plants in particular is closely connected with the environment, and difficulties with mass-producing them for supplies within a defined period of time result in limitations to their use in the pharmaceutical market or as functional materials. One approach to overcoming these limitations is production through structural identification and synthesis of extracted active substances.

Ideally, all synthesis would occur at the same stage. Problems with this include difficulties with synthesizing structurally complex substance and declining yield rates. A recent approach to addressing this problem has been to make an accurate assessment of the pharmacology and effects once the effective components have been isolated through precursors or other means, allowing for a determination of their development potential as new pharmaceutical or functional materials.

1.5.5   Safety of Biomaterials

Once the efficacy of the final effective component is establishment, safety testing (toxicity and pharmacological safety) is conducted to obtain a permit for manufacturing the product for clinical testing. Typically, ordinary researchers conduct general pharmacology and acute toxicity testing on their own; once the possibilities have been confirmed, toxicity and pharmacological testing is then commissioned from a trustworthy institution.

Genetically engineered materials such as insulin, growth hormone, and interferons require more accurate safety determinations; even in the case of naturally derived materials, the issue of provenance is extremely important. For fermentation products such as *Escherichia coli*, testing must be conducted for other microbes and external toxins.

1.5.6 Manufacturing of Biomaterials

Once the final effective component's efficacy and safety have been ensured, the next step is manufacturing. The most important step in the manufacturing design process is assessment of the effective component's characteristics, including its physical, chemical, and biological properties. Biological properties to be tested include pharmacology, efficacy, affected regions, dosage, blood concentration, absorption distribution, absorption regions, absorption rate, metabolism and excretion, biological half-life, and side effects. Physical properties to be tested include solubility and safety. Once the medicine has been produced, its characteristics are assessed. Areas included in this assessment are biological characteristics (absorption and excretion), sensory characteristics (taste, smell, and color), physiochemical characteristics, longitudinal safety, and the presence of foreign matter and microbe contaminants. After the medicine's characteristics have been assessed, the production process begins, and a final assessment is made on the experimental product, including clinical product.

1.5.7 Clinical Testing of the Final Product

Once pre-clinical testing of toxicity and pharmacology has been completed and a permit has been issued by the Minister of Health and Welfare, production manufacturing approval for clinical testing is received and a clinical product is manufactured. Clinical testing follows designated guidelines: safety is demonstrated through pre-clinical testing, and a trial drug that is expected to have clinical efficacy is used in three phases of, I, II and III, testing.

In the first testing phase, the trial drug is given to a human subject and examined for the first time based on the pre-clinical test results. A healthy subject is used to confirm the drug's safety, and safe amounts and maximum permissible doses are estimated to provide data for the second phase of testing. Pharmacokinetic research is also conducted for the drug to collect basic data on its absorption, metabolism, and excretion.

In the second testing phase, the trial drug is administered to patients for the first time to verify its efficacy and safety and examine its pharmacokinetic effects. Dosage quantities, intervals, and period are determined for the third phase of testing.

In the third testing phase, quantity and usage findings from the efficacy and safety testing of the trial drug in the second stage are used for an expanded range of clinic testing subjects in order to affirm the drug's effects and safety and identity any infrequent side effects. The final goal is to establish efficacy, effects, usage methods and amounts, and any areas for caution in view of the effectiveness and safety of the trial drug.

1.6 Chapter Summary and Conclusion

Marine bioresource are unlimited, but utilization to mankind is limited. Marine based bio industry plays an important role in produce several commercial products such as polysaccharides, proteins, peptides, small molecules, and lipids for various applications. Marine Biotechnology plays an significant role in our daily day life, and it is an emerging area of field in the past five decades. Marine biotechnological applications are widely utilized in food, pharmaceutical, cosmeceutical and energy applications. Some of the environmental applications such as bio fouling, removal of spilled oil is being overcome through some of the biotechnological methods. Production of novel chemicals with different biotechnological tools increases its production and can meet the consumer's expectation.

References

Borowitzka, M. A. (2013). High-value products from microalgae—Their development and commercialisation. *Journal of Applied Phycology, 25*, 743–756.

Brown, M., Jeffrey, S., Volkman, J., & Dunstan, G. (1997). Nutritional properties of microalgae for mariculture. *Aquaculture, 151*, 315–331.

Carroll, J., & Crews, P. (2009). Macromarines: A selective account of the potential of marine sponges, molluscs, soft corals and tunicates as a source of therapeutically important molecular structures. In D. B. Antony & M. S. Butler (Eds.), *Natural product chemistry for drug discovery* (pp. 174–214). Cambridge, UK.: The Royal Society of Chemistry.

Fusetani, N. (2000). *Drugs from the Sea*. Karger Medical and Scientific Publishers.

Harino, H., Yamamoto, Y., Eguchi, S., Kurokawa, Y., Arai, T., Ohji, M., et al. (2007). Concentrations of antifouling biocides in sediment and mussel samples collected from Otsuchi Bay, Japan. *Archives of Environmental Contamination and Toxicology, 52*, 179.

Harvey, S., Elashvili, I., Valdes, J. J., Kamely, D., & Chakrabarty, A. (1990). Enhanced removal of Exxon Valdez spilled oil from Alaskan gravel by a microbial surfactant. *Nature Biotechnology, 8*, 228.

Jha, R. K., & Zi-Rong, X. (2004). Biomedical compounds from marine organisms. *Marine Drugs, 2*, 123–146.

Jiao, G., Yu, G., Zhang, J., & Ewart, H. S. (2011). Chemical structures and bioactivities of sulfated polysaccharides from marine algae. *Marine Drugs, 9*, 196–223.

Kim, D., Yang, G.-G., Min, S., & Koh, C.-H. (2014). Social and ecological impacts of the Hebei Spirit oil spill on the west coast of Korea: Implications for compensation and recovery. *Ocean and Coastal Management, 102*, 533–544.

Kim, S.-K., & Lee, C.-G. (2015). *Marine bioenergy: Trends and developments*. CRC Press.

Kim, S.-K., & Venkatesan, J. (2015). Introduction to marine biotechnology. In S.-K. Kim (Ed.), *Springer handbook of marine biotechnology* (pp. 1–10). Berlin: Springer.

Lal, R. (2006). Managing soils for feeding a global population of 10 billion. *Journal of the Science of Food and Agriculture, 86*, 2273–2284.

Lee, S.-B., Cho, S.-J., Lee, S.-Y., Paek, K.-H., Kim, J.-A., & Chang, J.-H. (2009). Present status and prospects of marine chemical bioindustries. *KSBB Journal, 24*, 495–507.

Medini, D., Donati, C., Tettelin, H., Masignani, V., & Rappuoli, R. (2005). The microbial pan-genome. *Current Opinion in Genetics and Development, 15*, 589–594.

Murosaki, T., Noguchi, T., Hashimoto, K., Kakugo, A., Kurokawa, T., Saito, J., et al. (2009). Antifouling properties of tough gels against barnacles in a long-term marine environment experiment. *Biofouling, 25*, 657–666.

Notoya, M. (2010). Production of biofuel by macroalgae with preservation of marine resources and environment. In *Seaweeds and their role in globally changing environments* (pp. 217–228). Berlin: Springer.

Salazar, A., Hackney, R., & Howells, J. (2003). The strategic impact of internet technology in biotechnology and pharmaceutical firms: Insights from a knowledge management perspective. *Information Technology and Management, 4,* 289–301.

Schultz, M., Bendick, J., Holm, E., & Hertel, W. (2011). Economic impact of biofouling on a naval surface ship. *Biofouling, 27,* 87–98.

Silverman, H. G., & Roberto, F. F. (2007). Understanding marine mussel adhesion. *Marine Biotechnology, 9,* 661–681.

Van Hamme, J. D., Singh, A., & Ward, O. P. (2003). Recent advances in petroleum microbiology. *Microbiology and Molecular Biology Reviews, 67,* 503–549.

Venkatesan, J., & Kim, S.-K. (2015). Marine biomaterials. In *Springer handbook of marine biotechnology* (pp. 1195–1215). Berlin: Springer.

Verma, N. M., Mehrotra, S., Shukla, A., & Mishra, B. N. (2010). Prospective of biodiesel production utilizing microalgae as the cell factories: A comprehensive discussion. *African Journal of Biotechnology, 9,* 1402–1411.

Villa-Carvajal, M., Catalá, M., Barreno, E., & Tornero Martos, A. (2014). Bioproduction of bioactive compounds: Screening of bioproduction conditions of free-living microalgae and lichen symbionts, 10th International Conference on Renewable Resources & Biorefineries.

Wijesekara, I., Pangestuti, R., & Kim, S.-K. (2011). Biological activities and potential health benefits of sulfated polysaccharides derived from marine algae. *Carbohydrate Polymers, 84,* 14–21.

Introduction to Molecular Biology

2

Contents

To understand marine biotechnology, it is necessary to understand the genes that determine an organism's appearance and functions. This chapter will include a brief explanation of the genetics that directly informs biotechnology, as well how genetics has been applied to biotechnology.

2.1 What Are Genes?

Before Gregor Mendel established the concept of genes, it was believed that certain traits were passed on to offspring through "blending inheritance," or an intercombination of the genetic properties present in the sperm and egg. At the same time, the regularities of that transmission process and the mechanism of transmission long remained unknown.

© Springer Nature Switzerland AG 2019
S.-K. Kim, *Essentials of Marine Biotechnology*,
https://doi.org/10.1007/978-3-030-20944-5_2

In the mid-19th century, Mendel conducted cross-breeding experiments using pea plants in a monastery garden in the Austrian city of Brünn (today Brno). Through statistical analysis of his observation findings on the separation of traits, he elucidated a mode of inheritance rooted in the concept of the gene. The pea plants that Mendel used for his experiments (*Pisum sativum*) self-fertilized through a process whereby the pistil received flower from a stamen within the same flower. It was therefore possible to obtain genetically fixed organisms over several generations of breeding with almost no natural crossing, thereby maintaining a pure lineage. Peas were also easy to cultivate and had a short growth period, bearing fruit easily and proving simple to hybridize artificially.

Moreover, the seven genetic traits that Mendel focused on in his research were controlled by simple genes on different chromosomes. This resulted in clear genetic separation in later breeding generations, a fact that proved most suitable for the discovery of genetic laws. In the process, he refuted the hitherto accepted notion of blending inheritance and proved that the traits were governed by certain factors (genes). He used the term "dominant" to refer to traits found in the first generation of cross-breeding and "recessive" for traits that did not appear. This phenomenon became known as the "law of dominance," whereby dominant traits alone appear in the first hybridized generation.

Mendel used pea plants to conduct his research on modes of inheritance. Cross-breeding of a pair bearing conflicting traits, such as a plant bearing round seeds on one bearing wrinkled seeds, would have a first generation (F_1) consisting solely of round seeds. Those seeds were then sown and raised on their own, and the 7324 second-generation (F_2) hybridized seeds obtained through self-fertilization (the pistil's receipt of pollen from the same flower's stamen) were examined. A total of 5474 seeds were found to be round and 1850 wrinkled, giving a ratio of 2.96 to one. Cross-breeding experiments between plants with yellow and green cotyledons resulted in an F_1 consisting entirely of yellow cotyledons, while the ratio among the 8023 F_2 plants was 6022 yellow to 2001 green, or 3.01 to one (see Fig. 2.1). The cross-breeding experiments for the seven traits thus showed that the F_2 generation always showed parental traits at a ratio of roughly 3:1 (Bateson and Mendel 2013; Mendel et al. 1993; Orel 1996).

From these experimental findings, Mendel found that only one parental trait appeared in the first hybridized generation, while the other trait did not appear at all. He referred to the trait that did appear as dominant and the trait that did not appear as recessive. He also hypothesized a "genetic factor" that was transmitted for parents to offspring to cause the trait's appearance, which he indicated by means of a symbol. For instance, he would refer to the dominant of two conflicting traits with a capital "A" and the recessive genetic factor (hereafter "gene") with a lower-case "a."

As shown in Fig. 2.2, parental somatic cells contain two genes for each trait. When a male or female gamete such as pollen of an ovum is produced, however, these separate and are each distributed as a distinct gamete. The fertilized egg resulting from the combination of male and female gametes—the offpsring's somatic cell, in other words—contain a set of two genes, or one each from the

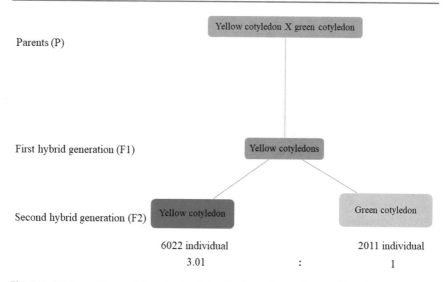

Parents (P)

First hybrid generation (F1)

Second hybrid generation (F2)

Yellow cotyledon X green cotyledon

Yellow cotyledons

Yellow cotyledon

Green cotyledon

6022 individual

3.01   :   1

2011 individual

Fig. 2.1 Modes of transmission in the hybridization of pea plants with yellow and green cotyledons

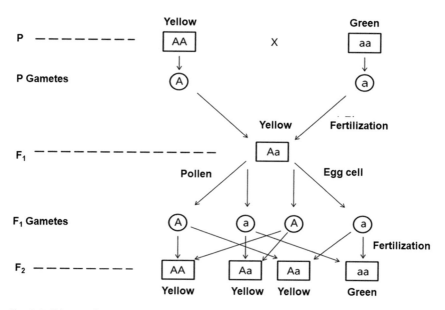

Fig. 2.2 Diagram for gene-based experimental findings in Fig. 2.1

mother and father. Combination of the genes results in an F_1 generation consisting entire of "Aa" pairs, in which all organisms manifest the dominant trait A. In the following F_2 generation, the gene pairs are AA:Aa:aa at a ratio of 1:2:1. At the same time, Mendel found that the ratio of organisms with the dominant yellow cotyledons and the recessive green cotyledons was 3:1.

Mendel published his findings in 1865 in a paper titled *Experiments on Plant Hybridization*, but contemporary biologists remained unaware of its true importance (Mendel 1996). It was not until 1900—a full 35 years after its publication by Mendel—that Mendel's rules were rediscovered independently by three different people: De Vries in the Netherlands, Correns in Germany, and Tshermark in Austria.

Perhaps it was the extreme originality of Mendel's ideas—their characterization of complex traits of organisms as being passed to offspring from parents by way of genes, and the use of abstract symbols to represent them—that prevented his contemporaries from readily understanding them. Fulling understanding the gene, however, required proof that it was more than simply an abstract symbol, but a comprehensible entity in and of itself. At least three questions thus remained to be answered: where genes existed in the cell, how they manifested their traits, and what chemicals they consisted of.

2.2 Where Do Cells Exist in the Body?

Mendel was a small boy when he learned that all organisms were made up of microscopic, sack-shaped components known as cells. It was later determined that cells proliferated through fission, but research in that area had only just begun when Mendel conducted his hybridization experiments with pea plants. Understanding the question of where genes exist in the cell requires an understanding of the cells themselves and their fission process.

For most organisms, cells are surrounded by a membrane and consist of a gelatinous cytoplasm containing a gene-bearing nucleus and organelles such as the mitochondrion, endoplasmic reticulum, and chloroplasts (in plant cells). These are surrounded by a very thin cell membrane. Plant cells contain a cell wall made of cellulose outside the cell membrane (see Fig. 2.3). When the cell divides, the nucleus first splits in two, after which the cell separates into two parts.

Prior to division, the resting nucleus appears as a kind of mesh or particle. Once division begins, however, the nucleus itself transforms into a long, slender, threadlike chromosome. The chromosome is made up of a combination of DNA and protein known as a chromonema. The proteins are necessary to maintain the chromonema's shape and regulate gene activity.

The number of chromosomes is fixed for different organism types: 46 in the case of human beings and 14 for pea plants. In higher organisms, each somatic cell receives one chromosome each from its mother and father, resulting in a homologous chromosome pairing one chromosome each from both parents. These pairs of chromosomes are called "diploid" (2n), while sex cells and gametes are called "haploid" because they carry only half the number of somatic cell chromosomes. When the cells that form the body (somatic cells) undergo division, each chromosome first divides in two vertically for distribution to two new cells; each resulting cell has the same number and same form of chromosomes (see Fig. 2.4a).

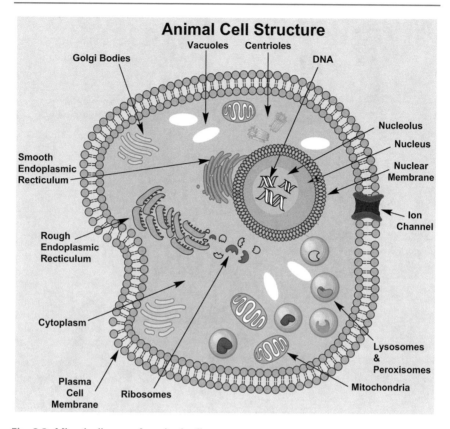

Fig. 2.3 Mimetic diagrams for animal cells

For the formation of ova, pollen, sperm, and other gametes, the division process is slightly different (see Fig. 2.4b).

When the ovum nucleus that is the source of gametes transforms into a chromosome, homologous chromosomes first attach to form what appears outwardly to be a single chromosome. The homologous chromosome then divides once again to form two separate nuclei as the cell divides in two. If the parent cell's number of chromosomes is 2n, then the number of chromosomes for the two cells formed through initial division in half that, or n. Another division then occurs, so that four gametes with n chromosomes are formed from a single ovum. Because the number of chromosomes is halved when gametes are formed in this way, the division process is known as "meiosis."

Fertilization in both plants and animals consists of the combination of male and female gametes with this halved number of chromosomes n, so that the new number of chromosomes is a multiple or sum 2n. The somatic cell of an organism that has matured from a fertilized egg thus contains 2n chromosomes, half of them from the

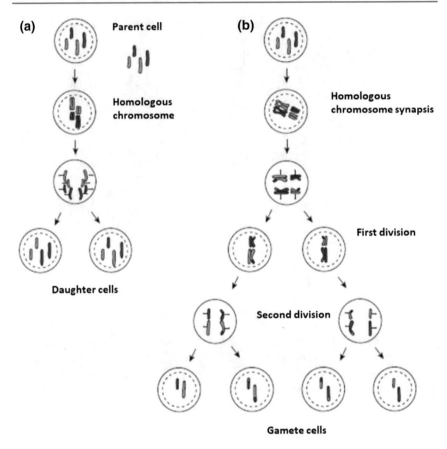

Fig. 2.4 Mitosis (**a**) and meiosis (**b**) for somatic cells, where 2n = 4

father and half from the mother. For homologous chromosomes, one set comes from the father and one set from the mother.

The genetic activity conceived by Mendel can therefore be seen to correspond fully with the chromosomal activity in meiosis. As shown in Fig. 2.5, overlaying the gene process on the chromosomal process shows no inconsistencies whatsoever in the genetic explanation. It is for this reason that many early in the 20th century began conceiving of the gene as existing on the chromosome. In the 1930s, the U.S. geneticist Morgan used fruit flies (*Drosophila melanogaster*) to show the location of many genes on the four (n) chromosomes, including those responsible for eye color—thus creating a chromosome map of genes. Through this research, it was learned that genes occupy a defined position on the chromosome (Morgan et al. 1925).

Chromosome maps have been studied from a number of different organisms, including corn and human beings. Artificial mutations have played an important role in such research. Studies of the genes responsible for eye color in fruit flies, for

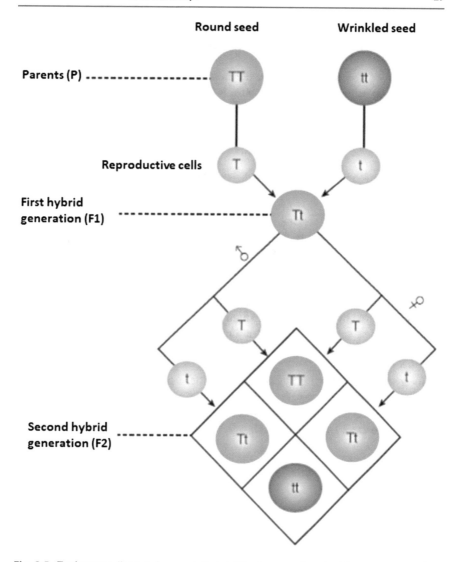

Fig. 2.5 Explanatory diagram for mendel's hybridization experiment with genes presumed to exist on chromosomes

example, have shown that wild organisms are almost exclusively red-eyed and therefore cannot be hybridized. If a mutation with white eyes can be obtained, however, and if that mutation is a genetically stable trait appearing across several generations rather than one alone, it becomes possible to study the alleles through cross-breeding. Mutations can be found in several traits besides eye color, including eye and wing shape, but are exceedingly rare in the natural world. Through the use of appropriate amounts of X-ray, ultraviolet, or other radiation, however, it may be possible to artificially raise the incidence of mutations by hundreds of times. It is a method currently used in research in microbiology and various other areas.

- For the purpose of convenience, the number of chromosomes has been set a 2n = 2 (the actual value for pea plants is 2n = 14; see Fig. 2.2).

2.3 Genes and Phenotypic Expression

As described above, genetic research has consisted chiefly of hybridization and cytological approaches concerning the position of genes on the chromosome for various traits. The question of how genes actually function to express traits has proven more difficult, however. A breakthrough finally came in the 1940s through research by Biddle and Tatum using the red bread mold *Neurospora crassa*.

Wild strains of *N. crassa* grow in minimal media containing mineral salts, sugars, and vitamins. Arginine auxotrophs, however, cannot grow without addition of the amino acid arginine to the minimal medium. Because the wild strains contain all the enzymes needed for amino acid synthesis, they are capable of absorbing the material supplied by the minimal medium to produce any amino acid. The arginine auxotrophs lack enzymes from the arginine synthesis pathway and are therefore incapable of synthesis on their own, so that they are unable to grow without an arginine supply.

Inorganic salts and sugars absorbed from the medium do not transform immediately into arginine. They first form glutamic acid as a raw material, passing through the seven types of intermediate materials listed here before arginine is finally synthesized. Because the pathway is so complex, it has been simplified into the diagram below.

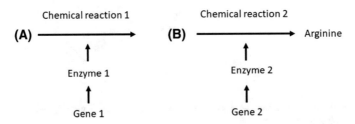

Raw material A transforms into B through the first chemical reaction, after which B becomes arginine through the second. The final product of arginine does not form if the catalysts for the two reactions, Enzyme 1 and Enzyme 2, are not present. For the arginine auxotrophic mutants, however, there exist a first type I that grows without the addition of arginine if intermediate material B is added to the minimal medium and a second type II that does not grow in B and requires arginine. The difference can be explained as follows: because Mutant I lacks Enzyme 1, the first chemical reaction does not take place and B is not synthesized. Because Enzyme 2 is present, however, the second chemical reaction can take place with the addition of B to the minimal medium, and arginine can be synthesized and developed.

Mutant II, in contrast, lacks Enzyme 2, which means that arginine cannot be synthesized even if B is added. Because both types are genetically stable, a mutation in Gene 1 governing the formation of Enzyme 1 results in failure to synthesize Enzyme 1 in Mutant I, whereas a mutation in Gene 2 governing the formation of Enzyme 2 results in failure to synthesize Enzyme 2 in Mutant II. In other words, the enzymes that catalyze each chemical reaction are governed by different genes. This is the famous "one gene-one enzyme hypothesis," which holds that a single gene is involved in formation of only one type of enzyme, governing its particularities and influencing the phenotype. If a change occurs in some gene, it becomes impossible to produce that enzyme; even if it is produced, it is incomplete. As a result, the reaction stops and the target material is not synthesized. If that material is a red pigment in the eye, it manifests in the visible trait of white eyes. In the case of arginine, the result is the physiological trait of bread mold being unable to grow in a minimal medium. Because the enzyme is a protein, a single gene is also seen as governing the formation of specific proteins. The explanation can be summarized as follows:

Gene $\rightarrow$ Enzyme $\rightarrow$ Chemical reaction $\rightarrow$ Product
              (Protein)   (Phenotypic expression)

The idea of a gene governing the formation of proteins means that it contains a diagram bearing information for protein synthesis. This means that if any mutation occurs in the gene, it results in irregularities in the protein diagram, so that protein cither is not formed or is formed but does not function normally. The resulting organism bears an altered trait; in other words, it is a mutant. Because that mutation is the result of a genetic change, it is also passed down to offspring.

2.4   What Is Nucleic Acid?

Explorations into the nature of the gene as a chemical began around more or less the same time as Mendel's research in a field—biochemistry—that bore no connection whatsoever to genes. In 1896, the Swiss biochemist Miescher was researching pus, a then poorly understood material extracted from white blood cells (Miescher and Schmiedeberg 1896). Because it contained nitrogen and phosphorus, it was a new material that different from a protein. The same material was also found in salmon milt (the white mass through which a male fish secretes sperm within its stomach) and yeast. Contained with the cell's nucleus, it was named "nuclein." Its main components were subsequently found to be nitrogenous bases, pentose (a monosaccharide consisting of five carbon atoms), and phosphoric acid. Because its pH was acidic, it became known as nucleic acid (Miescher and Schmiedeberg 1896).

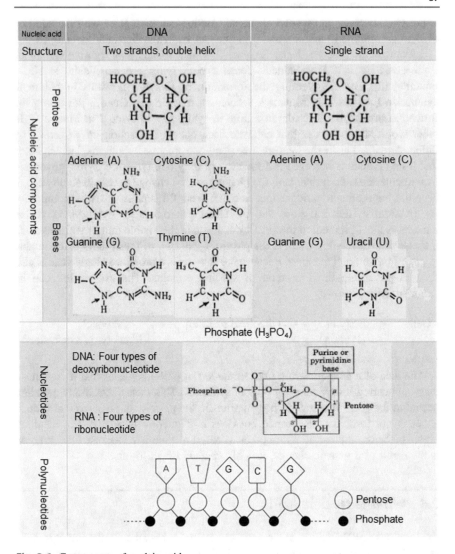

Fig. 2.6 Components of nucleic acid

The chemical structure of nucleic acid was identified in the 1940s (see Fig. 2.6). Because there are two types of pentose—ribose and deoxyribose—the nucleic acid containing ribose was called ribonucleic acid (RNA), and the nucleic acid containing deoxyribose was called deoxyribonucleic acid (DNA). For animals and plants alike, DNA was found to exist within the cell's nucleus and RNA in the nucleus or the cytoplasm. DNA and RNA both consist of four different bases. The bases adenine (A), guanine (G), and cytosine (C) are shared by both, but DNA contain thymine (T) while RNA contains uracil (U). The bases, pentose, and phosphoric acid form the basic unit of nucleic acid, namely the nucleotide, which exists in four types depending on differences in the nucleic acid's bases (Watson and Crick 1953).

Nucleic acid is a polynucleotide, or a polymer consisting of several nucleotides linked together in a chain. While this chemical structure was at least partially understood, the question of what function the material had and how it related to genes remained to be solved.

2.5   The Gene Entity

Proof that nucleic acid stored genetic information came through a 1928 experiment in trait transformation by Frederick Griffith. Griffith applied heat to kill pathogenic pneumococcus bacteria responsible for pneumonia in mice. After mixing it with non-pathogenic pneumococcus, he discovered the non-pathogenic strain had become pathogenic. He concluded that the pathogenic strain contained a trait transformation factor, which had entered the non-pathogenic strain and caused a change in its properties. The nature of this trait transformation factor remained uncertain for many years until 1944, when O. T. Avery and co-researchers purified the factor from pathogenic pneumococcus and identified it as deoxyribonucleic acid (DNA; see Table 2.1) (Griffith 1934; Avery et al. 1944).

After Avery's experiment, questions arose as to whether DNA was a universal genetic material, or whether proteins could also be responsible for genetic information. The answer to the question came from A. H. Hershey and M. Chase. In 1952, they performed an experiment using a T2 phage virus infecting *E. coli* bacteria. Using isotopes, they marked the DNA with ^{32}P and the protein with ^{35}S and infected bacteria (Fig. 2.7). Only the DNA marked with ^{32}P from the T2 phage was injected into the *E. coli*; the protein was not introduced. After around 20 min, they found that many offspring phages had been created and released from DNA information within the bacterium. From this, they concluded that DNA, not proteins, was responsible for genetic information (Hershey et al. 1953).

Later, it was proven that DNA carried genetic information not only in prokaryotes but also in eukaryotes. More recently, a general practice has arisen of inserting chemically synthesized DNA into cells to create corresponding gene products.

Table 2.1   Avery's experimental findings

Bacteria transplanted to medium	Form and ratios of resulting colonies
Live R type bacteria	R:S = 1,000,000:1[a]
Killed S type bacteria	O
Mixture of live R type bacteria and killed S type bacteria	R:S = 100:1

[a]S is believed to have arisen as a mutation of R

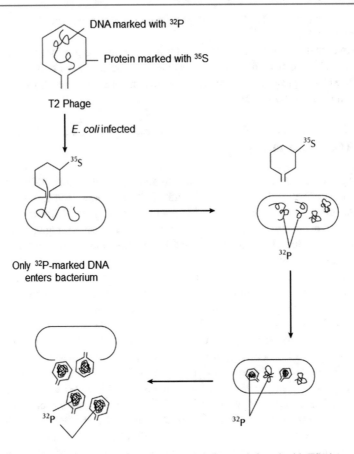

DNA marked with ^{32}P

Protein marked with ^{35}S

T2 Phage

E. coli infected

^{35}S

^{35}S

Only ^{32}P-marked DNA
enters bacterium

^{32}P

^{32}P

^{32}P

^{32}P

Fig. 2.7 Hershey and Chase's experiment. When *E. coli* were infected with T2 phages bearing DNA marked with ^{32}P and proteins marked with ^{35}S, only the DNA marked with ^{32}P entered the bacterium, while the ^{35}S-marked proteins did not

2.6 DNA's Double Helix Structure

After DNA was found to be the material for genes, scientists around the world competed to elucidate its molecular structure. In 1950, E. Chargaff analyzed the formation of DNA bases in various organisms and found that while it differed from one organism to the next, the amount of adenine was always equal to the amount of thymine and the amount of guanine to the amount of cytosine (Chargaff 1950). Around the same time, Wilkins learned that DNA extracted from different organisms showed similar X-ray diffraction, and that DNA was a very long, thin molecule that coiled into a spiral. He also discovered that all DNA molecules showed a rod-like structure measuring around 2 nm across, and that consisted of repeating units measuring 0.34 nm between base pairs and 3.4 nm between loops (Chargaff 2012).

In 1953, Watson and Crick proposed the double helix model for DNA based on Chargaff's observations and the X-ray diffraction data. Its characteristics were as follows:

1. Chains consisting of two strands of polynucleotides twist in opposing directions. In other words, if one strand followed a 5′ → 3′ pattern, the other would follow a 3′ → 5′ pattern.
2. The two strands' chain is maintained through hydrogen bonds between adenine and thymine and between guanine and cytosine.

As shown in Fig. 2.8, two hydrogen bonds exist between A and T and three between C and G. Because A-T and C-G base pairs are almost exactly the same size, the diameter of the double helix is maintained at a fixed length. Other arrangements of base pairs besides C-G and A-T would not be appropriate to form a double helix: A and G are both purine bases that would be too large for a base pair between them to fit in the helix, while a C-T base pair would slip too easily out of the helix, preventing formation of a stable hydrogen bond. The existence of only A-T and C-G as base pairs was also consistent with Chargaff's discovery that adenine and thymine quantities and cytosine and guanine quantities were always equal. The two bases forming a pair are described as complementary, as are the two polynucleotide chains forming the double helix.

3. The two-stranded polynucleotide chain has a single axis and winds to the right.
4. The diameter of the helix is 2 nm with a distance of 0.34 nm between base pairs, values consistent with the X-ray study observations. The distance between helix loops is 3.4 nm, or ten nucleotides.
5. The double helix structure of DNA was confirmed to be correct in numerous subsequent experiments. Once the double helix structural model was recognized as correct, the replication machinery of DNA became the subject of speculation, and it became clear that the genetic code was determined by the arrangement of nucleotides (Figs. 2.9 and 2.10).

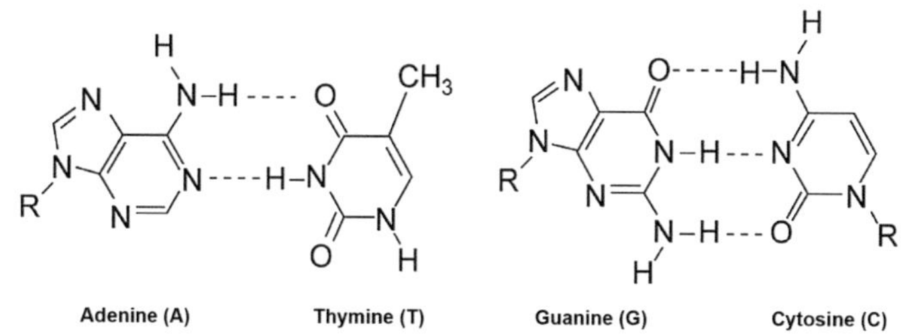

Fig. 2.8 Hydrogen bonds between DNA bases

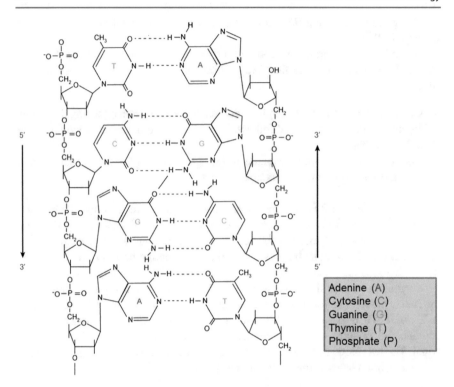

Fig. 2.9 Base pairs in DNA's two-stranded polynucleotide chain

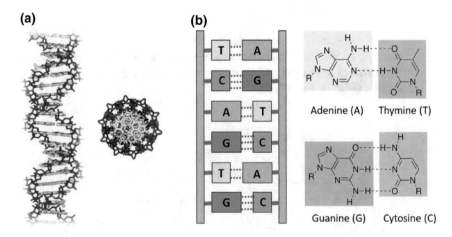

Fig. 2.10 Mimetic diagram of the DNA double helix structure. The figure is adopted from Wang (2018)

2.7 DNA Autoreproduction

Prior to division, the cell's DNA is replicated in its entirety and splits apart, resulting in the formation of two identical daughter cells (the name given to the cells produced through cell division). If the same DNA is not replicated, the result will be different kinds of cells. In 1940, Linus Pauling and M. Delbrück speculated that the surface of the genetic material provided a template, creating molecules of complementary shape that then became their own template, resulting in the creation of a material identical to the original (Pauling and Delbrück 1940). It was against this backdrop that Watson and Crick presented their double helix model for DNA in 1953, after which speculation as to DNA's replication mechanism was based on that model.

Because DNA consists of genes, it is located in the cell nucleus and is a component of the chromosome. When the cell divides, chromosomes are split equally and distributed to two cells, both of which must have the same DNA both in quantitative and qualitative terms. To ensure this, DNA must be precisely replicated prior to division to produce two identical molecules. This is known as DNA autoreproduction, where the double helix serves a structure to allow this process to occur (Fig. 2.11).

When autoreproduction begins, the hydrogen bonds between bases are first broken apart by helicase and two strands of main chains separate. In the gap between them gather the four nucleotides that make up DNA. These form complementary pairs with the bases from each separated chain, which are linked through hydrogen bonds. The sugar components of each nucleotide are subsequently linked with phosphate and a long chain is formed. Obviously, several enzymes, including those related to DNA synthesis, are involved in this synthesis process.

As a result of this, a new, complementary polynucleotide chain is synthesized along the entire length of each original DNA strand, resulting ultimately in the formation of two DNA molecules with the exact same base pair arrangement. In other words, the two existing DNA strands function as templatess for synthesis of a new chain. Because each of the two newly synthesized DNA molecules contains a strand from the original DNA chain that served as a template, this is referred to as "semiconservative replication." Watson and Cricks' double helix structure was thus capable of explaining the various properties and functions observed for DNA to date, and it was immediately recognized as the prevailing theory.

E. coli DNA consists of around 4.2 million nucleotide pairs, measuring approximately 1.4 mm in length and consisting of a ring with its two ends connected (Galau et al. 1974). These are packed tightly within the cell, which measures around 0.0007 mm across and 0.001–0.004 mm in length. At the same time, it has no cell nucleus surrounding it. This form of organism possessing DNA but no nuclear membrane is known as a prokaryote; examples include bacteria and blue–green algae. In other organisms, proteins known as histones combine with DNA molecules. These are divided into set numbers; when division occurs, they cluster to

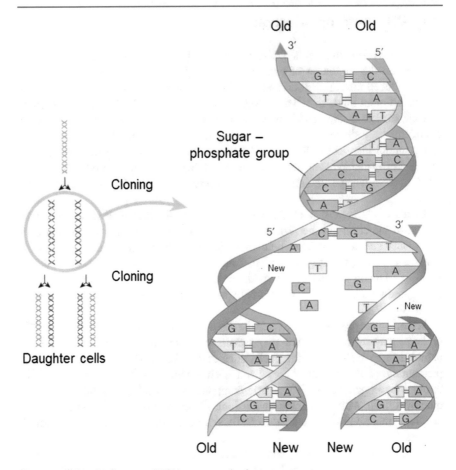

Fig. 2.11 Mimetic diagram of DNA autoreproduction

form thick and short chromosomes. When division is not occurring, the entirety is surrounded by a nuclear membrane; because a cell nucleus is present, such organisms are known as eukaryotes. The total number of nucleotide pairs in the human chromosome is estimated to be around three billion.

2.8 Plasmids

In addition to the DNA carrying the genes for all traits (sometimes called the bacterial chromosome), *E. coli* cells also contain a few separate pieces of DNA that are much smaller, at around one-millionth the scale. The genes that determine the *E. coli*'s sex or confer resistance to antibiotics are very small "mini-genes," ring-shaped DNA that replicate and proliferate independently from the main DNA and are passed down to offspring when the cell divides. While not essential for the cell's survival, they dominate different types of traits from the main DNA. These genes are called plasmids.

Eukaryote mitochondria and chloroplasts (see Fig. 2.3) also contain several ring-shaped pieces of mini-DNA. In addition to information for the proteins that form the mitochondria and chloroplasts, this DNA has also been found to carry information for leaf spots and male sterility, a condition in which pollen is not formed in flowers. Mitochondria and chloroplasts (for plants) are contained in egg cells; since only the nucleus is received from the male gamete during fertilization, DNA for these organelles contained in fertilized eggs and the somatic cells that develop from them is thought to come entirely from the mother. These plasmid-determined traits that are passed down from the mother thus do not conform to Mendel's laws, a phenomenon known as "cytoplasmic inheritance" or "maternal inheritance."

2.9   What Is Genetic Information?

So what is this diagram for proteins contained in DNA as genetic information? Before considering this question, it is first necessary to understand proteins.

Proteins (also known as polypeptides) are polymer compounds formed as water molecules leave amino acids and peptide bonds form between the acids, linking them into a chain. The number of amino acids formed is typically greater than 100 (see Figs. 2.12 and 2.13). Varieties include fibrous proteins like the keratin in hair and the fibroin in silk thread, in which peptides are linked through several S-S or hydrogen bonds and arrayed regularly in a single direction to form a long chain, and globular proteins such as enzyme proteins, in which peptide chains are bent and overlap into a round general form.

Twenty types of amino acids combine to form proteins. While the same amino acid may repeat several times, the sequence of amino acid combination differs between proteins. This sequence of amino acids is called the protein's primary structure.

In addition to the primary peptide bonds among their amino acids, the unique shapes of natural proteins are also maintained through secondary ionic, hydrogen, and S–S bonds between proteins molecules as a result of the many free amino groups and free carboxyl groups existing within those molecules. Those three-dimensional structures, however, are often transformed under the influence of physical actions such as heating, freezing, high pressure, and ultraviolet ray exposure and by chemical effects from acids, bases, organic solvents, and fibers. This transformation of structure in natural substances is known as denaturation.

In most cases, denaturation of natural globular proteins results in unfolding, which causes a sharp increase in the number of active sulfhydryl groups, amino groups, and carboxyl groups that had not been exposed in the original natural protein. Secondary bonds then form between these active groups, forming an insoluble protein. Because the biochemical activity of proteins such as enzymes and

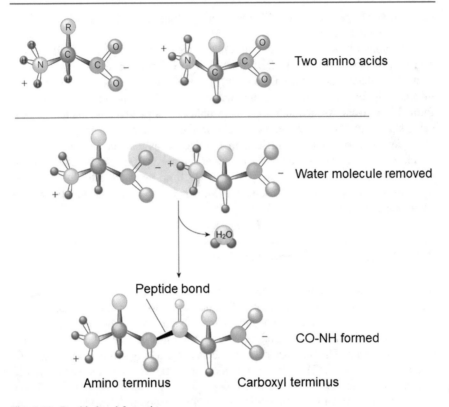

Fig. 2.12 Peptide bond formation

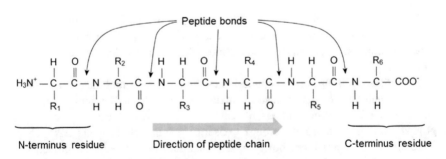

Fig. 2.13 Small peptide showing the direction of the peptide chain

hormones is determined by their three-dimensional structure, changes in that structure often result in them losing this activity.

Crucially, the molecule's three-dimensional structure and functions are established automatically, and the specific type of protein designated, once the protein's primary structure is given. The reference to a "diagram" in DNA concerns this primary structure of the protein, or its amino acid combination sequence.

Whereas DNA is contained entirely within the cell's nucleus, however, protein synthesis takes place on ribosomes contained in the cytoplasm, which requires the DNA diagram to be copied precisely within the nucleus and transported to the ribosomes. This role is performed by RNA.

In its principles, the mechanism is the same as for DNA autoreplication. Genetic information is not transcribed directly from DNA to proteins; rather, it is first transcribed to mRNA. One strand of DNA serves as the information chain here, with mRNA created to complement its base sequence. The enzyme that catalyzes this reaction is DNA-dependent RNA polymerase. When mRNA synthesis occurs, the base matched to adenine (A) from the DNA is not thymine (T), but uracil (U). Once the necessary information has been transcribed, the mRNA exits the nucleus and enters the cytoplasm, while the DNA returns to its original double helix shape. The name "mRNA," referring to the RNA that transports the protein diagram, is short for "messenger RNA." Other types of RNA include transfer RNA (tRNA), which carries the amino acids that form the raw material for proteins to the ribosome within the cytoplasm, and ribosomal RNA (rRMA), which is a component of the ribosomes. All of them are polynucleotide strands formed with DNA as a template (Figs. 2.14 and 2.15).

The question of how amino acid combination sequences form through DNA molecules is a very intriguing one. It is very likely that names of the amino acids are somehow encoded rather than written individually in sequence. In 1954, George Gamow found that in the four kinds of bases that form DNA, one sequence of three bases corresponded to a single amino acid (Gamow 1954). If a single base coded for a single amino acid, it would be impossible to designate more than four; if two bases coded for one amino acid, the result would be 16 types (4×4), which would not be enough to account for the 20 existing types. If three bases coded for a single amino acid, however, there would be 64 possible types ($4 \times 4 \times 4$), which would be more than adequate even if several different codes existed for the same amino acid. Indeed, genetic experiments found codons to be composed of three bases, with each codon designating an amino acid. These three-base codons are also known as "triplets."

The first amino acid to be "decoded" was phenylalanine. Its code was a series of three adenine bases (AAA) in DNA, and three consecutive uracil bases (UUU) in the messenger RNA transcribing it. Decoding of all 64 types was complete by around 1966, or 100 years after Mendel. Typically, the term "genetic code" refers to the base triplets transcribed to mRNA. Table 2.2 shows gene codes based on comparison of the DNA base sequence and the amino acid sequence in the resulting proteins.

In this table, the sequences UAG, UAA, and UGA do not code for any amino acid; when these units appear, they indicate the end of protein synthesis. The sequence AUG is a start codon that initiates reading of the code, but codes for methionine if it appears once reading has begun. As can be seen from the genetic code table, nearly all amino acids correspond to multiple codons, a phenomenon referred to as codon degeneracy. The number of codons conforms to the frequency with which their amino acids appear in proteins.

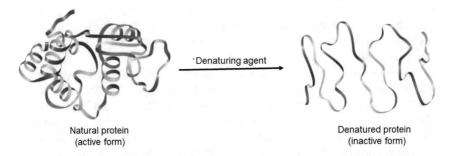

Natural protein
(active form)

Denatured protein
(inactive form)

Fig. 2.14 Denaturation of a protein (The natural three-dimensional structure is sometimes restored when the denaturation conditions are removed)

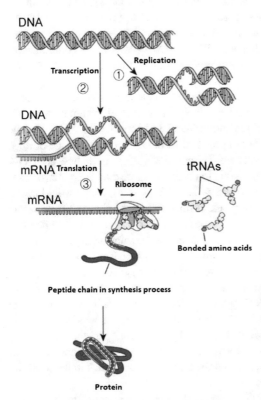

1. Replication

In DNA replication, two DNA molecules identical to the first are created, passing extremely accurate genetic information on to daughter cells.

2. Transcription

The DNA base sequence is converted into a complementary base sequence within a single-stranded mRNA molecule.

3. Translation

A three-base codon on the mRNA molecule corresponds to a specific amino acid, providing the amino acid sequence for protein generation. Codons are recognized by tRNA carrying the appropriate amino acids. Ribosomes are the "machinery" used to synthesize proteins.

Fig. 2.15 Replication, transcription, and translation of genetic information (This process is identical in general for both prokaryotes and eukaryotes)

The next question is how proteins are synthesized according to the information contained on mRNA. The mechanism by which proteins are formed on ribosome from mRNA information has been elucidated chiefly through the *E. coli*, and it is on

Table 2.2 The standard genetic code

First position (5'-end)	Second position				Third position (3'-end)
	U	C	A	G	
U	UUU Phe	UCU Ser	UAU Tyr	UGU Cys	U
	UUC Phe	UCC Ser	UAC Tyr	UGC Cys	C
	UUA Leu	UCA Ser	UAA Stop	UGA Stop	A
	UUG Leu	UCG Ser	UAG Stop	UGG Trp	G
C	CUU Leu	CCU Pro	CAU His	CGU Arg	U
	CUC Leu	CCC Pro	CAC His	CGC Arg	C
	CUA Leu	CCA Pro	CAA Gln	CGA Arg	A
	CUG Leu	CCG Pro	CAG Gln	CGG Arg	G
A	AUU Ile	ACU Thr	AAU Asn	AGU Ser	U
	AUC Ile	ACC Thr	AAC Asn	AGC Ser	C
	AUA Ile	ACA Thr	AAA Lys	AGA Arg	A
	AUG Met[a]	ACG Thr	AAG Lys	AGG Arg	G
G	GUU Val	GCU Ala	GAU Asp	GGU Gly	U
	GUC Val	GCC Ala	GAC Asp	GGC Gly	C
	GUA Val	GCA Ala	GAA Glu	GGA Gly	A
	GUG Val	GCG Ala	GAG Glu	GGG Gly	G

[a]AUG forms part of the initiation signal as well as coding internal methionine residues

the *E. coli* that this explanation will focus. The synthesis process is divided into initiation, elongation, and termination stages (Fig. 2.16).

(1) Initiation: Before protein synthesis begins, the ribosome dissociates into subunits (a). These small subunits bond near the AUG start codon on the mRNA (see Fig. 2.15). The AUG start codon then bonds with formylmethionyl-tRNA (fMet-tRNA), which has CAU as an anticodon, to form an initiation complex (c). The initiation complex then associates the larger subunit to form a complete ribosome (d). The location where the fMet-tRNA bonds on the ribosome is called the peptidyl site (P-site).

(2) Elongation: The next codon on the mRNA (GCU in the figure) is associated by the second aminoacyl-tRNA, here with IGC as its corresponding code unit (e). It should be noted that the amino acid does not associates directly with the mRNA, but uses tRNA as an adaptor. Aminoacyl-tRNA synthetase serves to combine each amino acid with its specific tRNA. The location where the aminoacyl-tRNA bonds is known as the aminoacyl site (A-site). The peptide bond then forms between formylmethionine (fMet) and alanine with the enzyme peptidyl transferase (f). Under these conditions, the empty tRNA migrates to the P-site and the fMet-tRNA from the A-site to the P-site, and the ribosome moves the equivalent of one codon down along the mRNA chain. A third piece of aminoacyl-tRNA (here phenylalanine-tRNA) then joins the next codon (here UCC) (g). Steps (e) through (g) are repeated and the peptide chain grows gradually longer (h). Termination: Once a stop codon (UAA, UGA, or UAG) is encountered, there is no longer any tRNA with a complementary codon, and peptide chain synthesis is terminated (i). At this point,

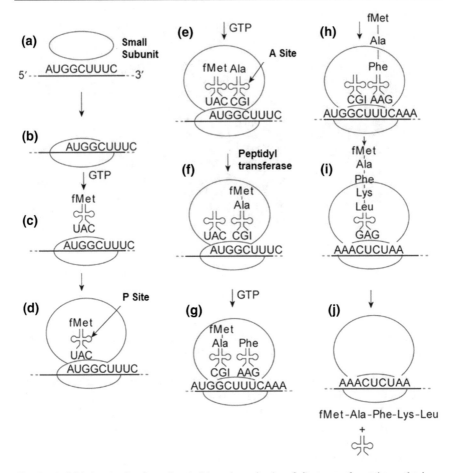

Fig. 2.16 Initiation (**a–d**), elongation (**e–h**), and termination (**i–j**) stages of protein synthesis

a releasing factor that recognizes the stop codons begins to operate, and the completed peptide chain exits the ribosome (j). Following step j, the formyl group (–CHO) is removed by a specific enzyme. In most cases, terminal methionine is also removed.

Once synthesized, the protein chain exits the ribosome and bends automatically to form a characteristic three-dimensional structure. Having served its purpose, the mRNA is immediately decomposed. Ribosomes in the body perform protein synthesis efficiently by forming several bonds with a single molecule of mRNA. A cluster of ribosomes attached to mRNA is known as a polysome. The rate of protein synthesis is such that five seconds is adequate to produce a polypeptide consisting of 100 residues. Figure 2.17 shows a mimetic illustration of the direction of flow for genetic information contained in DNA.

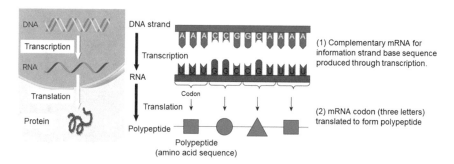

Fig. 2.17 Relationship diagram for the DNA information chain, mRNA, and codons

Genetic information is determined by the DNA base sequence; replication involves the creation of a strand with identical information. Genetic information is not transmitted directly from the DNA to the protein, but is first transcribed to mRNA. Here, one strand of DNA serves as the information chain from which complementary mRNA is formed.

It is sequences of three bases in mRNA that serve as the code for specific amino acids, information that is translated to form polypeptides. In general, this process of protein formation is identical for prokaryotes and eukaryotes; James Watson referred to it as the "central dogma." The later discovery of viruses with genetic information stored as RNA, which is then transcribed to DNA through the enzyme reverse transcriptase, resulted in the addition of an arrow pointing in the opposite direction, whereby DNA is synthesized from RNA (Watson 2012).

DNA thus contains diagrams for every protein that a cell or organism is capable of synthesizing. *E. coli* cell includes around 3000 types of proteins, and its DNA carries information for every one of them. Cells can use this information to synthesize proteins as needed, but are incapable of synthesizing proteins for which no information is contained in the DNA. DNA is also precisely replicated prior to cell division (see Fig. 2.10) and distributed equally to two daughter cells, such that for multicellular organisms, cells in any given part of the body contain the same information. That information is then passed down to offspring through gametes such as sperm and egg cells. Fundamentally, the aforementioned functions of the genes and genetic code apply to all organisms.

2.10   Why RNA Uses Thymine Instead of Uracil

The reason that DNA uses thymine instead of uracil long remained a mystery. Recently, however, it was found why the absence of uracil in DNA is advantageous. The cytosine in DNA naturally becomes uracil through deamination; the resulting uracil then forms a pair with adenine, creating an AU base pair instead of the original GC and causing a mutation. In such cases, uracil that is not present in

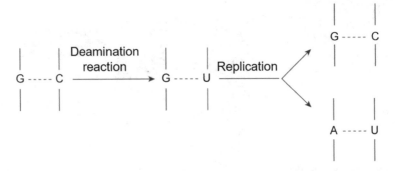

Fig. 2.18 Repair of a GU base pair. The uracil group from DNA is removed and its place filled by cytosine

DNA is recognized through aDNA conservation mechanism, and mutations through cytosine deamination can be repaired to their original shape (Fig. 2.18) (Lesk 1969).

First, uracil-DNA glycosidic bond hydrolyzes the glycosidic bond between deoxyribose and the uracil formed through the cytosine deamination. The DNA structure remains unimpaired, but one base (uracil) is removed; the resulting empty space is capped the AP-site (apurinic or apyrimidinic site). The AP-site is then recognized by AP endonuclease, and a nick is made in the structure near the absent base site. DNA polymerase I makes a nick in remaining deoxyribose phosphate that does not possess bases, and cytosine is introduce to pair with guanine in the complementary strand. The gap is finally filled by the enzyme DNA ligase, and the DNA is restored to its original shape (Fig. 2.19).

Unlike uracil, thymine possesses a methyl group at its C_5 site, while the DNA repair enzyme uracil-DNA–glycosylase (UDG) removes only the uracil, and not the thymine, from DNA. The DNA repair mechanism thus recognizes the methyl group-possessing thymine as a normal DNA base and the methyl group-lacking uracil as an attachment to DNA, which allows it to prevent the GC-to-AU mutations that take place due to cytosine's natural deamination. In other words, the reason that DNA possesses thymine as a normal base and does not use uracil appears to be that this increases the fidelity of genetic information transmission. Unlike DNA, RNA does not undergo repair, and uses the methyl group-lacking uracil as a normal base.

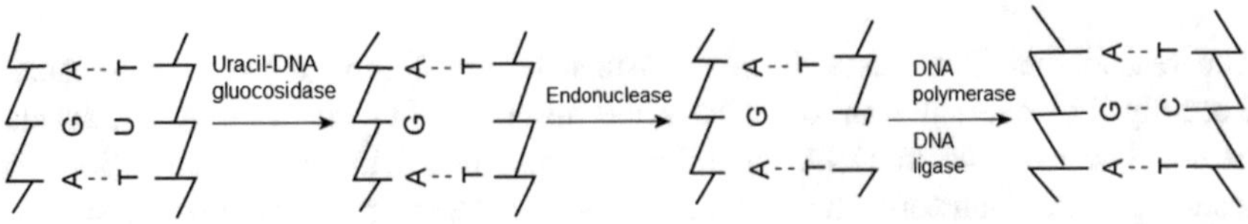

Fig. 2.19 Repair of a GU base pair. The DNA uracil group is eliminated and its space filled with cytosine

2.11 Restriction Endonuclease: Nomenclature, Types, and Characteristics

Bacteria possess restriction enzymes that nick and eliminate genetic information entering the cell in virus or DNA form. In 1970, H. O. Smith, and D. Nathans succeeded for the first time in purifying a restriction enzyme. More than 300 have been discovered since then.

One restriction enzyme is deoxyribonuclease, which recognizes DNA's specific base sequence and passes on the DNA chain. It is through this enzyme that long and thin chromosomal DNA is cut into various fragments, allowing specific pieces to be obtained. For the names of restriction enzymes, a capital letter representing the genus of the bacterium producing it is used, followed by two lower case letters from the species; the three letters are represented in italics. When the origin is listed, the name of the strain or plasmid is written afterwards. When two or more enzymes have been isolated from the same strain, Roman numerals are added to distinguish between them. For example, an example isolated from *Eschericia coli*, which possesses the antibiotic resistance factor R, would be labeled *Eco* RI, while examples isolated from *Haemophilus influenza* include *Hind* I, *Hind* II, and *Hind* III (Roberts et al. 2003).

For most bacteria, methylated bases within their DNA possess characteristic shapes, and a mechanism exists to distinguish and secrete different forms of DNA entering the cell from its own DNA. Modification methylase and restriction endonuclease are both part of this mechanism.

Modification methylase performs the role of producing the species's specific methylation form within the characteristically shortened base sequence in the host cell's DNA. This methylated base sequence is repeated several times in the host cell DNA. The methyl group in this sequence remains in its state throughout the cell's lifespan. In contrast, restriction endonuclease clips two different DNA chains when this characteristic base pair is not methylated.

When DNA becomes introduced through infection of the bacteria by a phase, the restriction enzyme distinguished whether it is self-modeled or outside-modeled DNA. If it is not a modified version of the cell's own model, the DNA is cut and deactivated, preventing reproduction of the outside phage.

The restriction and modification phenomenon was first reported in 1952 by Bertani and Weigle; an example is presented in Fig. 2.20. When a lambda phage ($\lambda \cdot$C) growth on the E. coli's C strain infected its K strain, breeding was restricted and only around 2/10,000 of the infecting phage's plaque formed. When the K strain was then infected with the lambda phage ($\lambda \cdot$K) breeding on the K starin ($\lambda \cdot$K), however, efficient plaque formation occurred. When the $\lambda \cdot$K was bred on the C strain and used to infect the K strain, the plaque formation rate was again around 2/10,000, showing that the λ phage's host range was not the result of mutation. In the 1960s, Arber concluded, based on the fact that this phenomenon was accompanied by DNA cutting and that uncut DNA was modified by methylation, that this

Fig. 2.20 Restriction
modification by the *E. coli* λ
phage K strain (Number
indicates efficiency of plating)

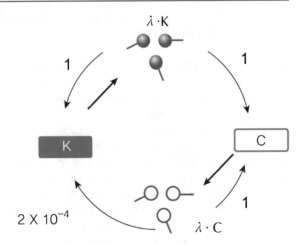

was restriction modification, and predicted the existence of a DNA breakdown
enzyme participating in the restriction (Wood 1966).

In performing their restriction, all restriction enzymes recognize sequences of
four to six bases and hydrolyze phosphodiester bonds. Two varieties of this can be
identified (Fig. 2.21).

The restriction enzyme *Hind* II produces DNA fragments with even ends by
cutting simultaneously at the same location on two strands from the center of the
recognition side (as indicated by the direction of the arrow). The resulting ends are
referred to as "flush" or "blunt." *Eco* RI, in contrast, recognizes and cuts palindromic sites on two strands of DNA, regardless of type. The resulting DNA
fragment ends have exposed portions of the strands with complementary nucleotide
sequences, allowing them to be joined easily through DNA ligase. Accordingly,
these ends are known as "cohesive" or "sticky." This ability to create ends that join
easily with foreign DNA is an important characteristic of restriction enzymes
(Sutcliffe 1978).

Restriction enzymes can be classified into types I, II, and III according to their
characteristics.

Type I restriction enzymes consist of molecular weight (Mw) 300,000–400,000
made up of various types of subunits, and possess modification methylase and
ATPase activity in addition to endonuclease activity. These enzymes require Mg^{2+},
ATP, and s-adenosylmethionine at the time of reaction, differ in the base sequences
and cutting sites that they recognize, and do not have define restriction sites.

Type II restriction enzymes require Mg^{2+} for their activity, but do not require
ATP or s-adenosylmethionine. They are also far smaller than Type I enzymes,
consisting of Mw 20,000–100,000. Type II enzymes recognize specific base
sequences within DNA and cut at defined sites. For example, the characteristic
Type II enzyme *Eco* RI cuts the ↓ section in the six-base sequence 5′-G↓AATTC-3′.
These base sequences differ between restriction enzymes, but enzymes such as *Pst* I
and *Sal* I recognize exactly the same sequences and cut at the same sites despite

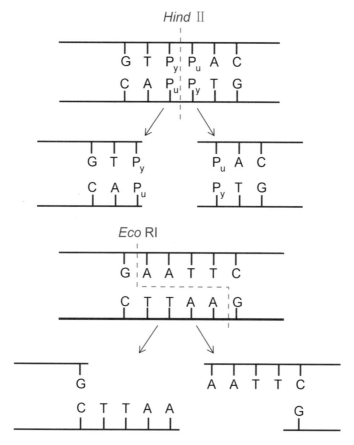

Fig. 2.21 Method of DNA cutting for the restriction enzymes *Hind* II and *Eco* RI

being obtained from different kinds of bacteria. Such enzymes are known as iso-schizomers. The bases recognized by Type II enzymes are specific sequences within a double strand of DNA, many of which have a palindromic structure with double-rotation symmetry. Among Type II enzymes are those that cut the same site on two strands of DNA to produce a flush end, and those for which the truncation site is located between the two strands, resulting in a cohesive end where either the 5′ or 3′ end protrudes. In either case, the 3′ end has an OH group and the 5′ a phosphate group, allowing for rejoining with DNA ligase.

Type III restriction enzymes possess endonuclease and methylase activity and require Mg^{2+} and ATP for DNA cutting. If ATP and *s*-adenosylmethionine are present, DNA methylation also occurs. Moreover, while Type III restriction enzymes recognize specific base sequences within the DNA molecule, they differ from Type II and cutting the DNA at a location somewhat distant from that sequence.

2.12   What Are Monoclonal Antibodies?

Because sexual reproduction through the fertilization of an egg by sperm involves recombination of the father and mother's chromosomes, the resulting second generation possesses mixed traits from both parents and is identical to neither of them. Asexual reproduction, in contrast, involves replication of the same chromosomes, which are distributed equally to two cells that are genetically identical. In other words, a single cell divides repeatedly to form a group of completely identical cells. The cells or individuals created from a single cell or individual through this form of asexual reproduction are known as clones, and their creation is referred to as cloning. An additional form of cloning in genetic engineering involves the isolation of two identical genes. The first experiments with animal cloning used frog eggs. Removal of the eggs' nucleus and insertion of a nucleus from the small intestine of a tadpole resulted in clones of completely identical frogs. Later experiments involved the cloning of mice. While human clones may appear in comics, this is actually a matter of semantics; not only is such a thing unnecessary, but no actual experiments to create it have ever been performed (Khazaeli et al. 1994; Groth and Scheidegger 1980).

The benefit of monoclonal cells is that creating them allows for research under the same conditions anywhere and at any time. For instance, the human cell as "HeLa" was extracted in 1951 from the U.S. uterine cancer patient Henrietta Lacke and bred into clones at cancer institutes all over the world. Cloning of a cell thus allows for mass production of identical materials in any location. Another example of this is the monoclonal antibody. Antibodies perform the role of bonding with and eliminating antigens, or foreign material such as bacteria or virus that enter an animal's body. The antibody is a protein known as immunoglobulin, which is produced by specialized plasma cells in lymphocytes (B lymphocytes, to be specific). Clones exist for the million or so varieties of lymphocytes in the human body, but each lymphocyte clone reacts only to one form of antigen. When an antigen invades, a lymphocyte clone carrying a corresponding antibody on its surface reproduces and specializes into plasma cells, which produce large volumes of antibodies. Because there are already many clones producing different antibodies within the body, many different varieties are already included in examples obtained from the blood. These are known as polyclonal antibodies. In this case, the term monoclonal antibody refers to a single kind of antibody produced by a single lymphocyte clone (or, to be more precise, the antibody that reactions only with a single antigenic determinant). In 1975, C. Milstein and G. Köhler developed a method for producing monoclonal antibodies. As shown in Fig. 2.22, a mouse was first given immunity with antigens. The mouse was then made to produce large volumes of antibodies for two antigens, after which the spleen (which contains the lymphocytes) was extracted. The lymphocytes from the spleen were subsequently fused with cancerous myeloma cells from the bone marrow. Two methods are used for fusing two types of cells, one in which polyethylene glycol is used to destroy part of the membrane and one in which electrical stimulation is applied.

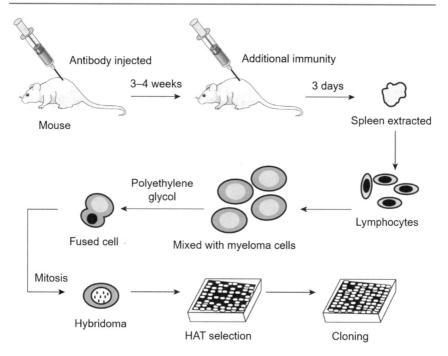

Fig. 2.22 Making monoclonal antibodies

The "hybridoma" cell resulting from the fusion of lymphocyte and myeloma cell possess properties of both the lymphocyte, which produces antibodies, and the cell, which can reproduce infinitely. Cloning of a hybridoma that produces antibodies for a certain antigen results in cells that only produce monoclonal antibodies. Benefits of monoclonal antibodies include the fact that they can be created identically anywhere in the world from the same hybridoma, and that because they are monoclonal, it is easy to interpret their activity. Currently, monoclonal antibodies are being put to use in self-pregnancy testing, which examines for a hormone known as human chorionic gonadotrophin (hCG) that is present in the urine of a pregnant woman. When the testing strip is placed in the urine, monoclonal antibodies attached to it bind with the hormone in the urine and change the strip's color. The antibodies are also being used for various other clinic diagnosis purposes, including cancer and allergy kits (Goding 1996).

2.13   DNA Sequencing and Marine Omics

DNA sequencing is a process of identify the nucleic acid sequence, otherwise, order of nucleotide in DNA. Any methods or any technology helps to determine the four important bases, adenine, guanine and thymine. Knowledge of DNA sequencing technology is playing major role in medical diagnostics, forensic medicine, medical biotechnology and system biology. Fast and accurate sequencing of DNA can be attained with modern instrument based technology. Next generation techniques and marine omics (genomics, metagenomics, glycomics, transcriptomics, metabolomics, nutrigenomics, pharmacogenomics, bioinformatics, lipidomics, and toxic genomics have been extensively discussed in the previous published book "Marine Omics: Principles and applications" book for further reading (Kim 2016).

2.14 Chapter Summary and Conclusion

The field of marine biotechnology includes many unfamiliar terms and marine organism names, as well as some species of marine plants and animals that do not exist in Korea. Korean students may also encounter difficulties in understanding newly minted academic terminology that does not exist in Korean, or does so only to an inadequate degree.

As a general science, the study of marine biotechnology requires first and foremost an understanding of such basic disciplines as biology, genetics, biochemistry, microbiology, organic chemistry, and fisheries science.

To assist in understanding marine biotechnology, this chapter attempted to focus on various basic aspects of genetics and other related areas. Because of page limitations, an explanation of related terminology has been provided in the Appendix for reference.

References

Avery, O. T., MacLeod, C. M., Lederberg, J., Dubos, R., & McCarty, M. (1944). Symposium February 2, 1979. *The Journal of Experimental Medicine, 79*(2), 137–158.

Bateson, W., & Mendel, G. 2013. *Mendel's principles of heredity*. Courier Corporation.

Chargaff, E. (1950). Chemical specificity of nucleic acids and mechanism of their enzymatic degradation. *Experientia, 6*(6), 201–209.

Chargaff, E. (2012). *The nucleic acids*. Elsevier.

Galau, G. A., Britten, R. J., & Davidson, E. H. (1974). A measurement of the sequence complexity of polysomal messenger RNA in sea urchin embryos. *Cell, 2*(1), 9–21.

Gamow, G. (1954). Possible relation between deoxyribonucleic acid and protein structures. *Nature, 173*(4398), 318.

Goding, J. W. (1996). *Monoclonal antibodies: Principles and practice*. Elsevier.

Griffith, F. (1934). The serological classification of *Streptococcus pyogenes*. *The Journal of Hygiene, 34*(4), 542.

Groth, S. F. d. S., & Scheidegger, D. (1980). Production of monoclonal antibodies: Strategy and tactics. *Journal of Immunological Methods, 35*(1–2), 1–21.

Hershey, A. D., Dixon, J., & Chase, M. (1953). Nucleic acid economy in bacteria infected with bacteriophage T2: I. Purine and pyrimidine composition. *The Journal of General Physiology, 36*(6), 777–789.

Khazaeli, M., Conry, R. M., & LoBuglio, A. F. (1994). Human immune response to monoclonal antibodies. *Journal of Immunotherapy with Emphasis on Tumor Immunology: Official Journal of the Society for Biological Therapy, 15*(1), 42–52.

Kim, S.-K. (2016). *Marine OMICS: Principles and applications*. CRC Press.

Lesk, A. M. (1969). Why does DNA contain thymine and RNA uracil? *Journal of Theoretical Biology, 22*(3), 537–540.

Mendel, G. (1996). Experiments in plant hybridization (1865). *Verhandlungen des naturforschenden Vereins Brünn.* Accessed on January 1, 2013. Available online: www.mendelweb.org/Mendel.html.

Mendel, G., Corcos, A. F., & Monaghan, F. V. (1993). *Gregor mendel's experiments on plant hybrids: A guided study.* Rutgers University Press.

Miescher, F., & Schmiedeberg, O. (1896). Physiologisch-chemische Untersuchungen über die Lachsmilch. *Archiv für experimentelle Pathologie und Pharmakologie, 37*(2–3), 100–155.

Morgan, T. H., Bridges, C. & Sturtevant, A. (1925). The genetics of *Drosophila melanogaster. Biblphia Genet, 2*(1–262).

Orel, V. (1996). *Gregor Mendel: The first geneticist.* USA: Oxford University Press.

Pauling, L., & Delbrück, M. (1940). The nature of the intermolecular forces operative in biological processes. *Science, 92*(2378), 77–79.

Roberts, R. J., Belfort, M., Bestor, T., Bhagwat, A. S., Bickle, T. A., Bitinaite, J., et al. (2003). A nomenclature for restriction enzymes, DNA methyltransferases, homing endonucleases and their genes. *Nucleic Acids Research, 31*(7), 1805–1812.

Sutcliffe, J. G. (1978). pBR322 restriction map derived from the DNA sequence: accurate DNA size markers up to 4361 nucleotide pairs long. *Nucleic Acids Research, 5*(8), 2721–2728.

Wang, K. (2018). DNA-based single-molecule electronics: From concept to function. *Journal of functional Biomaterials, 9*(1), 8.

Watson, J. (2012). *The double helix.* UK: Hachette.

Watson, J. D., & Crick, F. H. (1953). The structure of DNA. In *Cold Spring Harbor symposia on quantitative biology* (pp. 123–131). Cold Spring Harbor Laboratory Press.

Wood, W. B. (1966). Host specificity of DNA produced by *Escherichia coli*: Bacterial mutations affecting the restriction and modification of DNA. *Journal of Molecular Biology, 16*(1), 118-IN3.

Fish Genetics

3

Contents

3.1 Gene Expression in Fish and Other Eukaryotes

Cells have a characteristic structure that is appropriate for performing their individual functions. Sperm cells, for example, have powerful flagella that allow them to swim through a female's reproductive tract and find the egg. Nerve cells are elongated in shape in order to communicate signals between distant parts of the body. In this way, the different cells in the human body generally support the living individual in its activities (Chalfie et al. 1994).

What is responsible for this sort of diversity in cells? Do the cells possess different genes? All somatic cells (that is, every cell apart from reproductive cells) in our bodies are the result of repeated somatic cell division that begins in the fertilized egg. Because somatic cell division results in exact replications of the genome, the many cells in our bodies all possess the same genome as the original fertilized egg. So what accounts for the differences among cells? The only way for a cell bearing the same genetic information to develop into various cells possessing different structures and functions is through regulation of gene activity.

© Springer Nature Switzerland AG 2019
S.-K. Kim, *Essentials of Marine Biotechnology*,
https://doi.org/10.1007/978-3-030-20944-5_3

In the cell, regulators somehow activate certain genes while leaving others inactive. In other words, each cell must undergo a process of cellular differentiation (structural or functional specialization of cells). This process of specialization is directed by gene regulation.

What does it actually mean for genes to activate or deactivate? Genes determine base sequences in specific messenger RNA (mRNA), which then determines the sequence of amino acids in proteins. Activated genes are transcribed to RNA, and the resulting message is translated into a specific protein. The term "gene expression" refers to the entire process of genetic information flowing from gene to protein (Schena et al. 1995).

Unlike simple unicellular organisms such as bacteria or blue-green algae, eukaryotes are cells with a nucleus enclosed in a membrane and specialized organelles that perform essential functions.

Genes in eukaryotes include "housekeeping genes," or genes that are permanently expressed structurally and code for proteins that are necessary for the cell's general functioning, and "luxury genes," which are expressed on a more or less permanent basis in cells with different specialization, but characteristically in cells such as hemoglobin or immunoglobin that are particularly specialized due to an induced reaction.

Most genes in eukaryotes are classified as either exons or introns. When the gene undergoes phenotypic expression, introns typically become linked to nearby exons to undergo RNA transcription. Alternatively, the exon of the RNA molecule assumes the same sequence and directionality as the original gene during post-transcription RNA processing in the nucleus, while the intron becomes loop-shaped RNA and is removed. Typically, the 5′ end of the intron is a GU sequence and the 3′ end an AG sequence, in what is known as Chambon's rule. A relatively shared base sequence exists for the mRNA precursor at the splicing area, or the boundary section between the exon and intron parts.

RNA splicing mechanisms differ according to the type of intron included in the RNA (precursor). In the case of splicing of the mRNA precursor encoded by the eukaryote nucleus, a splicing complex is first created by the mRNA precursor and small nuclear RNA (SnRNA). Simultaneously, a loop-shaped intron/exon intermediate RNA form is created by a 2′–5′ bond between the non-expressed portion's 5′-phosphoric acid end created by truncation at the intron/exon boundary (the 5′ splicing site) and the adenosine ribose-2′–OH at the molecular position within the intron (stage 1). This is followed by the simultaneous joining of the two exon and formation of the loop-shaped intron portion (stage 2). Both stages are phosphoric acid ester exchange reactions that take place due to 2′–OH (stage 1) or 3′–OH at the truncated exon end (stage 2; see Fig. 3.1) (Black 2003).

7-methylguanosine (cap) joins the mRNA's 5′ end, which consists solely of the exon portion. Addition of a poly-A sequence chain at the 3′ end results in formation of mature RNA. This mature RNA migrates into the cytoplasm (outside the nucleus), and as the mRNA's three-nucleotide sequences designate a single amino acid, proteins are formed on the ribosome with the help of tRNA (Perry et al. 1987).

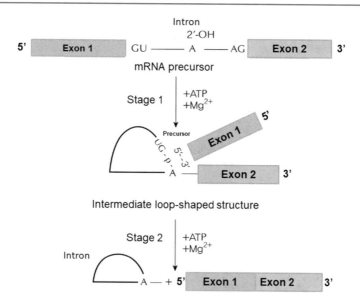

Fig. 3.1 mRNA precursor splicing reaction pathway in the nucleus

The optional expression of genes may be visualized through recently developed technology known as a DNA microarray (Fig. 3.2a). A DNA microarray is a glass slide that localizes thousands or more types of single-stranded DNA fragments arranged on an array (lattice). Each DNA fragment is obtained from a single gene, so that a single microarray contains thousands of genes' worth of DNA.

To use a microarray, researchers first assemble all mRNA expressed in a particular cell ①. As the mRNA is mixed with the viral enzyme reverse transcriptase, this results in ② synthesis of complementary DNA for each mRNA. This complementary DNA (cDNA) is synthesized with nucleotides marked with fluorescent substances. Small quantities of the fluorescent-marked cDNA mixture are then ③ mixed with thousands of kinds of single-stranded DNA in the microarray. If a molecule in the cDNA mixture complements a piece of DNA at a particular position in the microarray, that cDNA molecule binds with its corresponding DNA fragment and remains fixed at that location. Once the unbound cDNA molecules are washed off ④, the cDNA molecules remaining on the microarray become fluorescent (Fig. 3.2b). Observing the patterns in the fluorescent areas allows researchers to determine what genes have been turned on and off in the cell (Schena et al. 1995).

Microarray experiments have allowed researchers to observe differences in gene expression between different tissues and tissues in humans with different health conditions. It is a powerful new technology that has opened new horizons for gene regulation research.

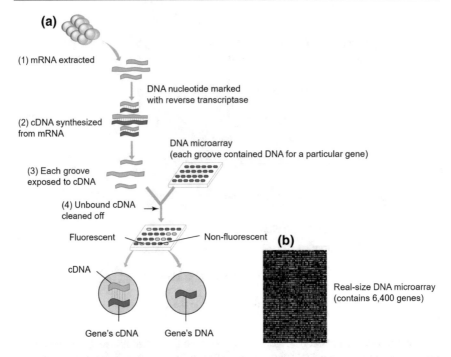

Fig. 3.2 DNA microarray. **a** To visualize the extent of gene expression, researchers assemble mRNA from a semicell to create fluorescent-marked cDNA. This cDNA is then applied to a DNA microarray with many types of fixed genetic DNA fragments. Once the unbound cDNA is washed away, the remaining fluorescent points indicate genes that are expressed in the cell used. **b** Microarrays are used to analyze many different types of gene activity simultaneously

3.2   Recombinant DNA Technology

In 1946, the U.S. geneticists J. Lederberg and E. Tatum, conducted experiments with the *E. coli* that showed the possibility of combination between two genes originating from different bacteria. This phenomenon was previously thought to occur only in sexual reproduction among eukaryotes. As a result of their findings, the two researchers explored the new field of bacterial genetics, and *E. coli* became the best-understood organism in the field at the molecular level. It was *E. coli* that brought about developments in recombinant DNA technology (i.e., techniques for combining genes originating in different organisms) during the 1970s (Young et al. 1985).

Gene recombination technology is widely used today to transform DNA in various cells for practical purposes. Through manipulation of genes in bacteria, scientists have succeeded in producing large quantities of useful substances ranging from anti-cancer agents to pesticides. It has also become possible for genes to be

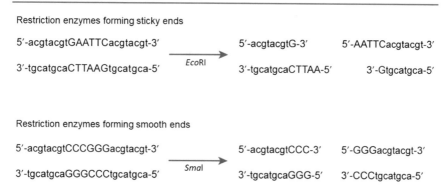

Fig. 3.3 Restriction enzymes responsible for truncating DNA and their truncated portions

transmitted from bacteria to plants or animals. These applications have ushered in advancements in practical biotechnology involving the use of biological organisms.

The discovery of restriction enzymes truncating double-stranded DNA ushered in rapid developments in recombinant DNA technology. Bacterial enzymes known as restriction enzymes perform the role of "scissors" in creating recombinant DNA. Most restriction enzymes recognize and clip specific sequences of bases consisting of three to eight DNA nucleotides. Types of truncation include smooth ends, in which a single strand is left without a "tail" as though clipped with scissors, and sticky ends, in which a chain with a complementary base sequence is left exposed at the 5′ or 3′ position. Over 300 types of restriction enzymes have been marketed to date (Fig. 3.3).

Once truncated by restriction enzymes, the DNA double helix can be connected with ligase. The use of restriction enzymes and ligase allows for joining of genes from entirely different origins. In 1973, S. N. Cohen of Stanford University and Boyer H. W. of the University of California perfected a genetic manipulation technique using circular cytoplasmic factor DNA (plasmids) to insert foreign DNA and developing a vector capable of amplifying it within the host cell. A variety of different vectors have since been developed, including E. coli, Bacillus subtilis (hay bacillus), and yeast (Yi 2011).

Vectors include a foreign gene insertion portion and functions (replication origins) for autoreplication within the host cells. They also possess optional indicators (drug-resistant indicators or lacZ′ genes) that allow them to distinguish between vector-possessing host cells and other cells. Establishment of this technology has permitted the amplification of large numbers of specific DNA fragments. Figure 3.4 shows an example of cloning using circular DNA (plasmid) vectors. In this process, genome DNA is partially truncated with the restriction enzyme Eco RI, and the same enzyme is used to truncate the vector DNA (pBR$_{322}$) and join it with ligase. With E. coli, the recombinant DNA is subjected to transformation, and a drug-resistant colony is selected (Aoki 2000).

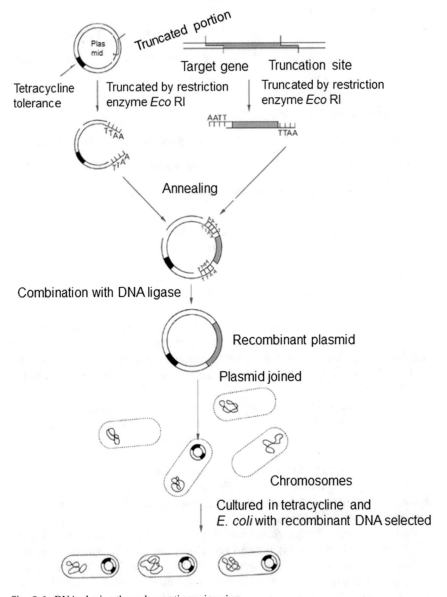

Fig. 3.4 DNA cloning through genetic engineering

3.3 Cloning of a Target Gene

The term "clone" refers to groups of organisms arising from asexual reproduction and organisms with the same genetic composition. "Cloning" refers to the creation of such groups of individual with identical genetic composition. Gene cloning is thus a phenomenon of separating out a particular gene and increasing its number by insertion in a prokaryote or eukaryote cell. It is therefore possible to use cells with the right gene for some purpose to create large quantities of the proteins that it encodes or to produce large quantities of the gene itself.

Several methods of gene cloning exist. First, there is the method of using restriction endonuclease to truncate the genome DNA in some organism to the length of approximately one gene. The recombinant DNA procedures through joining of that DNA fragment to a vector is introduced to *E. coli* or a similar host cell to function as a genome library, or mRNA is extracted from a donor and reverse transcriptase is used to synthesize complementary DNA (cDNA) to create a cDNA library. After that, a specific DNA probe is used to select and separate cells with the target gene through the genome library or cDNA by means of hybridization. Target cDNA may also be screened with antibodies.

Recently, target DNA has been cloned from organisms besides fish. Base sequences from the best preserved of these regions serve as a reference for creating primer, with the polymerase chain reaction (PCR) technique used to proliferate some or all of the cDNA or gene DNA region from the cDNA or genome library.

When the amino acid sequence for the protein produced by target DNA is at least partially determined, that sequence can serve as a reference for creating multi-primer, allowing for multiplication of the gene through PCR. Discovered by Kary Mullis in 1986, PCR is a technique by which DNA fragment becomes amplified. The principle behind PCR is simple. The process requires a particular form of DNA polymerase obtained from prokaryotes living in hot spring water. Unlike other proteins, the enzyme is capable of withstanding the high-temperature treatment needed to separate DNA strands in the PCR process. The DNA sample to be amplified is mixed with this characteristic polymerase, nucleotide monomers, and small pieces of DNA to serve as primers in DNA replication. Heat treatment of the mixture results in the DNA's replication, and as two new strands of DNA form, that new DNA undergoes its own separation and replication due to the heat in a repeating process. The result has been to make gene cloning extremely simple, as large numbers of DNA molecules can be obtained in a short time from a DNA fragment with the same base sequence (Fig. 3.5) (Ríus et al. 1998).

PCR is in wide use today, with applications not only in gene cloning but also in fish diagnosis, species and genus identification, and diagnosis of microbial infections.

Figure 3.6 shows the various gene cloning methods described above. Progress in gene manipulation has been accelerating, and staggering advancements continued to be made in the field.

Region of target DNA to be amplified

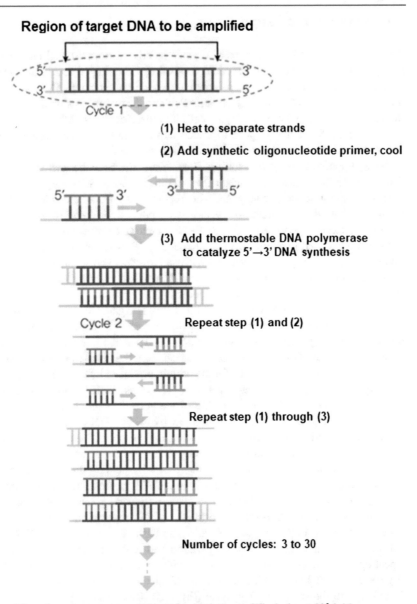

(1) Heat to separate strands

(2) Add synthetic oligonucleotide primer, cool

(3) Add thermostable DNA polymerase to catalyze 5'→3' DNA synthesis

Cycle 2   **Repeat step (1) and (2)**

Repeat step (1) through (3)

Number of cycles: 3 to 30

After 25 cycles, the target sequence has been amplified about 10^6 fold

Fig. 3.5 DNA amplification through the PCR process. The PCR procedure has three steps, DNA strands are (1) are separated by heating, then (2) annealed to an excess of short synthetic DNA primers that flank the region to be amplified; (3) new DNA is synthesized by polymerization. The three steps are repeated for 25 or 30 cycles. The thermostable DNA polymerase *Taq*I (from *Thermus aquatics*, a bacterial species that grows in hot springs) is not denatured by the heating steps

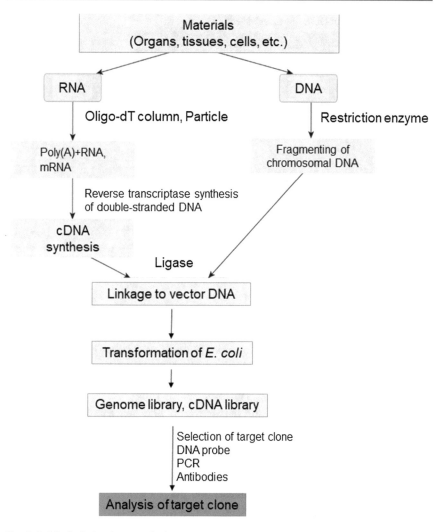

Fig. 3.6 Manipulation in gene cloning

3.4 Fish Genes

Cloning of various fish genes, including those associated with production of bioactive substances, have resulted in the identification of their structure, speculation on their functions from amino acids, and gene functions. Table 3.1 shows the number of fish base sequences registered with the Gene Bank database as of January 1, 2013. The largest number come from the Channel catfish (*Ictalurus punctatus*) and Rainbow trout (*Oncorhynchus mykiss*) followed by Zebrafish (*Danio rerio*),

Table 3.1 Number of DNA genes registered

Species	Number registered (as of 01.01.2013)	Species	Number registered (as of 01.01.2013)
Zebra fish (*Danio rerio*)	53,558	Rainbow trout (*Oncorhynchus mykiss*)	117,120
Kili fish (*Fundulus heteroclitus*)	5440	Nile tilapia (*Oreochromis niloticus*)	17,426
Cod fish (*Gadus morhua*)	41,275	Cyprinodont (*Oryzias latipes*)	21,803
Stickleback (*Gasterosteus aculeatus*)	16,728	Minnow (*Pimephales promelas*)	20,664
Blue cat fish (*Ictalurus furcatus*)	17,357	Salman (*Salmo salar*)	29,820
Channel cat fish (*Ictalurus punctatus*)	187,480	Swell fish (*Takifugu rubripes*)	3800

Cod (*Gadus morhua*) and Salmon (*Salmo salar*). Expressed sequence tag (EST) interpretation has allowed for elucidation of many fish genes. EST interpretation is a process of randomly selecting a clone from the cDNA library, setting a base sequence from the 3′ or 5′ end, and comparing that sequence with amino acid sequences registered in a database (Wittbrodt et al. 1998).

EST interpretation is also helpful in discovering new genes and interpreting the characteristic genes expressed in various tissues and organs. Body mapping offers data categorized by the organs and tissues that express genes; by analyzing the mRNA types and expression frequency in a single organ or tissue and comparing the data with other organs and tissues, it is possible to estimate whether the gene in question is a housekeeping gene expressed in many different organs or a characteristic gene of that organ. This process was used to build LIFESED, a human cDNA database that includes information on gene expression frequency, and possible commercialization of genome information, such as database access sharing, is under consideration (Heikkila et al. 1982).

As of January 1, 2013, the zebrafish accounted for the largest number of clones input through EST interpretation of fish, followed by the Japanese rice fish (Table 3.2).

Genomic analysis of fish has taken place at a comparable rate to human or rice genome analysis, but the full-scale effort launched in 1999 (chiefly in the U.S.) to conduct genomic or EST analysis of the zebrafish offers good news for those involved in analyzing the genes of the fish that serve as food resources for us. With the advancements to come in applied research to understand and improve fish through gene analysis, genomic analysis should also be performed for Korea's local fish species.

Table 3.2 Number of clones registered through dbEST interpretation

Organism types	Number registered (as of 01.01.2013)
Total registered	74,186,692
Human; *Homo sapiens*	8,704,790
Mouse + chicken; *Mus musculus + domesticus*	4,853,570
Zebra fish; *Danio rerio*	1,488,275
Asian rice; *Oryza sativa*	1,253,557
Rat; *Rattus norvegicus* sp.	1,162,136
Vinegar fly; *Drosophila melanogaster*	821,005
Asiatic rice fish; *Oryzias latipes*	666,891
Zel worm; *Caenorhabditis elegans*	396,687
Speckles fish; *Ictalurus punctatus*	354,516
Rainbrow trout; *Oncorhynchus mykiss*	287,564
Nile tilapia; *Oreochromis niloticus*	120,991
Color carp; *Cyprinus carpio*	34,316
Atlantic flat fish; *Hippoglossus hippoglossus*	20,836
Bastard halibut; *Paralichthys olivaceus*	15,234
American flat fish; *Pleuronecte americanus*	1483

Many clones have already been analyzed through EST for Japanese medaka, eels, and flounder. This section will introduce some of the clones obtained through EST analysis of flounder livers, spleens, and leukocytes.

For a total of 1374 clones, including 350 in the flounder's liver, 41 in its spleen, and 983 in its leukocytes, sequences averaging 700 bases have been set from their 3′ or 5′ end and their homologues sought in database sequences. Of the 1374 clones, a total of 787 showed homology with previously registered genes, while the remaining 587 were as yet unknown (Table 3.3).

The 787 clones with homologs in the registered gene sequences include some with identical base sequences. A total of 416 were ultimately identified, including 86 in the liver, 17 in the spleen, and 313 in the leukocytes. These genes were categorized into those involved in cell division, factors in intercellular information and signal transmission, those involved in cell structure and movement, factors in cell and organism defense, those involved in gene and protein expression, and those involved in energy metabolism (Table 3.4). Particularly interesting were the many genes found with involvement in organism defense and immunity. In addition to the clones exhibiting homology, clone base sequences without homology were registered in the Gene Bank database, allowing them to be searched at any time.

Because of a lack of equalization when the cDNA library was created, large quantities of gelatinase-b, ribosomal protein L23, collagenase, β-actin, brain myosine II isoform, and osteoclast-activating factor were detected in the leukocytes. These may be said to constitute the most expressed genes among those in the flounder leukocytes. In the liver, large amounts of apolipoprotein and complement C3 were found, along with cystatin.

Table 3.3 Number of clones analysed in the leukocytes, liver, and spleen (Nam et al. 2000)

Number of clones analyzed	1374
Leukocytes	983
Liver	350
Spleen	41
Number of ESTs analyzed	1792 base sequences
Leukocytes	1242 ESTs
Liver	493 ESTs
Spleen	57 ESTs
Clones with visible homology	787 (57.3%)
Clones without visible homology	587 (42.7%)
Number of different clones with visible homology	416
Leukocytes	313
Liver	86
Spleen	17
Total number of bases analyzed	1,254,858 base pairs (bp)
Average number of bases set per EST	700 base pairs (bp)

Table 3.4 Number of genes expressed in the flounder's leukocytes, liver, and spleen as determined by EST analysis

Cell function	Leukocytes	Liver	Spleen
Cell division	8	3	1
Intercellular information/signal transmission factors	46	2	0
Cell structure and movement	28	3	2
Cell/organism defense factors	21	8	2
Gene/protein expression	83	27	7
Energy metabolism	33	9	1
Other	94	34	4
Total	313	86	17

3.5   Fish Gene Structures

This section introduces the fish genes for globin, hormones, and transferrin, which are currently the subject of analysis.

A. Globin Genes

The hemoglobin found in fish blood performs a very important role in respiratory physiology. Hemoglobin is a conjugated protein formed as two α-globin and two β-globin chains are joined and each of the four globins combines with a heme.

While amino acid sequences had previously been analyzed for the α-globins and β-globins in carp, goldfish, and rainbow trout, among other species, Takeshita et al. succeeded in 1984 in extracting mRNA from carp blood and using reverse transcriptase to create a cDNA library and detect α-globins cDNA among the clones. This led to further research into carp globin genes, which found that α-globin and β-globin genes were composed of three exon sections and two intron sections. The upper 5′ region of the globin gene for all α-globins and β-globins was also count to include a -100 region, ATA box, and CCAAT box, along with a promoter region of gene DNA, which is essential for RNA polymerase to attach and initiate transcription. The lower 3′ region was found to include a poly(A) addition signal (Chen et al. 1997).

Cloning was performed for eight α-globin genes and nine β-globin genes with different base sequences in the same carp. The total length of the three open reading frames (ORFs) coding for amino acids was found to be 429 base pairs (bp) for α-globin and 441 (441) base pairs (bp) for β-globin (Tables 3.5 and 3.6).

Table 3.5 Number of α-globin genes from the same carp (Aoki and Miyata 1997)

	ORF1[a]	IVS1[b]	ORF2	IVS2	ORF3	ORF 길이
No. 1 α	95 bp	428 bp	208 bp	96 bp	126 bp	429 bp
No. 2 α	95 bp	96 bp	208 bp	112 bp	126 bp	429 bp
No. 3 α	95 bp	152 bp	208 bp	96 bp	126 bp	429 bp
No. 4 α	95 bp	183 bp	208 bp	107 bp	126 bp	429 bp
No. 5 α	95 bp	159 bp	208 bp	385 bp	126 bp	429 bp
No. 6 α	95 bp	172 bp	208 bp	99 bp	126 bp	429 bp
No. 7 α	95 bp	152 bp	208 bp	92 bp	126 bp	429 bp
No. 8 α			208 bp	96 bp	126 bp	

[a]Open reading frame
[b]Intervening sequence

Table 3.6 Number of β-globin genes from the same carp (Aoki and Miyata 1997)

	ORF1[a]	IVS1[b]	ORF2	IVS2	ORF3	ORF 길이
No. 1 β	92 bp	102 bp	223 bp	120 bp	129 bp	444 bp
No. 2 β	92 bp	133 bp	223 bp	109 bp	126 bp	441 bp
No. 3 β	92 bp	101 bp	223 bp	112 bp	129 bp	444 bp
No. 4 β	92 bp	106 bp	223 bp	122 bp	129 bp	444 bp
No. 5 β	92 bp	98 bp	223 bp	121 bp	129 bp	444 bp
No. 6 β	92 bp	101 bp	223 bp	121 bp	129 bp	444 bp
No. 7 β	92 bp	100 bp	223 bp	121 bp	129 bp	444 bp
No. 8 β	92 bp	154 bp	223 bp	114 bp	129 bp	441 bp
No. 9 β	92 bp	94 bp	223 bp	121 bp	129 bp	444 bp

[a]Open reading frame
[b]Intervening sequence

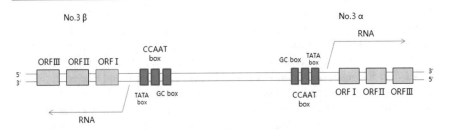

Fig. 3.7 Structure of α-globin and β-globin genes in carp

Like human globin genes, the findings suggested that carp globin genes form multigene families. In mammals, α-globins and β-globins are encoded in different chromosomes; in the case of humans, the α-globin family is located on chromosome 14 and the β-globin family on chromosome 11.

Among fish, α-globin and β-globin genes are encoded on the same chromosome, with the α-globin encoded in the opposite direction facing the β-globin gene and upper region (see Fig. 3.7) (Miyata and Aoki 1997). In humans, hemoglobin proteins have been found to differ in globin protein composition for embryos, fetuses, and adults and to be expressed sequentially from each upper region gene of the α-globin and β-globin families. In fish, the α-globin and β-globin genes that have been clones can be categorized into three groups, which are respectively found to be expressed in embryos, fry, and adult fish. In other words, α-globin and β-globin genes are converted with the growth process in fish as well.

B. Growth Hormone Genes

Growth hormones exist in minute quantities in the pituitary gland and are thus very difficult substances to produce in mass quantities in an organism. Recent developments in genetic engineering techniques have helped make mass-production a reality, however. Salmon growth hormone genes have been cloned and recombinant genes introduced into *E. coli*, allowing for the production of large amounts of salmon growth hormone within it (Sekine et al. 1985). The cloning of growth hormone genes in various fish through this process has allowed identification of the structure of those genes (Du et al. 1992; Rand-weaver et al. 1933).

Growth hormones produced in the pituitary gland function in the liver to promote the synthesis and secretion of insulin-like growth factor-1 (IGF-I), which functions in turn to promote growth. IGF-I is a growth factor consisting of polypeptides with a molecular weight of 7500 and a similar structure to insulin. While it functions similarly to insulin in serum, it is not controlled by insulin antibodies. In addition to mediating the proliferation of cartilage cells and growth hormone activity in protein biosynthesis, it also exhibits similar physiological functions to insulin. It exists in high concentrations in plasma but is mostly deactivated as it combines with binding proteins.

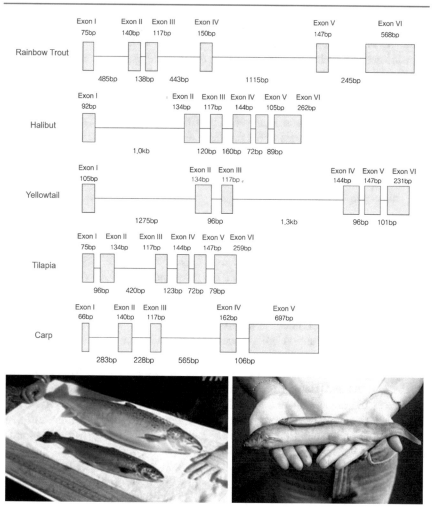

Fig. 3.8 Structure of fish growth hormone genes and photos refers to super salmon (L) and super mud loach (R) created with growth hormone genes

In salmonidae fish, growth hormones function not only in the liver but also in the gills, performing a role in adjusting osmotic pressure to increase seawater adaptability. Growth hormone genes in mammals and carp consist of five exon and four intron; in rainbow trout, yellowtails, flounder, and tilapia, they consist of six exon and five intron (Fig. 3.8). Growth hormone genes differ in length among species: for tilapia, they are roughly 1.7 Kb, which is similar to a human being, while in Atlantic carp, rainbow trout, and yellowtails, growth hormone genes are longer than in mammals, measuring around 3.9 Kb from start to stop codon (Chen and Powers 1990).

Among growth hormone genes, those for prolactin gene and somatolactin gene show the same number of expressed sites. While no homologues are observed for expression site I, amino acid sequence homologues did appear for exon II to V, and the discovery of a transcription regulation area in the upper region of exon I is thought to indicate common ancestry.

Kawauchi et al. found that growth promotion effects actually did occur from introduction of growth hormone genes into fish and abalone. Through inserting these growth hormone genes into a fertilized egg, they were able to create a "super salmon" that grew to large sizes in a short time. At Pukyong National University, Kim et al. created a "super mud loach".

C. Transferrin Genes

Transferrin has a molecular weight of around 75,000 and is a kind of β-globin. It is an iron-carrying glycoprotein: two of molecules absorbed into serum bond with iron ions (Fe^{3+}), supplying cells with the iron they need for proliferation and hemoglobin production by means of transferrin receptors. Over 99% of iron in serum binds with transferrin. Normally, around one-third of transferrin binds with iron; the remaining quantity that does not bind with iron is referred to as unsaturated iron binding capacity (UIBC) (Liu et al. 2010).

Alongside insulin and selenious acid, transferrin is one of the three necessary additives to the serum medium for animal cell cultures. It is also an important active substance as an accelerator factor for scleromere differentiation. Transferrin and similarly structured lactoferrin are involved in immune function regulation and regulate cytokine secretion and the differentiation and proliferation of lymphocytes. Administration to rainbow trout has been found to result in higher levels of phagocytosis and chemiluminescence by means of macrophage and eosinophilic leukocyte receptors.

Isozyme polymorphism analysis of coho salmon transferrin showed the existence of three types (A, B, and C). Coho salmon possessing the C variety were found to be strongly resistant to bacterial kidney disease (BKD) (Suzumoto et al. 1977). It remains unknown whether transferrin acts directly on pathogenic bacteria, or whether there exists a gene with resistance to the BKD bacterium *Renibacterium salmoninarum* in genomes of the C type. It is necessary, however, to create new fish that are resistant to infection through the use of genetic engineering methods rather than the traditional hybridization breeding and selection approach. By cloning and analyzing transferrin genes in Japanese rice fish, salmon, flounder, and other fish, Takashima discovered that the length of the protein open reading frame (ORF) coding for transferrin proteins, though slightly different from one species to the next, formed a sequence of 2061–2073 bp (687–691 amino acids). Structures for this transferrin family are known to be similar in the front portion (N-lobe) and back portion (C-lobe) of the peptide chain and to be created through gene duplication.

Japanese rice fish transferrin genes, like those in human beings, consist of 17 exon and 16 intron. The length of the exon is no different than in humans, but the intron are characteristically very short. The amino acids needed for iron bonding

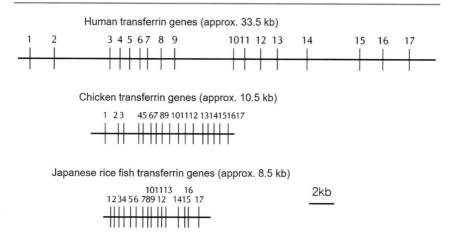

Fig. 3.9 Transferrin gene lengths in Japanese rich fish, chickens, and humans

and the cysteine needed for disulfide bonds are preserved in fixed locations in the ORF. In the upper 5' region, a transcription factor binding region and enhancer region exist near the promoter region (Fig. 3.9) (Mikawa et al. 1996).

Analysis of amino acid sequences in ten Salmonidae fishes (including *Oncorhynchus*, *Salverinus*, and *Salmo*) showed a high level of homology at over 84%. Homology of over 90% was found between members of the same genus, and characteristic amino acid sequences were identified. These amino acids are believed to have undergone transposition following the splitting of genera. Dendrogram visualization based on the amino acid sequences showed categorizations using transferrin amino acid sequences as an indicator to be valid due to their similarity in those produced according to tradition methods of fish classification (Lee et al. 1998).

3.6 New Analyses of Fish Genomes

3.6.1 Importance of Fish Genome Analysis

The term "genome" is a portmanteau of "gene" and "chromosome," referring to the entirety of genetic material (genetic information) used to regulate an organism's biological phenomena. Genomic research makes use of biological phenomena, analyzing the full base sequence of DNA within a cell (containing all the necessary information for the organism's basic functioning) and the genes contained within it. The 2001 identification of the human genome offered the first step towards understanding the design of all life forms. By understanding the base sequence of DNA within the genome, it became possible for us to identify the roles of specific parts and analyze the functions of genes represented in that sequence. Genomic research has also enabled us to understand the changes and relationships of innumerable genes, and genome analysis has allowed for an understanding of specific biological phenomena (Hudson et al. 1980; Venkatesh et al. 2007).

In terms of fish genomics, complete genome analysis, first for the puffer fish Fugu rubripes and then for model zebrafish and medaka, has been attempted as a means of understanding human gene functions. Around the world, recent genomic research has been conducted on fish varieties with high economic or industrial value and applied in their breeding or industry use. In China, for example, analysis of the Cynoglossus semilaevis genome was completed in 2010, and an investigation of the genes responsible for sex determination was applied towards the development of a fast-growing variety. In Japan, analysis of the Thunnus thynnus genome was completed in 2011; where the resulting information has been used in the development of farmed strains with outstanding growth, feed efficiency, texture, and disease-resistance properties; management of tuna resources, including accurate place-of-origin determinations; establishment of a place-of-origin tracking system from farm to dinner table; and development of functional food and pharmaceutical products with special properties such as DHA accumulation.

In South Korea, Paralichthys olivaceus is one of the most preferred varieties of fish for sashimi. In 2012, its genome was decoded for the first time by the team of Dr. Kim of the National Institute of Fisheries Sciences, allowing for its use in studies on disease resistance, flavor, texture, and sex determination. Genomic research has also been conducted to explain biological phenomena: in 2011, Norwegian researchers spearheaded an effort to analyze the full Gadus morhua genome, identifying the genetic factors accounting for the fish's unique immune system allowing it to survive under highly varying temperatures. As this shows, genomic research can be used in a broad range of areas, including understanding biological phenomena, improving strains, and developing biomaterials.

3.6.2 Trends in New Genomic Analysis Technology

The double helix structure of DNA was first identified in the 1950s, while the first base sequence analysis technology was developed in the 1970s by Frederick Sanger and Walter Gilbert. In the 1980s, the development of the polymerase chain reaction (PCR) technique led to huge advancements in gene analysis technology. Sanger's method was ultimately used to identify the full human genome sequence in the early 2000s. Since 2007, the development of new base sequence analysis technologies distinct from the Sanger approach has resulted in the coinage of the term "Next Generation Sequencing" (NGS); at present, second- and third-generation NGS techniques have been developed and are in active use. These NGS analysis techniques are capable of reading large quantities of base sequences and have ushered in historic changes in terms of time and cost (König et al. 2012).

Examples of second-generation analysis techniques include Roche's GS FLX, Ilumina's Solexa and Hiseq 2000, and ABI'S SOLID. Third-generation technologies have offered a new paradigm to address shortcomings in existing forms of NGS. Their biggest difference from second-generation technologies is their ability

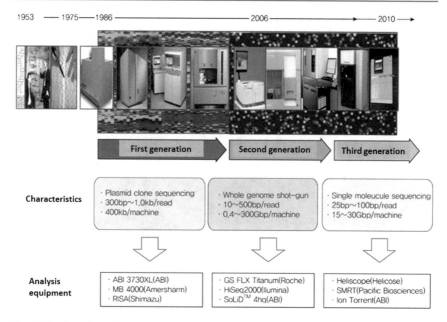

Fig. 3.10 Evolution of base sequence analysis techniques

to analyze base sequences using single molecules, potentially obviating the DNA amplification process and reducing the money and time needed.

Major examples of third-generation analysis technology include SMRT by Pacific Biosciences and ION Torrent by ABI. These techniques are capable of reading sequences of 20 billion bases, or seven times the length of the human genome (3 billion bases). Rapid advancements in genome analysis techniques have served as a driving force for an astronomical increase in base sequence information, while developments in high-throughput techniques have enabled analysis of the entire genome sequence. Various other methods of decoding and processing vast amounts of biological information have further enabled analysis and decoding of genomic information (Fig. 3.10).

3.6.3 Genome Decoding Process

The process of decoding the genome of an organism for which no prior genomic information exists is known as de novo sequencing of a reference genome. The process of reference genome decoding depends on the organism under analysis or the researcher performing it. This section will outline the strategy used in decoding the Paralichthys olivaceus genome (Agris 2004; Gunderson et al. 2004; Murat et al. 2012).

First, next-generation sequencing (NGS) is used to produce large quantities of base sequence data, which is subjected to de novo assembling with a Newbler

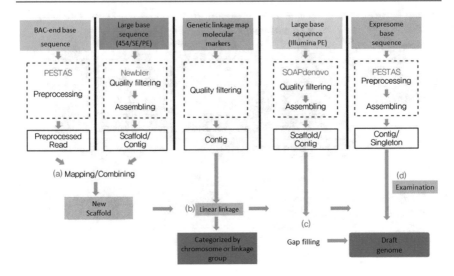

Fig. 3.11 Genome decoding process

program to create a contig (collection of overlapping and continuous cloned DNA) and scaffold. Second, a bacterial artificial chromosome (BAC) library is used to generate BAC end sequencing (BES) data, which is integrated with the earlier data to produce a new contig and scaffold. The resulting scaffold is matched with molecular markers used for a genetic linkage map, after which it is arranged in sequence by chromosome, and large base sequence analysis or a BAC cloning 3D pooling system is used to fill in the scaffold's gaps and connect between scaffolds. The genetic information produced to date is tested with expressed sequence tags (EST), after which a draft genome is completed. To predict the genes existing within the genome, a prediction program and DNA base sequence and EST data are used for annotation (Fig. 3.11).

3.6.4   Future of Fish Genomics

The recent development of techniques for analyzing large base sequences has led to efforts to decode the genomes of economically and academically important fish varieties. In the fishing industry, decoding of fish genomes is extremely important in three regards: for understanding biological phenomena, use of genome information in selective breeding, and efficient classification of biological resources.

Advancements in large-scale analysis techniques have resulted in the development of various techniques for identifying the functions of large numbers of genes. The combination of functional genomics and genome information means more opportunities for understanding various biological phenomena, and particularly for the efficient identification of useful gene functions with strong industry usage potential. Also, because biological phenomena are regulated systematically by networks of genes rather than individual genes, genome information and functional genome techniques may be used to efficiently gain an understanding of these networks (Crollius and Weissenbach 2005).

Two of the drawbacks of traditional breeding approaches are the large time and money investments involved. Molecular biology techniques have been developed to address these issues, and recent attention has focused on molecular breeding selection techniques that distinguish according to differences in trait-related DNA base sequences. Molecular marker development is very costly, however, and may not be possible in cases where there is inadequate genetic polymorphism among breeding lines. Genomic information may therefore be used not only for simpler molecular marker development, but also as a basis for developing new molecular biology techniques with superior efficiency.

One potential example is the use of genomic information itself as a substitute for molecular markers or genetic maps. This conceptual approach has recently become known as "genomic breeding," which is expected to establish itself as a core method in molecular breeding in the future. As a reflection of the growing importance of biological resources, the world's advanced economies have been focusing their energies on gathering, preserving, increasing, and using large amounts of biological resources. It may prove possible to use large-scale genome based sequence analysis to develop molecular markers for the categorization or selection of useful genetic resources, or to apply the genome sequence information itself in developing genome-based genetic resource classification and selection methods. As this shows, genomic information is expected to emerge not only as a high-tech means of overcoming the limitations seen in fishery science and the key techniques of resource classification, selection, and improvement, but also as a new paradigm in the field of fishery resource, where the development of genome-derived techniques usher in a new chapter in the discipline's advancement (Cossins and Crawford 2005; Allendorf et al. 2010; Palti 2011).

3.7 Chapter Summary and Conclusion

The ichthyologist Nelson has said that 24,618 species of fish are known to exist on Earth, and that the 14,652 inhabiting the oceans account for over half those in existence.

A recent issue of the science journal Nature featured the rather shocking claim that intensifying destruction of marine ecosystems around the world had resulted in 90% of large fish species disappearing from the waters in the past 50 years. To see the abundance of fish in the market and the growing numbers of sashimi restaurants, one might be under the impression that there are still many fish in the sea. Yet countries around the world have been raising the alarm since the 1990s over the severe depletion of fish resources.

According to data from the World Conservation Monitoring Centre (WCMC), rising water temperatures as a result of global warming, advancements in fishing technology, and large-scale overfishing have resulted in the disappearance of around one-third of large species such as tuna, sharks, and swordfish, while catches have dwindled to around one-tenth past sizes. Some have predicted that if the current devastation is left to continue, most of the fish species in the seas will be gone by around 2048.

Marine products account for 16% of the animal protein needed for human survival, and fish account for over 80% of marine products; thus the problem is quite serious. Addressing this global trend and satisfying the public's demands will require maximal improvements in productivity and increases in hatchery yields through the development of intensive, state-of-the-art technology.

Recently, efforts have been made worldwide to produce outstanding, high value-added varieties through the application of genetic engineering techniques to maximize productivity per unit of effort over a short period of time. A particularly great amount of research has focused on the fish varieties with the highest economic value among marine products. The world's advanced economies have already entered an age of untrammeled competition in global freshwater and oceanic gene resources and biotechnology.

Expressed sequence tag (EST) techniques and cDNA chip techniques based on a library of fish cDNA have already been applied to analysis of fish genetic structure, tracking of mutants, gene marking, and the restoration of ecosystems through the preservation of various species.

Rapid advancements in genetically modified fish are expected to usher in a major revolution in aquaculture. In the medium and long term, genetic transformation technology may be used to increase economic value-added for marine products in general.

References

Agris, P. F. (2004). Decoding the genome: A modified view. *Nucleic Acids Research, 32*(1), 223–238.

Allendorf, F. W., Hohenlohe, P. A., & Luikart, G. (2010). Genomics and the future of conservation genetics. *Nature Reviews Genetics, 11*(10), 697.

Aoki, T. (2000). Gene of fish. In F. Takashita (Ed.), The next generation of fisheries biotechnology (pp. 1–15). Tokyo, Japan: Seizando-shoten Pubilishing Co.

Aoki, T., & Miyata, M. (1997). Globin gene. In T. Aoki, et al. (Eds.), DNA of fish: Geneticapproach of fish gene molecule (pp. 158–200).

Black, D. L. (2003). Mechanisms of alternative pre-messenger RNA splicing. *Annual Review of Biochemistry, 72*(1), 291–336.

Chalfie, M., Tu, Y., Euskirchen, G., Ward, W. W., & Prasher, D. C. (1994). Green fluorescent protein as a marker for gene expression. *Science, 263*(5148), 802–805.

Chen, L., DeVries, A. L., & Cheng, C.-H. C. (1997). Evolution of antifreeze glycoprotein gene from a trypsinogen gene in Antarctic notothenioid fish. *Proceedings of the National Academy of Sciences, 94*(8), 3811–3816.

Chen, T., & Powers, D. (1990). Transgenic fish. *Trends in Biotechnology, 8*, 209–215.

Cossins, A. R., & Crawford, D. L. (2005). Fish as models for environmental genomics. *Nature Reviews Genetics, 6*(4), 324.

Crollius, H. R., & Weissenbach, J. (2005). Fish genomics and biology. *Genome Research, 15*(12), 1675–1682.

Du, S. J., Gong, Z., Fletcher, G. L., Shears, M. A., King, M. J., Idler, D. R., et al. (1992). Growth enhancement in transgenic Atlantic salmon by the use of an "all fish" chimeric growth hormone gene construct. *Nature Biotechnology, 10*(2), 176.

Gunderson, K. L., Kruglyak, S., Graige, M. S., Garcia, F., Kermani, B. G., Zhao, C., et al. (2004). Decoding randomly ordered DNA arrays. *Genome Research, 14*(5), 870–877.

Heikkila, J., Schultz, G., Iatrou, K., & Gedamu, L. (1982). Expression of a set of fish genes following heat or metal ion exposure. *Journal of Biological Chemistry, 257*(20), 12000–12005.

Hudson, A. P., Cuny, G., Cortadas, J., Haschemeyer, A. E., & Bernardi, G. (1980). An analysis of fish genomes by density gradient centrifugation. *European Journal of Biochemistry, 112*(2), 203–210.

König, J., Zarnack, K., Luscombe, N. M., & Ule, J. (2012). Protein–RNA interactions: New genomic technologies and perspectives. *Nature Reviews Genetics, 13*(2), 77.

Lee, J. Y., Tada, T., Hirono, I., & Aoki, T. (1998). Molecular cloning and evolution of transferrin cDNAs in salmonids. *Molecular Marine Biology and Biotechnology, 7*(4), 287–293.

Liu, H., Takano, T., Abernathy, J., Wang, S., Sha, Z., Jiang, Y., et al. (2010). Structure and expression of transferrin gene of channel catfish, *Ictalurus punctatus*. *Fish & Shellfish Immunology, 28*(1), 159–166.

Mikawa, N., Hirono, I., & Aoki, T. (1996). Structure of medaka transferrin gene and its 5'-flanking region. *Molecular Marine Biology and Biotechnology, 5*(3), 225–9.

Miyata, M., & Aoki, T. (1997). Head-to-head linkage of carp α- and β-globin genes. *Biochimica et Biophysica Acta (BBA) - Gene Structure and Expression, 1354*(2), 127–133.

Murat, F., Peer, Y. V. D., & Salse, J. (2012). Decoding plant and animal genome plasticity from differential paleo-evolutionary patterns and processes. *Genome Biology and Evolution, 4*(9), 917–928.

Nam, B. H., Yamamoto, E., Hirono, I., & Aoki, T. (2000) A survey of expressed genes in the leukocytes of Japanese flounder, Paralichthys olivaceus, infected with Hirame rhabdovirus. *Developmental & Comparative Immunology, 24*(1), 13–24.

Palti, Y. (2011). Toll-like receptors in bony fish: From genomics to function. *Developmental and Comparative Immunology, 35*(12), 1263–1272.

Perry, K. L., Watkins, K. P., & Agabian, N. (1987). Trypanosome mRNAs have unusual "cap 4" structures acquired by addition of a spliced leader. *Proceedings of the National Academy of Sciences, 84*(23), 8190–8194.

Ríus, C., Smith, J. D., Almendro, N., Langa, C., Botella, L. M., Marchuk, D. A., et al. (1998). Cloning of the promoter region of human endoglin, the target gene for hereditary hemorrhagic telangiectasia type 1. *Blood, 92*(12), 4677–4690.

Rand-weaver, M., Kawachi, H., & Ono, M. (1993). Evolution of structure of the growth hormone and prolactin family. In M. P. Schrebman, C. G. Scanes & P. K. Pang (Eds.), *The endocrinology of growth development and metabolism in vertibrates* (pp. 13–42). Academic Press: New York.

Schena, M., Shalon, D., Davis, R. W., & Brown, P. O. (1995). Quantitative monitoring of gene expression patterns with a complementary DNA microarray. *Science, 270*(5235), 467–470.

Sekine, S., Mizukami, T., Nishi, T., Kuwana, Y., Saito, A., Sato, M., Itoh, S., & Kawauchi, H. (1985). Cloning and expression of cDNA for salmon growth hormone in Escherichia coli. *Proceedings of the National Academy of Sciences, 82*(13), 4306–4310.

Suzumoto, B. K., Schreck, C. B., & McIntyre, J. D. (1977). Relative Resistances of Three Transferrin Genotypes of Coho Salmon (Oncorhynchus kisutch) and their hematological responses to bacterial kidney disease. *Journal of the Fisheries Research Board of Canada, 34*(1), 1–8.

Venkatesh, B., Kirkness, E. F., Loh, Y.-H., Halpern, A. L., Lee, A. P., Johnson, J., et al. (2007). Survey sequencing and comparative analysis of the elephant shark (*Callorhinchus milii*) genome. *PLoS Biology, 5*(4), e101.

Wittbrodt, J., Meyer, A., & Schartl, M. (1998). More genes in fish? *BioEssays, 20*(6), 511–515.

Yi, D. (2011). Who owns what? Private ownership and the public interest in recombinant DNA technology in the 1970s. *Isis, 102*(3), 446–474.

Young, R. A., Bloom, B. R., Grosskinsky, C. M., Ivanyi, J., Thomas, D., & Davis, R. W. (1985). Dissection of *Mycobacterium tuberculosis* antigens using recombinant DNA. *Proceedings of the National Academy of Sciences, 82*(9), 2583–2587.

Fish Breeding and Biotechnology

4

Contents

4.1 The Global Food Shortage and Genetic Engineering

On October 12, 1990, at a hospital in Sarajevo, Bosnia, then United Nations Secretary-General Kofi Annan declared the birth of the six-billionth human being. It had taken 12 years for the world's population to increase by one billion people from five to six billion, and 13 years before that to climb from four to five billion. As this shows, the rate of population increase had accelerated enough to shorten the time that it took for the population to grow by one billion people. Increases in population are inevitably accompanied by food production issues; methods that have been used to date to increase food production include expansions of farmland area, use of chemical fertilizers and pesticides, and cultivation of high-yield grain varieties. The amount of land available for us to use is limited, however, and issues with the safety of chemical fertilizers and pesticides have called attention to constraints in terms of their use to increase food production. For this reason, breeding scientists have turned to genetic engineering methods for more efficient production of new varieties (Cassman and Liska 2007; Long et al. 2015).

Genetic engineering has undergone astonishing developments since Watson and Crick first unlocked the secret of the genes in DNA in 1953. Artificial organisms have been cloned, as with the sheep "Dolly" produced in 1996. Since then, artificially cloned organisms such as the sheep "Dolly" (produced in 1996) have been produced: a cloned monkey and cow followed in 1997 and 1998, respectively, while South

© Springer Nature Switzerland AG 2019
S.-K. Kim, *Essentials of Marine Biotechnology*,
https://doi.org/10.1007/978-3-030-20944-5_4

Korea succeeded in cloning a Korean beef cow in early 1999. In a larger sense, genetic engineering is technology that takes advantage of the functions of organisms; along with genetic manipulation techniques, other key technologies in biotechnology include those for cell fusion, mass cell culturing, and bioreactors. Genetic engineering has brought advancements in medicine and biology and contributed greatly to human welfare, but it is also widely used in the areas of aging, cancer and immunity research, mass production of growth hormones and non-polluting pesticides, crops that do not require pesticide use, and a range of other basic research, industry, and livestock agriculture areas (Evans et al. 1999; Shiels et al. 1999).

From their initial focus on microorganisms, developments in genetic engineering technology have expanded in application to plants and animals. As studies expanded to the as yet unexplored field of the oceans, this led the creation of the new field known as marine biotechnology. Marine biotechnology is complex bioengineering technology that applies other state-of-the-art techniques to marine organisms and ecosystems; in simple terms, it is research on cell tissue culturing, cell manipulation, genetic recombination, and bioengineering that involves marine organisms.

Studies in genetic engineering that use ocean animals have drawn a great deal of notice in recent years. Especially active areas involve the manipulation of fish chromosomes and the use of gene manipulation to produce transgenic fish. The technology is now available to produce a female-only population through pressure treatment of an egg fertilized with ultraviolet-treated sperm. The development of such techniques for producing females are of great help in promoting value added in the fishery industry through mass production of salmon or herring roe. Also considered promising for the industry is technology for triploid or quadruploid fish, which has made a great contribution to the development of fish of large sizes—as seen with the triploid "super-loach" (Morse 1984; West 2005).

Breeding is a matter of creating and maintaining or propagating groups of organisms with genetic traits that are beneficial for humans; typically, it is conducted with farming populations. Breeding of agricultural and livestock products is relatively well advanced, and improved strains are currently being marketed. In contrast, breeding has been lagging for the fish and shellfish that are central to marine products; presently, nearly all seafood that is consumed is wild.

Fish and shellfish, which typically require water, are difficult to raise or propagate. Conventional breeding methods of generational raising, selection, and breeding cannot be used with them, and because they are so diverse and mass caught, there has been little interest in breeding per se. While farming technologies have gradually improved, however, catches have declined in recent years due to the 200 nautical mile economic zone issue, among other factors. Hopes for farming are growing due to resource depletion, and improving farming efficiency has become an urgent matter. Potential demand for fish and shellfish as health or luxury foods has also recently grown. Under these circumstances, the production of revolutionary farming strains with superior economic traits has become the focus of some anticipation (Morrison 1993).

In contrast, countries in Africa and elsewhere in Asia are still faced with a severe shortage of animal protein resources. Many of these regions have vast inland waters or ocean areas that go almost entirely unused. It would not be an overstatement to say that achieving a revolutionary increase in food self-sufficiency hinges on advancements in fish and shellfish aquaculture. One means of addressing this would be through the breeding of species that can be cultivated easily and are capable of withstanding particular environmental conditions, such as local water temperature and quality and salt concentrations.

In that sense, the specific goals of fisheries breeding—its aims and the fish involved—have changed with the times even as general interest in fisheries breeding has grown. For instance, while it was once thought that fast-growing lines were better, recent examples involving color carp and whales have actually required dwarf lines that do not grow. Even when the goal is to produce tasty fish for eating, the genetic characteristics demanded of the fish vary according to how it is eaten.

For this reason, researchers involved in fish breeding must make active use of biotechnology to accelerate breeding methods.

4.2 Chromosome Manipulation

Chromosome manipulation techniques originated in agriculture with the use of varieties with naturally or artificially modified chromosome numbers for breeding (i.e., the creation of superior or improved varieties through hybridization). Typically, organisms that engage in sexual reproduction (reproduction in which organisms are sexually specialized into female and male forms) are diploid (2n), meaning that they possess pairs of identical chromosomes. Humans possess 46 chromosomes, or 23 pairs (2n = 46). One comes from the maternal line by way of the egg, while the other comes from the paternal line via the sperm. When the number of chromosomes is altered, however, the result is a new, never-before-seen organism. This manipulation of chromosomes to create new organisms is known as chromosome set manipulation. Chromosome set manipulation is widely used today, as it not only can be performed through simple methods but has the advantage of not requiring special or expensive equipment (Vorobjev et al. 1993; Arai 2001; Endo 2007).

For improvement of varieties through chromosome manipulation, polyploids are typically used. As mentioned previously, an organism's chromosomes exist in diploid (2n) form, with one gene each received from the mother and father. Some organisms, however, exist in quadruploid (4n) or sextuploid (6n) form, with their number of chromosomes doubled or tripled; others are haploid (n), meaning that they have only half the number of chromosomes. The term "polyploid" refers to organisms with multiples of the number of chromosomes, while "polyploidy" refers to the state in which polyploids exist. Autopolyploids are organisms that have three or more of the same genome (chromosome pairs). The term "aneuploidy" describes cases in which part of the chromosome is lost, such as 2_{n+1} or 2_{n-1} (Oshiro 1990).

The term "polyploid" is typically applied in cases of whole number multiples of genomes; depending on the number of genomes, they may be haploid (n), diploid (2n), triploid (3n), and so forth. The majority of fish and shellfish are thus diploid.

Why are polyploids used in breeding? Since they have more chromosomes and genes than diploids, they also yield more genetic products, resulting in large organisms. In the case of odd-number polyploids, clonal propagation appears to have developed as a means of survival because gamete pairs cannot be formed. A representative example of this is the triploid seedless watermelon. While ordinary watermelons are diploid, treatment with colchicine to check the formation of spindle fibers (microtubules) during meiosis (the characteristic form of cell division to create reproductive cells, or eggs and sperm) resulted in the development of 2n sex cells, in which the number of chromosomes has not been halved. Subsequent pollination of the polyploid sex cells (2n) and diploid sex cells (n) results in the formation of a triploid (3n). Because the inability of chromosomes to pair during meiosis prevents the formation of normal cells, odd-number polyploids are sterile. The seedless watermelon is one such result.

Quadruploids can sometimes bear especially large flowers or fruit, resulting in large yields. In 1936, Blakeslee and Avery found that treatment of seed sprouts with a solution of colchicine (a type of alkaloid contained in colchicum, a perennial of the lily order) resulted in multiples of the chromosome set, most of them quadruploid. This method holds great significance for breeding science, as it can be used to produce large pumpkins, radishes with large roots, cotton with long fibers, high-yield leaf tobacco, vitamin-rich tomatoes, and large-blossomed flower-of-the-hours. Quadruploids can also be obtained with *Quercus variabilis* and other forest plants.

Examples of well-known crops for which polyploids are used are bananas (triploid), peanuts and potatoes (quadruploid), and sweet potatoes (sextuploid). These are larger in form than their diploid counterparts because the chromosomes replicated in the somatic cell division process do not divide.

While active breeding with polyploids has taken place in agriculture, chromosome manipulation for marine organisms was introduced very late. Indeed, serious application has only taken place since the 1980s, and its history is relatively brief.

The situation today in research and development of chromosome manipulation for marine organisms is one of a near total absence of such techniques for marine microalgae. In the case of marine plants, cultured cells from the kelp class have been used to attempt chromosome set manipulation through apospory and apogamy. In terms of marine animals, chromosome manipulation methods have been established for several varieties of fish from the salmon and flatfish families and for several kinds of shellfish, including pearl and other oysters and abalone. Production of female-only populations is currently under way, along with the commercialization of grains with superior growth properties, such as triploid shellfish.

The following are the four basis techniques used in chromosome manipulation, which can be combined to obtain the genetic variomes seen in the application examples.

4.2.1   Basic Techniques

A. Multiplication

Because cell division is preceded by g., replication of the DNA (chromosomes) , the quantity of cellular DNA (number of chromosomes) is doubled if the division can be artificially blocked. Typically, an egg with two genomes is produced by checking the maturation division process that occurs in the egg after oviposition and fertilization (the second maturation division for fish, the first or second for shellfish and crustaceans). This process of halting division is sometimes called polar body emission prevention, as one side of the daughter cell during division (the polar body) is very small and appears to be in a process of emission from the other side. Doubling can also be achieved by preventing the subsequent first cleavage (division of the somatic cell), although the success rate is low. Methods used in doubling include chemical (e.g., through immersion in colchicine or cytochalasin B solution), physical (e.heating, cooling, and pressure), and biological (e.g., hybridization of particular combinations) types (Purdom 1983; Thorgaard 1983; Isaacs et al. 2011).

B. Parthenogenesis

Fertilization in which one gamete nucleus is treated with radiation such as ultraviolet or gamma rays, or chemicals such as toluidine blue, results in development occurring in only the untreated gamete nucleus. Development in the egg nucleus alone is known as gynogenesis, while development in the sperm nucleus alone is known as androgenesis; both require artificial fertilization. Naturally, the genetic traits of the undeveloped nucleus are not transmitted.

Detailed research has considered the question of why gynogenesis should occur when this treatment is applied to the sperm. Two methods are believed to exist.

C. Aneuploidy

An aneuploid is an organism in which the number of chromosomes is not a whole number multiple of the genome. Artificial production of aneuploids follows the same approach as parthenogenesis, but involves fertilization with a gamete treated under weak conditions so that at least part of the chromosome remains undestroyed, or artificial insemination with aneuploidy sperm obtained from a triploid male. An aneuploidy embryo may be obtained when an over-ripened egg is fertilized artificially after losing part of its chromosomes (Hassold and Hunt 2001).

D. Hybridization

If we take the letters "A" and "B" to represent individual genomes, such that the parents' somatic cells have karyotypes of 2A and 2B, fertilization of an egg (A) and sperm (B) obtained from different species results in a hybrid possessing a new karyotype A + B. While the karyotype does not change when the exact reverse is

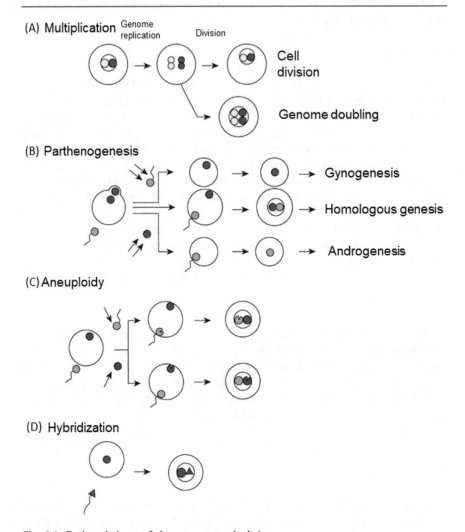

Fig. 4.1 Basic techniques of chromosome manipulation

done, the genes within the cytoplasm are slightly altered. This approached cannot be used when the species are too far apart. Some cases of heterosis are observed, in which the first hybrid generation shows superior traits to the parents, although the reason for this is not clearly understood. The only way to learn which mixtures produce stronger hybrids is to actually make them. The techniques described in this section are illustrated in Fig. 4.1 (Thomas 1980; Oshiro 1990).

4.2.2 Example of Application

Genome Units

Parthenogenesis, Diploid, (B) + (A) Because the egg or sperm nucleus possesses only one genome, even successful parthenogenesis results in a haploid embryo. The haploid is severely deformed and cannot survive, so multiplication treatment is used to produce a diploid soon after genesis (Baldacci et al. 1980).

Gynogenesis, Diploid The fish's stomach is pressed lightly and the roe extracted. The roe is halted in the middle of the second maturation division. When artificial fertilization is performed, the division is re-started and a second polar body is emitted. Fertilization is performed with sperm treated with ultraviolet rays, and low- or high-temperature or high-pressure treatment is used to block the second polar body emission (Fig. 4.2). The result is offspring that acquires only female genetic traits. Gynogenesis methods have yet to be established for shellfish or crustaceans.

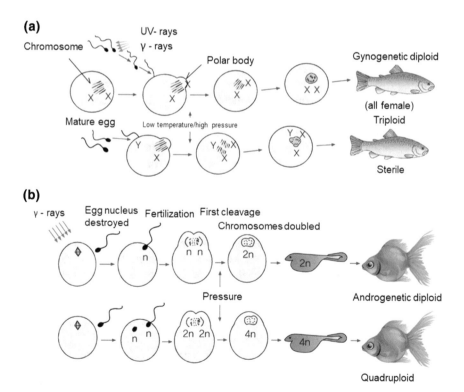

Fig. 4.2 Breeding fish with chromosome manipulation **a** making gynogenetic diploids and triploids **b** making androgenetic diploids and quadruploids (Oshiro 1990)

Because inadequate ultraviolet exposure may result in the sperm nucleus not being destroyed and ordinary fertilization and genesis occurring, it is important to examine whether the fish is indeed gynogenetic. As a first step, both parents are thoroughly exposed to the rays. Varieties of sperm may be used that achieve fertilization but without viability in the resulting embryo, or fertilization may involve a male sperm with a dominant gene controlling for a trait that can be easily distinguished outwardly and a contrasting recessive gene.

The embryos resulting when loach sperm is applied to goldfish eggs all die particularly easily. Impregnation with loach sperm exposed to ultraviolet rays results in a goldfish haploid that displays the characteristic deformities of the haploid (haploid syndrome) and is incapable of survival. When multiplication is applied to create a diploid, the result is a beautiful goldfish resembling a female.

In fish as in humans and insects, many species appear to have female homo sex determination mechanisms (XX for females and XY for males, where X and Y are the sex chromosomes or genes for sex determination). Males are only produced when two forms of sperm are present with an X and Y chromosome and fertilization takes place with the Y sperm. Because gynogenetic offspring inherit only the genetic traits of the mother, all of them would be expected to be female. In many cases, females have greater economic value than males as foodstuffs.

Koreans tend to prefer fish roe (such as salmon or cod) or ovaries, and growth is typically superior for females compared to males for some varieties such as halibut. In the cases of salmon and trout, males have a muddy-colored body as a secondary sexual characteristic, which markedly reduced their commercial value. With aquarium fish, the female body type is preferred, with its fleshiness and bulging stomach. In seedling production, it is possible to have far fewer males than females, but the typical scenario is a shortage of females.

Several means exist for controlling fish sex, but gynogenesis is applied to various species because it can be achieved through a single manipulation just after fertilization. Gynogenetic seedlings, however, cannot be used as is for industry purposes, as viability is low and various deformities tend to arise as a result of a higher degree of inbreeding (as with incestuous hybridization) or the effects of treatment. Gynogenetic fish are thus treated with hormones to make them functionally male at a point shortly after birth when their sex is indeterminate. Genetically, the resulting fish are all female and produce only X sperm. This method is expected to be actively adopted in the future, as it is possible to achieve entirely female seedling through normal hybridization of these males and females.

At this point, "false males" are created continuously to maintain a line from entirely female seedlings without relying on gynogenesis. Females are always selected from the most genetically superior for experiments in breeding outstanding pedigrees (see Fig. 4.3) (Oshiro 1990).

In fact, examination of sex ratios for gynogenetic fish shows three categories (Oshiro 1990): varieties in which all examples are female as expected, such as rainbow trout, coho, silver carp, grass carp, catfish, and loach; varieties in which roughly half or more are male, such as halibut and righteye flounders; and varieties in which males are present at a fixed percentage or not present at all within the same

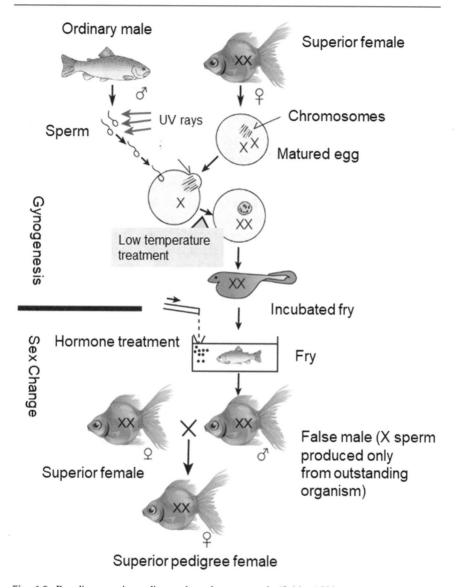

Fig. 4.3 Breeding superior pedigrees through gynogenesis (Oshiro 1990)

species, such as zebrafish and goldfish. The comparative figure of triploid and diploid of giant pacific oyster (*Crassostrea gigas*). The effect of triploidization is remarkable with edible portion growing larger as compared to diploid.

In diploid organisms, one or two different genes involved in specific traits occupy the same position on the pairing chromosome. The former is referred to as homozygosity, the latter as heterozygosity. Among organisms descending from a common ancestor, however, the likely of inheriting the same genes is high, and repeated inbreeding results in the positions of all genes approximating a homozygotic organism. At the same time, a group of clones is achieved in which all organisms are nearly identical genetically. Homozygotic clones are referred to as a "pure line." Several repetitions of gynogenesis were expected to result in establishment of a pure line in the generation faster than with inbreeding. If no crossing occurs at all between homologous chromosomes at a synapsis during maturation division, blocking of second polar body emission to form a gynogenetic diploid results in the position of heterosynaptic genes becoming homosynaptic, producing an organism that is homosynaptic for all genetic positions (homosynaptic organism).

For the second gynogenetic generation from these female parents, homosynaptic groups (clones) appear that always have the same genes at the same gene locations. Indeed, crossing occurs between homologous chromosomes, and the position of hetero genes at those locations does not become homo. The rate of crossing is different for each area, and because it is defined more or less precisely, the rate at which hetero gene positions become homo as a result of gynogenesis varies greatly by position. Depending on the genetic trait, it can therefore be difficult to obtain homo genes for that trait through gynogenesis.

4.3   Cell Manipulation

Cells are structural and functional units for nearly every organism on the planet. A biological organism is, in a sense, a society of cells. Cells typically measure 10–30 μm and are surrounded by a membrane, with one nucleus and cytoplasm inside (Fig. 4.4). Cell biology, which seeks to use the cell as a basis for explicating biological phenomena, was established as a field in the early 19th century. Subsequent advancements led to the development of various sophisticated techniques for manipulating these tiny cells. Some of these techniques may be applied as is to the field of biotechnological breeding.

The chromosome manipulation described in the preceding section may be characterized as cell manipulation as well, in that it involves gamete cells. The academic context is slightly different, however, and actual techniques applied with larger fish consist mainly of manipulation of individual organisms by holding them by the armpit or grabbing them barehanded to squeeze out roe or sperm; roe, which measure 1 mm or more in diameter, are handled in the dozens to thousands, while sperm is handled in units of several cubic centimetre. Since the cell manipulation described below does not typically require sophisticated techniques, it has been divided into several categories (Oshiro 1990).

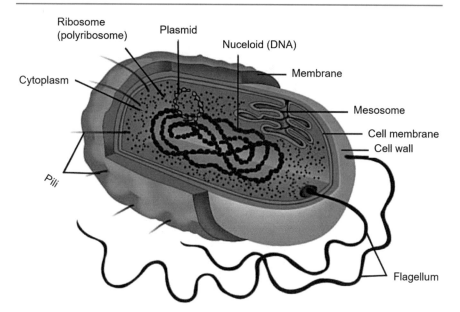

Fig. 4.4 Structure of a cell

4.3.1 Basic Techniques

A wide range of cell manipulation techniques have been developed over the history of cell biology, but those corresponding to the following four basic goals may be applicable for breeding (see Fig. 4.5). Each of them is closely connected the characteristic properties exhibited by the cell as a biological unit, and acquisition of them requires some degree of professional temperament in terms of understanding the cell completely and becoming one with it. These days, however, the development of special devices for the different purposes has made it possible for even a non-expert to manipulate cells with relative ease.

A. Culturing

Cells may be propagated in a suitable culturing medium, or cell propagation factors or growth factors may be added to a culture medium to maintain these functions, in addition to salts to control osmotic pressure and nutrients such as amino acids and vitamins. Recent methods have been applied to culture and produce large quantities of animal cells for useful substances that cannot be produced with E. coli, including complex proteins such as gonadotopic hormones and monoclonal antibodies.

(1) Proliferation

(2) Division

(3) Mixture

(4) Insertion

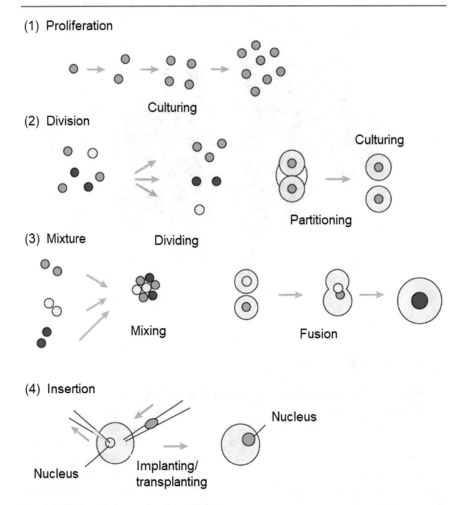

Fig. 4.5 Basic techniques of cell manipulation

B. Division and Partitioning

Technology for separating cells when there is a mixture of several different forms of cells in a suspension includes special devices that select particular cells based on subtle differences in size (the mesh technique), density (centrifugal separation), surface charge (carrier-free electrophorosis), and content or dyeing properties (cell sorting).

In holoblastic cleavage, where cleavage occurs throughout the cytoplasm of a fertilized egg, an organism may arise at the blastomeres at two to four cell stages, allowing for the production of genetically identical twins or quadruplets. Among land animals, successful examples have already taken place with (livestock) mammals.

Among marine animals, shellfish, crustaceans, and sea urchins have total cleavage eggs, yet almost no experimentation has been done with them. Roe are meroblastic, and while a single organism does not appear from a single blastomere, the separated blastomere can be used in addition with the technology described in ③ to produce a chimera.

C. Mixture and Fusion

This refers to techniques for mixing cells. Creating mixed conditions while maintaining boundaries between different cells is relatively easy, and ordinary not especially meaningful. Mixture of genetically different blastomeres from different fertilized eggs early on in genesis, however, can produce a chimera.

At the same time, it is also possible to produce new hybrid cells by inducing adhesion, membrane fusion, cytoplasm mixing, or nuclear unification between cells. This is known as "cell fusion," with chemicals such as polyethylene glycol or electrical shocks typically used to induce it in cells.

D. Transplantation and Amalganation

This refers to techniques for introducing something into the cell or removing something from it. Micropipettes may be used to extract relatively large organelles such as nuclei and insert the nucleus from another cell (transplantation) or introduce minute quantities of various materials into the cell (Oshiro 1990).

4.3.2 Applications

The following are some of the breeding applications of these basic techniques.

A. Gamete Selection and Improvement

Cell separation techniques may be applied to separate sperm in hetero male animals that are believed to exist in two different varieties. Fertilization of each allows for the production of offspring that can be distinguished between female and male.

When sex chromosomes are of markedly different sizes, X sperm carrying X chromosomes have a different weight (density) from Y sperm carrying Y chromosomes, and can therefore be separated by centrifuge. Taking advantage of differing charges on the sperm surface, Takashi Oshiro used a non-carrying cataphoresis device to separate X and Y sperm in loach and obtain seedlings that were either entirely female or entirely male. Once techniques are discovered for dyeing two varieties of living sperm, it will be possible to separate them by cell fractionator. Another method involves refining antigens characteristic of the sexes to produce antibodies and separate out one sex by killing only the sperm, egg, or embryo carrying those antigens.

Once a cell has been fused with two or more sperm, it becomes possible to create various polyploids through fertilization. By fertilizing a sperm that has had genetic material introduced through electroporation, it is possible to insert it relatively easily into nuclear DNA within the embryo (Singh 1994; Terán and Singh 2009).

B. Creating Clones (Hetero)

As described previously, whatever clones are produced from continued inbreeding or gynogenesis are not genetic copies of the initial parent organism. Also, because most gene locations are subject to homozygosis, recessive genes possessed by the parents often manifest new activities. In such cases, there is a high probability of undesirable traits appearing in addition to desirable ones. Today, if one has a genetically superior organism and wishes to make another organism with a very similar genetic makeup (a genetic clone, that is), the only option available at the current stage is nuclear transplantation (Fig. 4.6).

Another method that is limited to female clones involves performing some kind of artificial manipulation on egg cells in the organism's gonads to allow for maturation division and the formation and parthenogenesis of eggs while completely omitting the initial division involving the synapsis (connections between identical chromosomes during the cell's meiosis) and crossing over (partial exchanges that occur between homologous chromosomes during meiosis) of homologous chromosomes. Elucidation of various factors controlling maturation division of egg cells has enabled application of this approach today (Oshiro 1990).

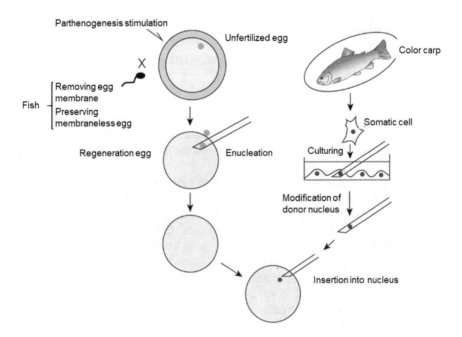

Fig. 4.6 Cloning through nuclear transplantation

C. Creating Nucleus/Cytoplasm Hybrids

The products of nuclear transplantation may be viewed more precisely as nucleus/cytoplasm hybrids than as a clone population. Successful nuclear transplantations were carried out with amphibians in Great Britain in the 1950s and 1960s and with fish in China in the 1980s. In a series of nuclear transplantation experiments conducted with fish, the organisms produced from eggs that had their nuclei extracted and were transplanted with somatic cell nuclei from different species showed not only traits from the species that provided the nucleus, but also some traits from the species providing the eggs. From this, cytoplasmic genes (genetic expression outside the nucleus in the cytoplasm—in the mitochondria, for example) are thought to be involved. In some cases, however, a portion of the ookaryon chromosomes remains due to incomplete micropipette extraction of the nucleus. In China, however, the organism produced was described as a nucleus/cytoplasm hybrid to emphasize the effects of heterosis (a phenomenon in which the first hybrid generation grows to be larger than its parents) due to nucleus and cytoplasm interaction (Mukai et al. 1978; Becraft and Taylor 1989).

D. Chimeras

Blasts may also be combined to produce chimeras (Fig. 4.7). The word "chimera" refers to a mythological creature described by the ancients as a beast with a lion's head, a goat's body, and a dragon's tail. Many other mythical beasts have

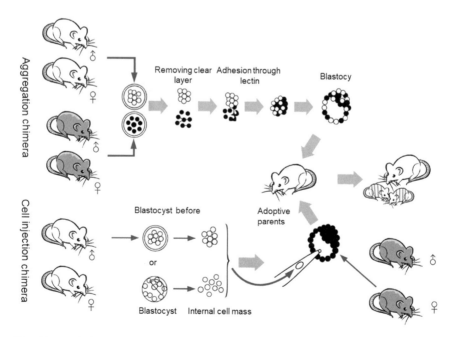

Fig. 4.7 Producing an aggregation chimera and a cell injection chimera

also been created by selecting and combining superior body parts from different species, such as the sphinx, sirens, mermaids, and centaurs. Typically, the term chimera refers to an organism composed of two or more kinds of genetically different cells. In some cases, the different cells are mixed in a mosaic structure; in others, cells of the same time assemble to form specific parts of the body.

While approaches such as cutting are already in widespread use in plant breeding, it has also recently become possible to perform something similar with eggs or embryos in the midst of genesis through microsurgery under a microscope. In terms of breeding science value, it is now possible to assemble superior functions from different animals in one body, just as the ancestor of modern humans sought to do. Contrary to their expectations, it is also possible to combine only the most inferior parts of each body, as in the case of heterosis breeding. In terms of fish, might it not be possible to create a "one fish, two flavors" organism that is halibut on its left side and sole on its right, or a hermaphrodite fish with both seminal glands and ovaries that is capable of reproducing on its own (Fig. 4.8) (Oshiro 1990; Taniguchi 2000).

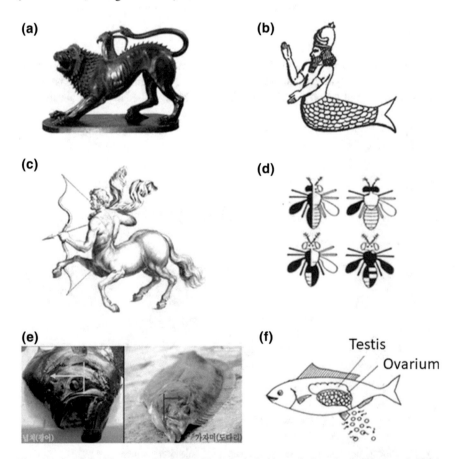

Fig. 4.8 Possible chimera applications **a** Mythological chimera **b** Assyrian half-human, half-fish deity **c** Centaur **d** Male-female fruit fly mosaic **e** Left half flounder, right half sole **f** Hermaphrodite fish

E. Making Transgenic Organisms

The first step in making transgenic organisms is to create an organism with a new genome through cell fusion. Unlike with the chimera, all cells in the organism may have the same genetic composition. If it is not possible to obtain both gametes of different species, or if fertilization is physically impossible because the sperm cannot pass through the micropyle, one may fuse somatic cells from the two species and culture them to select hybrid cells with stable nuclei (genomes). While it is possible to reproduce an organism from one cell in plants, it is typically not possible with animals (including fish). The organism is thus produced by transplanting the hybrid cell nucleus into an egg with its nucleus previous removed and applying a parthenogenic stimulus (Krimpenfort and Berns 1992; Regal 1994; Weeks et al. 2010).

Figure 4.9 shows the example of an organism produced by fusing cells from a tropical fish that prefers high water temperatures and is strongly resistant to disease with those from a salmon, which prefers low temperatures and is migratory, and transplanting them into eggs from the loach, nuclei from which are readily available. As a result, it may be possible to create not a migratory fish that prefers the cold waters of the north (such as the fish caught in Russia), but a new, large and strongly disease-resistant form of fish that migrates to grow in warm southern waters before returning to Japan.

Another example of application would be the culturing of fish embryonic cells that have all capabilities but have not yet specialized, with different genes introduced and the chimera production method applied; introduction into the organism's embryonic cells would result in at least a portion specializing into reproductive

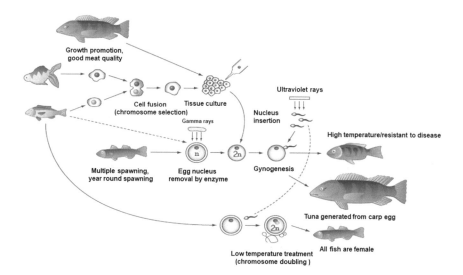

Fig. 4.9 Creating new biological resources through cell fusion, nuclear transplantation, and gynogenesis

cells. Organisms produced from the reproductive cells with these newly introduced genes would be transgenic in the following generation.

F. Producing Beneficial Materials (Cell Breeding)

Examples of obtaining useful materials through culturing of somewhat specialized fish or shellfish cells include the secretion of a nacreous layer in oyster mantle cells and the production of maturity inducing substances in the follicle cells of starfish ovaries. For fish, experiments have been conducted to produce peptide hormones through culturing of pituitary gland cells. More possibilities exist besides these. It would also be beneficial not simply to take advantage of secretion cells that have undergone specialization, but to fuse them with cells from tumors such as myelomas to produce hybrid cells with reproduction capabilities, which could then be cultured for the effective production of useful substances. Possibilities are also emerging for the use of hybridomas from mouse lymphocytes to produce monoclonal antibodies for specific fish antigen proteins, which could be used for clinical testing and research (Oshiro 1990).

4.4 Gene Manipulation

At root, breeding is a matter of assembling genes that are beneficial to humans within a particular organism or population. The traditional approach of selection and cross-breeding and the techniques used to date for chromosome and cell manipulation follow a somewhat general breeding approach of simply assembling chromosomes (the "cassette tape") or chromosome sets (O X cassette collections) that have the target genes. As a result, the collection comes to include songs that are unnecessary or even undesirable in addition to the desired ones. In contrast, gene manipulation, or the extraction of only the desired genes for introduction into the genome of a fish that one is seeking to improve, could be considered the equivalent of selecting the songs one likes and copying them to or splicing together a master tape. The resulting fish with new genes introduced are called transgenic fish, or simply hosts (Old and Primrose 1981).

In the natural world, regulation of expression (as with the puff observed in the fruit fly's saliva chromosome, where a puff is a discontinuously swollen structure observed in specific parts of a multi-stranded chromosome), amplification (as with ribosome RNA genes during the formation of fish and amphibian eggs), recombination (as with the reconstitution of antibody genes in lymphocytes), and transfer (as with transposons and transposable elements) at the level of the single gene occur frequently and are often essential to survival.

Artificial gene manipulation began in 1944 with Avery et al., who transferred DNA extracted from the S variety of pathogenic *Streptococcus pneumoniae* through culturing of bacteria from the non-toxic R variety. Transformation through introduction of such DNA fragments (genes) has already been performed successfully in several species, including higher-order plants and animals. As various more genes are expected to be discovered and studied through the biological world, these too will be introduced and applied to breeding across species boundaries.

4.4.1 Basic Techniques

Genes were first understood as hypothetical particles, proposed by Mendel in 1865 to explain genetic phenomena and their principles. Today, molecular genetics and genetic engineering have achieved rapid advancements, and it has become possible for to actually obtain genes and to improve and use them freely.

As noted previously, genes are large molecules within the double helix structure of DNA, which exist in the same form in all somatic cell nuclei. As is widely known, a single gene, depending on the sequence of bases that compose it, designates and possesses a diagram for the structure of specific proteins and RNA produced by the cell.

Within chromosomal DNA, genes are located at specific intervals. They may be understood as similar to stations on a railway or oases in a desert. Internally, genes have a complex structure; the portions that are finally translated into proteins are dispersed and are known as expressed sites (structure sequences), while the untranslated portions between the expression sites are known as non-expressed sites (intervening sequences). Within the nucleus, all information—including both types—is transcribed to RNA, after which only the expressed sites, and not the non-expressed sites, are spliced together to form messenger RNA (mRNA). This mRNA is then transported to the cytoplasm and translated into proteins. A single complete gene also includes a portion known as an enhancer, a promoter gene that is actually transcribed and is necessary for expression to take place (Fig. 4.10).

Fish breeding may be conceived as consisting of two stages: modification of the gene to be introduced and introduction of the gene to the egg cell. The following section explains the basic techniques needed for each stage.

A. Modification of Foreign Genes

1. *Separation*

Following removal of proteins from the tissue extract with enzymes and phenol, the nucleic acid (DNA and RNA) is precipitated with alcohol. The RNA may be removed with enzymes to obtain DNA. mRNA may be separated into specific end sequences (poly(A)) through chromatography (Williams et al. 1991).

2. *Dissociation and Truncation*

Dissociation and recombination of the DNA double helix can be performed easily through acid and base treatment. Various enzymes are used as "scissors" to cut the nucleic acid chain itself. These enzymes include DNase (or RNase), which truncate whatever DNA (or RNA) they come across, and a number of restriction

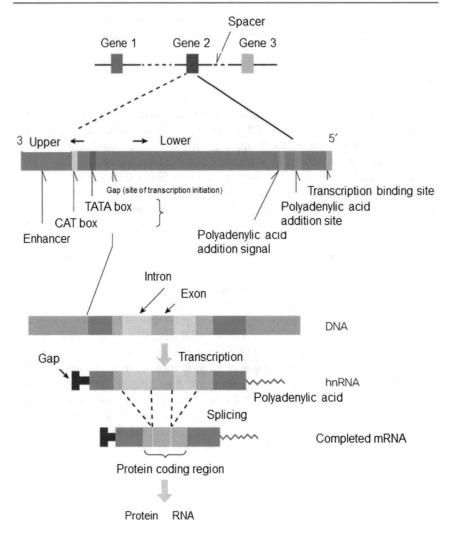

Fig. 4.10 Gene structure

enzymes that recognize four or six specific base sequences and only truncate there (Fig. 4.11). Restriction enzymes include those that clip the two DNA strands such that the truncation surfaces are parallel, and those that cut them so that they intersect; the latter results in a "sticky" surface at the time of DNA adhesion. Restriction enzymes remove target genes or portions of genes from the genome (Surmacz et al. 1995; Ishikawa 1985).

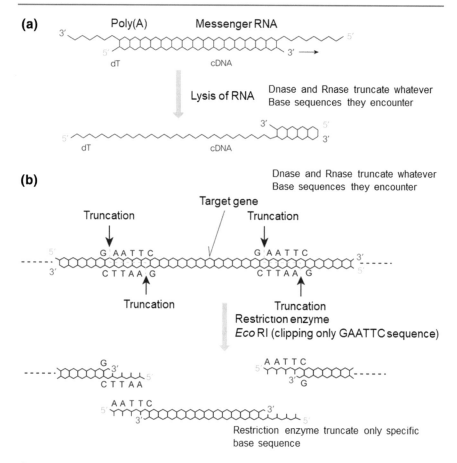

Fig. 4.11 Basic techniques of gene manipulation (1): clipping

3. *Adhesion*

The next stage requires some "adhesive" to connect nucleic acid. Two approaches used for this are DNA ligase, which uses the adhesive surface to connect DNA, and RNA ligase, which connects RNA. Genes or portions of genes removed through (2) in the former are connected by promoters or vectors (Fig. 4.12).

4. *Transcription and Cloning*

Also essential is manipulation that freely "clones" information possessed by the nucleic acids: from DNA to DNA (DNA polymerase), from RNA to RNA (RNA replicase), from DNA to RNA (RNA polymerase), and from RNA to DNA (reverse transcriptase; Fig. 4.13). Reverse transcriptase in particular is essential when forming complementary DNA (cDNA) with information transcribed from

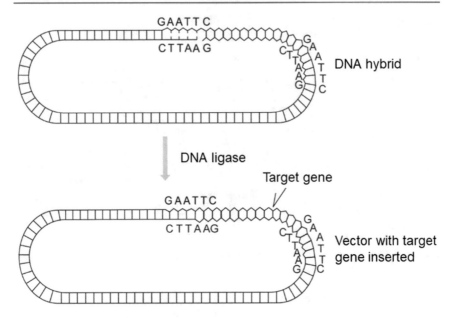

Fig. 4.12 Basic techniques of gene manipulation (2): connecting with "adhesives"

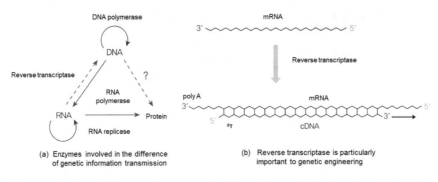

Fig. 4.13 Basic techniques of gene manipulation (3): multiplication through cloning (Honjo 1986)

messenger RNA (mRNA). DNA replication is typically carried out by attaching it to a vector (such as a virus or plasmid) and introducing it into bacillus bacteria (Fig. 4.11).

As with creating a music or video cassette, the organization and "editing" of the genetic information that must be passed along requires a minimal set of suitable tools: the glue, the scissors, and the replication device.

5. *Hybridization*

The double strand of DNA consists of a series of set base pairs; if one strand has a certain sequence, the corresponding sequence connected to it has been determined. It is also possible to dissociate and combine the double strand easily. To take advantage of this property and investigate a particular DNA sequence (gene), one simply needs to mark single-chain DNA that is already known with a radioactive isotope, and use a probe to choose the combining part of the unknown single DNA chain.

B. Components of Introduction

1. *Inserting*

Various approaches exist for introducing genes into an egg cell, including the calcium phosphate, microinjection, cell chimera, cell fusion, and electroporation methods. A portion of the gene introduced into the cell happens to be incorporated into the genome, where it is replicated and distributed to cells throughout the body with the remaining genome at the time of cell division. To date, microinjection of a highly concentrated gene solution near the nucleus of the fertilized egg appears to be the surest approach for introduction to the genome.

Before, it was thought that foreign genes had to be attached to a vector such as a virus, plasmid, or transposon to increase the efficiency of introduction into the gg genome, but such introduction also takes place effectively through the genes alone. The use of ring-shaped rather than linear chains for the DNA fragments is also reported to be more effective.

The success or failing of foreign gene introduction can be examined through hybridization, with the entire DNA of the resulting fry separated and the gene or its main sequence probes.

2. *Promoting Expression*

As a following step, it is important to induce actual expression of the gene that has been introduced into the fish (host) genome. In other words, it must be transcribed to mRNA and translated into a specific protein on the ribosome so that the host's properties are transformed. Typically, the resulting mRNA or proteins are examined to determine whether expression has taken place. It is also necessary to ensure expression in the proper part (tissue) of the host body and the proper time. Additionally, it may be possible for humans to control gene activity by inducing expression only when a specific kind of signal (expression stimulus) has been sent from outside. In this case, the question of how to treat the introduced gene becomes important. Combination with a suitable promoter is essential when promoting or controlling expression. Often, the promoter possessed by the gene itself is simply used as it is.

To induce powerful expression without specifying a time or site, an effective approach is to use a virus such as simian virus 40 (SV40), which is isolated from monkeys, or respiratory syncytial virus (RSV). Powerful promoters that allow expression to be controlled may also be used from metallothionein or heat shock protein genes.

3. *Transmission to Later Generations (Establishing a Transgenic Line)*

One introduced to a host genome, the gene is transmitted to future generations through eggs of sperm. Not all of the offspring will possess the gene, however. The genes are not introduced simultaneously at the same sites in two homologous chromosomes; rather, gametes with and without the genes form after meiosis. Transfer of genes microinjected into a fertilized egg typically takes place after the second cell stage; often, cells in the parents' bodies (including pre-meiosis reproductive cells) are separated into those with and without the genes, which form a kind of mosaic.

To establish a line where foreign genes are constantly being transmitted to all descendants, it is necessary to first create organisms in which those genes form a homo junction on the chromosome. These may be obtained through gynogenesis in the initial transgenic female, or similarly through androgenesis in the male.

4.4.2   Examples of Applications

The previous sections have given a sequential account of the basic techniques needed for fish breeding. In contrast with chromosome or cell manipulation, these basic techniques hold no significance at all on their own. Rather, they are mere constituents in the technical system of the single greatest example of application, namely the creation of transgenic fish.

For this reason, the explanation in this section will consider the introduction of tuna growth hormone genes into a rainbow trout as an example of application (Fig. 4.14). The purpose of this research is to significantly promote the trout's growth and create a fish that is as large and flavorful as tuna through marine culture.

The entire system consists of modification of growth hormone genes, a process that consists of three stages.

First, a specific base sequence is located in the growth hormsone gene, and a probe is created with a complementary oligonucleotide (a DNA fragment consisting of dozens of bases). This requires extraction and purification of growth hormone (a protein molecule) from the tuna's pituitary gland (an organ for hormone production) and analysis of its amino acid sequence. By estimating the corresponding DNA base sequence on the genetic code, it is possible to identify the major sequences of the tuna's growth hormone gene. A very specific sequence is then selected and a DNA fragment consisting of 20 to 50 bases is artificially synthesized with a synthesizer. Once it has been labeled with ^{32}P, it is now a probe.

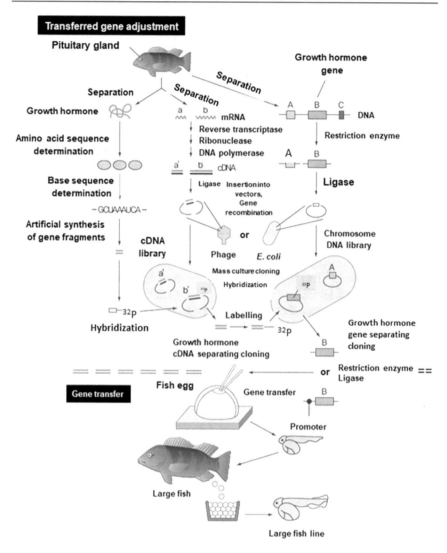

Fig. 4.14 Introducing a useful gene into fish eggs

The second step is formation of complementary DNA (cDNA) for the growth hormone gene. mRNA is separated from the tuna's pituitary gland, which is the site of prolific growth hormone production and secretion, and an enzyme reaction is used to create cDNA replicating that information. This sort of connecting of information components from each gene does not exist in nature. The cDNA is attached to a plasmid or other vector and introduced into a phage or colon bacterium for large-scale culturing and cloning. Once obtained, cDNA clones can be preserved; because it comes from various genes active in the pituitary gland at the time of mRNA extraction, this is known as a cDNA library. A synthetic probe is used to select only the growth hormone cDNA for cloning.

The third step is extraction of the growth hormone gene itself. Within the egg cell, there is a strong probability that the entire introduced gene will be expressed, rather than the cDNA, which lacks intron. First, the entire genome DNA is extracted from body tissues such as corpuscles or muscles polymeric state, and restriction enzymes are used to separate it into fragments, each of which is recombined with a bacillus bacterium or phage and cloned. This is known as a chromosomal DNA library, as it is believed to contain all genes on the chromosome. A more precise method of selecting the growth hormone gene involves the use of a cDNA probe containing a sequence complementary to the entire information section. Cloned growth hormone genes may be microinjected into the egg cell as it is or in combination with a suitable promoter. The gene's introduction into the host rainbow trout genome, its expression, and its transmission to future generations must be individually verified.

Next, we will consider what genes may be introduced to produce particular kinds of fish. In traditional cross-breeding approaches, available genes have been limited to those within species that are the same or closely related. With gene introduction, it is possible for any gene from any organism to be introduced according to a set method. In other words, it becomes possible to use the genetic resources of all 20,000 or so species of fish, or of the millions of species of organisms living on Earth today.

In some cases, genes already possessed by the fish, or something similar, are introduced and their activities amplified with an emphasis on specific beneficial or economic properties (such as growth hormone genes). In others, entirely new genes are introduced to transfer an outstanding capability specific to one fish intact to another (as with genes for flounder anti-freezing proteins or carp α-globin). Both are crucial for breeding.

Introduction of growth hormone genes was touched off by research conducted in 1982, when Richard Palmiter and other U.S. researchers introduced rat growth hormone genes to a fertilized mouse egg to create a "super mouse" nearly twice the weight of a normal mouse. In the future, genes from fish will be used rather than genes from previously cloned mammals. This approach is beneficial in terms of regulating gene expression in the host and the physiological activity of the final peptide hormone product.

Anti-freezing proteins in the blood of flounders allow them to live in Arctic Ocean where their blood would otherwise freeze. Attempts are now under way to introduce those genes into salmon to increase the range of regions in which they can be farmed.

Salmon and trout are vulnerable to oxygen deficiency, whereas carp are stronger. This stems from differences in the structure of their hemoglobin (a protein that transports oxygen in the blood), and in its globin molecule subunits. By introducing carp globin genes into rainbow trout, it may be possible to significantly increase the latter's respiratory capabilities—allowing them to be raised in non-flowing waters —and produce rainbow trout that are easy transported.

Thus far, we have seen how genes introduced into the genome are transmitted to future generations. It is also possible to observe the expression of those genes.

To achieve significant improvement in a rainbow trout's respiratory capabilities, one first (a) removes α-globin from a carp's red blood cells. (b) The entire DNA of rainbow trout fry incubated in the recipient egg is subjected to analysis. (c) The carp gene is introduced into the rainbow trout genome at a high rate of probability. (d) The historic creation of a new species: a transgenic rainbow trout with carp genes.

The following section will discuss some of the genes that will need to be introduced into fish for breeding purposes.

① Physiologically active peptide genes: Various hormones, hypothalamus substances to promote or suppress their secretion, genes for polymer factors directly controlling hormone activities.

② Metabolic enzyme genes: Includes genes for Q-enzyme (enzyme for shifting part of the α-1,4 glycan chain to the C6 position), which converts from the Embden-Meyerhof to the Warburg-Dickens pathway and from the progestin production pathway to the corticoid production pathway and metabolic system.

③ Genes controlling innate behaviors: Includes genes for migration, elasticity, flavor, and spawning.

④ Asynaptic genes: Eggs are formed through division of ordinary somatic cells, and clones of parents are produced through gynogenesis.

⑤ Other: Genes for controlling virus or germ proliferation and conferring disease tolerance.

For gene introduction to become part of the mainstream of future breeding, however, an abundant store of knowledge will be needed on various genetic traits and the genes that govern them. In the field of fish, such basic research is still very scant, and future development is anticipated for this field (Oshiro 1990).

4.5 Chapter Summary and Conclusion

The oceans are growing in importance as a source of high-level protein resources and a storehouse of food for future humankind. Recently, however, severe difficulties are predicted for these supplies due to deteriorating international fishing conditions and environmental pollution, including coastal contamination and reclamation. Limits have already become apparent in the ability of current catches to satisfy marine product demand. This environment of unlimited competition coincides with full-scale globalization and openness in farming and marine products with the recent launch of the World Trade Organization (WTO) system. As a result, it has become crucially important to explore needs in terms of establishing international marine industry competitiveness and promoting national welfare, to strategically development technology to contribute to addressing national food issues, and to devise a strategy for marine industry exports through advancement into the global market. Adapting to this worldwide trend and meeting national demand urgently requires maximal productivity improvements through intensive high technology development, as well as an increase in farming yields. Among the basic conditions for farming production, production of seedlings (fry) is considered most important in terms of establishing the raw materials for production. Production of high-quality seedlings is also viewed as a shortcut to maximizing farming productivity improvements.

Recently, efforts have been under way worldwide to produce superior, high value-added varieties, using genetic engineering techniques to maximize the productivity of unit efforts over a short time frame. Much research effort has been focused on fish, which have the highest economic value among marine products. Additionally, the introduction of foreign genes from various farmed fish varieties has recently been observed in laboratories in various advanced economies, with investigations under way on gene expression and transmission to future generations.

This field is thus seen as having great potential for development as it enjoys support from the rapidly advancing fields of molecular biology and genetic engineering.

References

Arai, K. (2001). Genetic improvement of aquaculture finfish species by chromosome manipulation techniques in Japan. In *Reproductive biotechnology in finfish aquaculture* (pp. 205–228). Amsterdam: Elsevier.

Baldacci, G., de Zamaroczy, M., & Bernardi, G. (1980). Excision sites in the GC cluters of the mitochondrial genome of yeast. *FEBS Letters, 114*(2), 234–236.

Becraft, P., & Taylor, G. (1989). Effects of nucleus, cytoplasm and male sterile nucleus-cytoplasm combinations on callus initiation in anther culture of wheat. *Euphytica, 44*(3), 235–240.

Cassman, K. G., & Liska, A. J. (2007). Food and fuel for all: Realistic or foolish? *Biofuels, Bioproducts and Biorefining: Innovation for a Sustainable Economy, 1*(1), 18–23.

Endo, T. (2007). The gametocidal chromosome as a tool for chromosome manipulation in wheat. *Chromosome Research, 15*(1), 67–75.

Evans, M. J., Gurer, C., Loike, J. D., Wilmut, I., Schnieke, A. E., & Schon, E. A. (1999). Mitochondrial DNA genotypes in nuclear transfer-derived cloned sheep. *Nature Genetics, 23*(1), 90.

Hassold, T., & Hunt, P. (2001). To err (meiotically) is human: The genesis of human aneuploidy. *Nature Reviews Genetics, 2*(4), 280.

Honjo, T. (1986). *Life image express by gene* (p. 204). Tokyo, Japan, Blue Backs: Kodansha Publishing Co.

Isaacs, F. J., Carr, P. A., Wang, H. H., Lajoie, M. J., Sterling, B., Kraal, L., et al. (2011). Precise manipulation of chromosomes in vivo enables genome-wide codon replacement. *Science, 333*(6040), 348–353.

Ishikawa, T. (1985). *Molecular evolution* (pp. 185–189). Tokyo, Japan: Shokabo Publishing Co.

Krimpenfort, P. J., & Berns, A. J. (1992). Transgenic mice depleted in mature t-cells and methods for making transgenic mice. Google Patents.

Long, S. P., Marshall-Colon, A., & Zhu, X.-G. (2015). Meeting the global food demand of the future by engineering crop photosynthesis and yield potential. *Cell, 161*(1), 56–66.

Morrison, C. (1993). Fish and shellfish. In *Frozen food technology* (pp. 196–236).

Morse, D. E. (1984). Biochemical and genetic engineering for improved production of abalones and other valuable molluscs. *Aquaculture, 39*(1–4), 263–282.

Mukai, Y., Maan, S. S., Panayotov, I., & Tsunewaki, K. (1978). Comparative studies of the nucleus-cytoplasm hybrids of wheat produced by three research groups. Indian Society of Genetics and Plant Breeding.

Old, R. W., & Primrose, S. B. (1981). Principles of gene manipulation: An introduction to genetic engineering. Univ of California Press.

Oshiro, T. (1990). Fish breeding and high technology. In *Biotechnology and high technology of fisheries* (pp. 31–64). Tokyo, Japan: Seizando-Shoten Publishing Co.

Purdom, C. (1983). Genetic engineering by the manipulation of chromosomes. *Aquaculture, 33*(1–4), 287–300.

Regal, P. J. (1994). Scientific principles for ecologically based risk assessment of transgenic organisms. *Molecular Ecology, 3*(1), 5–13.

Shiels, P. G., Kind, A. J., Campbell, K. H., Waddington, D., Wilmut, I., Colman, A., et al. (1999). Analysis of telomere lengths in cloned sheep. *Nature, 399*(6734), 316.

Singh, S. P. (1994). Gamete selection for simultaneous improvement of multiple traits in common bean. *Crop Science, 34*(2), 352–355.

Surmacz, E., Sell, C., Swantek, J., Kato, H., Roberts, C. T., Jr., LeRoith, D., et al. (1995). Dissociation of mitogenesis and transforming activity by C-terminal truncation of the insulin-like growth factor-I receptor. *Experimental Cell Research, 218*(1), 370–380.

Taniguchi, N. (2000). Genetic diversity of fishes and DNA markers. In F. Takashita (Ed.), *The next generation of fisheries biotechnology* (pp. 43–53). Tokyo, Japan: Seizando-shoten publishing Co.

Terán, H., & Singh, S. P. (2009). Gamete selection for improving physiological resistance to white mold in common bean. *Euphytica, 167*(3), 271–280.

Thomas, P. S. (1980). Hybridization of denatured RNA and small DNA fragments transferred to nitrocellulose. *Proceedings of the National Academy of Sciences, 77*(9), 5201–5205.

Thorgaard, G. H. (1983). 8 chromosome set manipulation and sex control in fish. In *Fish physiology* (Vol. 9, pp. 405–434). Amsterdam: Elsevier.

Vorobjev, I. A., Liang, H., Wright, W. H., & Berns, M. W. (1993). Optical trapping for chromosome manipulation: A wavelength dependence of induced chromosome bridges. *Biophysical Journal, 64*(2), 533–538.

Weeks, D. P., Wang, X.-Z., & Herman, P. L. (2010). Methods and materials for making and using transgenic dicamba-degrading organisms. Google Patents.

West, C. (2005). Economics and ethics in the genetic engineering of animals. *Harvard Journal of Law & Technology, 19,* 413.

Williams, R. S., Johnston, S. A., Riedy, M., DeVit, M. J., McElligott, S. G., & Sanford, J. C. (1991). Introduction of foreign genes into tissues of living mice by DNA-coated microprojectiles. *Proceedings of the National Academy of Sciences, 88*(7), 2726–2730.

	Layout: **T1 Color**	Book ID: **462027_1_En**	Book ISBN: **978-3-030-20943-8**
	Chapter No.: **5**	Date: **23-9-2021** Time: **3:43 pm**	Page: **109/144**

Genetic Diversity and DNA Markers in Fish

5

Contents

5.1   The Meaning of Genetic Diversity

Among organisms, genetic diversity is a quantitative representation of the diversity existing within and between species that is passed on to descendants through genetic mechanisms, or the differences in homologous genes within and between species. An example would be to compare genetic base sequences and use differing base ratios for measurement. The origins of genetic diversity lie in the various mutations that arise in species, such as base transposition, deletion, and repetition or gene transformation, recombination, and transfer (Liu and Cordes 2004).

As a category, genetic diversity encompasses both fluctuation within a population and differentiation between populations. Fluctuations occur over the short and long term through the balance between mutation supply and loss within a given natural population. The genetic mutations within a wild population are understood

The original version of this chapter was revised: belated author corrections have been incorporated. The correction to this chapter is available at
https://doi.org/10.1007/978-3-030-20944-5_12

S.-K. Kim, *Essentials of Marine Biotechnology*,
https://doi.org/10.1007/978-3-030-20944-5_5

39 to be of great significance in relation to the survival capabilities of individuals and
40 the group as a whole.

41 Genetic diversity is thus also an indicator of population health for various
42 organisms. It is understood to be of great importance from the standpoint of
43 long-term development of the farming, forestry, and fishing industries, which
44 depend on organism production, and it also possesses great significance for industry
45 in terms of its being essential in providing a breeding stock for improvements in
46 marine aquaculture.

47 Why should we be interested in biodiversity? The decline in diversity that we are
48 bearing witness to today is both a direct and an indirect product of human activities.
49 Diversity has been diminishing by the day due to deliberate hunting and harvesting,
50 habitat destruction from development activities, and indirect causes such as global
51 environmental changes and oceanic pollution. Reduced diversity may lead to a
52 collapse in interdependent ecosystem relationships, resulting in the disappearance
53 of species of organism that may hold as yet unknown benefits for us. An excellent
54 example is the Pacific yew (*Taxus brevifolia*). Native to the west coast of the U.S.,
55 this tree previously received little attention due to its slow growth and small size.
56 Today, it has an important role in providing raw material for the powerful
57 anti-cancer medication Taxol. Declining species diversity, and the resulting
58 destruction to ecosystems, can also be a source of both spiritual and psychological
59 damage to human beings.

60 To address these issues, the United Nations Food and Agriculture Organization
61 (FAO) moved in June 1992 to establish the Biodiversity Convention, an agreement
62 with the aim of preserving diversity among species. The document affirmed the
63 need for genetic preservation of wild fish and shellfish preservations and assessment
64 of pure stocks from a breeding perspective, as well as the importance of developing
65 highly sensitive DNA markers as a means of studying them (Toledo and Burlin-
66 game 2006).

67 Population size is critical for species survival. Rare species living in special
68 habitat environments are often threatened with extinction due to declines in their
69 population numbers from loss of those habitats. Typically, genetic diversity is
70 reduced in populations of fewer than 100 organisms due to genetic drift, inbreeding,
71 and diminished genetic flow. Many research studies have shown that at least
72 100,000 organisms must exist within a population for it to be capable of long-term
73 preservation. The number of organisms that can produce the next generation of
74 gametes at an equal level of probability within a population is normally referred to
75 the effective population size (N_e), which is always smaller than the absolute pop-
76 ulation size N. N_e may be calculated as $4(N_m N_f)/(N_m + N_f)$, where N_m is the number
77 of males and N_f the number of females.

78 Effective population size is also affected by fluctuations in absolute population
79 size from one generation to the next.

80 For surviving organisms from an endangered species, a population may be
81 established through capture and breeding programs. Genetic diversity may be
82 reduced by bottlenecks (temporary reductions in population size) or the founder
83 principle (the emergence of new populations as a result of migration by small

numbers of organisms). Genetic influence is even more severe in small and isolated populations (ecological corridors). Genetic drift occurring in populations with small effective sizes results in a serious loss of genetic diversity. Genetic drift is a random phenomenon, and selective survival of genes cannot be expected. Inbreeding also occurs at a high rate among small populations, increasing the percentage of homozygotes within them. The inbreeding constant F may be expressed in the equation $F = (2pq - H)/2pq$, where H is the actual frequency of heterozygotes within the population. As this shows, in populations that are reduced enough for genetic drift to occur, H declines with each generation in proportion to effective population size. F increases as a result.

What effects does inbreeding have on a population's long-term survival? Self-fertilizing plant population exhibit a high level of homozygosity and a relatively low level of genetic mutation, but considerable differences appear between populations. Descendants produced through inbreeding show lower adaptability and survival rates than species not produced through inbreeding, a phenomenon referred to as "inbreeding depression."

The reason for this has to do with an increase in homozygosity for harmful genes. The number of harmful alleles within a population is referred to as its genetic load. Harmful genes expressed phenotypically after inbreeding may be removed due to their failure to adapt and their low survival capabilities, while isolation and fragmentation of populations reduces gene flow, and thus genetic diversity.

5.2   Genetic Diversity and Marine Organism Resources

Because mariculture fisheries uses a small number of parents to produce the next generation, changes in genetic composition and declining variability are often observed as a result of inbreeding or the bottleneck effect. While measures to prevent this have recently been adopted at marine fish seedling production farms, some unexpected outcomes have arisen. These unwitting genetic changes that arise in artificial seedling populations must therefore be monitored through genetic markers, which act as a sort of "ID (Jensen and Fenical 1996)."

To date, there have been no cases or examples of seedlings for fishing industry having a negative genetic influence on wild types. This may simply be because no means exist for monitoring such an influence. As the fishing industry is dependent on wild types for a production base, measures that take wild type genetic preservation into account are needed, and a hurried effort is currently under way to develop highly sensitive DNA markers that may be used on fish species subject to genetic change due to artificial seedlings, along with tracking surveys for released populations.

From a biodiversity preservation standpoint, targets for preservation consist of a single local group or clade within a single species. One concern is that if a different clade is introduced and released, hybridization with the native group could occur, resulting in loss of the native group.

	Layout: **T1 Color**	Book ID: **462027_1_En**	Book ISBN: **978-3-030-20943-8**
	Chapter No.: **5**	Date: **23-9-2021** Time: **3:43 pm**	Page: **112/144**

Recently, it has been become possible to use DNA polymorph (possessing multiple alleles) markers to detect and identify previously unknown local clades of marine fish species. Since resource management focuses on local groups and clades, genetic marker-based monitoring systems for these populations must be adjusted to acquire parent species and produce and release seedlings.

In contrast with fishing, fish farming involves artificial management of organisms, with genetic improvements carried out as in agriculture to meet breeding goals related to various producer or customer demands. Explication of the improved varieties' genetic characteristics and preservation and management of the new varieties are important tasks for this field. When a new variety is developed, the spatial, labor, and economic requirements for the line's preservation and management increase, and costs may increase substantially. If mixed raising becomes possible through the employment of genetic markers in these improved varieties and lines, however, the maintenance costs and labor requirements for management of the parents and preservation of the line may decrease significantly down the road.

5.3 What Are Genetic Markers?

Genetic markers refer to instances in which a gene's phenotype can be easily distinguished and used to identify organisms or cells possessing it. Genetic markers may be used as probes to identify the nucleus, chromosomes, or a gene's position. They may be used as indicators in assessing the genetic diversity of a fish or shellfish population that is the subject of attention, or in determining intraspecific geographic groups. In other words, they allow for direct observation of genetic information and estimation of genetic relationships between organisms.

Markers commonly used in population genetics include body color, shape, special features, blood type, proteins (isozymes, etc.), and defined DNA regions. In all cases, polymorphs are recognized as a result of alleles. Effectiveness as a marker depends on the degree of diversity (Giblett 1969; Queller and Goodnight 1989; Williams et al. 1990; Kidd 1993; Sunnucks 2000).

Isozyme polymorphs, which have come to assume an important role as DNA markers over the years, exist within DNA's structural domain. Because they perform functions for the organism, they are associated in many cases with single gene loci; even when multiple gene loci exist, variability is somewhat lacking. For this reason, there are limits to their applicability in detecting differences between local populations shortly after isolation from their habitat environment. In population genetics research, the amount of genetic information is determined by the quantity and quality (sensitivity) of genetic markers. This has naturally led to a desire to develop more sensitive markers than isozymes, and recent years have seen rapid advancements in research using DNA base sequence polymorphs that include regions exhibiting high variability.

166   Among DNA base sequences, high-variability intron have been in frequent use
167 recently, including D-loop regions of mitochondrial DNA, nuclear DNA min-
168 isatellites (sections of non-genetic regions with large numbers of repeating
169 sequence units), sites (DNA fingerprints), and microsatellites (sections of
170 non-genetic regions with small numbers of repeating sequence units). Since these
171 sites have a great number of accumulated genetic polymorphs, they may be used as
172 genetic markers in identifying as yet undetected local populations (clades) with a
173 very low level of differentiation, assessing genetic variability, and clearly identi-
174 fying changes in genetic diversity for artificially produced selective populations or
175 genetically manipulated populations, as well as increases in the inbreeding
176 coefficient.

177   As an example, the number of bases in fish DNA may range from the hundreds
178 of millions to billions per genome. Various genetic individual mutations are
179 included in that sequence. Mutations in an organism's DNA accumulate in greater
180 numbers on non-coding introns than on coding exons. Non-genetic regions include
181 repeating sequences of bases in the single to double digits; when the number of
182 repeating base units for the site is large, it is known as a minisatellite, while
183 microsatellites are sites where the basic number is small (Fig. 5.1). Extranuclear
184 mitochondrial DNA also contains exon and intron, where genetic individual
185 mutations accumulate in greater numbers.

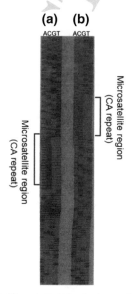

A repeated CA sequence (microsatellite region) can be clearly seen. Base sequence information on either
side of the repeated CA sequence can be used to design a PCR primer to multiply that region. Primers have been
designed for detection of Pma1 in image (a) on the left and Pma2 in image (b) on the right.

Fig. 5.1 DNA base sequence ladder in a red seabream *Pagrus major* nucleus

186   The names and characteristics of DNA polymorph markers differ by the DNA
187 region that they are detecting. DNA polymorph detection methods may be classified
188 into those in which the reason for polymorphism stems from base substitution
189 (restriction fragment length polymorphism, RFLP) and those in which it stems from
190 differences in the number of unit sequence repetitions (variable number tandem
191 repeats, VNTR). They may also be classified according to whether the polymerase
192 chain reaction is used (Table 5.1).
193   While chromosomal DNA typically possesses a fixed base sequence per
194 organism, closer observation shows subtle differences within organisms of the same
195 species. These are known as DNA polymorphs, which can be easily detected
196 through Southern blotting because of differences in the lengths of truncation

Table 5.1 Polymorphism and applicability of fish DNA polymorph markers

DNA polymorph detection method	Polymorphism indicator		Variability and applicability
	Level of mutation	Specialization between populations	
RFLP method			
Nuclear DNA	Band sharing	Band sharing	Low variability: Phylogenetic classification
Mitochondrial DNA	Haplotype gene diversity	Base substitution rate	High variability: Phylogenetic classification
PCR-RFLP method			
Nuclear DNA (AFLP method)	Band sharing	Band sharing	Medium variability: Phylogenetic classification, lineage analysis
Mitochondrial DNA (D-loop region)	Haplotype gene diversity	Base substitution rate	High variability: Phylogenetic classification, lineage analysis
VNTR method			
Minisatellite DNA polymorph gene location method	Band sharing	Band sharing	High variability: Quantifying mutation level, proving clonality
Minisatellite DNA single gene location method	Number of alleles	Heterozygosity rate	High variability: Subpopulation analysis, quantifying population mixture rate and degree of inbreeding
Microsatellite DNA single gene location method	Number of alleles	Heterozygosity rate	High variability: Subpopulation analysis, quantifying population mixture rate and degree of inbreeding
RAPD method	Band sharing	Band sharing	Low-high variability: lineage analysis, species identification
PCR/Base sequence method	Base substitution rate	Base substitution rate	Low-high variability: Quantifying mutation level, phylogenetic classification

fragments produced when they are expressed to restriction enzymes, a phenomenon known as restriction fragment length polymorphisms (RFLP). RFLP is exhibited randomly throughout chromosomal DNA and transmitted along a Mendelian model. Accordingly, several different RFLP expression patterns may be combined to distinguish organisms, lines, populations, and species.

Animal DNA includes series-type repetition sequences that are prone to changes in frequency through repetition. Restriction enzyme fragments including these sequences exhibit RFLP according to differences in the number of sequence repetitions. These forms are known as variable tandem pair repeats (VTPR).

With RFLP, each organism exhibits different numbers of repeats as pairs of 10–20 bases showing the same sequence at a given gene site are repeatedly established, producing differing forms of mutations.

Polymorphism refers to diversity in the forms and traits exhibited by organisms within the same species (Taniguchi 2000).

5.4   DNA Polymorph Detection Methods

The following is a brief explanation of the chief DNA polymorph detection methods (Table 5.1).

(1) DNA Extraction: VNTR (as in DNA fingerprinting), which is a DNA polymorph detection method that does not use PCR, requires large volumes of highly pure DNA.

(2) Blood is often used as a sample in this case. It is subjected to manipulation including extraction with phenol, purification, and precipitation with ethanol to extract DNA from the blood cell nuclei.

With other methods of DNA polymorph detection, PCR may be used to amplify DNA during the manipulation process, which means that a small amount of DNA will suffice. This results in a simpler DNA detection method that allows for extraction and processing from multiple organisms. (Taniguchi and Takag 1997)

(3) Variable number of tandem repeat (VNTR) method

Minisatellite region polymorphs (Fig. 5.2): These may be divided into multiple gene locus and single gene locus methods, depending on the probe used. Probes used in minisatellite region polymorph detection include M13 phage DNA (GAGGGTG GNGGNTCT) and the human-derived minisatellite 33.15 (AGAGGTGGGCAGG TGG). (Vassart 1987; Jeffreys et al. 1985)

These methods require a sample with large quantities of highly pure DNA, with around 60 h of electrophoresis following restriction enzyme truncation. After electrophoresis comes additional manipulation, including Southern blotting, hybridization with a probe marked with a radioactive isotope, and band detection through autoradiography. A drawback of this approach is that the manipulation sequence is somewhat complex and the analytical method demanding, with a total process lasting seven days (Fig. 5.2). Time reductions and simplification of the

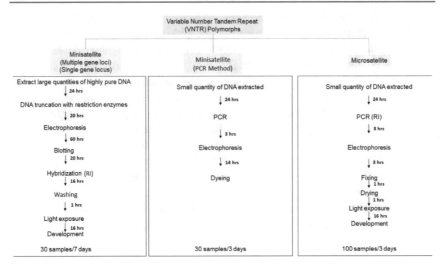

Fig. 5.2 Flow chart for VNTR detection method

197 experimental manipulations for the single gene locus method are currently being
198 achieved through development of a primer set to amplify probes through PCR.

199 **Microsatellite region polymorphs**: In VNTR-based approaches, primer sets are
200 designed so that a repeating GT or CA sequence can be found in the middle and
201 used to amplify the microsatellite region through PCR. Because this approach uses
202 PCR, a small DNA sample is sufficient. Accordingly, fin fragments may be col-
203 lected as tissue for DNA extraction, and the approach can be applied to samples
204 involving young fish, obviating any need to kill the fish used in the experiment.
205 Because the single gene locus method allows accurate yet simple identification of
206 alleles at individual gene loci with molecular markers, comparison of films is
207 possible, and such markers are also optimal for population analysis.

208   Once a primer has been developed for one fish species that is being researched, it
209 can also be used for related species in the same genus. It cannot be applied for more
210 distantly related species, requiring development of different primers for different
211 species. In the case of microsatellites and the single gene locus method, a primer
212 must be developed, and the process requires somewhat complex and specialized
213 facilities, equipment, and technology compared to other methods (Fig. 5.3).

214 **Primer Design**: Microsatellite region polymorph detection requires a PCR primer
215 for the DNA sequence that includes the repeating CA or GT sequence.

216   DNA from the experimental fish's DNA is truncated with restriction enzymes,
217 and the DNA sequence containing the repeating CA or GT sequence is cloned with
218 plasmids (Fig. 5.3). The repeating CA or GT sequence within the DNA sequence
219 read by the sequencer is confirmed, and the primer is designed so that this portion is
220 located in the middle.

Fig. 5.3 Flow chart for microsatellite gene locus detection method

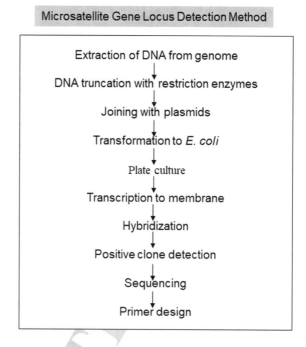

(4) Restricted fragment length polymorphisms (RFLP)

Mitochondrial DNA D-loop region polymorph (Fig. 5.4): All DNA, including mitochrondrial DNA, is extracted and the target region is amplified through PCR. After restriction enzyme truncation, DNA fragments are separated by electrophoresis, dyed with ethidium bromide, and exposed to UV rays (with a transilluminator) to be recorded by Polaroid camera or other means. Because this approach uses PCR, no radioactive isotope is needed, and polymorph detection can be carried out simply in an ordinary biological laboratory without special facilities or equipment (Fig. 5.4; Ishida et al. 1994).

For a PCR primer, a base sequence is designed so that the D-loop region is effectively embedded in the center.

Amplified fragment length polymorphisms (AFLP): A small quantity of nuclear DNA is truncated with restriction enzymes and an adapter sequence is attached to the fragment end to serve as a primer. The primer, which includes the restriction enzyme recognition sequence, is used for PCR processing. Because so many amplified fragments are present, PCR is conducted twice using a primer including a random (facultative) base. Placement of DNA size markers at either end during electrophoresis allows for easy identification of bands and comparison between different films. This is an excellent technology that combines the reproducibility of RFLP with the simplicity of PCR, allowing for processing of multiple samples.

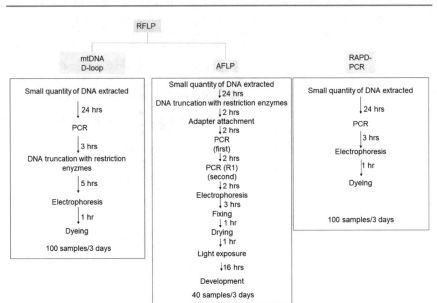

Fig. 5.4 Flow charts for RFLP and RAPD detection methods

241 (5) Random amplified polymorphic DNA (RAPD): The RAPD method is a form
242   of mutation detection using a random primer and PCR. It allows for mutation
243   detention even with very small quantities of DNA from a sample fish.
244   A primer is sought from among many possible candidates for use in poly-
245   morph detection, and the level of genetic mutation is quantified using a band
246   sharing index (BSI). As this method does not use RI, it allows for convenient
247   and simple detection in the laboratory, but in many instances variability is
248   lacking because it does not specifically seek out and multiply sites with large
249   numbers of DNA mutations. Another drawback is that the results are prone to
250   being affected by PCR conditions (Martin et al. 1992).

5.5 Genetic Analysis Through DNA Markers

5.5.1 Mitochondrial DNA Haplotype Polymorphs

254 The term "haplo" refers to organisms that form single units among the chromosomal
255 pairs represented in somatic cells. In molecular evolution research, a haplo group (the
256 representation of various gene types on a single chromosome) represents a large
257 group of haplotypes, or a series of alleles at a specific location on the chromosome
258 (Williams et al. 1990, 1995; Excoffier et al. 1992; Tingey and del Tufo 1993).
259   In human genetics, the most commonly researched haplo groups are the Y
260 chromosome DNA haplo group and the mitochondrial DNA (mtDNA) haplo group,

which may be used in defining a genetic group. Y-DNA is transmitted only through the paternal line, while mtDNA is passed on only through the maternal line. This also means that a haplotype is a collection of statistically related single nucleotide polymorphisms (SNPs) on a single chromosome. By examining these relationships and the allelomorphic character of the haplogroup block, it is possible to see other polymorphism regions in the area, information that is highly used in researching the background of various diseases.

Because mitochondrial DNA has a higher rate of mutation than nuclear DNA (around ten times as large), some individuals may have a sequence of recurring, independently arising mutations at the same site, which can cause some difficulties in analysis. In other words, if there is a probability that one base in a sequence of Y-DNA within the nucleus will mutate is one in 100,000 years, then for mtDNA such a mutation may occur around once in 10,000 years. This means that quite a few recurring mutations may have occurred independently in the same mtDNA base locus in various continents over the past few dozen millennia. Because mtDNA does not cross like the Y chromosome, however, it is possibility to classify old sequences by analyzing lines according to haplotype.

The degree of polymorphism in a restriction enzyme fragment at a substitution region in mtDNA amplified through PCR may be obtained through nucleon diversity, the extent of RFLP mixing ratio through haplo diversity using the equation $h = 2n(1 - \sum X_i^2)/(2n - 1)$. In this case, X_i refers to the frequency of nucleons or haplos in the same, while n is the number of individual in the sample.

Number of base substitutions and net base substitutions: Serving as indicators of genetic specialization between groups, the number of base substitutions and net base substitutions may be obtained through the following formulas. The number of base substitutions between two haplos in the same population is shown by $d_x = (n_x/n_x - 1) \sum X_i X_j d_{ij}$. Here, n_x is the number of samples, while d_{ij} represents the number of base substitutions per site for haplos i and j. The number of base substitutions arising between the DNA haplo and Y-DNA haplo from populations X and Y and its average is shown by $d_{xy} = \sum X_{i\ yj\ dij}$. Here, dij is the number of base substitutions between haplo i taken from population X and haplo j taken from population Y.

Net base substitutions between the two populations can be found through the equation $d_A = d_{xy} - (d_x + d_y)/2$. The time since the two groups' division can be found by substituting $d_A = 2\lambda T$ into the equation $d_A = d_{xy} - (d_x + d_y)/2$ (Taniguchi 2000).

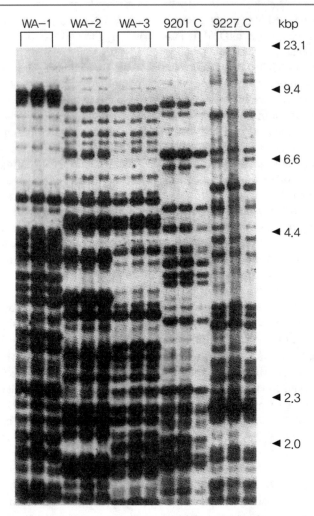

DNA fragment sizes are characteristically large. This marker enables proof of clonality
and distinguishing of clones

Fig. 5.5 Artificial clone DNA fingerprints for five ayu fish (microsatellite DNA polymorphs)

5.5.2 DNA Fingerprinting

Band sharing index (BSI) for organismal DNA fragments: In the DNA fingerprint
search method (DNA-FP), each band (allele) corresponds to a genetic pattern of
codominance, in which both alleles in a pair are fully expressed in heteromorphic
individuals. Because alleles at multiple gene loci are expressed within the same
mobile phase, they cannot determine the gene type for each gene locus (Fig. 5.5). It

is therefore impossible to estimate the frequency of the necessary allele or the hybridization rate during population analysis (Lynch 1990; Vos et al. 1995).

Accordingly, this approach uses the percentage of total bands that are shared by the two individual under comparison as a band sharing index (BSI), which represents the genetic similarity between the organisms. The average sharing value is also used as an indicator of population variability and relatedness between groups. The BSI for individuals' DNA fragments (bands) can be calculated with the following equation:

$$\mathrm{BSI} = 2 \times N_{ab}/(N_a + N_b)$$

N_{ab}: Number of DNA fragments (bands) shared by individuals a and b; N_a, N_b: Number of DNA fragments (bands) found in individuals a and b.

Average BSI within and between populations: Normally, a BSI is calculated for all combinations of organisms within a population, and average BSIs are obtained within and between populations. The average BSI within a population is higher when the population consists of organisms that are closely related, and lower in the opposite situation. For a population where the inbreeding coefficient F is equal to one, the average BSI for that population will obviously have a maximum value of one. Wild populations, however, tend to have low values.

The average BSI between populations is an indicator of the genetic flexibility between those populations. Typically, BSI values are calculated for all combinations between populations in the same autoradiography, and an average BSI between populations is obtained. For wild populations with a very high level of diversity within species, the average BSI within and between populations will sometimes show little difference.

Advantages and problems of BSI: Due to the large size of the fragments in DNA-FP (minisatellites), it may be difficult to determine whether bands are different or identical, and the possibility of subjective judgments cannot be ruled out. Also, technical factors in the DNA-FP detection process are such that some transformation in the number of bands produced is unavoidable. For this reason, the rule in BSI is to make calculations and compare data within the same autoradiography, and the inability to compare data from different films remains a technical limitation (Takagi et al. 1995).

5.5.3   Gene Amplified Fragment Length Polymorphism (AFLP) and Fingerprinting

Allele frequency and approximate heterozygosity rate: In the case of nuclear DNA AFLP fingerprints (AFLP-FP), fragment sizes are small and bands can be established with size markers, allowing for comparison of data from different films. AFLP is also used as a genetic map marker because of the high reproducibility of its band detection.

343   When used in population structure analysis, this method can identify the fre-
344 quency of a gene and the rate of heterozygosity with it. If a particular band is
345 positive, the corresponding gene is determined to be either homo or hetero, and if it
346 is negative it is understood to be a conjugating outside that allele. The AFLP
347 method is thus considered to be a superior assessment method to the microsatellite
348 DNA polymorph BSI in terms of evaluation of a population's genetic diversity and
349 reproducibility of band detection. One problem with this approach is that the rate of
350 polymorph gene loci and heterozygosity may be overestimated. Also, while the
351 possibility of some of the appearing bands being mitochondrial DNA fragments
352 cannot be ruled out, the percentage appears to be negligible when compared with
353 the nuclear genome (Taniguchi 2000).

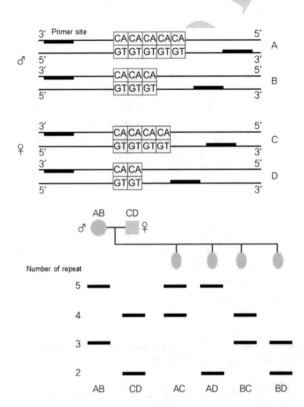

The upper portion shows the locations of the primer and homologous parent chromosomes with
different repeating CA sequence numbers, while the lower portion shows the movement
pattern for PCR fragments in the young produced by these parents.

Fig. 5.6 Mimetic diagram for microsatellite DNA polymorph mode of inheritance in red
seabream

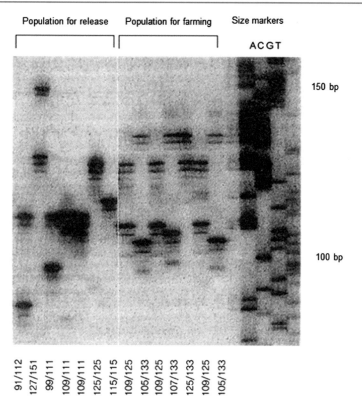

Comparison of patterns for artificial seedlings for release and artificial seedlings for farming. The four on the right are size markers.

Fig. 5.7 Microsatellite DNA polymorphs

5.5.4 Microsatellite DNA Polymorphs

Gene locus and alleles: In terms of microsatellite DNA (sites in non-coding regions where the number of repeating base sequence units is small) and minisatellite DNA (where the number is large), DNA fragments are referred to as gene loci and alleles. The terms are used for convenience because the mode of inheritance for the fragments follows Mendelian principles regardless of the presence of absence of gene functioning in the detected site (Figs. 5.6 and 5.7). The situation is thus similar to isozymes in terms of being an indicator for level of mutation and for differentiation between populations (Fahima et al. 1998; Schlötterer 2000; Norris et al. 2001; Taniguchi 2000).

Genetic variability within populations: This is assessed with the average number of alleles per gene locus or the rate of polymorph gene loci and average heterozygosity rate. For microsatellite DNA polymorphs with large numbers of

	Layout: **T1 Color**	Book ID: **462027_1_En**	Book ISBN: **978-3-030-20943-8**
	Chapter No.: **5**	Date: **23-9-2021** Time: **3:43 pm**	Page: **124/144**

366 low-frequency alleles, the effective number of alleles ($1/(1 - H_e)$) per gene locus is
367 used.

Hardy-Weinberg equilibrium assessment: With microsatellite DNA polymorphs,
calibration with the Markovnikov chain method is used because of the large number
of alleles present. In such cases, use of computer software (such as Arlequine) is
essential; by testing for each gene locus in a population, or several gene loci all at
once, it may be determined whether or not the sample comes from a Mendelian
population.

Calculation may also be performed by selecting a few major alleles and conducting a chi-squared test for predominance in the difference between the number of
genotypes observed and the absolute value.

**Genetic differentiation between populations and cladograms for multiple
populations**: The extent of genetic differentiation between populations may be
assessed using the genetic differentiation index (Gst) and the genetic distance
between populations (D). Computer software known as PHYLIP (Phylogeny
Inference Package) is used to calculate genetic distance and create cladograms.

Simple population mixture, genetic permeation, and inbreeding: In cases that
deviate significantly from Hardy-Weinberg equilibrium, it becomes necessary to
presume mixing between different populations, inbreeding, and crossing between
different populations. The ratio between the observed heterozygosity rate (H_o) and
the expected rate (H_e) is assessed through a fixed index ($F = 1 - H_o/H_e$).

Advantages and problems: The single locus method (single probe, or a probe that
corresponds to only one location in the genome) can be used to calculate the genetic
parameters of a population, and marker genes have been praised for their sensitivity
in fish clade analysis and monitoring of genetic changes in artificial seedling
populations. In the case of microsatellite DNA, polymorphism is very high, but
effective use of marker genes requires numerous samples (80–100) in a single
sample group. Care is also necessary in analysis due to the presence of so-called
"Nar" genes among the alleles, where mutations in the primer section result in PCR
products not forming.

Because there are no groups that fully satisfy the conditions for a Mendelian
population, very few populations in their natural state conform to Hardy-Weinberg
equilibrium. The gene pool is subject to change due to mutation, natural selection,
genetic drift, migration, and segregation, among other factors. Because this
microevolution process is so slow, however, populations appear to be nearly
achieving equilibrium. The frequency of allelomorphic genes and genotype
expression may therefore be estimated and applied to studies of population evolution and public health (Taniguchi 2000).

5.5.5 Computer Software Use in Population Analysis

As described above, high-sensitivity DNA polymorphs markers also have large numbers of alleles or genotypes at a single gene locus, which severely complicates calculations of Hardy-Weinberg equilibrium of population heterogeneity testing in population genetics analysis. Due to low allele frequency, it may also be necessary to prepare a large number of samples per sample group to use these markers effectively. Given this use, use of population analysis computer software is essential. Fortunately, such software may be downloaded online (Nishita 1998).

5.6 Research Applying DNA Markers

5.6.1 DNA Polymorph Marker Sensitivity and Use

When a marker is described as sensitive, this typically means a high level of polymorphism in the marker gene. Among the DNA polymorphs serving as new genetic markers, some allow for differentiation of organisms; these are known as DNA fingerprints. In studies involving the release of fish and shellfish, some of the organisms released are marked through attachment of a label or clipping of a fin. DNA fingerprinting offers the convenience of allowing all artificial seedlings to be marked automatically at the stage of parent genotype determination (Saunders 2005).

In some cases, even these high-sensitivity markers cannot be called necessarily appropriate in distinguishing different species or making determinations of relatedness among multiple species, as they are subject to influence from non-statistical similarities in allele frequency.

When protein and DNA polymorphs are compared for usefulness as gene markers for population analysis, isoenzymes are still seen as highly trustworthy in terms of gene identification accuracy and data stability, and highly useful for certain organisms (such as freshwater organisms) where small populations can be segregated easily. It is not rare for sibling species that cannot be distinguished by outward form to be identified through isozyme markers.

In contrast, mitochondrial DNA D-loop region polymorphs exhibit great diversity and are highly useful in making determinations on genetic differentiation in a population and relatedness between populations and in estimating evolution periods. At the same time, DNA D-loop region polymorphs have disadvantages in terms of the characteristic rises in inbreeding and homozygosity in sexually reproducing populations and are unable to provide effective information in distinguishing simple population mixing from genetic mixing.

The single locus method for minisatellites and microsatellites (VNTR) is sure to play a major role in analysis of fish population structure going ahead, as it shows both outstanding reproducibility and a high degree of sensitivity. In particular, the high degree of usability in microsatellite DNA polymorphs, which have a far higher

443 number of potential gene loci than minisatellite DNA polymorphs, is certain to
444 result in new information emerging (Taniguchi 1996).

445 ## 5.6.2 Genetic Diversity and Population Structure in Wild
446 ## Populations

447 Genetic diversity: To date, DNA polymorph analysis has been carried out for the
448 full range of regions in many varieties of fish, including the herring, black bass,
449 trout, ayu, blue spotted snapper, and pacific yellowfin tuna, in accordance with
450 restriction enzyme fragment length polymorphism (RFLP). With the introduction of
451 PCR, however, recent RFLP analysis has been conducted for amplified
452 high-mutation DNA D-loop regions. Mutations are detected more clearly in these
453 DNA loop regions than in functional regions, with applications in population
454 analysis (Frankham 1996; Yuhiro 1997)

455 The high level of haplotype diversity in DNA D-loop regions is observable in
456 wild populations of red seabream. Twenty-seven varieties of haplotype have been
457 detected, resulting from combinations of five different RFLPs, with a very high
458 diversity level of 0.905. Wild species population structure analysis using DNA
459 haplotype polymorphs as genetic markers has also been conducted for other ben-
460 eficial fish species besides red seabream (Tabata and Mizuta 1997).

461 Development of microsatellite DNA polymorph marker detection primers has
462 enabled detection of multiple gene loci in many wild fish species (Takagi et al.
463 1997). In terms of the diversity of microsatellite DNA gene loci, comparison of the
464 average number of alleles per gene locus (Fig. 5.8) and heterozygosity rate clearly
465 shows greater sensitivity as a molecular gene marker than the values of isoenzyme
466 markers (Fig. 5.9). With regard to genetic diversity in wild red seabream popula-
467 tions, those of the Northern Pacific have been found to exhibit higher values than
468 those of the Southern Pacific.

469 Genetic differentiation between wild populations: Given the great diversity of
470 microsatellite DNA markers, one may expect them to exhibit great sensitivity
471 (applicability) in analysis of population structure, including genetic differentiation
472 between local fish populations (Tabata and Mizuta 1997; Tabata et al. 1997).
473 Microsatellite DNA markers remain in the development stages technically, how-
474 ever, and examples of application are few and far between. Primers are required for
475 microsatellite DNA marker detection, and have already been developed for the
476 Japanese seabass, three-spined stickleback, rainbow trout, Atlantic salmon, and red
477 sea bream; among these, population analysis research has been carried out for the
478 Atlantic salmon, trout, and red seabream. Ongoing development and data analysis
479 are under way for important fishery and cultured fish species such as the ayu, carp,
480 black skipjack tuna, amberjack, and olive flounder.

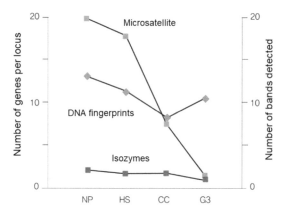

NP = wild population, HS= seedlings for release, CC= seedlings for farming, G3 = third generation of polar body emission-blocked gynogenesis. Microsatellites and isoenzymes are assessed by average number of alleles per gene, while DNA fingerprints are assessed by number of bands appearing

Fig. 5.8 Genetic diversity comparison for red seabream with microsatellites, DNA fingerprints, and isozymes

5.6.3   Farming Population Variability and Line Differentiation

Mitochondrial DNA haplotype reduction: Farming populations of red seabream showed an average of 11 haplotypes by the seventh generation, a clear decrease from the average of 27 for wild populations.

While the haplotype diversity level decreased from 0.905 to 0.791, a high level of variability was maintained despite generational crossbreeding, and the inbreeding coefficient was observed to remain low. Consistent with findings from microsatellite DNA polymorphs, this shows the ability to detect changes in genetic diversity with great sensitivity (Avise et al. 1992; Tabata et al. 1997).

Decrease in average alleles: Artificial seedling development has been attempted for various fish species for farming and release. Because of the founder and bottleneck effects, steep changes in the genetic composition of artificial seedling populations and the level of mutation possession are predicted to occur.

In farming populations of red seabream, a rapid reduction in the number of alleles per microsatellite gene locus (Fig. 5.8) and a steady decline in the heterozygosity rate (Fig. 5.9) have been observed (Taniguchi and Takagi 1997). The rapid decline in the number of alleles per gene locus in red seabream farming seedling production may be attributed to the bottleneck effect. The steady decrease in the heterozygosity rate, for its part, means that other zygosity rates remain high, and the inbreeding coefficient remains low despite repeated generational crossbreeding. This example suggests sites producing red seabream farming seedlings have established the minimum number of parents to prevent a rapid rise in the inbreeding coefficient.

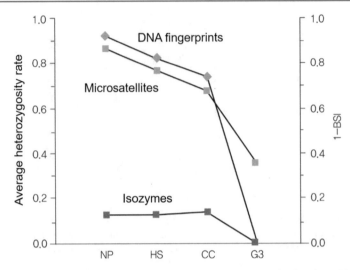

NP = wild population, HS= seedlings for release, CC= seedlings for farming, G3 = third
generation of polar body emission-blocked gynogenesis. Microsatellites and isoenzymes
are assessed by average heterozygosity rate, while DNA fingerprints are assessed
by band sharing index (BSI).

Fig. 5.9 Genetic diversity comparison for red seabream using microsatellites, DNA fingerprints,
and isoenzymes as indicators

5.6.4 Genetic Diversity in Chromosomally Manipulated Fish

Artificial gynogenetic diploids: BSI comparison for normal ayu diploids,
cleavage-blocked gynogenetic diploids, and second-generation clones within the
same film has shown the clones to possess the maximum BSI value of one, while
normal diploids had a low value of 0.527. The BSI for cleavage-blocked gynogenetic diploids was even lower at 0.314. The inbreeding coefficient F equaled one
for the cleavage-blocked gynogenetic diploids and around 0.5 for polar body
emission-blocked gynogenetic diploids, albeit with a sharper difference between
organisms in the cleavage-blocked case because the genotypes separated. The
number of bands appearing also declined as a result of joining of similar bands.
This theoretical prediction has been tested through average BSI within populations
and the number of bands detected (Arai 2001; Takagi et al. 1993).

In the case of microsatellite DNA polymorphs, the heterozygosity rate for
cleavage-blocked gynogenetic diploids is reported to be zero; for polar body
emission-blocked gynogenetic diploids, it is reported to range between 0 and 1.0,
depending on the distance between the marker gene and centromere. The fact that
this characteristic arises due to crossing between homologous chromosomes during
meiosis may be used to estimate the distance between G and C.

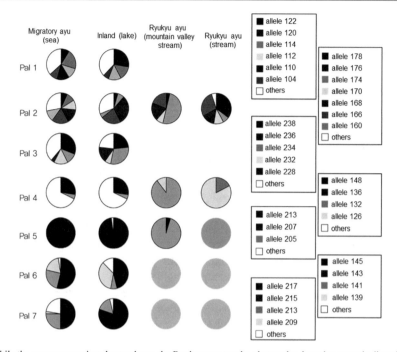

While the common ayu is polymorphous, the Ryukyu ayu can be observed to have lost genetic diversity.

Fig. 5.10   Microsatellite DNA polymorphs for the ayu: Pal7 gene locus

Proof of clonality: The question of whether DNA-FP changes as a result of technology or other factors is a matter of great interest. Identical patterns have been exhibited by artificially produced clone populations (ayu, olive flounder) and crucian carp from the same embryo (Fig. 5.5). The possibility of factors besides genes affecting the difference between organisms has been shown to be nonexistent as long as the organisms are being compared within the same film (Hulata 2001).

Separation of environmental variance and genetic variance due to mixed breeding of multiple clonal lines: Precise estimation of genetic variance and environment variance in quantitative traits through variance analysis requires multiple produced lines to be raised in the same environment. Attempts have been made to use DNA markers to distinguish lines for a more precise calculation than genetic or environmental variation (Taniguchi et al. 1996).

5.6.5   Assessing Genetic Diversity in Endangered Species

For the red seabream and ayu, the wild population variability index (1-BSI) is a very high 0.7–0.9. In the case of the endangered Ryukyu ayu, however, the variability index has fallen to around 0.5, and genetic homogenization has been

538   suggested. This reduction in genetic diversity can be understood more clearly
539   through the use of microsatellite DNA markers (Fig. 5.10) (Mace and Lande 1991;
540   Seki et al. 1995).

541   ### 5.6.6   Inbreeding Coefficient Estimation

542   In inbreeding populations, heterozygote frequency declines. If H_o is taken to rep-
543   resent the heterozygosity rate under conditions of random mating within a wild
544   population and F_t is the inbreeding coefficient when artificial seedling production is
545   carried out through t generations, then the heterozygosity rate after t generations is
546   shown by the equation $H_t = H_o \cdot (1 - F_t)$ (Li and Horvitz 1953).

547   While it is not possible to estimate gene loci in DNA fingerprinting, each band
548   appears in accordance with a Mendelian mode of inheritance. The average number
549   of bands appearance thus declines in inbred populations due to reduced hetero
550   frequency. Indeed, examination of the number of bands appearing in a specific
551   section of the membrane in DNA fingerprinting has shown a definite decline in the
552   cleavage-blocked gynogenetic diploids ($F = 1$) and clonal populations.

553   Because microsatellite DNA polymorphs enable measurement of the heterozy-
554   gosity rate at each gene locus, the inbreeding coefficient F for generation t can be

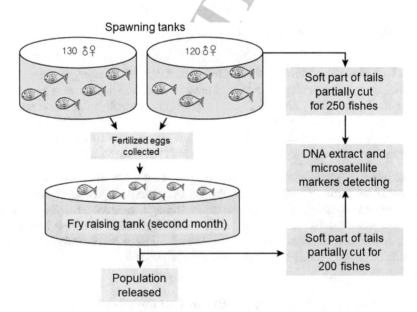

After the end of the spawning season, part of the parents' fin is cut to extract DNA.
Fry are sampled just before release (two months after incubation), and a portion of
their fin is cut to extract DNA.

Fig. 5.11 Flow chart for artificial seeding labeling with DNA markers

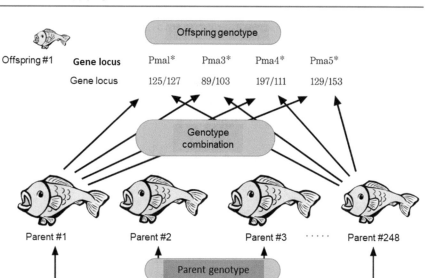

Genes in the fry at the top can be seen to come from parents #1 and #248.

Fig. 5.12 Parent-offspring determinations made by computer following detection of parent and offspring DNA polymorphs to determine genotype

estimated by substituting the heterozygosity rates for wild and artificial seedling populations into the equation $F_t = (H_o - H_t)/H_o$ (Taniguchi 2000).

5.6.7   Diagonosis Parentage and Estimating Effective Size of Artificial Seedling Populations

With microsatellite DNA polymorphs, all artificial seedlings are automatically marked at the stage during which the parental genotype is determined (Fig. 5.11). The parents' and offspring's genotypes can also be compared for every gene locus (Fig. 5.12). For artificial seedling production, this may be applied in specifying the parents directly involved in reproduction (Fig. 5.13), examination of the number of organisms and parent-offspring correlations for various traits, and tracking studies for seedlings that have been released (Taniguchi 2000).

This has allowed for actual measurement of the approximate effective population size for artificial seedlings. In the production of artificial red seabream seedlings using 250 parents (Fig. 5.11), the number of parents directly involved in reproduction was found to be 87. After correction for error due to imbalance between the male-female ratio and number of offspring per fair, the effective population size

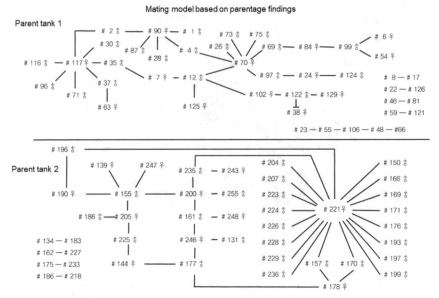

Mating model based on parentage findings

In many cases, a single female has mated with multiple males, which is consistent with actual observed spawning activities.

Fig. 5.13 Mating model developed from parentage confirmation findings

N_e was estimated to be 63.7, or slightly larger than the N_e standard proposed by the FAO (Taniguchi 2000).

5.6.8   Tracking Studies for Released Fish

With microsatellite DNA polymorphs, the large number of alleles means that the number of genotypes within a single gene locus is very large. When multiple gene loci are combined, genotype frequency decreases greatly. The likelihood of the same form being expressed in two individuals can be calculated with the equation $I = \sum_i p_i^4 + \sum_i \sum_j > i (2 p_i\, p_j)^2$. In this case, p_i and p_j refer to the frequency of the ith and jth alleles. For artificial red seabream seedling production, the observed I value for alleles at five gene loci ranges between 3.31×10^{-9} and 5.76×10^{-9}. This means that the likelihood of two random organisms having an identical genotype is one in 175 million (Taniguchi 2000).

Because a determination of the genotype of parents used in producing artificial red seabream seedlings allows for distinguishing of organisms, they can be released into natural waters. After time has passed, wild specimens are collected, and if organisms are found to have a genotype that is likely to have come from an artificial seedling, it can be judged in statistical terms to have been a released specimen. This

588 method of tracking released fish has the potential to provide useful information on
589 the effects of releases from cultured fisheries industry.

5.6.9   Future Tasks

591 Natural fishery resources have been declining from year to year due to environ-
592 mental destruction and overfishing. Meanwhile, expectations for fishery production
593 as a protein source continue to grow. Under these conditions, the need to preserve
594 genetic diversity is likely to only grow. It is therefore essential to quickly provide
595 data to serve as a standard for future preservation of wild populations in rivers,
596 lakes, and various marine ecosystems, including estuaries, intertidal zones, and
597 shallow and deep ocean waters.
598   Monitoring systems using genetic markers must be introduced in farming
599 (brackish water) to preserve and manage improved lines and in aquiculture to
600 prevent unintended genetic changes from arising in the artificial seedling production
601 process.
602   More and more species are becoming increasing rare or threatened with
603 extinction. This extinction may be hastened if fish are collected and killed to
604 determine the current state of their genetic diversity. Avoiding the sacrifice of these
605 fish will request detection of genetic markers using only small samples.
606   For populations that have already died out, considered should be given to the
607 question of whether it would suffice to transplant some organisms from a surviving
608 population to restore the local population, and to designing guidelines for founding
609 populations. The need for genetic research to restore such environments (ecosys-
610 tems) is likely to only grow.
611   None of these tasks can be carried out with DNA polymorphs and other genetic
612 markers. Continued research in the development and simplification of genetic
613 marker detection technology will be needed in the days ahead.

5.7   New Marine Breeding with DNA Markers

616 As rapid advances continue to be made in studies of genome information for human
617 beings and many other organisms, new information is being amassed on genetic
618 information, or the structure and functions of the genome at the molecular level.
619   Developments in marine farming technology have enabled quantitative pro-
620 duction of many marine creatures, including the salmon, trout, olive flounder, red
621 seabream, and kuruma prawn. As quantitative production technologies have
622 enabled mass production, efforts are now under way to achieve qualitative
623 improvements, or improvements to economic properties in terms of disease resis-
624 tance, high growth, and meat quality. While active breeding of cattle, horses, swine,
625 and chickens has resulted in the development of multiple new varieties over the
626 centuries, successes with improvements in farmed fish varieties that have recently

627 become amenable to mass production are few and far between. Reasons for this
628 include the short timespan over which mass production of marine organisms
629 through farming has been possible, as well as the concentration of efforts to date on
630 achieving stable quantitative production. Consumers today are demanding quali-
631 tative improvements to farmed fish, in the hopes of enjoying safer and more deli-
632 cious farmed fish that is also amenable to stable mass production.

633 Qualitative improvements are achieved through breeding (strain improvements),
634 which in turn is made possible through an understanding of genetic character. As
635 noted in the previous section, genome research at the molecular level has resulted in
636 many achievements in terms of genetic information. One of these that may be
637 described as historic in its implications is the method known as quantitative trait
638 locus (QLT) analysis, in which DNA markers can be used to explore as yet
639 unknown gene loci that govern or are connected with phenotype. While QLT
640 analysis is ultimately a matter of using positional cloning to locate responsible
641 genes, the discovery of DNA markers associated with phenotype would also enable
642 their use in breeding.

643 This method stands to increase efficiency to an unimaginable degree in breeding
644 that has previously been reliant solely on phenotypes. It is the subject of great
645 attention in modern science, where the new breeding approach is referred to as
646 marker assisted selection (MAS). In terms of use of genetic information, two
647 approaches exist. One is the so-called "hard path," which involves actively
648 manipulating genes to alter organisms and artificially produce useful materials for
649 application in production or treatment. The other is the "soft path," which involves
650 using genetic information to take maximum advantage of organisms' genetic
651 capabilities without actually manipulating their genes. Selective breeding through
652 genetic markers falls into this category: because the fish's genetic information can
653 used when selecting parents, allowing for selection through recombination, greater
654 efficiency can be achieved than in current selective breeding, where parents have
655 been selected based on phenotypic information. In terms of breeding methods,
656 MAS is no different from current statistically based selected breeding, but the
657 genetic information involved in economically related traits can provide a means of
658 precisely selecting genetically superior organisms (Okamoto 2000).

659 ### 5.7.1 Reasons for Using Genetic Marker Selective Breeding
660 ### in Marine Fisheries Products

661 As noted previously, the achievement of stable quantitative production of farmed
662 fish has led consumers in the contemporary environment of widespread farmed fish
663 use to demand safer and more delicious seafood. Producers of farmed fish similarly
664 would like to create varieties that meet these consumer demands. By using fish
665 species that are resistant to disease and can adapt readily to temperature changes,
666 they can reduce the incidence of disease and farm without the use of medicines or
667 antibiotics, allowing them to meet consumer demand for safe food products.
668 Consumer demands can also be met with high-quality fish and reductions in

production costs through the use of fish with excellent growth properties, a difference that is ultimately reflected in the market price (Liu and Cordes 2004).

The tool traditionally used in breeding was a phenotypic record based on naked-eye observation of fish phenotypes; this record was subsequently used in combination with statistical technology to assess genetic capabilities. In theoretical terms, however, this phenotypic record-based genetic capability assessment approach was rooted in the idea that innumerable different genes are involved in specific traits, and each of these genes exhibits the same effects. Genetic capability assessments in traditional breeding have thus been a matter of evaluating based solely on phenotypic information, while remaining ignorant about the "black box" that is the collection of genes actually influencing phenotype.

Recently, rapid advancements in DNA technology have uncovered information about the chief genes influencing major economically related characteristics. The various genetic markers developed through DNA technology may actually be involved in phenotype, but it is arguably the QTLs associated with those markers that are involved. A genetic map is currently being created for QTL analysis, and associations with traits in standard line populations are being researched to confirm the QTLs on the map.

This approach is expected to help in commercialization of new varieties, as it is possible to use genetic markers as indicators to select the desired genetic treatments, maintaining only the target genes while also preserving genetic diversity. Selective breeding with DNA markers is a matter of adding DNA marker information to the traditional phenotypic record to increase the accuracy of genetic capability assessments and boost efficiency in genetic improvement (Okamoto 2000).

5.7.2   Selective Breeding with Genetic Markers: An Overview

While DNA markers are being isolated and stockpiled, lines are subjected to linkage analysis (transmission of two or more allelomorphs together; alleles [genes governing allelomorphic properties] arise because they are located on the same chromosome and do not following independent Mendelian principles in transmission) to find DNA markers linked to useful genetic properties. If DNA markers linked to such properties are found, it becomes possible to distinguish the properties through genotype (using DNA markers) rather than phenotype, allowing for determinations as to whether the organism possesses that trait (Gjedrem 1983; Okamoto 2000).

A. DNA Markers and Gene Maps

The strategy for learning where genetic traits lie on the genome can be easily understood if we think of a puzzle. The pieces of the puzzle correspond to DNA fragments, while the DNA marker is like a symbol showing the characteristics of the puzzle piece (Fig. 5.14). Hidden in these pieces are genetic traits that are useful in marine animal breeding, such as those determining fast growth or related to disease resistance. The proteins or genes determining useful genetic traits are not

| Layout: **T1 Color** | Book ID: **462027_1_En** | Book ISBN: **978-3-030-20943-8** |
| Chapter No.: **5** | Date: **23-9-2021** Time: **3:43 pm** | Page: **136/144** |

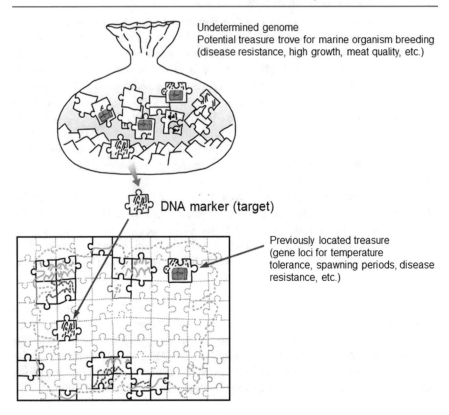

Fig. 5.14 Schematic diagram for the positional cloning method (visualization of the puzzle piece analogy)

709   separated like puzzle pieces, however, and it is almost impossible to identify their
710   functions directly. Even if the genome is differentiated and analyzed as though it
711   consists of puzzle pieces, each piece has its own characteristics, and the overall
712   picture can be completed as the pieces are fitted together.

713   In terms of the creation of a genetic map using DNA markers, we may imagine
714   something like fitting puzzle pieces together. The resulting genetic map of rela-
715   tionships among gene marker—a virtual framework map, in other words—serves a
716   tool for efficiently finding the gene markers linked to useful traits. Because multiple
717   genetic markers are displayed on a single chromosome, we can investigate using
718   genetic markers representing different chromosomes. If one of these markers is
719   linked to a useful trait, we can focus on examining markers on the same chromo-
720   somes, and thereby find the genetic markers most strongly linked to the trait.

721   Not having a genetic map means that all available genetic markers must be used
722   for investigation. Because the entire genome must be examined to find the markers
723   with the best linkages to useful traits, a tremendous number of markers have to be
724   used, and the probability of finding the desired gene is slight.

	Layout: **T1 Color**	Book ID: **462027_1_En**	Book ISBN: **978-3-030-20943-8**
	Chapter No.: **5**	Date: **23-9-2021** Time: **3:43 pm**	Page: **137/144**

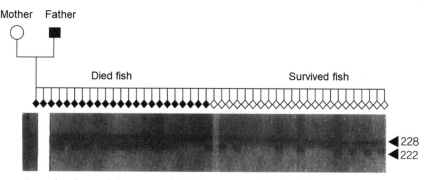

It can be observed that many within the group of offspring with allele loci from the paternal line (lower band, size 222) died.

Fig. 5.15 Actual example of a viral disease resistance/sensitivity linkage analysis in rainbow trout

B. Linkage Analysis in Standard Lines

Genetic markers may be used to visually distinguish male and female genetic loci in a particular region of the genome. If, for example, a gene locus determining disease resistance is located on near a DNA marker, with a gene for disease resistance on a chromosome from either the paternal or maternal line, the disease-resistant fish can be mated with a disease-sensitive fish to produce offspring of four types, with combinations of genes coming from the disease-resistant and disease-sensitive fish. Among these, only organisms with the disease-resistant gene locus from disease-resistant parents will exhibit disease resistance (Fig. 5.15). Because each individual DNA marker has its own electrophoresis pattern (genotype), parent-offspring correspondence in the DNA marker genotype and hereditary of disease resistance must be sought in a standard line.

Once a correspondence has been found between the genotype and a gene for disease resistance, the DNA marker can be described as linked to the disease resistance trait. In other words, we will have found a DNA marker that can be used to detect the disease resistance trait (Fig. 5.15). Finding a linkage between the DNA marker and the disease resistance trait is not a simple matter, however. Indeed, the target trait and DNA marker are often on different chromosomes, and no linkage at all exists in such cases. For linkage, the target trait and DNA marker must at least be on the same chromosome. In such cases, greater distances between the two mean that the recombination rate increased due to the effects of crossover during meiosis.

The resulting findings can be used to find DNA markers that are closer to the target trait, and ultimately to discover those with a strong linkage. The genetic map, which as previously mentioned shows relationships between DNA markers, offers some help in this process. As was noted before, once even a single DNA marker linked to a target trait has been found, it is simply a matter of focusing on DNA markers on the same chromosome to find one with a stronger linkage. Without a

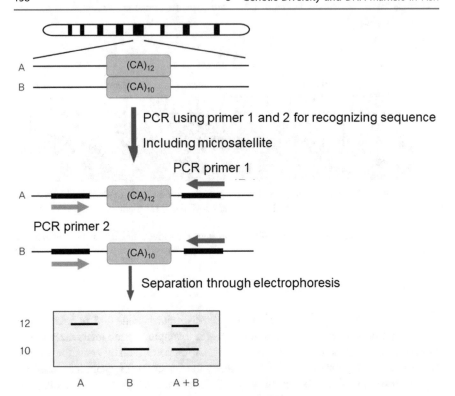

Fig. 5.16 Principles of the microsatellite marker detection method

genetic map, one must establish correspondences between DNA marker and line at random, which is not a very efficient approach. As was mentioned in the case of DNA markers, it is crucial to establish a framework map showing positional relationships among markers (Okamoto 2000).

5.7.3 Current State of Research Toward Selective Marine Breeding with Markers

This section introduces research toward the use of DNA markers in marine breeding, as well as fish species for which microsatellite markers (Fig. 5.16) are being isolated. While these markers are still for the most part used for research analysis (resources prediction), they may also be applicable to marine breeding in the future (Xu and Crouch 2008).

First, research toward marine breeding will be examined. Researchers in five European countries (Norway, Ireland, the United Kingdom, Denmark, and France) and Canada have launched the SALMAP project with the goal of improving quantitative traits (QT) in fish from the salmon family, including meat quality, disease resistance, and growth. Three hundred microsatellite markers each have been isolated in three species of fish (rainbow trout, Atlantic salmon, and brown trout) and used to draft genetic maps, with line analysis conducted to investigate quantitative train loci (QLTs). Researchers in Spain and the U.S. are also conducting their own independent research on fish from the salmon family. DNA marker isolation has been performed for pictus catfish and striped bass in the U.S., European carp in the Netherlands, and olive flounder in Japan to achieve variety improvements (Bentzen et al. 1991; Hallerman and Beckmann 1988).

In the case of kuruma prawns (Australia and Thailand), isolation of microsatellite markers is difficult. For breeding purposes, attempts are being made to create a gene map for breeding applications using amplified fragment length polymorphism (AFLP), although this approach is less generalizable than one using microsatellite markers (Okamoto 2000).

The next section will focus on fish that are not being directly targeted for breeding at present, but which have microsatellite markers that could potentially be used for breeding in the future.

In the case of tilapia, U.S. researchers have isolated and reported around 60 microsatellite markers in an effort to explain separation and differentiation in organisms in the cichlid tribe. Genetic map creation efforts are also under way. Microsatellite markers have similarly been isolated for the European plaice (Spain), oyster (Australian), Indian carp (India), red seabream (Japan), and ayu (also Japan).

Creating a genetic map involves large research costs, however, and maps cannot be created for all species. In the case of livestock, commonality with markers can be found in cattle, sheep, and goats. Similarly, effective map creation should be carried out through examination of commonalities among markers from fish species.

To date, no reports have been published in academic journals on the discovery of useful gene loci for marine organism breeding in line investigations using linkage analysis. Similar findings in other areas, including human beings, are the focus of great attention, and hopes for their discovery in the area of marine breeding are high.

Isolation and stockpiling of markers and development of genetic maps are poised to take place at a rapid rate in the future. Preparations are also being carried out for analysis of genomes in rainbow trout and other fish from the salmon family. The delay may simply be a matter of examining what traits consumers and producers desire in their fish and where the fish possessing them live (or are being raised). Fortunately, many research institutions in Japan have been developing clonal lines for rainbow trout, olive flounder, and other fish, which are being used for marine organism breeding. A broader investigation of their characteristics to identify the markers linked to them through linkage analysis (line analysis) in assessing clonal traits will soon allow the use of those markers to practically achieve varieties fixed with only the target trait. Linkage analysis with DNA markers has also produced significant results in marine breeding, and a global consensus will be necessary on matters such as the scale of economic value from those results, as well as DNA marker patenting issues.

5.7.4   Analyzing Unknown Gene Loci Associated with Beneficial Marine Fish Species

Fish in the salmon family have been the most closely analyzed of useful marine species. As these species are also popular worldwide, they are expected to play a leading role among fish species in the future (Limborg et al. 2012).

In rainbow trout, albinism is a dominant trait. (Albinism dominance is a rarity among organisms in general.) The locus for the gene responsible for it (i.e., the gene's position) has been ascertained. This also means that a DNA marker has been discovered to identify the location where the responsible gene is present. Among quantitative traits that are difficult to analyze (i.e., traits for which the phenotype is governed by a complex mixture of genes), gene loci associated with temperature tolerance, spawning times, and resistance or sensitivity to infectious pancreatic necrosis (IPN) have been found.

The achievement of these results is due in larger part to the creation of a (still insufficient) genetic map for the rainbow trout. The finding demonstrates the effectiveness of DNA markers and linkage analysis, and anticipation is growing for selective breeding methods using DNA markers linked to useful traits. As more and more DNA markers are stockpiled, analyses will focus on a greater number of economically valuable traits, and DNA markers linked to those traits will soon be discovered and used in selective breeding.

5.7.5   DNA Marker Identification of Gene Loci Associated with Conditions in Human Beings and Other Animals

An almost uncountable range of diseases and conditions exist for which genes have been isolated and identified in humans using DNA markers, including Duchenne muscular dystrophy, familial adenomatous polyposis (FAP), Huntington's chorea, familiar Alzheimer's disease, retinoblastoma, osteitis fibrosa cystica, viral tumors, type 1 diabetes, ankylosing spondylitis, juvenile breast cancer, and obesity. Some of these cannot be explained in terms of Mendelian inheritance or polygenic theory, which has led to a constant stream of new findings (Andersson 2001).

Other organisms besides humans include livestock such as cattle, sheep, swine, and chickens and grains such as rice and wheat, most of which have been subject to genome analysis for industry or commercial purposes. Efforts are being made to use as yet unknown genes governing economically valuable traits as information for breeding.

Genes (including gene loci) have been specially identified for milk production yields and milk fat quantity and meat quality (brisket) in cattle; for reduced fat (pork acidification), non-elastic meat, *E. coli* resistance, number of vertebrae, and growth in swine; for resistance to the Marek's disease virus in chickens; and for rice blight resistance and germination periods in rice.

Zebrafish have also been used in studied focused on hereditary diseases in humans. In particular, a gene responsible for a hereditary disease known as

	Layout: **T1 Color**	Book ID: **462027_1_En**		Book ISBN: **978-3-030-20943-8**
	Chapter No.: **5**	Date: **23-9-2021**	Time: **3:43 pm**	Page: **141/144**

"one-eyed pinhead" has been identified. This condition is characterized by a single eye and poor development in the head region. A disease with similar symptoms also occurs in humans, but because so few examples can be found, fish are being used as models for human beings in an example of analysis that is gaining major attention (Sakamoto et al. 1999).

Genetic analysis with DNA markers certainly provides an excellent approach to marine organism breeding. Once a DNA marker map for the entire genome has been completed, this method can be used in analyzing various useful genetic traits in the same species.

This section explained about the need for marker use in selected marine organism breeding, as well as an overview of its methods and the current state and results of research in the field. A basic account of achievements already made in human beings and other organisms was also provided. While the use of DNA markers in selective breeding of marine organisms began relatively recently, the achievements made with plants and livestock animals to date offer proof of its efficacy (Williams et al. 1990).

5.8 Chapter Summary and Conclusion

Comparison of protein polymorphs and DNA polymorphs in terms of utility as genetic markers for population analysis shows that isozymes are extremely reliable in terms of accuracy of genetic identification and consistency of data. While some sensitivity issues do exist, isozymes are highly useful in certain instances.

D-loop region polymorphs from mitochondrial DNA are very diverse and useful in determinations of genetic differentiation and relatedness in populations and estimation of evolution times. Drawbacks include issues with the inbreeding and increasing homozygosity rate characteristic of sexually reproducing populations and their inability to provide effective information in distinguishing between simple mixing and genetic mixing within a population.

As explained previously, single gene loci from minisatellites and microsatellites, which combine the excellent reproducibility and high sensitivity of isozymes, are certain to play an important role in the future. Microsatellite DNA polymorphs in particular, which have a far greater number of usable gene loci than minisatellite DNA polymorphs, are extremely useful and are expected to serve as a source for new findings. Drawbacks to these VNTRs include the complex and costly nature of the genetic polymorph detection method, the relative scarcity of data to date, and the inability to compare findings with those from other markers. Such issues are likely to be resolved at some point as the approach is applied with new technology.

From a biodiversity preservation standpoint, the focus is on preserving local populations and groups within a single species. Use of the latest DNA polymorph markers will allow detection and identification of hitherto unknown local populations. Research management can be focused on populations as basic units of reproduction. In terms of fish farming, these markers will need to be used in

903   monitoring systems for local population preservation, and efforts to acquire
parents and release seedlings will need to be carried out.

Like agriculture, fish farming subjects organisms to artificial management.
Future genetic improvements will be focused on breeding goals to meet various
demands. Important tasks in this field include explication of the genetic characteristics of improved varieties and preservation and management of these new
varieties. Development of new varieties is associated with rising costs due to the
labor, expenses, and site needed for line preservation and management.

Because the lines in question are labeled with genetic markers from birth, they
can be raised in mixed groups. In this case, it suffices to have a small number of
separate tanks for the parents to preserve their line.

As this chapter has shown, DNA polymorphs appear poised to play an even
more important role in marine farming research and industry as a new form of
genetic marker.

References

Andersson, L. (2001). Genetic dissection of phenotypic diversity in farm animals. *Nature Reviews Genetics, 2*(2), 130.

Arai, K. (2001). Genetic improvement of aquaculture finfish species by chromosome manipulation techniques in Japan. In *Reproductive biotechnology in finfish aquaculture* (pp. 205–228). Elsevier.

Avise, J. C., Bowen, B. W., Lamb, T., Meylan, A. B., & Bermingham, E. (1992). Mitochondrial DNA evolution at a turtle's pace: Evidence for low genetic variability and reduced microevolutionary rate in the Testudines. *Molecular Biology and Evolution, 9*(3), 457–473.

Bentzen, P., Harris, A. S., & Wright, J. M. (1991). Cloning of hypervariable minisatellite and simple sequence microsatellite repeats for DNA fingerprinting of important aquacultural species of Salmonids and Tilapia. In T. Burke, G. Dolf, A. J. Jeffreys & R. Wolff (Eds.), *DNA Fingerprinting: Approaches and Applications* (pp. 243–262). Basel/Switzertland: Birkhauser Verlag.

Excoffier, L., Smouse, P. E., & Quattro, J. M. (1992). Analysis of molecular variance inferred from metric distances among DNA haplotypes: Application to human mitochondrial DNA restriction data. *Genetics, 131*(2), 479–491.

Fahima, T., Röder, M., Grama, A., & Nevo, E. (1998). Microsatellite DNA polymorphism divergence in *Triticum dicoccoides* accessions highly resistant to yellow rust. *Theoretical and Applied Genetics, 96*(2), 187–195.

Frankham, R. (1996). Relationship of genetic variation to population size in wildlife. *Conservation Biology, 10*(6), 1500–1508.

Giblett, E. R. (1969). *Genetic markers in human blood*. Blackwell Scientific Oxford.

Gjedrem, T. (1983). Genetic variation in quantitative traits and selective breeding in fish and shellfish. *Aquaculture, 33*(1–4), 51–72.

Hallerman, E. M., & Beckmann, J. S. (1988). DNA-level polymorphism as a tool in fisheries science. *Canadian Journal of Fisheries and Aquatic Sciences, 45*, 1075–1087.

Hulata, G. (2001). Genetic manipulations in aquaculture: A review of stock improvement by classical and modern technologies. *Genetica, 111*(1–3), 155–173.

Ishida, N., Hasegawa, T., Takeda, K., Sakagami, M., Onishi, A., Inumaru, S., et al. (1994). Polymorphic sequence in the D-loop region of equine mitochondrial DNA. *Animal Genetics, 25*(4), 215–221.

Jeffreys, A. J., Wilson, V., & Thein, S. L. (1985). Hypervariable 'minisatellite' regions in human DNA. *Nature, 314*(6006), 67–73.

Jensen, P. R., & Fenical, W. (1996). Marine bacterial diversity as a resource for novel microbial products. *Journal of Industrial Microbiology, 17*(5–6), 346–351.

Keim, P., Price, L., Klevytska, A., Smith, K., Schupp, J., Okinaka, R., et al. (2000). Multiple-locus variable-number tandem repeat analysis reveals genetic relationships within Bacillus anthracis. *Journal of Bacteriology, 182*(10), 2928–2936.

Kidd, K. K. (1993). Associations of disease with genetic markers: Deja vu all over again. *American Journal of Medical Genetics, 48*(2), 71–73.

Li, C., & Horvitz, D. (1953). Some methods of estimating the inbreeding coefficient. *American Journal of Human Genetics, 5*(2), 107.

Limborg, M. T., Helyar, S. J., De Bruyn, M., Taylor, M. I., Nielsen, E. E., Ogden, R., et al. (2012). Environmental selection on transcriptome-derived SNPs in a high gene flow marine fish, the Atlantic herring (*Clupea harengus*). *Molecular Ecology, 21*(15), 3686–3703.

Liu, Z. J., & Cordes, J. (2004). DNA marker technologies and their applications in aquaculture genetics. *Aquaculture, 238*(1–4), 1–37.

Lynch, M. (1990). The similarity index and DNA fingerprinting. *Molecular Biology and Evolution, 7*(5), 478–484.

Mace, G. M., & Lande, R. (1991). Assessing extinction threats: Toward a reevaluation of IUCN threatened species categories. *Conservation Biology, 5*(2), 148–157.

Martin, A. P., Humphreys, R., & Palumbi, S. R. (1992). Population Genetic Structure of the Armorhead, in the North Pacific Ocean: Application of the Polymerase Chain Reaction to Fisheries problems. *Canadian Journal of Fisheries and Aquatic Sciences, 49*(11), 2386–2391.

Nishita (1998). Population analysis method by mitochondria DNA analysis. *Fisheries breeding, 26,* 81–100.

Norris, D. E., Shurtleff, A. C., Touré, Y. T., & Lanzaro, G. C. (2001). Microsatellite DNA polymorphism and heterozygosity among field and laboratory populations of *Anopheles gambiae* ss (Diptera: Culicidae). *Journal of Medical Entomology, 38*(2), 336–340.

Okamoto, N., (2000). New Marine breeding using DNA markers. In F. Takashita (Ed.), *The next generation of fisheries biotechnology* (pp. 42–52). Tokyo, Japan: Seizando-Shoten Publishing Co.

Perez-Enriquez, R., & Taniguchi, N. (1999). Genetic Structure of Red Sea Bream (Pagrus major) Population off Japan and the Southwest Pacific, Using Microsatellite DNA Markers. *Fisheries science, 65*(1), 23–30.

Perez-Enriquez, R., Takagi, M., & Taniguchi, N. (1999). Genetic variability and pedigree tracing of a hatchery-reared stock of red sea bream (Pagrus major) used for stock enhancement, based on microsatellite DNA markers. *Aquaculture, 173*(1–4), 413–423.

Queller, D. C., & Goodnight, K. F. (1989). Estimating relatedness using genetic markers. *Evolution, 43*(2), 258–275.

Sakamoto, T., Okamoto, N., Ikeda, Y., Nakamura, Y., & Sato, T. (1994). Dinucleotide-repeat polymorphism in DNA of rainbow trout, Oncorhynchus mykiss, and its application in fisheries science. *Journal of Fish Biology, 44,* 1093–1096 (1994).

Sakamoto, T., Okamoto, N., & Ikeda, Y. (1996). Application of PCR Primer Pairs from Rainbow Trout to Detect Polymorphisms of CA Repeat DNA Loci in Five Confamilial Species. *Fisheries science, 62*(4), 552–555.

Sakamoto, T., Danzmann, R. G., Okamoto, N., Ferguson, M. M., & Ihssen, P. E. (1999). Linkage analysis of quantitative trait loci associated with spawning time in rainbow trout (Oncorhynchus mykiss). *Aquaculture, 173*(1–4), 33–43.

Saunders, G. W. (2005). Applying DNA barcoding to red macroalgae: A preliminary appraisal holds promise for future applications. *Philosophical Transactions of the Royal Society of London B: Biological Sciences, 360*(1462), 1879–1888.

Schlötterer, C. (2000). Evolutionary dynamics of microsatellite DNA. *Chromosoma, 109*(6), 365–371.

Seki, S., Takagi, M., & Taniguchi, N. (1995) *Fish Breeding, 43,* 97–102 (1995).

Sunnucks, P. (2000). Efficient genetic markers for population biology. *Trends in Ecology and Evolution, 15*(5), 199–203.

Tabata, K., & Mizuta, A. (1997). RFLP Analysis of the mtDNA D-loop Region in Red Sea Bream Pagrus major Population from Four Locations of Western Japan. *Fisheries science, 63*(2), 211–217.

Tabata, K., Kishioka, H., Takagi, M., Mizuta, A., & Taniguchi, N. (1997). Genetic Diversity of Five Strains of Red Sea Bream Pagrus major by RFLP Analysis of the mtDNA D-loop Region. *Fisheries science, 63*(3), 344–348.

Takagi, M., et al. (1993). DNA fingerprints of chromosome manipulation ayu. *Fish Breeding, 19*, 45–53.

Takagi, M., Taniguchi, N., Yamasaki, M., & Tsujimura, A. (1995). Identification of Clones Induced by Chromosome Manipulation in Ayu Plecoglossus altivelis by DNA Fingerprinting with RI and Non-RI Labelled Probes. *Fisheries science, 61*(6), 909–914.

Takagi, M., Taniguchi, N., Cook, D., & Doyle, R. W. (1997). Isolation and Characterization of Microsatellite Loci from Red Sea Bream Pagrus major and Detection in Closely Related Species. *Fisheries science, 63*(2), 199–204.

Taniguchi, N. (1996). Analysis of the Fish Genome and Its Necessity. G-C Mapping of Chromosome Manipulated Fish Using Hyper-Variable DNA Markers. *Nippon Suisan Gakkaishi, 62*(4), 685–686.

Taniguchi, N. (2000). Genetic diversity of fishes and DNA markers. In F. Takashita (Ed.), *The next generation offisheries biotechnology* (pp. 43–53). Tokyo, Japan: Seizando-shoten publishing Co.

Taniguchi, N., & Takagi, M., (1997). Diversity analysis of DNA Polymorphism and fish population. In T. Aoki, F. Takashima & T. Hirano (Eds.), *DNA of fish* (pp. 117–137). Tokyo, Japan: Seizando-Shoten Publishing Co.

Taniguchi, N. Takagi, M. & Matsumoto, S. (1997). Genetic evaluation of quantitative traits of hatchery stocks for aquaculture in red sea bream. *Bulletin National Research Institute of Aquaculture Supplement 3*, 35–41.

Taniguchi, N., Yamasaki, M., Takagi, M., & Tsujimura, A. (1996). Genetic and environmental variances of body size and morphological traits in communally reared clonal lines from gynogenetic diploid ayu, Plecoglossus altivelis. *Aquaculture, 140*(4), 333–341.

Tingey, S. V., & del Tufo, J. P. (1993). Genetic analysis with random amplified polymorphic DNA markers. *Plant Physiology, 101*(2), 349.

Toledo, Á., & Burlingame, B. (2006). Biodiversity and nutrition: A common path toward global food security and sustainable development. *Journal of Food Composition and Analysis, 19*(6–7), 477–483.

Vassart, G., Georges, M., Monsieur, R., Brocas, H., Lequarre, A., & Christophe, D. (1987). A sequence in M13 phage detects hypervariable minisatellites in human and animal DNA. *Science, 235*(4789), 683–684.

Vos, P., Hogers, R., Bleeker, M., Reijans, M., Lee, T. v. d., Hornes, M., et al. (1995). AFLP: A new technique for DNA fingerprinting. *Nucleic Acids Research, 23*(21), 4407–4414.

Williams, J. G., Kubelik, A. R., Livak, K. J., Rafalski, J. A., & Tingey, S. V. (1990). DNA polymorphisms amplified by arbitrary primers are useful as genetic markers. *Nucleic Acids Research, 18*(22), 6531–6535.

Williams, J. G., Hanafey, M. K., Rafalski, J. A., & Tingey, S. V. (1995). Genetic analysis using random amplified polymorphic DNA markers. In *Recombinant DNA Methodology II* (pp. 849–884). Elsevier.

Wirgin, I. I., & Maceda, L. (1991). Development and use of striped bass-specific RFLP probes. *Journal of Fish Biology 39*, 159–167.

Xu, Y., & Crouch, J. H. (2008). Marker-assisted selection in plant breeding: From publications to practice. *Crop Science, 48*(2), 391–407.

Zhang J., Talbot, W. S., & Schier, A. F. (1998). Positional cloning identifies zebrafish one-eyed pinhead as a permissive EGF-related ligand required during gastrulation. *Cell, 92*, 241–251.

Seaweed Biotechnology

6

Contents

6.1 Introduction

The term seaweed typically refers to cryptograms that grow in seawater and do not bloom. They include green, brown, and red algae, which are visible to the unaided eye and survive by attaching to other objects.

Because water is the chief setting for their growth, they differ considerably from land organisms in their development temperatures and photosynthetic properties. Even seaweeds belonging to the same group often possess unique biogenic substances and metabolic characteristics. Seaweeds are a promising source of energy and bioactive substances that will make them key, alongside plankton, in resolving future food issues. Recently, research has been conducted toward applying marine biotechnology—an area where notable advancements have been made—to seaweeds to develop useful breeds or produce useful substances. Despite these efforts, however, the biochemical properties of green, brown, red algae differ due to their belonging to

---

The original version of this chapter was revised: belated author corrections have been incorporated. The correction to this chapter is available at
https://doi.org/10.1007/978-3-030-20944-5_12

© Springer Nature Switzerland AG 2019, corrected publication 2021   145
S.-K. Kim, *Essentials of Marine Biotechnology*,
https://doi.org/10.1007/978-3-030-20944-5_6

different evolutionary lines, and many problems remain unresolvable with the understanding hitherto gained from land-based plant life (Mazarrasa et al. 2013).

Biotechnology, a field where outstanding recent advancements have been achieved in the area of microorganisms, has produced many new findings as its scope has broadened from animals to higher-order plants. In manufacturing, tissue culturing of seaweeds was attempted in the early 1950s. Initially, alcohol or sodium hypochlorite solution was used in various attempts to obtain aseptic tissue, but the results were not spectacular. The later emergence of antibiotic substances and technical improvements led in the 1980s to developments through prolific cutting-edge biotechnology research in countries such as Japan, the United States, and China (Wheeler et al. 1979; Tseng 2001).

The field of seaweed biotechnology includes a number of different areas, including tissue culture, callus induction, protoplast production and regeneration, cell fusion, and gene manipulation. At the same time, a number of issues remain to be solved, including the failure to achieve full consistency in terminology. Nevertheless, the technology is very important in opening up new possibilities for the use of seaweed not only as a food resource but also for mass-production of useful substances or for bioactive substances and biotechnological resources, and greater development is expected for the field going ahead.

South Korea has a long history of using seaweeds such as laver (*Porphyra suborbiculata*), sea mustard (*Undaria pinnatifida*), kelp (*Laminaria japonica*), green laver (*Enteromorpha* spp.), gulf weed (*Sargassum fusiforme*), and fusiforme (*Hizika fusiformis*) for food, and these varieties have been the subject of active farming efforts. In Europe, seaweeds have long been used for fertilizer and as a source of iodine and soda ash for glass-making (Kim et al. 2000).

In addition to their use for food, seaweeds today are used around the world as source algae for the extraction of useful substances such as agar, alginic acid, and carrageenan. In the case of agar, red algae such as agar-agar and kkosiraegi (*Gracilariaceae*) are used as chief sources. Alginic acid and carrageenan are widely used in the food processing and cosmetics industries as fixatives and viscosity agents and for enzymes. Brown algae such sea tangle (*Laminariaceae*) and giant kelp, rhubard (*Eisenia bicyclis*), and gamtae (*Ecklonia cava*) are used for alginic acid, while red algae such as *Gloiopeltis tenax*, *Chondrus ocellatus*, *Gracilaria verrucosa,*, *Chondracanthus tenellus* and Gigartinales are used as sources for carrageenan. Seaweeds are also very important as food for useful marine creatures such as turban shells, sea urchins, and abalone.

6.2 Mass Production of Seaweeds

The seaweeds used for the purposes described above were initially collected from the sea, where they grew naturally. Farming techniques gradually developed to provide stable mass production to meet rising demand. Representative cases in South Korea include the farming of laver, sea mustard, and kelp. Farming of these food seaweeds also takes place today in China and Japan. Laver farming has been attempted recently in the United States, Canada, and New Zealand as well; in the U.S., giant kelp is farmed not as a food source, but as a source of alginic acid or material for methane fermentation. In the Philippines, *Gracilaria verrucosa* is farmed for export as a carrageenan source (Lüning and Pang 2003; Kraan 2013).

Recently, giant kelp has been the focus of continued attempts in the U.S. to mass-produce seaweed as a methane fermentation resource for alternative bioenergy development. In Japan, ocean farming has been considered for ongoing mass production of kelp and other highly productive large varieties of brown algae. These efforts would entail establishing large ocean farms on the coast for year-round production and supply of seaweed as a methane fermentation resource.

The ocean farming facilities used in Japan cover an area measuring 8.05 km × 5.12 km, or a total of 41.216 km^2. This area contains 28 unit farms (each measuring 850 m × 980 m), where kelp and other large brown algae are grown separately or together. In each unit, seedlings grown with different land production times are transplanted and farmed for a suitable length of time, after which those that have become particularly large are automatically harvested by harvest vessels. These vessels are essential for automatic cultivation, but for the vessels to fully demonstrate their capabilities, farming facilities must be in place with the structures to respond to them efficiently.

Once harvested, seaweed is transported to methane fermentation facilities and used as a fermentation substrate. For reasons of economy, useful ingredients are extracted prior to fermentation. These useful ingredients include pigments, alginic acid, water-soluble alginate, and various anti-cancer and anti-bacterial agents. Consideration is also given to the recovery of useful ingredients from methane fermentation residue, which include iodine, vitamins, and eicosapentaenoic acid (EPA). Vitamin B$_{12}$ and EPA are known to be of particular physiologically and nutritionally significance for young fish and may be used as food additives for farmed fish (Pereira and Yarish 2008; Ugwu et al. 2008).

6.3 Tissue Culture

Various methods have been considered for the mass production of useful seaweeds. One of these is mass seedling production through the application of tissue culture techniques. Achieving this first requires the sterilization and successful culturing of seaweed tissue. Because the seaweed surface is rich in mucilage, and because large numbers of tiny flora and fauna adhere to the surface or penetrate the cell layer close to the surface (particularly in naturally growing seaweeds), acquiring aseptic tissue is no easy feat. Some sterilization methods that have been in use since early on include pipette flushing (most often used for single-cell organisms such as zoospores and regular spores), agar plate washing, UV ray exposure, ultrasound application, antibiotic application, and germicide through the use of alcohol, iodine, or chlorine solution. These methods are used in combination rather than individually (Aguirre-Lipperheide et al. 1995; Kumar et al. 2004; Chen and Tayler 1978; Saga et al. 1982).

In the case of seaweed collected from the ocean, any attached organisms visible to the naked eye would be removed, and the seaweed would then be subjected to ultrasound treatment to remove smaller organisms. Next, it would be cultured for two days in a solution containing various antibiotic substances; iodine and chlorine would be effective means for final germicidal treatment. These techniques require considerable skill, and are unfortunately not something that anyone could apply to achieve sterilization. In the case of seaweeds emitting zoospores, the single-stage selection method shown in Fig. 6.1 is a relatively simple and reproducible method that can be used to achieve a sterilization rate of 90% or more.

The following is an outline of the sequence in the single-stage selection method for seaweeds in the kelp order.

(1) Scrub mature seaweed (sporophyte) with gauze and rinse several times in sterilized seaweed before preparing fragments of around 1 cm.
(2) Place ten of these fragments in a flask containing a solution of 100 ml of antibiotic substances and keep refrigerated for two days.

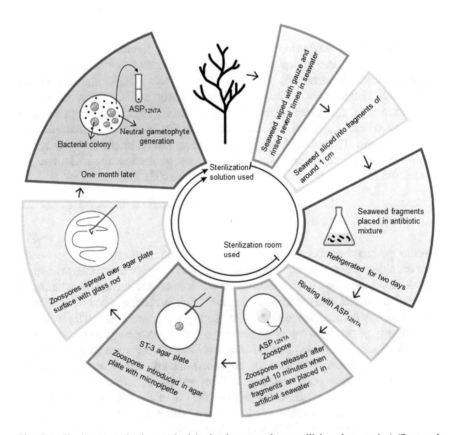

Fig. 6.1 Single-stage selection method (a simple approach to sterilizing algae strains) (Saga and Sakai 1982)

(3) Rinse one fragment in sterilized artificial seawater and place in a laboratory dish of sterilized artificial seawater. After 10 min or so, many of the sporophytes will be released.

(4) Draw out sporophytes with a micropipette, place two to three droplets on ST3 agar medium, and spread with glass rod.

(5) After one month, the sporling will become a cluster of branching filaments measuring around 1 mm in diameter, and the mixed bacteria will form a colony measuring 3–55 mm. Place the seaweed that is not contaminated with bacteria into a test tube containing 10 ml of $ASP_{12}NTA$ medium.

(6) A month later, the seaweed will becoming a microcallus measuring around 3 mm in diameter. For the culturing of this microcallus, sterilization can be verified through the use of ST3, ASP-B1, or other search media.

Research has already reported successful sterilization of 90% or more with this single-stage selection method.

Medium use is crucially important for successful seaweed tissue culturing. Varieties of medium used for seaweed culturing include enriched media, which use a natural seawater base enriched with nutrient salts and other essential nutrients, and synthetic media, which use artificial seawater prepared using chemicals alone and enriched with nutrient salts and other essential nutrients (Table 6.1).

Chief examples of enriched media include the Erd-schreiber (ES) medium, which has long been used for seaweed culturing. Recent examples include PES, which has often been used for culturing of green and brown algae; PESI, which is often used for brown algae culturing; and ESS, which was developed for seaweed seed production and is useful for lavers, kelps, and sea lettuces. Examples of synthetic media include ASP, which developed by Provasoli et al. in the U.S.; ASP_1 and ASP_7 have been used frequently for green algae, ASP_2 and ASP_6 for red algae, and ASP_{12} for brown algae. The ASS_1 media was developed as a basic medium for seaweed tissue culturing and can used to produce excellent results with green, red, and brown algae; it is also used as a preservation medium for sterilized strains (Tables 6.2 and 6.3).

Some room for improvement of these media remains, as the current cases are as yet inadequate for completion of the life cycle or smooth completion of morphogenesis.

Examples of success with tissue culturing of large seaweed include Chem et al. in Canada, who in 1978 cultured a tissue fragment measuring several millimeters taken from the center of the red alga *Chrondrus crispus*; three months later, this had grown into an alga measuring several centimeters. That same year, Sega et al. in Japan used the kelp *Laminaria angustata* for tissue culturing and succeeded in producing cloned kelp using cells separated from callus tissue.

Tissue culturing has been attempted with sterilized seaweeds through the methods described above, but in all cases they can be introduced to an agar medium to obtain a callus and the cells derived from it. A callus is an amorphous cluster of cells that appears when part of a parent organism is removed and cultured; under certain conditions, it does not exhibit differentiability.

Table 6.1 Developing crude seawater solution for major seaweeds

	ES[a]	PES[b]	PESI[b]	ESS[c]
Distilled water	–	100 ml	100 ml	60 ml
Seawater	100 ml	–	–	–
Sodium nitrate (NaNo$_3$)	10 mg	350 mg	350 mg	350 mg
Na$_2$·glycerol phosphate	–	50 mg	50 mg	80 mg
Disodium phosphate (Na$_2$HPO$_4$.12 H$_2$O)	2 mg	–	–	–
Iron (as EDTA, mole ratio 1:1)	–	2.5 mg	2.5 mg	–
Iron sequestrene	–	–	–	40 mg
Mixed metal solution PII[d]	–	25 ml	25 ml	40 ml
Vitamin B$_{12}$	–	10 μg	–	–
Thiamine	–	0.5 mg	–	–
Biotin	–	5 μg	–	–
ESS vitamin mixed solution[e]	–	–	–	1 ml
Soil extract	5 ml	–	–	–
Tris	–	500 mg	500 mg	–
HEPES	–	–	–	1 g
Potassium iodide (KI)	–	–	100 μg	–
pH	–	7.8	7.8	7.8

[a]ES solution should be diluted after sterilization
[b]2 ml added to 100 ml of seawater after filtering of undiluted PES, PESI solution
[c]1 ml added to 100 ml of seawater after filtering of undiluted ESS solution
[d]See note on Table 6.2
[e]Mixing of ESS vitamin solution (in 1 ml): 10 μg vitamin B$_{12}$, 10 μg biotin, 1 mg thiamine hydrochloride, 1 mg nicotinic acid, 1 mg calcium pantothenate, 100 μg p-aminobenzoic acid, 10 mg inositol, 1 mg thymine

Typically, calluses are induced by the placement of <u>nicked</u> tissue between a solid (such as agar, carrageenan, or dampened filter paper) and air saturated with sea-water vapor. Callus induction has been performed successfully for many seaweeds. Table 6.4 shows some of the important varieties of red, brown, and green algae for which tissue culturing has been reported.

In this section, a large brown alga will be used as an example for tissue culturing. The following is the sequence followed by Sega et al. using the kelp *Laminaria angustata* (Saga and Sakai 1982).

(1) Thoroughly rinse both ends of section from seaweed stalk and cut to lengths of around 5 cm.
(2) Soak both ends of the fragment in 100% ethanol and heat slightly with alcohol lamp before cutting to around 5 mm each.
(3) Bore with cork borer (approximate 4 mm in dimeter) and cut to thickness of around 2 mm.
(4) Introduce the resulting disk-shaped fragments into ASP$_{12}$NTA agar medium (50 ml).

Table 6.2 Developing synthetic medium for major seaweeds (in 100 ml)

	ASP_1	ASP_2	ASP_6	ASP_7	ASP_{12}	KDX	ASS_1
Sodium chloride (NaCl)	2.4 g	1.8 g	2.4 g	2.5 g	2.8 g	1.9 g	2.5 g
Sodium sulfate (Na_2SO_4)	–	–	–	–	–	0.32 g	–
Magnesium sulfate ($MgSO4.7H2O$)	0.6 g	0.5 g	0.8 g	0.9 g	0.7 g	–	1.0 g
Magnesium chloride ($MgCl_2.7H_2O$)	0.45 g	–	–	–	0.4 g	0.87 g	–
Potassium chloride (KCl)	60 mg	60 mg	70 mg	70 mg	70 mg	–	70 mg
Calcium chloride ($CaCl_2$)	40 mg	10 mg	15 mg	30 mg	40 mg	50 mg	30 mg
Sodium nitrate ($NaNO_3$)	10 mg	5 mg	30 mg	5 mg	10 mg	8.5 mg	10 mg
Ammonium sulfate [$(NH4)_2SO_4)$]	–	–	–	–	–	0.66 mg	–
Dipotassium phosphate (K_2HPO_4)	2 mg	0.5 mg	–	–	–	–	–
Tripotassium phosphate (K_3PO_4)	–	–	–	–	1 mg	–	–
Monosodium phosphate ($Na_2HPO4.H2O$)	–	–	–	–	–	0.7 g	–
Sodium silicate ($Na_2SiO_3.9H_2O$)	2.5 mg	15 mg	7 mg	7 mg	15 mg	–	–
Sodium carbonate ($NaCO_3.H_2O$)	–	3 mg	–	–	–	–	–
Sodium bicarbonate ($NaHCO_3$)	–	–	–	–	–	8.8 mg	10 mg
Sodium glycerophosphate	–	–	10 mg	2 mg	1 mg	0.14 mg	2 mg
Iron [Fe (as Cl^-)]	–	50 μg	–	–	–	–	–
Metal mixture P II[a]	1 ml	3 ml	–	3 ml	1 ml	–	–
Metal mixture S II[b]	–	–	–	–	1 ml	–	–
Metal mixture P8[c]	–	–	1 ml	–	–	–	–
Metal mixture (ASS)[d]	–	–	–	–	–	–	1 ml
Vitamin B_{12}	0.02 μg	0.2 μg	0.05 μg	0.1 μg	0.02 μg	–	–
Biotin	–	–	–	–	0.1 μg	–	–
Thiamine hydrochloride	–	–	–	–	10 μg	–	–
Mixed vitamin solution S3[e]	–	1 ml	–	1 ml	–	–	–
Mixed vitamin solution 8A[f]	0.05 ml	–	0.1 ml	–	–	–	–
Mixed vitamin solution (ASS)[g]	–	–	–	–	–	–	0.1 ml
KDS solution[h]	–	–	–	–	–	1 ml	–
KDTM solution[i]	–	–	–	–	–	1 ml	–
Triaminomethane	0.1 g	0.1 g	0.1 g	0.1 g	0.1 g	–	–

(continued)

Table 6.2 (continued)

	ASP_1	ASP_2	ASP_6	ASP_7	ASP_{12}	KDX	ASS_1
Glycylglycine	–	–	–	–	–	75 mg	–
HEPES[j]	–	–	–	–	–	–	100 mg
pH	7.6	7.8	7.6	7.8–8.0	7.8–80	8.3–8.4	8.0

[a]Preparation for metal mixture P II (in 1 ml): 1 mg Na_2·EDTA, 0.01 mg Fe (as Cl^-), 0.2 mg B (as H_3BO_3), 0.04 mg Mn (as Cl^-), 0.005 mg Zn (as Cl^-), 0.001 mg Co (as Cl^-)

[b]Preparation for metal mixture S II (in 1 ml): 1 mg Br (as Na^+), 0.2 mg Sr (as Cl^-), 0.02 mg Rb (as Cl^-), 0.02 mg Li (as Cl^-), 0.05 mg Mo (as Na^+), 0.001 mg I (as K^+)

[c]Preparation for metal mixture P8 (in 1 ml): 3 mg Na_3-versenol, 0.2 mg Fe (as Cl^-), 0.2 mg B (as H_3BO_3), 0.1 mg Mn (as Cl^-), 0.05 mg Zn (as Cl^-), 0.01 mg Co (as Cl^-), 0.05 mg Mo (as Na^+), 0.002 mg Cu (as Cl^-)

[d]Preparation for metal mixture for ASS (in 1 ml): 100 µg Fe (as Fe-sequestrene), 100 µg B (as H_3BO_3), 100 µg Mn (as Cl^-), 10 µgZn (as Cl^-), 1 µg Co (as Cl^-), 10 µg Mo (as Na_2MoO_4), 1 µg Cu (as Cl^-), 1 mg Br (as K^+), 100 µg Sr (as Cl^-), 10 µg Rb (as Cl^-), 10 µg Li (as Cl^-), 1 µg I (as K^+)

[e]Preparation for vitamin mixture S3 (in 1 ml): 0.05 mg thiamine hydrochloride, 0.01 mg niacin, 0.01 mg calcium pantothenate, 1 µg p-aminobenzoic acid, 0.1 µg biotin, 0.5 mg inositol, 0.2 µg folate, 0.3 mg thymine

[f]Preparation for vitamin mixture 8A (in 1 ml): 0.2 mg thiamine hydrochloride, 0.1 mg nicotinic acid, 0.04 mg putresine-2HCl, 0.1 mg calcium pantothenate, 5 µg riboflavin, 0.04 mg pyridoxine 2-HCl, 0.02 mg pyridoxamine-2HCl, 0.01 mg p-aminobenzoic acid, 0.5 µg biotin, 0.5 mg choline citrate, 1 mg inositol, 0.8 mg thymine, 0.26 mg orotic acid, 0.05 µg vitamin B_{12}, 0.2 µg folinic acid, 2.5 µg folate

[g]Preparation for ASS vitamin mixture (in 1 ml): 100 µg thiamine hydrochloride, 100 µg nicotinic acid, 10 µg putresine-2HCl, 100 µg calcium pantothenate, 10 µg riboflavin, 10 µg pyridoxine-2HCl, 10 µg pyridoxamine-2Hcl, 10 µg p-aminobenzoic acid, 1 µg biotin, 1 mg inositol, 100 µg choline citrate, 100 µg thymine, 100 µg orotic acid, 1 µg cyanocobalamin, 1 µg folate, 0.1 µg folinic acid

[h]Preparation for KDS solution (in 1 ml): 7.84 mg KBr, 54.2 mg KCl, 1.95 mg $SrCl_2 \cdot 6H_2O$, 0.01 µg cyanocobalamin, 0.05 µg biotin, 10.0 µg thiamine hydrochloride

[i]Preparation for KDTM solution (in 1 ml): 668.4 µg H·EDTA, 1.14 mg H_3BO_4, 199.0 µg $FeSO_4·7H_2O$, 3.9 µg $CuSO_4·5H_2O$, 12.6 µg Na_2MoO_4, 36.0 µg $MnCl_2·4H_2O$, 44.0 µg $ZnSO_4·7H_2O$, 4.0 µg $CoCl_2·6H_2O$, 2.3 µg NH_4VO_3, 3.9 µg KI

[j]HEPES buffer solution: 4-(2-hydroxyethyl)-1-piperazineethane sulfonic acid

The resulting disk-shaped tissue fragments are nearly 100% sterilized, and after one to two months calluses will form on them. Figure 6.2 shows the tissue culturing method used for the brown alga *Ecklonia stolonifera* (from the kelp order).

While medium preparation is the most important part of tissue culturing, recent examination of physical conditions such as light, osmotic pressure, and agar concentration has resulted in callus induction being accelerated in the red algae *Chrondia crassicaulus*, *Laurencia intermedia* and *Grateloupia filicina* when agar concentrations approach 0.3–1.5%. Temperature and light intensity were first found to have a large influence in the late 1990s (Table 6.5).

Table 6.3 Preparation of sterile detection media for major seaweeds

	STP	ST$_3$	ESS$_{B1}$
Seawater	750 ml	700 ml	900 ml
Distilled water	200 ml	250 ml	70 ml
Undistilled ESS solution[a]	–	–	10 ml
Soil extract	50 ml	50 ml	25 ml
Sodium nitrate (NaNO$_3$)	200 mg	50 mg	–
Dipotassium phosphate (K$_2$HPO$_4$)	10 mg	–	–
Na$_2$-glycerol phosphate	–	10 mg	–
Bacto peptone	–	–	5 mg
Hy-case (Scheffielde Chemical)	200 mg	20 mg	
Casein lysates	–	–	5 mg
Yeast rate (Difco)	200 mg	–	–
Yeast extract	–	10 mg	5 mg
Malt lysates	–	–	5 mg
Liver (Oxo L25, Oxo Ltd)	–	20 mg	–
Beef extract	–	–	5 mg
Vitamin B$_{12}$	–	0.1 μg	–
Vitamin mixture 8A[b]	1 ml	1 ml	–
Sucrose	1 g	–	–
Glucose	2 g	–	–
Carbon source mixture II[c]	–	20 ml	5 ml
Na H-glutamic acid	0.5 g	–	–
D, L-alanine	100 mg	–	–
Glycine	100 mg	–	–
Glycylglycine	–	400 mg	–
Agar	(4 g)	(4 g)	(10 g)
pH	7.5–7.6	7.9	7.8

[a]See Table 6.1

[b]Preparation for vitamin mixture 8A (in 1 ml): 0.2 mg thiamine hydrochloride, 0.1 mg nicotinic acid, 0.04 mg putresine-2HCl, 0.1 mg calcium pantothenate, 5 μg riboflavin, 0.04 mg pyridoxine-2HCl, 0.02 mg pyridoxamine-2HCl, 0.01 mg p-aminobenzoic acid, 0.5 μg biotin, 0.5 mg choline citrate, 1 mg inositol, 0.8 mg thymine, 0.26 mg orotic acid, 0.05 μg vitamin B$_{12}$, 0.2 μg folinic acid, 2.5 μg folate

[c]1 ml includes 1 mg glycine, 1 mg D, L-alanine, 1 mg L-asparagine, 2 mg sodium acetate, 2 mg glucose, and 2 mg L-glutamic acid

Callus cells typically multiply under relatively low temperatures and light intensity, but differentiation into sporophytes under relatively high temperatures and light intensity was found in culturing results using leaf body tissue from kelp. This was later found to be related to rising or falling melanosis within the cells (Fig. 6.3) (Borowitzka et al. 2009).

Table 6.4 Seaweeds currently used for tissue culturing (Polne-Fuller and Gibor 1987; Polne-Fuller 1988; Amano 1990)

Green algae	Brown algae	Red algae
Bryopsis plumosa	Dictyosiphon foeniculaceus	Porphyra ochotensis
Ulva linza (Green laver)	Ecklonia stolonifera	Porphyra yezoensis (Nori)
Ulva intestinalis (Green laver)	Laminariaceae (Kelp)	Agardhiella subulata
Kornmannia leptoderma (Green laver)	Laminaria japonica (Japanese seatangle)	Chondrus crispus
Monostroma angicava (Green laver)	Laminaria angustata	Gelidiaceae
Ulva taniata (Green laver)	Laminaria saccharina	Gelidium amansii (Long limu)
	Macrocystis pyrifera (Giant kelp)	Pterocladiella capillacea
	Undaria pinnatifida	Gelidium divaricatum
	Sargassum heterophyllum	Gigartina exasperata
	Chondracanthus tenellus	Gigartinaceae
		Gracilaria deblis
		Gracilaria epihippisora
		Gracilaria tikvahiae

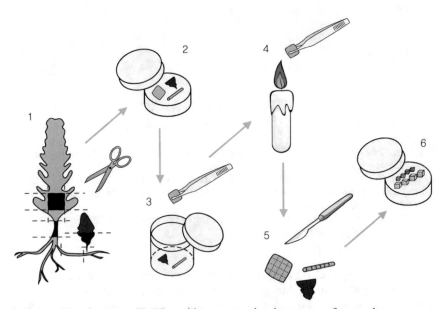

(1) Cut tissue fragment. (2) Wipe with paper towel and remove surface portion.
(3) Dip in 100% ethanol for 1 to 2 seconds. (4) Sear surface with burner.
(5) Cut to 3–4mm. (6) Place on agar medium.

Fig. 6.2 Tissue culturing sequence for *Ecklonia stolonifera* (Aruga 1990)

Table 6.5 Suitable temperatures and light intensities for callus proliferation and sporophyte differentiation in gonidium tissue from seaweeds in kelp order. (1) Cut tissue fragment. (2) Wipe with paper towel and remove surface portion. (3) Dip in 100% ethanol for 1 to 2 s. (4) Sear surface with burner. (5) Cut to 3–4 mm. (6) Place on agar medium   (Notoya 2000)

Species	Callus cell multiplication	Leaf body cell differentiation
Agarum cribrosum	10 °C, 10–80 μ mol m^{-2}s^{-1}	–
Costaria costata	15 °C, 10–40 μ mol m^{-2}s^{-1}	15 °C, 80 μ mol m^{-2}s^{-1}
		20 °C, 40–80 μ mol m^{-2}s^{-1}
	15 °C, 10–20 μ mol m^{-2}s^{-1}	10 °C, 80 μ mol m^{-2}s^{-1}
		15 °C, 40–80 μ mol m^{-2}s^{-1}
		20 °C, 10–80 μ mol m^{-2}s^{-1}
Ecklonia cava	15 °C, 30–32 μ mol m^{-2}s^{-1}	15 °C, 30–32 μ mol m^{-2}s^{-1}
	20 °C, 10–20 μ mol m^{-2}s^{-1}	15 °C, 40–80 μ mol m^{-2}s^{-1}
		20 °C, 40–80 μ mol m^{-2}s^{-1}
		25 °C, 10–80 μ mol m^{-2}s^{-1}
	15 °C, 10–20 μ mol m^{-2}s^{-1}	20 °C, 80 μ mol m^{-2}s^{-1}
		20 °C, 10–80 μ mol m^{-2}s^{-1}
Ecklonia stolonifera	10 °C, 10–40 μ mol m^{-2}s^{-1}	15 °C, 80 μ mol m^{-2}s^{-1}
	15 °C, 10–40 μ mol m^{-2}s^{-1}	20 °C, 20–80 μ mol m^{-2}s^{-1}
		25 °C, 10–40 μ mol m^{-2}s^{-1}
Eisenia bicyclis	20 °C, 10 μ mol m^{-2}s^{-1}	15 °C, 20–80 μ mol m^{-2}s^{-1}
		20 °C, 20–80 μ mol m^{-2}s^{-1}
		25 °C, 10–80 μ mol m^{-2}s^{-1}
	20 °C, 10 μ mol m^{-2}s^{-1}	10 °C, 40–80 μ mol m^{-2}s^{-1}
		15 °C, 40–80 μ mol m^{-2}s^{-1}
		20 °C, 20–80 μ mol m^{-2}s^{-1}
		25 °C, 20–80 μ mol m^{-2}s^{-1}
Kjellmaniella crassifolia	10 °C, 10 μ mol m^{-2}s^{-1}	10 °C, 80 μ mol m^{-2}s^{-1}
Undaria pinnatifida	15 °C, 10 μ mol m^{-2}s^{-1}	10 °C, 80 μ mol m^{-2}s^{-1}
		15 °C, 20–80 μ mol m^{-2}s^{-1}
		20 °C, 20–40 μ mol m^{-2}s^{-1}
	15 °C, 10 μ mol m^{-2}s^{-1}	10 °C, 80 μ mol m^{-2}s^{-1}
		15 °C, 80 μ mol m^{-2}s^{-1}
		20 °C, 40 μ mol m^{-2}s^{-1}

Also drawing attention is the use of auxins and cytokines that mediate plant growth in the culturing of seaweed tissue. Indole acetic acid and kinetine have proven effective when mixed in sea lettuces (a form of brown algae), as have phenyl acetic acid, zeatin, 2,4,5-trichlorophenoxy acetic acid, 6-benzylaminopurine, indole acetic acid, and kinetine when mixed with the red alga *Agardhiella*. Research on the

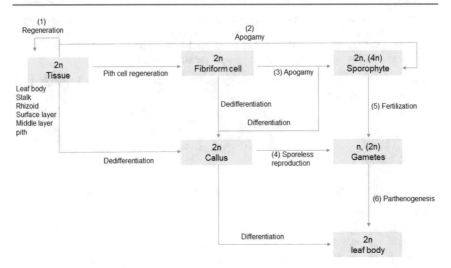

(1) Tissue regeneration (no change in generations, nuclear phase) (2) Subtrahend
from quadruploid sporophyte to diploid sporophyte (nuclear phase changes, no
generation change) (3) Differentiation from fibriform callus cells to sporophytes
(nuclear phase change may or may not occur, no generation change) (4) Differentiation
from callus cell to gamete (nuclear phase change may or may not occur, generation
change occurs), (5) Nuclear phase duplication through fertilization (generation change),
(6) Parthenogenesis (generation change).

Fig. 6.3 Various forms of cell differentiation in tissue culturing (Notoya 2000)

use of plant growth regulation substances remains sparse, but is expected to develop
further going ahead.

Several forms of cell differentiation have been reported in tissue culturing. In
some cases, differentiation occurs beyond generations and the nuclear phase is
changed (Fig. 6.4). For example, sporophyte cells may be of a haploid 2n generation, but seaweeds differentiated from the tissue form an n generation of gametes.
In this case, the appearance of a reproductive cell may be merely external, and the
cell will be diploid without any change in the nuclear phase. In other cases, the cells
become haploid through subtraction. The mechanism and causes of this differentiation remain completely unknown.

In many cases, seaweed tissue culturing has the explicit aim of breeding varieties
that are important for the use of marine organism resources. In the case of *Ptero-
cladiella capillacea* (a red alga that provides a source for agar), calluses have been
induced to extract and isolate polysaccharides. The polysaccharides included the
typical agarose that is the main polysaccharide in agar, but sulfate and methyl group
contents were reported to have declined slightly from their natural levels. Development is also under way for various self-sterile mutant strains of the most

Fig. 6.4 *Ulva fasciata*
Protoplasts

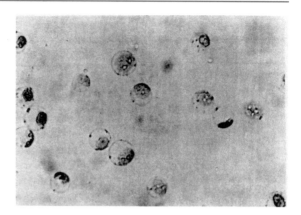

important agar source, *Gelidium amansii*, with tissue culturing techniques used to achieve improved varieties. The advantage of tissue culturing lies in the use of the resulting calluses, or the cells isolated from the calluses, for so-called cloning propagation, producing numerous clones with the same traits as parents without genetic information and acquiring numerous seeds from a particular superior parent strain (Amano 1990).

6.4   Protoplast Production in Seaweed Cells

Among the various organisms in the population, some survive while others are removed. This kind of selection (winnowing) or artificial hybridization has been used for a very long time in the fields of crop and livestock breeding. Recently, the strong desire to produce hybrids or new strains not accessible through previous artificial hybridization or selection has drawn attention to rapid advancements in genetic engineering and cell engineering technology. Creating somatic cell hybrids and applying advanced gene manipulation technology to breeding requires protoplasts, or cells from which the plant cell wall has been removed.

The term "protoplast" is used to refer the protoplasmic and non-protoplasmic cell material surrounded by a cell membrane or protoplasm membrane that is obtained through the use of enzymes to remove the cell wall in plant tissue or calluses.

Protoplasts may be obtained physically, depending on the species of seaweed. Protoplasts are an essential part of breeding by cell fusion or gene manipulation, and may also serve as important resources in explaining the cell wall formation mechanism and examining cell differentiation and nutrition.

Seaweed protoplasts were first obtained in 1979 from the green alga *Ulva intestinalis* (Millner et al. 1979). This was followed by successful examples of the use of enzymes from land-based plants to produce protoplasts for many green algae. Protoplast development and production for brown and red algae came relatively later due to differences in the polysaccharides making up the (chiefly cellulose) cell wall in all land plants and the glycosaminoglycans existing between cells.

For example, protoplasts were first obtained by breaking down alginic acid or fucoidans in brown algae for the seaweed polysaccharides porphyran, xylan, and mannan in laver, which is a red alga. Enzymes used to dissolve the cell wall include those obtained from the digestive organs of marine animals that consume seaweed (such as abalone, sea urchins, and sea slugs) and those secreted by fungi and bacteria with seaweed decomposition capabilities. Each of these enzymes has its own characteristics and may be used independently, but more often several enzymes are used in combination.

As an example, this chapter will consider the protoplast separation method used with the red alga *Porphyra yezoensis*. Once sterilized with antibiotics and iodine solution, the seaweed is treated for 10 min with protease to break down proteins on the cell wall surface, after which the seaweed is cut to fragments of 2–3 mm in length. Around 0.5–1 g of these fragments are placed for two to four hours in a cell wall lyase solution consisting of 2% powdered abalone midgut gland treated with acetone, 2% macerozyme, 0.3% mannitol, and 0.5 M dextran <u>and</u> potassium sulfate placed in seawater (pH 6.0, 20–22 °C). Once separated, the protoplasts are filtered in 20 μm nylon mash and placed in a centrifuge for 5 min at 1000 rpm before the enzyme extract is removed.

Cellulose, pectinase, and dolicellase are added to the cell wall lyase solution for the brown algae. In the case of green alga cell wall lyase solution, the cellulase currently on the market is often sufficient. Table 6.6 shows the important

Table 6.6 Algae for which protoplasts have been isolated (Notoya 2000)

Green algae	Brown algae	Red algae
Ulva intestinalis	*Laminaria japonica*	Bangiaceae
Gloiopeltis tenas	*Undaria pinatipida*	– *Porphyra lanceolate*
Ulva linza	*Macrocystis pyrifera*	– *Porphyra tenera*
Enteromorpha compressa	Dictyotaceae	– *Porphyra suborbiculata*
Enteromorpha prolifera	– *Dictyopteris prolifera*	– *Porphyra yezoensis*
Monostroma angicava	– *Dictyota dichotoma*	– *Porphyra ochotensis*
Monostroma nitidum	*Scytosiphon lomentaria*	*Meristotheca papulosa*
Kornmannia leptoderma	*Colpomenia bullosa*	*Gratelopia turuturu*
Ulvaceae	*Hizikia fusiformis*	*Grateloupia filicina*
– *Ulva lactuca*	Sargassaceae	*Kallymenia crassiuscula*
– *Ulva pertusa*	– *Sargassum filipendula*	Gigartinacea
– *Ulva fasciata*	– *Sargassum hemiphyllum*	– *Gracilaria tikvahiae*
– *Ulva conglobata*	– *Sargassum muticum*	– *Gracilaria incurvata*
– *Ulva taniata*		Gelidiaceae
Bryopsis plumose		– *Gelidium robustum*
Bryopsis maxima		– *Gelidium divaricatum*
Derbesia marina		

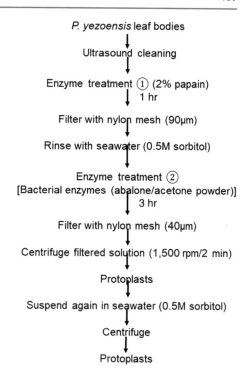

Fig. 6.5 Sequence for
protoplast production from
P. yezoensis

P. yezoensis leaf bodies

↓

Ultrasound cleaning

↓

Enzyme treatment ① (2% papain)
↓ 1 hr

Filter with nylon mesh (90μm)

↓

Rinse with seawater (0.5M sorbitol)

↓

Enzyme treatment ②
[Bacterial enzymes (abalone/acetone powder)]
↓ 3 hr

Filter with nylon mesh (40μm)

↓

Centrifuge filtered solution (1,500 rpm/2 min)

↓

Protoplasts

↓

Suspend again in seawater (0.5M sorbitol)

↓

Centrifuge

↓

Protoplasts

protoplasts that have been reported to date, while Fig. 6.5 shows protoplasts from the green alga *Ulva fasciata*. Most are derived from somatic cells, but in the case of the fucales genus, large protoplast quantities of 95% or more can be obtained from zygotes (diploid cells forming from the combination of pistil gametes) and are currently being used in research on cell wall biosynthesis (Aruga 1990).

When cultured for 10 h in PES medium at a photosensitivity of 19.5 $\mu E^{-1} m^{-2}$ and for 14 h in a dark room, *P. yezoensis* protoplasts grow from callus-like tissue into normal leaf bodies through cell wall regeneration and repeated cell division. In the cases of other red algae, regeneration is difficult. Green alga protoplasts regenerate relatively easily and grow into normal organisms, but this is very difficult to accomplish with brown algae (Cheney et al. 1986; Le Gall et al. 1990; Garcia-Reina et al. 1991; Aguirre-Lipperheide et al. 1995; Reddy et al. 2008).

In some instances, protoplasts do not regenerate even when apparently normal *P. yezoensis* protoplasts have been obtained, due to protein denaturation from the protease treatment conditions in the first stage.

For this reason, the enzymes used in protoplast production must be examined beforehand. Attempts are currently under way to use protoplasts from this process to induce calluses in *L. japonica* for suspension culturing of the cells isolated from these calluses. It has also become possible to fix laver and sea lettuce protoplasts with beads made from polysaccharides such as alginic acid and agarose so that calluses can be induced within the beads and a leaf body grown on the bead's surface.

Protoplast isolation has also been attempted with leaf bodies from the red alga *P. yezoensis*. Protoplast production through the use of abalone or sea slug digestive enzymes, or with a combination of sea slug digestive enzymes combined with dolicellulase, has not been successful, but it has been possible to isolate protoplasts effectively through the use of extracellular enzymes from bacteria separated from decomposed laver. The sequence consists of the following steps (Fig. 6.6):

1. Apply ultrasound to *P. yezoensis* leaf body to remove materials adhering to surface.
2. Transfer leaf body to seawater (pH 6.0) containing 2% papain and let sit for one hour at room temperature.
3. Filter with nylon mesh (90 μm) and rinse remaining tissue fragments in seawater containing 0.5 M sorbitol.
4. Transfer to solution (pH 6.0 or 7.0) of bacteria-derived enzymes (or mixture of abalone and acetone powder with same) and let sit for three hours at room temperature.
5. Filter with 40 μm nylon mesh and place filtered solution in centrifuge for 2 min at 1500 rpm to recover protoplasts.
6. Suspend and centrifuge recovered protoplasts again in seawater with 0.5 M sorbitol and rinse.

With this method, 10^5–10^6 protoplasts may be obtained from 1 g of *P. yezoensis* leaf bodies (raw weight; see Fig. 6.7).

If produced very freshly, protoplasts obtained through method described above will be able to regenerate through culturing under optimal conditions. Green alga protoplasts are typically reported to reproduce well, with normal organisms forming through cell well formation and cell division. Apart from some varieties of the red alga laver, however, regeneration remains very difficult for red and brown algae. For this reason, many are anticipating the development of protoplasm production

Fig. 6.6 *P. yezoensis* Protoplasts

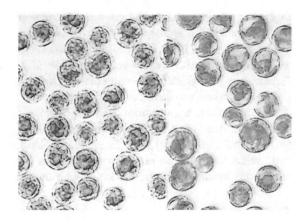

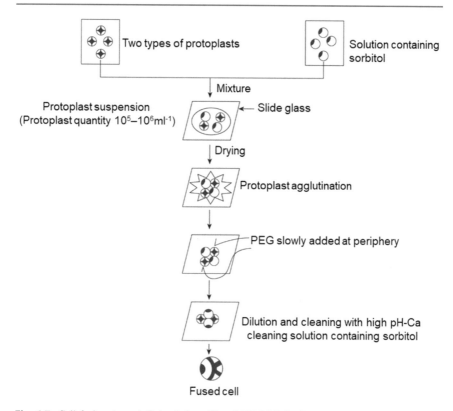

Fig. 6.7 Cell fusion through Polyethylene Glycol (PEG Method)

methods with higher survival rates or improvements in protoplast culturing methods. Examinations of protein growth using protoplasts from sea lettuce, kelp, and laver have found very few proteins to exist in the cell wall.

Protoplasts are also used for research on the biosynthesis and structure of seaweed polysaccharides and the production of aromatic constituents. Issues that will have to be addressed in future tissue culturing include identification and isolation of seaweed growth and differentiation factors, a more precise understanding of the cell wall's components and structure, and development of standardization techniques for cell wall zymolysis (Amano 1990).

6.5 Cell Fusion

Cell fusion is a very useful technique for improving strains of seaweeds that are difficult or impossible to cross-breed. Methods used for protoplast cell fusion in higher plants include the sodium nitrate ($NaNO_3$), high pH-high calcium, polyethylene glycol (PEG), dextran, electricpulse, and centrifugal force approaches;

among these, the high pH-high calcium, PEG, and electricpulse methods have been shown to be reproducible. Recent years have seen greater available of electrical cell fusion devices using electrofusion, or fusion through electrical stimulation. The pomato, created from the fusion of potato and tomato protoplasts, may be the best-known example of cell fusion in higher plants. Many other hybrids have also been produced through cell fusion recently, most notably with members of the nightshade family (Cheney et al. 1986).

Examples of cell fusion use in seaweed remain rare, with only minor reports from studies of freshwater algae. A recent successful and reproducible example in seaweeds was obtained from a syncytium produced through the PEG and electro-fusion methods. It remains difficult, however, to distinguish whether complete fusion has occurred even within seaweeds from the same genus and species, or between different species and families. Another issue concerns whether the new cell resulting from fusion will be capable of developing into a new seaweed organism. Development of new techniques for syncytium (hybrid) culturing and selection may become an important issue in the future (Waaland 1975).

Let us consider an actual example of seaweed cell fusion using protoplasts. Figure 6.8 shows the principles of cell fusion through the PEG method. Even with this fusion method, cell fusion occurs randomly, and cells of different species do not fuse at a 1:1 ratio. The fusion rate is around 10% for the same species and 15% for different species, but because fusion occurs within the same species or with two or more cells, target cells must be selected from these for culturing. In some cases, traits from both types may not be expressed in the fused cell, and the target trait may not be expressed. Many issues remain to be addressed in the selection of fused cells and subsequent regeneration through culturing, and continued research is expected to result in functional improvements going ahead.

As an example of the PEG method, cell fusion involving the red alga laver and the green alga sea lettuce occurs according to the following sequence (Fig. 6.8).

1. Protoplasts are separated from laver (a red alga) and sea lettuce (a green alga).
2. Protoplasts are mixed at a 1:1 ratio in a Tris-sorbitol buffer solution and adjusted to 10 per milliliter.
3. Two to three droplets of the mixture are transferred to an agar-coated glass slide and allowed to naturally draw for several minutes.
4. After verification that the protoplasts have agglutinated through drying, 0.1 ml of PEG is slowly added. [Preparation of PEG: 25% polyethylene glycol 6000, 0.7 M sorbitol, 0.1 M calcium chloride ($CaCl_2$), 0.1 M Tris, pH 10.0]
5. Rinse slowly with Tris-sorbitol buffer solution.

In this case, the protoplasts of laver (a red alga) will assume a reddish tinge and the protoplasts of sea lettuce (a green alga) a green tinge due to differences in pigments, which makes it easy to verify by microscope whether protoplast fusion has occurred. Because the red and green alga fusion occurs between very distance species, complete fusion of protoplasts will not be easy to achieve. It will also be difficult to conclude that fusion has occurred for any product (Reddy et al. 2008).

Fig. 6.8 Developing a new strain through cell fusion

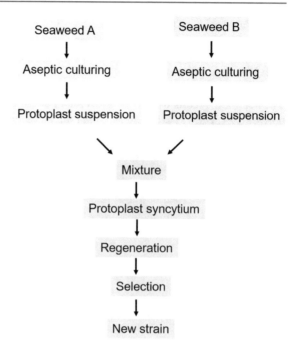

6.6 Gene Manipulation

Protoplast production, tissue culturing, and cell fusion have already reached the commercialization stage in the biotechnology of higher plants, and mass production through gene manipulation is under way. In contrast, the only example of genetic manipulation in algae concerns the ability to achieve transformation in the fresh-water unicellular green alga *Chlamydomonas* through the use of enzyme-derived plasmids as a vector; no examples at all have taken place for marine algae. At the same time, advancements are likely to be made in the future thanks to rapid progress in two areas essential to genetic manipulation, namely nucleic acid isolation and base sequence analysis (Hulata 2001).

In terms of DNA and RNA isolation, sperm DNA has been isolated from the alga *Porphyra yezoensis*, and leaf body DNA from the brown alga *Macrocystis*, and methods have been established for removing the polysaccharide polyphenol and isolating mRNA from lavers (red algae) and the green alga *Ulva intestinalis*. Restriction enzyme fragments of leaf body DNA from the green alga *Bryopsis*, the brown alga *Laminaria*, and the red alga *Porphyra yezoensis* have also been analyzed for use in seaweed line classification. Complete ribosomal RNA base sequences have been determined for four species of the red alga laver.

Nucleic acid base sequences will similarly be identified for other useful marine algae going ahead. Genetic manipulation also requires the development of vectors to transport foreign DNA into algae suited to specific purposes. Typical vectors are viral DNA and plasmids.

The success or failure of DNA experimentation can be seen as dependent upon whether good vectors can be found. Viruses are reported to exist for two to three brown alga species such as *Chorda*, but tumor-like cell reproduction is known to be taking place already through bacteria and fungi for 34 other varieties of seaweed. The possibility of developing bacterial plasmids or viruses to serve as vectors for seaweeds as they do for higher plants is raising hopes for the production of new strains through genetic recombination technology.

As yet, no new strain of transformed seaweed has been commercialized from cell fusion or genetic recombination. New strains of useful seaweeds are very likely to be produced with biotechnology in the near future in countries that are heavily dependent on them. For such new strains, it is essential to fully examine beforehand what effect their farming in natural waters will have on the ecosystem.

6.7 Future Tasks

Protoplast production and cell fusion in seaweeds has recently become the subject of very active research. For the sake of future development, however, research will need to involve not simply experimentation with protoplast production and cell fusion, but proceeding toward some target in terms of which traits to combine to produce new forms of seaweed.

Figure 6.9 depicts the process of selecting a new strain suited to a particular target from the hybrid seaweeds produced through separation and fusion of protoplasts from two varieties of seaweed (A and B) and direct or indirect (callus) regeneration from the fused cells.

While the sequence is quite simple, many technical issues remain to be solved at its individual stages, including protoplast isolation, protoplast fusion, and regeneration of fused cells. Many seaweeds have complex life cycles, and leaf bodies have both haploid and diploid nuclear phases. Because protoplasts from haploid seaweeds are haploid (n) and protoplasts from diploid seaweeds are diploid (2n), the fusion combination must be considered before any cell fusion experimentation. Another potential issue concerns whether the cytoplasm alone has fused or the nuclei as well; determining whether these have fused or not is not a simple matter.

In seaweed cell fusion, the goal must be clear, and knowledge of the seaweed's genes is necessary for advancement to occur. At present, however, knowledge of seaweed genes is scant, and basic research in this field is an area that will require development in the future.

Recent years have seen advancements in genetic manipulation technology, most notably in research areas involving microbes. This has led to the first consideration of genetic manipulation for breeding of higher plants. Many issues remain to be

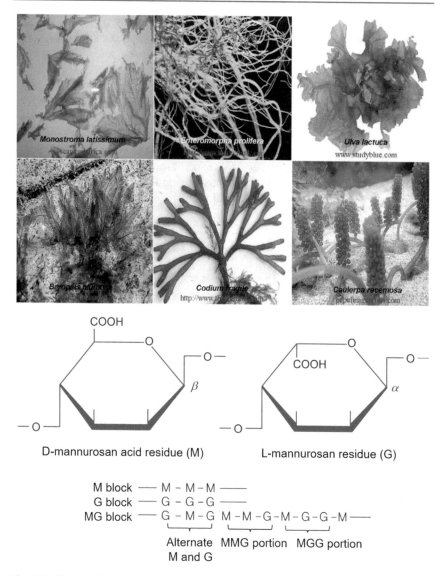

Fig. 6.9 The morphology of some representative species of green seaweed in living. Figure from (Wang et al. 2014)

solved in the future, including host and vector development and establishment of transformation lines.

While research in areas involving microbes and higher plants has been developing by the day, however, research into seaweeds has fallen markedly behind.

One approach under consideration for seaweed breeding with genetic technology involves isolating protoplasts from aseptically cultured seaweed tissue and using a suitable vector for transformation. This, however, will require swift development of related technology and techniques, including not only the DNA handling methods that are fundamental to genetic manipulation but also hosts and vector lines, vector introduction   methods,   selection   methods   for   transformed   cells,   and post-transformation cell management methods.

Future developments are anticipated in farming technology (using biotechnology to improve the genetic traits of useful seaweeds as desired for food production) and research on the production of energy and useful substances with seaweed biomass, specifically through the artificial formation of seaweed beds. Wide-ranging basic to applied research will be required on the use of cell fusion or genetic manipulation to produce seaweeds that can be put to use for human welfare, including highly productive and highly disease-resistant varieties and those that can serve as especially productive sources for certain useful substances.

6.8   Industrial Applications of Seaweeds

6.8.1   General Composition of Seaweeds

Table 6.7 shows the general combination of leading green, brown, and red algae. As it indicates, while the organisms consist of roughly 60–90% water, the remaining components once water is removed consist mostly of carbohydrates. Sugars in particular are the most important component, typically accounting for 50% of dry matter.

After carbohydrates, the next largest component is ash, which accounts for as much as 40%, although considerable variation can be seen among species. Proteins are relatively scarce, accounting for 15% or less of dry matter, although they may exceed 20% in green and red algae. Protein content is especially high in the freshwater green algae freshwater sea lattus *Prasiola japonica* (31%) and in the red alga *Porphyra yezoensis* (around 40%). Lipid content is quite low: for many green and red algae it is less than 1%, although it is slightly higher in brown algae.

The constituents of seaweeds are strongly influence by a variety of factors, including not only physical and chemical conditions of ambient water such as temperature and nutrient salts but also sunlight, seasonal conditions, growth environment, and the different parts of the seaweed.

Because seaweeds grow in water, they may absorb dissolved substances from that water, and they may use the accumulated substances to synthesize necessary components. As such, they are subject to influence by the physical and chemical properties of ambient water. Moreover, the basal metabolism of seaweeds is photosynthesis, which means that amount of sunlight is also a large influence.

Seasonal changes in the general composition of seaweeds are not only subject to effects from seasonal fluctuations in ambient water, but also closely related to internal factors stemming from the seaweed's growth and reproductive cycle (Ito and Hori 1989; Marinho-Soriano et al. 2006; Manivannan et al. 2008; Gosch et al. 2012).

Table 6.7 General composition of seaweeds (% of anhydrides)

Type	Protein	Lipids	Carbohydrates		Ash
			Sugars	Fibers	
Green Algae					
Sea lattus (Ulvaceae)[+]	11.7	0.1	64.3	11.2	12.7
Sea lattus (Ulvaceae)[+]	24.8	0.1	51.4	7.6	16.1
Ulva pertusa	5.8	0.1	74.3	6.2	13.6
Sea lattus (*Ulva* sp.)	5.8	0.1	71.3	5.9	16.9
Sea staghorn (*Codium* sp.)[+]	13.3	0.5	60.2	16.8	9.2
Sea staghorn (*Codium* sp.)[+]	13.8	0.3	43.5	9.8	32.6
Fresh wáter sea lattus (*P. japonica*)	31.3	0.1	57.9	5.5	5.2
Brown Algae					
Sea tangle (*L. japonica*)[+]	8.9	1.3	57.9	9.1	22.8
Sea tangle (*L. japonica*)[+]	12.6	0.8	39.3	10.6	36.7
Sea mustard (*C. costata*)	15.1	2.1	38.5	14.1	30.2
Sea mustard (*U. pinnatipida*)[+]	13.6	2.2	31.1	15.2	37.9
Sea mustard (*U. pinnatipida*)[+]	14.3	1.9	40.0	8.5	35.3
Sea Oak (*E. bicyclis*)[+]	12.2	1.3	62.6	7.1	16.8
Sea Oak (*E. bicyclis*)[+]	3.7	2.0	72.7	7.2	14.4
Gulf weed (*M. myagroides*)	7.9	1.2	43.6	16.0	31.3
Fusiforme (*H. fusiformus*)[+]	10.1	0.8	32.6	17.2	39.3
Fusiforme (*H. fusiformus*)[+]	6	1.5	30.2	23.9	38.4
(*S. siliquastrum*	11.8	9.6	34.6	21.2	22.8
(*S. thunbergii*)	8.2	1.0	20.7	39.0	31.1
Red Algae					
(*Acanthopeltis* sp.)	16.5	0.4	64.1	11.6	7.4
Irish moss (*C. ocellatus*)	19.1	0.2	51.3	1.3	28.1
(*Gymnogongrus*)	27.2	0.2	48.9	3.3	20.4
(*G. deblis*)[+]	17.2	+	63.2	14.4	5.2
(*G. deblis*)[+]	24.9	0.1	61.8	6.7	6.5
(*E. humifusa*)	7.2	0.1	72.2	6.2	14.3
(*G. furcata*)	19.5	0.2	49.9	0.9	29.5
(*P. yezoensis*)	41.6	0.5	40.0	6.1	11.8

*Members of the same species differ in basic constituents due to different habitats
**Sugar represent value when other components are subtracted from 100.0
[+]: 0.05% or less

As an example, Table 6.8 shows the general composition of the leaves and stalks of *Undaria pinnatipida*. Lipid content in the leaves increases between the first halves of May and June before subsequently decreasing. Protein content rises from April to early May before beginning a rapid decline in late May; by July, when the growth period for naturally occurring *U. pinnatipida* ends, protein content is down to one-third of its peak level. Fiber exhibits a tendency to increase proportionally with the reproductive period. Alginic acid content begins to rise in late May before

Table 6.8 Seasonal changes in the general composition of sea mustard *Undaria pinnatipida* (% of dry mass)

Harvesting date	Part	Proteins	Lipids	Carbohydrates		Ash
				Alginic acid	Fiber	
10-Apr	Leaf	21.6	1.7	–	1.6	36.5
	Stalk	15.1	1.2	–	4.2	38.6
25-Apr	Leaf	21.1	1.6	24.4	1.2	34.1
	Stalk	11.6	0.8	20.7	4.9	40.9
10-May	Leaf	29.6	1.9	24.7	–	40.0
	Stalk	7.0	1.4	24.0	–	43.9
24-May	Leaf	11.3	2.3	27.9	2.4	36.2
	Stalk	–	1.2	22.9	5.8	43.0
9-Jun	Leaf	11.8	2.5	30.7	3.6	32.5
	Stalk	3.9	1.2	25.9	6.3	38.1
26-Jun	Leaf	10.9	2.1	26.9	3.0	36.6
	Stalk	4.6	1.0	25.3	6.6	49
10-Jul	Leaf	8.4	1.1	28.9	3.3	34.6
	Stalk	4.9	1.1	27.5	–	42.4

peaking in late June. Ash does not vary in any fixed way, and changes over time are not large. Stalks show lesser lipid, protein, and alginic acid content and greater ash and fiber content than leaves, but seasonal changes do not necessary coincide with those seen in the leaves.

6.8.2   Seaweed Polysaccharides

Polysaccharides represent the chief general components of seaweeds besides water. Polysaccharides can be classified into three types according to their distribution within the seaweed: cell wall polysaccharides, which form a small fibrous crystal structure at the outermost layer; glycosaminoglycans, which form an amorphous gel over them; and storage polysaccharides located within the cell.

Seaweeds inhabit waters with large salt concentrations and swift currents. Polysaccharides play a major role in helping them adapt to this growth environment. For example, the seaweed's cell wall, which is made of polysaccharides, is thicker than that of a land plant, but also flexible and resilient. Additionally, seaweeds possess glycosaminoglycans between their cells that land plants do not have. The constituent sugars of these polysaccharides include sugars with carboxyl groups or sulfate groups. Their role is to select and absorb or exchange ions within the seaweed to maintain water content at a fixed level.

Seaweeds thus possess glycosaminoglycans that land plants do not, as well as storage polysaccharides that are unique to them. In addition to D-glucose, D-galactose,   D-xylose,   L-rhamnose,   L-arabinose,   D-glucuronic   acid,

D-galacturonic acid, their constituent sugars include uronic acids such as L-fucose, D- and L-3,6-anhydrogalactose, 6-O-methyl-D-galactose, D-mannuronic acid, and L-glucuronic acid.

In addition to having different constituent sugars, seaweed polysaccharides also have more complex physical and chemical properties and structures than those of land plants.

Because of the large distance between seaweed lines, differences in seaweed classification and line location, form, and life patterns may be connected in some way with polysaccharide composition. This may be seen as similar to the way in which differences in pigments are used as classification and line indicators for seaweed. In the broad sense, seaweeds represent a population associated with 13 phyla, but the three that account for the largest number and the most frequent use are green, brown, and red algae (Table 6.9).

Table 6.9 Polysaccharides produced by seaweeds

Seaweed type	Skeletal polysaccharides (Cell wall)	Muco-polysaccharides (Between cells)	Storage polysaccharides
Green algae	Cellulose I (Valonia) and Cellulose II (*Ulva* sp.) β-1,3-xylan • Caulerpaceae, Bryopsidaceae, • Ricaniidae, • Ostreobiaceae β-1,4-mannan • Codiaceae, • Polyphysaceae	Sulfated xyloarabinogalactan, • *Cladophora* • *Chaetomorpha,* • *Caulerpa* • *Codium* Sulfated glucuronoxylorhamnan • *Ulvaceae,* • *Monostroma* Sulfated glucuronoxylorhamnogalactan • *Acetabularia*	Amylose Amylopectin
Brown algae	Cellulose II Hemicellulose	Alginic acid • *Undaria* • *Laminaria* • *Eisenia* • *Ecklonia* • Macrocystin Fucoidans • *Fucus*	Laminaran
Red algae	Cellulose II Hemicellulose β-1,3-mannan [Porphyra] β-1,4-xylan [Porphyra]	Agar • *Gelidiales* • *Gigartinales* • *Ceramiales* Carageenan • *Gigartinaceae* • *Solieriaceae* • *Phyllophoraceae* Porphyran • *Porphyra*	Red algae starch

Seaweed polysaccharides are divided into skeletal polysaccharides, viscous polysaccharides, and storage polysaccharides (Percival 1979; de Souza et al. 2007; Lahaye and Robic 2007; Costa et al. 2010).

A. Skeletal polysaccharides

The chief polysaccharide making up the seaweed's skeleton is cellulose, but constituents differ among seaweed species. Because seaweeds grow in sea water, they need the flexibility and resilience to withstand intense waves and currents.

(1) Green Algae

For green algae, three types of skeletal polysaccharides exist.

1. In sea lattus, *Ulva* sp., *Enteromorpha* sp., and *Valoniapsis* sp. (Valonia), X-ray analysis has shown the chief constituent to be true cellulose.
2. As observed in Cladophorales and Siphonocladales, they include the cellulose I of *Chaetomorpha*, with its parallel array of glucans and clear X-ray diffraction image, and the cellulose II of *Ulva* sp. and *Monostroma*, which has some glucan molecules in the opposite direction and an unclear diffraction image.
3. In place of cellulose with an unclear diffraction image, *Codium* and *Acetabularia ryukyuensis* polysaccharides include hemicellulose such as β-1,4-mannan, while *Caulerpa okamurai* and *Bryopsis plumose* include hemicelluloses such as β-1,3-xylan.

(2) Brown and Red Algae

Brown and red algae have a structure of cellulose II and hemicellulose. The structure in *Porphyra* is slightly different. True red alga classes (not including protozoans) have a dry mass of around 1–9% cellulose.

Xylan can be obtained through application of hydrochloric acid and sulfuric acid to *Porphyra umbilicalis*, *Laurencia pinatifida*, and *Rhodymenia palmate*. Structurally, these are intermediate between the xylan in land plants and green algae. Protozoan red alga classes have a structure formed with glycan and mannan rather than cellulose.

B. Mucopolysaccharide

The mucopolysaccharide that serve a replenishing function between cells also differ among seaweed species.

(1) Green algae

The mucopolysaccharide contained in green algae necessary include ester sulfates. They can be grouped into the following three categories by their constituent sugars:

1. Sulfated glucuronoxylorhamnan (present in *Ulva lactaca* and *Entermorpha compressa*)
2. Sulfated xyloarabinogalactan (present in *Codium* and *Caulerpa*)
3. Sulfated glucuronoxylorhamnogalactan (present in *Acetabularia).*

(2) Brown algae (McCandless and Craigie 1979).

These are polysaccharides unique to seaweeds, with a very broad range of usages.

1. Alginic acid: This is an acidic polysaccharide contained in *Undaria pinnatifida*, *Laminaria japonica*, *Eisenia bicyclis*, and *Ecklonia cava* and a copolymer polysaccharide consisting of uronic acids such as β-D-mannuronic acid and α-L-guluronic acid. Alginic acid is reported to only exist in seaweeds, but some species of bacteria, including *Azotobacter uinelandii* and *Pseudmonas aeruginosa* produce mucopolysaccharide that are similar to approximately 20% acetylated alginic acid.
2. Fucoidans: These are present in *Fucus*, *Laminaria*, and *Undaria*, with L-fucose and sulfated esters as their main components.
3. Sargassans: Present in *Sargassum limifolium*, these sulfated polysaccharides are very similar to fucoidans but differ in containing mannose.

(3) Red algae

Red algae contain numerous useful polysaccharides that are very widely used for industrial purposes (Stevenson and Furneaux 1991; Usov 2011).

1. Agar: Agar is contained in Gelidaceae and Kkosiraegi consists of agarose and agaropectin. Agarose consists of two kinds of agarobiose made up of D-galactose and 3,6-anhydro-L-galactose connected in a straight chain by an D-1,3 linkage.
2. Carrageenan: Present in *Eucheuma* and <u>*Chondrus*</u>, carrageenan possesses more D sulfate groups such as anhydrogalactose than agar. It consists of D-mannuronic acid and L-glucaric acid.
3. Alginic acid: Known to be present at a rate of around 2% in the red alga *Serraticardia maxima* and to not exist in brown algae, alginic acid was first found in red algae.
4. Funoran: Isolated from *Gloiopeltisfurcata*, funorans have D-galactose and 3,6-anhydrogalactose linked together alternately in a structure similar to agarose.
5. Porphyran: Found in *Porphyratenera* et al., porphyran has a structure similar to a combination of agar and carrageenan.

C. **Storage polysaccharides (Renn** 1997)

(1) Green algae

These polysaccharides are similar to the starch found in land plants and consist of around 16–27% amylose, with amylopectin accounting for the remainder.

(2) Brown algae

The chief example is laminaran (β-glucan), which is hot water soluble. Content is as high as 25% in *Laminaria*. Laminaran has been found to be synthesized from the photosynthensis product mannitol. In *Eisenia bicyclis*, laminaran has a molecular chain structure with β-D-glucopyanose connected by 1,3 and 1,6 linkages.

(3) Red algae

Known as red algae starch, these polysaccharides have been isolated from the Gelidiopsis *Furcedaria fastigiata* and the Constantinea *Constantinea sabutinea*. From their physical and chemical characteristics, they are believed to fall midway between the higher plant starch amylopectin and the glycogen found in animals. Pure red algae starch has been obtained through purification of *Serraticardia maxima*, which flourishes in areas that are profuse with calcareous algae due to seaweed withering on seaside rocks. Precursors for the red algae starch biosynthesis mechanism are reported to include depolymerized compounds that have become the subject of recent attention, including trehalose, floridoside, and isofloridoside.

6.8.3 Alginic Acid

Alginic acid is a polysaccharide contained in brown algae. The substance was named alginic acid after *alga*, from the Latin for "seaweed." Alginic acid was discovered by the Scottish scientist Stanford in 1884 while examining the chemical constituents of the brown alga *Ascophyllum*. After pre-treatment of brown algae with dilute hydrochloric acid, acidification through the addition of hydrochloric acid to a viscous solution extracted from an aqueous solution of alkaline chemicals such as sodium carbonate or sodium hydroxide results in the formation of white, fibrous precipitate, which can be dried to obtain alkali salts of alginic acid. Alginic acid is white and odorless and exhibits acidity due to a separated carboxyl group. The name alginic acid is used to refer to sodium alginate, which has an especially broad range of uses.

Alginic acid is present in all brown algae, albeit in different quantities. For an algae to serve as an alginic acid source for industrial purposes, it must meet several conditions: large volumes, mass colony formation, growth in environments where harvesting is easily performed, and not detracting from its functioning as a seaweed forest.

Table 6.10   Alginic acid content of brown algae (% of dry mass)

Algae type	Alginic acid (%)	Algae type	Alginic acid (%)
L. japonica[a]	17.05	U. pinnatifida[a]	27.06
L. japonica[a]	22.54	U. pinnatifida[a]	28.76
L. japonica Aresch[a]	25.43	E. bicyclis	17.87
L. japonica Aresch[a]	22.00	H. fusiformis[a]	20.48
Saccharina gyrata[a]	30.17	H. fusiformis[a]	20.15

[a]Same species inhabiting different environments

From this perspective, around two of ten brown algae orders are typologically capable of meeting these conditions, and around ten species are actually used.

As shown in Table 6.10, alginic acid consists of up to around 30% of brown alga dry mass, with variations according to brown alga type and time of harvesting. Brown algae containing large amounts of alginic acid are a favored food for algivorous fauna such as abalone, which characteristically possess alginase in their viscera from their juveline shell period (Yamada 2001).

A. **Alginic Acid Chemical Structure and Physico-Chemical Composition and Production**

As shown in Fig. 6.10, alginic acid consists of an M block (made up of D-mannurosan residue (M)), a G block consisting of D-guluronic acid residue, and MMG and MGG blocks with alternating residues. Ratios vary according to brown alga species, the part of the seaweed, the season, and other factors. The intensity of the M-to-G ratio of 1.5–2.5 appears as follows when compared with gels made with other metals.

$$\text{Lead} > \text{copper} > \text{cadmium} > \text{barium} > \text{strontium}$$
$$= \text{calcium} > \text{cobalt} > \text{nickel} > \text{zinc} > \text{manganese}$$

The difference in gel intensity is also closely tied to usage value.

Alginic acid is thus a macromolecule electrolyte that is also a weakly acidic positive ion exchanger. In particular, it exhibits selective exchange for iron (iii) and copper (ii). It forms insoluble salts with metals with a valence of (ii) or greater (apart from mercury) and soluble salts with other metals, resulting in a viscous solution. Molecular weight differs by seaweed type and harvesting date, but ranges between 200,000 and 2,000,000. Depolymerization occurs during cooking due to vinegar and heat exposure. When used as a material in processed food, alginic acid's viscosity must be reduced and its solvency increased through depolymerization of the natural substance. For this reason, alginic acid that is currently sold has its solubility increased through hydrolysis to depolymerize to an average molecular weight of around 50,000 ± 10,000. Derivatives are also in use, including propylene glycol alginate (PGA), which includes a propylene glycol group.

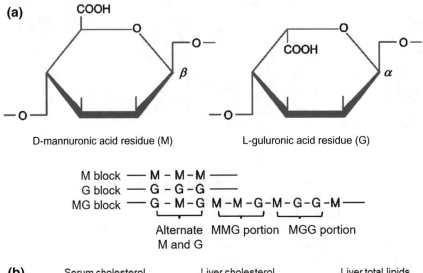

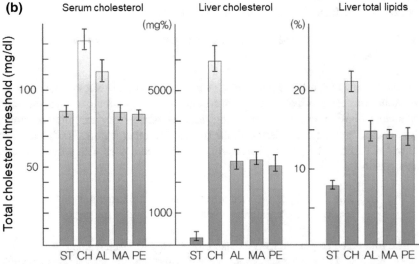

Fig. 6.10 (a) Mimetic diagram of alginic acid block structure (b) Blocking of serum and liver cholesterol levels by Sodium Alginate, Cutin, and Konjac Mannan. ST: zero-cholesterol food, CH: food with cholesterol, AL: 5% sodium alginate, MA: 5% konjac mannan, PE: 5% cutin

Seaweeds also contain insoluble salts such as calcium and magnesium, although these are solubilized through transposition with sodium or potassium with alkali treatment.

The first industrial production of alginic acid took place in the United States in 1926. Since then, it has been produced at large-scale factories in the United Kingdom, Norway, France, and elsewhere. Basic research began in Japan in 1928, serving as a basis for industrial production since 1938. Today, around 15 factories are in operation, including one each in the United Kingdom, United States, Norway, and Chile; two each in France and Japan; and various larger and smaller ones in China. Altogether, around 35,000 tons of alginic acid is being produced throughout the world.

The current Korean alginic acid market stands at around 3500 tons a year. Around 3300 tons are reportedly used for printing, 100 tons for food, and 60 tons for machine industry. All of it is produced by a single company (Rowley et al. 1999; Davis et al. 2003; Nickerson and Paulson 2004; Iwamoto et al. 2005; Augst et al. 2006; George and Abraham 2006; Klemmer et al. 2012; Lee and Mooney 2012).

Mimetic diagram of alginic acid block structure (Table 6.11).

B. **Uses of alginic acid**

Applications in Food-Related Areas: Alginic acid is recognized by the U.S. Food and Drug Administration (FDA) as safe for general food consumption, a standard that is also applied in South Korea.

Alginic acid is used in food because it possesses properties such as water retention, gelling, and thickening stabilization. As Table 6.12 shows, it has a broad range of uses in foods that take advantage of these characteristics, and is used in the forms of propylene glycol alginate, sodium alginate, and alginic acid. While it is not indicated in the table, alginic acid is also used for the membranes of artificial salmon roe, which is created by suspending encapsulation in alginic acid soda slime and lowering it into congealing solution from a suitable height. Additionally, alginic acid is used in producing fibrous proteins. Its fiber plasticity is used to spin protein mixtures and create fibrous protein products. Examples include the production of artificial fibrous meats mixed with proteins from sardines, krill, plasma, and casein.

(1) Applications in Medical-Related Areas

   1. Tooth mold framework production: As shown in Table 6.13, alginic acid is used when making molds for false teeth. Agar, a polysaccharide derived from red algae, may also be used as an impression agent for these molds. Table 6.13 shows the constituents of these impression agents. They are referred to as mixed impression materials. Agar impression agents are often used for the inner tooth mold and alginic acid impression agents for the outer mold, though they may also be used independently.
   2. Pharmaceutical coating: Alginic acid is used as a coating or styptic agent when treating ulcers in the esophagus, stomach, and duodenum.
   3. Surgical thread: Alginic acid is also used in stitches applied after surgery. In such cases, the stitches do not need to be removed. It is also used as a dispersing agent for refined components.

Table 6.11 Major uses of alginic acid and its derivatives

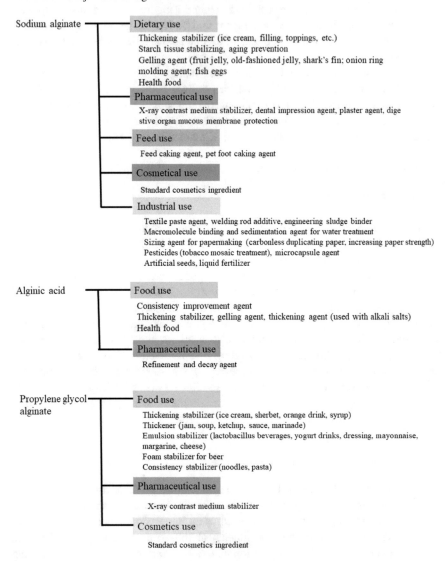

Sodium alginate

Dietary use
Thickening stabilizer (ice cream, filling, toppings, etc.)
Starch tissue stabilizing, aging prevention
Gelling agent (fruit jelly, old-fashioned jelly, shark's fin; onion ring
molding agent; fish eggs
Health food

Pharmaceutical use
X-ray contrast medium stabilizer, dental impression agent, plaster agent, dige
stive organ mucous membrane protection

Feed use
Feed caking agent, pet foot caking agent

Cosmetical use
Standard cosmetics ingredient

Industrial use
Textile paste agent, welding rod additive, engineering sludge binder
Macromolecule binding and sedimentation agent for water treatment
Sizing agent for papermaking (carbonless duplicating paper, increasing paper strength)
Pesticides (tobacco mosaic treatment), microcapsule agent
Artificial seeds, liquid fertilizer

Alginic acid

Food use
Consistency improvement agent
Thickening stabilizer, gelling agent, thickening agent (used with alkali salts)
Health food

Pharmaceutical use
Refinement and decay agent

Propylene glycol
alginate

Food use
Thickening stabilizer (ice cream, sherbet, orange drink, syrup)
Thickener (jam, soup, ketchup, sauce, marinade)
Emulsion stabilizer (lactobacillus beverages, yogurt drinks, dressing, mayonnaise,
margarine, cheese)
Foam stabilizer for beer
Consistency stabilizer (noodles, pasta)

Pharmaceutical use
X-ray contrast medium stabilizer

Cosmetics use
Standard cosmetics ingredient

(2) Applications in Textile Industry-Related Areas

In textile dyeing, color sometimes changes as gelling agents react with the fibers.
This rarely happens with alginic acid thickening agents. Sodium alginate paste is
idea for dye printing, making it ① suitable for textile printing. Other advantages
include ② its markedly low chemical reactivity, ③ its ability to maintain reliability
viscosity during processing, ④ its stability for all dye colors, ⑤ its ability to wash
off easily, and ⑥ its protection of expensive screens from scratching or staining.

Table 6.12 Recent uses of alginic acid in food (Ocio et al. 1996; Ocio et al. 1997; Pranoto et al. 2005; Bierhalz et al. 2012)

Property used	Major uses	Main types used
Aging prevention and improvement of noodle quality	Added to flour to prevent noodles from breaking; reinforcing noodle stripes; improving bread texture; improved texture, noodle stripes, and degreasing in instant noodles	Sodium alginate, alginic acid
Water holding capacity	Preventing dehydration in high-sugar foods, preventing dehydration in jams and ketchups	Propylene glycolester alginate (PGA), sodium alginate, alginic acid
Viscosity enhancer	Increasing noodle soup viscosity, dispersing and stabilizing solid components in marinade and jam, contributing elasticity to surimi gel, improving texture by adding viscosity to puddings and jellies, thickening and shaping dairy products	PGA, sodium alginate, alginic acid
Acid resistance	Preventing milk protein precipitation in lactobacillus beverages, dispersing solids in thick juices	PGA
Salt resistance	Thickening and dispersing agent for soy and other sauces	PGA
Emulsion stability	Maintaining overrun in ice cream and sherbet, maintaining emulsion state in mayonnaise and dressings	PGA, sodium alginate
Foam stability	Stabilizing foam in beer	PGA
Agglutination	Removing dregs, microbes from brewed beverages	Sodium alginate
Film plasticity	Preventing moisture evaporation from food surface coating, preventing moisture transfer when two types of food with increasing moisture content settle, improving preservability through noodle surface processing, adding luster, pre-coating for freezing of seafood	Sodium alginate
Heat resistance	Improving agar heat resistance, shaping and gelling of dairy products, adding heat resistance, fixing jelly contents	Sodium alginate
Gelling	Copying products fish eggs, shark's fin, truffles, and cheese, binding agent for reconstitution of livestock and fish meat; fixing flavoring agents; preventing deformation in old-fashioned jelly, agar	Sodium alginate

Table 6.13 Composition of agar and alginic acid impression agents

Ingredient	Percentage (%)
Agar impression agent composition	
Agar	12.5
Boric acid	0.2
Potassium sulfate	1.7
Alkylbenzoic acid	0.1
Water	85
Composition of impression agentalginic acid	
Potassium alginate	18
Potassium sulfate	14
Sodium phosphate	2
Zinc oxide additives • Silicate • Borate • Fluoride	10
Excipient, silicious matter	6
Water	50

(3) Applications in Industrial-Related Areas

Because alginic acid binds optionally with iron ions, it has long been used for boiler management and the removal of sediment and extraneous matter. Sodium alginate is used at concentrations of 1–3 ppm. Recently, however, the use of alginic acid salts has been threatened by the emergence of outstanding chelating agents that include surfactants.

(4) Bioreactor Immobilizing Agent

Alginic acid's properties as a cell inclusion agent are being put to use as an immobilization agent for bioreactors. Cells are suspended in a sodium alginate solution and added to a multivalent metal ion solution to form beads. Research has been reported on this method's use in producing ethyl alcohol and vinegar. Examples are shown in Table 6.14.

(5) Uses in Artificial Seeds

Alginic acid is used to embed adventitious embryos and buds propagated by tissue culturing into capsules that confer seed functions and can be sown directly in fields.

C. **Physiological Functions of Alginic Acid**

(1) Lowering Cholesterol

The cholesterol contained in the food that we consume can be encircled by a network of sodium alginate molecules and expelled from the body, thereby suppressing its absorption by the intestines and lowering cholesterol in the blood.

Table 6.14 Uses of cells immobilized with alginic acid (Smidsrod and Skjak-Brak 1990)

Cell	Product/Target
Bacteria	
• *Erwinia rhapontici*	Isomaltose
• *Pseudomonas denitrificans*	Drinking water
• *Zymomonas mobilis*	Ethanol
Blue-green algae	
• *Anabena* sp.	Ammonia
Filamentous bacteria	
• *Kluyveromyces bulgaricus*	Whey hydrolysis
• *Saccharomyces cerevisiae*	Ethanol
• *Saccharomyces bajanus*	Champagne
Green Algae	
• *Botryococcus braunii*	Hydrocarbons
Plant cells	
• *Chatharanthus roseus*	Alkaloids
• *Daucus carota*	Alkaloids
• Various other plants	Artificial seeds
• Plant protoplasts	Cell processing/microscope
Mammalian cells	
• Hybridomas	Monoclonal antibodies
• Islets of Langerhans	Insulin/transplantation
• Fibroblasts	Interferon-β
• Lymphomas	Interferon-α

Research has confirmed that raising white mice with food triggering hypercholesterolemia (including 1% cholesterol and 0.25% bile acid) combined with 5% sodium alginate showed effects in preventing rises in concentrations of serum cholesterol, cholesterol in the liver, and total lipids and fatty acids. PGA also has an effect in suppressing an increase in serum cholesterol, but that effect has been confirmed not to be present in propylene glycol itself. Reports have additionally shown that rats given food mixed with cholesterol showed a rise in their cholesterol level over the short term, but that the increase was suppressed (albeit weakly) compared to konjac mannan or pectin when 5% sodium alginate was mixed in (Jiménez-Escrig and Sánchez-Muniz 2000; Yoshida et al. 2004; Draget et al. Draget 2005).

To examine the effects of sodium alginate of different viscosities (high and low) on intestinal flora and serum lipids, male rats were given food containing 2% alginic acid salts for seven days. No significant effect was observed for the high viscosity condition, but pH in the appendix was markedly reduced in the low-viscosity condition, and serum triglyceride and cholesterol levels decreased (Fig. 6.10). The findings suggest that food containing sodium alginate of different viscosity has differ effects on intestinal flora and serum lipid levels.

(1) Reducing Hypertension

Hypertension is reported to result from excessive salt consumption. This results from contraction of the blood vessels as the balance between sodium and calcium

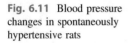

Fig. 6.11 Blood pressure changes in spontaneously hypertensive rats

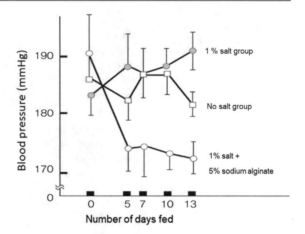

ions in the absorbed salts breaks down. As with other foods containing salts, when potassium alginate is absorbed in the body, potassium is removed from the alginates due to gastric acid within the stomach. Because the small intestine is weak alkaline, alginic acid transferred to the intestines combined with some of the sodium ions also consumed and expelled from the body as waste. This occurs due to the ion exchange reactions chemically characteristic of alginic acid that take place in digestive organs. Potassium ions removed from alginic acid are absorbed by the interests and drive out sodium from the blood. Alginic salts thus reduce blood pressure in two ways. As models for essential hypertension, spontaneously hypertensive rats (SHRs) have been developed. These rats, which enjoy salt and possess genes to increase blood pressure, were shown to have that increase blocked when given potassium alginate, regardless of their salt consumption (Fig. 6.11).

As shown in Fig. 6.12, this occurs because pH variation in the digestive organs results in ion exchange reactions; the positive ions that combined with alginic acid are separated due to the stomach's acidity and recombined with positive ions in the small intestine due to its non-alkalinity. Any excess sodium at this point bonds more easily than potassium; sodium absorption is reduced, and expulsion as feces is promoted. The potassium shed in the stomach also contributes to lowering blood pressure as it is absorbed in the small intestine.

(3) Enzyme Activation

Reports shows the consumption of sodium alginate with food to be associated with increased amylase and protease activity in the intestines.

(4) Intestinal Regulation

Alginic acid is known to be excellent for intestinal regulation due to its very high hydration and water retention capabilities, lubrication, and plasticity.

(5) Suppressing Cancer Cell Proliferation

The anti-cancer properties of polysaccharides first began drawing the attention of cancer researchers with a 1891 case in which modified toxins from suppurating

Fig. 6.12 Changes in alginic acid negative ion bonding in the digestive organs

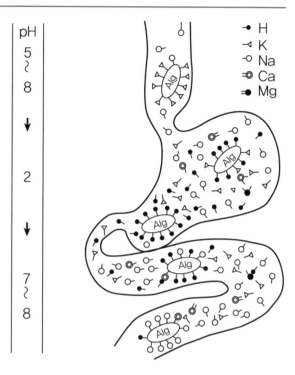

Streptococcus and *Serratia macescens* culturing filtrate were shown to have a marked effect when used with inoperable cancer patients, and with a 1936 study showing the effective ingredient to be polysaccharides contained with the bacteria. In terms of anti-cancer polysaccharides found in seaweeds, anti-cancer activity screening tests of various seaweed-derived polysaccharides in 1959 reported marked effects when alginic acid (prepared from kelp) and mesogloea (a polysaccharide from *Mesogloea divaricate*) were introduced within the abdominal cavities of cancerous mice (which had sarcoma 37 cells transplanted in them).

Injection of alginic acid refined from sargassum into the abdominal cavities of mice implanted with sarcoma 180 solid tumor cells resulted in a 50% reduction in growth for those cells. Proliferation of sarcoma solid tumor cells was also found to be suppressed when alginic acid marketed as reagent was introduced in the abdominal cavity (Yamada 2001).

D. Alginic Acid Oligosaccharide Functions

The alginic acid polysaccharides that are the source of alginic acid oligosaccharides are found in brown algae (most notably kelp, sea mustard, and hijiki). They are naturally acidic with a characteristic molecular weight of 2.7 million, and exist mostly between the seaweed's tissue cells in calcium salt form, keeping the seaweed flexible and tough. Alginic acid polysaccharide content is between 20 and 50% of the dry seaweed mass and consists of straight chain polymers made up of two types of uronic acids, namely D-mannuronic acid with a β-1,4 linkage (M) and L-guluronic acid with an α-1,4 linkage (G).

To date, the use of alginic acid polysaccharides has involved the sodium salt form's application to bread, noodles, and mayonnaise as a thickener and gelling and stabilizing agent. Recently, however, they have also been used in imitation shark's fin and caviar and as a foam improvement agent in beer. As can be seen from their use as a thickening agent, alginic acid polysaccharides are highly viscous and do not readily dissolve in water, which presents some limitations in their application as a food ingredient.

Recent years have seen the development and marketing of alginic acids reduced to a molecular mass of 50,000 $\pm$ 10,000 through heating hydrolysis under pressure (hereafter "depolymerized alginic acids"), with alginic acid polysaccharides found in foods designed for specific health benefits such as lowering of cholesterol and intestinal regulation. Alginic acid oligosaccharides are produced not through pressure and heat application to alginic acid polysaccharides, but through the use of naturally microbe-produced enzymes to break those polysaccharides down to a molecular weight below 1000. This section presents an introduction to the characteristics and physiological functions of alginic acid oligosaccharides (Chaki 2005; Kawada et al. 1999; Yoshida et al. 2004; Uno et al. 2006).

(1) Alginate Lyase-Producing Bacteria

Since the first reports by Waksman in the 1930s, research on the isolation of bacteria producing enzymes that break down alginic acid polysaccharides has been conducted on various subjects, including mollusks, seawater, the intestinal content of saltwater fish, land organisms, and decomposed brown seaweed. This section will examine the production and characteristics of alginic acid oligosaccharides using enzymes obtained from ocean bacteria.

(2) Characteristics of Alginic Acid Oligosaccharides

1. Structure: Alginic acid oligosaccharides may be identified as shown in Fig. 6.14 through placement in a DEAE-HPLC (Cosmogel DEAE glass, 8 × 75 mm), resulting the appearance of eight kinds of peaks (see Fig. 6.13; these have been labeled P1, P2, P3, P4, P5, P6, P7, and P8 by emission order) that can be subjected to MS and NMR analysis.

2. Nutritional Component Analysis Values: The analysis values for the nutritional components in alginic acid oligosaccharides show almost no change from the alginic acid polysaccharide sodium salts (hereafter alginic acid polysaccharides) that provide their material, as the reaction is a simple breaking down of those polysaccharides with enzymes (Table 6.15).

3. Physiochemical Properties: Table 6.16 shows physiochemical properties of alginic acid oligosaccharides. They show little difference from alginic acid polysaccharides in terms of taste, smell, or pH, but can be seen to have much lower viscosity and color acceptance. Viscosity in particular was greatly reduced from 810 mPa s (25 °C) in a 2% solution of alginic acid polysaccharides to 55 mPa s in a 75% solution of alginic acid oligosaccharides. The value is nearly identical to the 41 mPa s for a 75% sucrose solution, indicating that alginic acid oligosaccharides may be used as a

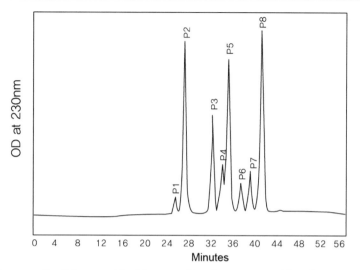

P1 : △ – G, P2 : △ – M, P3 : △ – G – G, P4 : △ – M – G, P5 : △ – M – M,

P6 : △ – G – G – G, P7 : △ – M – G – G, P8 : △ – M – G – M

Fig. 6.13 HPLC pattern for alginic acid polysaccharides

food product with only their water solubility and viscosity reduced and no change in nutritional composition.

(3) **Physiological Functions**

To date, reports on the physiological functions of the alginic acid polysaccharides that serve as a source for alginic acid oligosaccharides have indicated reductions in cholesterol and blocking of blood pressure increases. In the case of alginic acid oligosaccharides, a broad range of other research reports can be found on effects in multiplying *Bifidobacterium*, promoting growth in barley sprouts, improving solubility and stability in fish fibril proteins, strengthening gamma interferon (IFN-γ) production, checking immunoglobin E (IgE) production, and improving gelatin functions. This section introduces physiological functions that have been confirmed in alginic acid oligosaccharides to date.

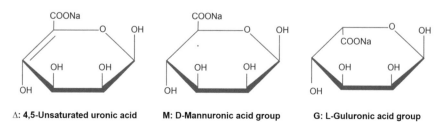

△: 4,5-Unsaturated uronic acid   M: D-Mannuronic acid group   G: L-Guluronic acid group

Fig. 6.14 Structure of alginic acid oligosaccharides

Table 6.15 Nutritional component values for alginic acid polysaccharides and oligosaccharides (per 100 g)

	Alginic acid oligosaccharides	Alginic acid polysaccharides
Water (g)	3.6	11.6
Protein (g)	0.6	0.3
Carbohydrates (g)	69.8	64.3[a]
Lipids (g)	0	0
Ash (g)	24.7	23.8
Salt (g)	0.8	-
Sodium (g)	10.7	-
Potassium (g)	0.2	-
Calcium (g)	0.06	-

–Not analyzed
[a]*Includes plant fibers*

Table 6.16 Physiochemical properties

	Alginic acid oligosaccharides	Alginic acid polysaccharides
Form	Powder	Powder
Color	White	White to light yellow
Color acceptance	0.14 (30% solution, OD420 nm)	0.26 (2% solution, OD420 nm)
Turbidity	0.02 (30% solution, OD660 nm)	0.13 (2% solution, OD660 nm)
Taste	None	None
Smell	None	None
pH	7.0 (2% solution, room temperature)	7.2 (2% solution, room temperature)
Viscosity (mPa s)	55 (75% solution, 25 °C)	810 (2% solution, 25 °C)

1. Checking Blood Pressure Increases

The effects of alginic acid oligosaccharides on blood pressure have been observed in seven-week-old male spontaneously hypertensive rats. An experimental group was given food mixed with 4.0% alginic acid polysaccharides for free consumption, while the control group was given food with 1.2% sodium chloride to provide equal sodium content. The breeding environment included room temperature of 23 ± 3 °C, moisture of 30–60%, 12 h each of light and darkness, and free access to water. Noninvasive blood pressure measurement conducted each week showed suppression of rising blood pressure in the experimental group by the fourth week after consumption began. By the eighth week, a significant continued suppression of the rise in blood pressure was observed (Fig. 6.15).

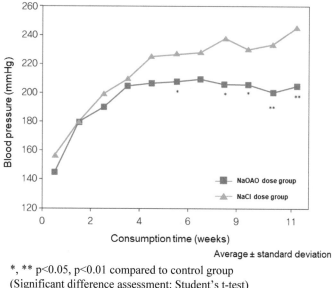

*, ** p<0.05, p<0.01 compared to control group
(Significant difference assessment: Student's t-test)

Fig. 6.15 Blood pressure increase suppression effects of alginic acid oligosaccharides

Another experiment conducted with the suppression of blood pressure increases using alginic acid oligosaccharide potassium salts and calcium chloride showed that those salts, like the alginic acid oligosaccharide sodium salts, also had an effect in blocking an increase in blood pressure. The findings show that the effect in checking increased blood pressure is not due to the metal ions (sodium and potassium), but is released to the sugar component (Hiura et al. 2001; Terakado et al. 2012; Moriya et al. 2013; Chaki 2005).

2. Effects on Human Blood Pressure

Since effects in suppressing increases in blood pressure have been observed in animal experiments, possible effects in lowering blood pressure in humans have also been investigated. In a preliminary experiment, 15 unmedicated mild hypertension patients (14 male and one female) were divided into an experimental group of 10 and a placebo group of five. In a double-blind study, the experimental group was instructed to take 10 grams of alginic acid oligosaccharides per day in beverage form for eight weeks. The results showed systolic blood pressure for the experimental group decreasing substantially from 153.5 ± 6.7 to 137.4 ± 11.7 mmHg after eight weeks, which was a significant difference from the placebo group. Blood pressure was also observed to continuously decline afterwards (Fig. 6.16). From this, it was determined that alginic acid oligosaccharides are also effective in lowering blood pressure in humans, and that 10 g represented an effective dose (Ueno et al. 2012).

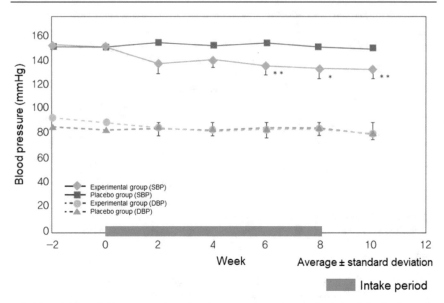

*, ** p<0.05, p<0.01 compared to placebo group (variation analysis by Scheffe showed significance level of 5% for both sides)

Fig. 6.16 Blood pressure changes while taking alginic acid oligosaccharides

In a subsequent study to examine an optimal amount, effects were observed in 48 experimental subjects who exhibited systolic blood pressure (SBP) of 130–159 mmHg and diastolic blood pressure (DBP) of 85–99 mmHg and were not taking medications that affected blood pressure. These were divided into four groups of 12 each receiving 2.5 g of alginic acid oligosaccharides per day (NaAO 2.5 g group), 5.0 g per day (NaAO 5.0 g group), 10 g per day (NaAO 10 g group), and a placebo (NaAO 0 g group). Findings for the three-month double blind comparative study showed a large standard deviation and unclear dosage reliability due to the small sample size of 12, but decreases or decreasing tendencies in blood pressure were observed for all groups that took alginic acid oligosaccharides (Chaki 2005).

3. Cholesterol Reduction Effects

Based on preliminary test showing lipid metabolism improvements in animals given alginic acid oligosaccharides, possible effects in improving cholesterol levels in human through alginic acid oligosaccharide consumption were examined, taking advantage of the fact that six out of ten experimental group members in the pre-liminary human blood pressure experiment related in (2) also had cholesterol above the threshold (220 mg/dl). Values were observed to decline from 258.5 ± 24.1 (mg/dl) at the consumption start date to 232.8 ± 15.2 (mg/dl) by the eighth week of consumption, a difference of 25 (mg/dl). This suggests the same cholesterol level

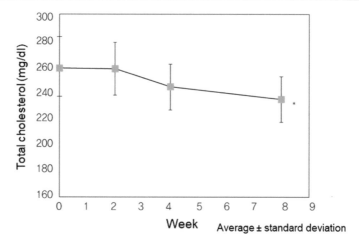

*: p<0.05 compared to level before consumption (corresponding t-test)

Fig. 6.17 Effects of alginic acid oligosaccharides on total cholesterol

improvements observed in animal testing also appear in humans (Fig. 6.17) (Brownlee et al. 2005; Chaki 2005).

4. Other Effects

In addition to the physiological effects of alginic acid oligosaccharides described above, other effects have been observed in the skin, including promoting the multiplication of epidermal growth factor (EGF)-dependent keratinocytes and vein endothelial growth factor (VEGF)-dependent epithelial cells in human blood vessels.

6.8.4   Fucoidans: Functions and Health Food Applications

A. What Are Fucoidans?

The term fucoidans refers to sulfated polysaccharides contained in brown algae that contain L-fucose among their constituent sugars. While fucoidan is typically present in all brown algae, the rate varies by location and extent of growth even within the same seaweeds. Fucoidans with different chemical structures are also found within seaweeds of the same species, and rates of fucoid presence are reported to differ markedly among different seaweed species. Fucoidan molecules thus encompass a wide variety of types differing in L-fucose linkage type, constituent sugars, degree of sulfation, and molecular weight. Nevertheless, these fucoidan polysaccharides are often treated as the same substances.

Failure to at least state the name of the source seaweed when discussing the structure and bioactivity of fucoidans may result in a profusion of differing structures or the belief in bioactivity that cannot be identified within fucoidans of a particular seaweed, which are ultimately detrimental to healthy development of the fucoidan industry.

B. Fucoidan Structure

While fucoidan structure has been reported in the literature as universal, this has generally been one of the average fucoidan structures derived from the brown alga *Fucus vesiculosus* as determined around 1950.

While the structure of fucoidans derived from *F. vesiculosus* has been subjected to repeated analysis since then, the principal chain chemical structure was only identified recently by Chevolot et al. (Fig. 6.18). The two structures shown in Fig. 6.18 differ in their L-fucose linkage site, although subsequent research has generally supported the second. The fucoidans from *F. vesiculosis* identified in 1913 were the first reported anywhere in the world and have since been studied by many researchers, but the same seaweed also contains numerous other kinds of fucoidans, which exhibit rather complex variation according to their chain structure and acetyl groups. It was thus only around 2001 that the correct main chain structure was determined. The chain structure has yet to be fully ascertained, but one structure of a different *F. vesiculosis* fucoidan has been elucidated by Sakai et al.

Sakai began research in 1994 to examine enzymes the break down fucoidans in various seaweeds. Before then, the typical practice had been to use physiochemical methods to determine an average structure for fucoidans. Sakai prepared various fucoidan lyases produced by ocean bacteria, which he used to break down fucoidans to determine chemical structure for the resulting oligosaccharides. While

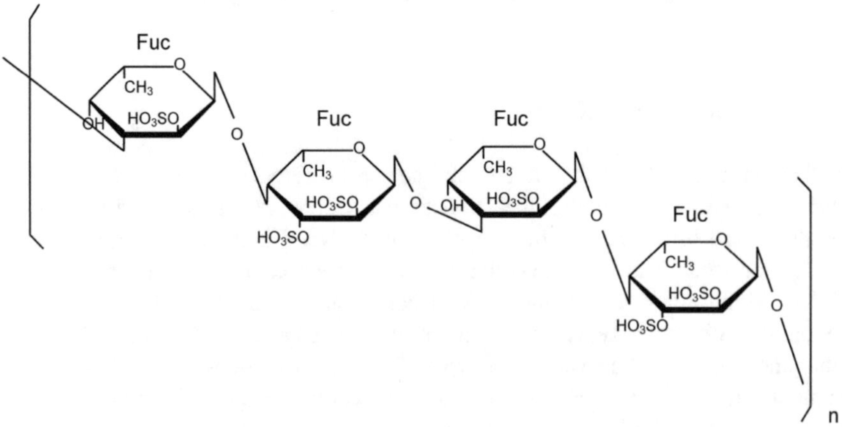

Fig. 6.18 Chemical structures (main chain) of derived sulfated fucoidans (Sakai and Kato 2005)

fucoidans had previously been thought to have complex structures, Sasaki's research showed the structures to actually be relatively simple and repetitive.

The main focus of the structural research by Sakai et al. was on fucoidans derived from Japanese kelp. To date, three types of fucoidans from Kagome seaweed have been broken down to identify enzymes providing oligosaccharides with iterative structural units, which have then been used to analyze the fucoidans' structure.

Figure 6.19a shows the structure of U-fucoidans (sulfated fucoglucuronomannans) from Kagome seaweed. The enzyme used to determine this structure was one that clips the α-D-mannosyl linkage of the main U-fucoidan chain by internal separation and breaks it down into three sugars of repeating structural units. Because clipping the sulfated L-fucose group from the U-fucoidan chain results in no breakdown occurring with this enzyme, the enzyme can be identified for certain as recognizing the sulfated L-fucose group on the U-fucoidan chain.

Figure 6.19b shows the structure of G-fucoidan (sulfated fucogalactan) from Kagome seaweed. The enzyme used to determine this structure is one that internally hydrolyzes the β-D-galactosyl group on the main G-fucoidan chain and breaks it down into six sugars of repeating structural units.

Figure 6.19c shows the structure of F-fucoidan (sulfated fucan) from Kagome seaweed. The enzyme used to determine this structure is one that internally hydrolyzes the β-L-fusoyl linkage in the main F-fucoidan chain. F-fucoidan contains portions where the chain's sulfated L-fucose fully exists, but the enzyme cannot be used to truncate in the vicinity. Comparison of F-fucoidan from Kagome

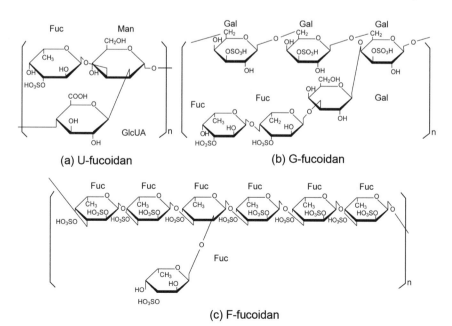

(a) U-fucoidan

(b) G-fucoidan

(c) F-fucoidan

Fig. 6.19 Structures of various derived fucoidans

seaweed and sulfated fugan from *F. vesiculosus* shows differences not only in the location of the L-fucose linkage but also in sulfation group density and linkage location (Figs. 6.18 and 6.19c) (Sakai and Kato 2005).

C. Manufacturing Fucoidans

Fucoidans are refined from brown algae, but because of various constraints on those manufactured as a food ingredient, most are produced through a process of seaweed extraction and simple fractionation. Because the seaweeds that typically are a source for fucoidans have fucoidan content of around 2–50% of dry weight, highly pure fucoidans can be produced through a relatively simple process.

Due to the highly acidic nature of the extraction process, the chemical structure of the fucoidans is destroyed. Sulfation groups are also reported to hydrolyze under alkaline conditions.

Figure 6.20 shows the process of manufacturing fucoidans from Kagome seaweed as a food ingredient.

The ultrafiltration process includes removal of materials that are harmful to health when consumed excessively, including arsenic, iodine, and sodium.

This process is a very important part of the manufacturing of fucoidan for food.

D. Fucoidan Bioactivity

Fucoidans are substances with many different bioactive properties, and the related literature is quite large. Nearly all of the fucoidans sold for research purposes are

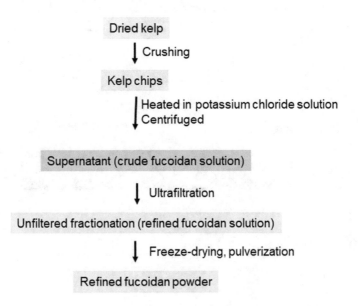

Fig. 6.20 Manufacturing process for kelp-derived fucoidans as a food ingredient

from *Fucus vesiculosus*, however, and research on the bioactivity of fucoidans currently sold as health food ingredients is lacking.

In particular, the bioactivity of orally administered fucoidans has received scant research attention. Proving the usefulness of fucoidans as a health food ingredient will require investigation of their bioactivity when administered orally.

This section will describe the bioactivity of fucoidans (particularly those from Kagome seaweed) when orally administered.

Because the sugar chain structure, sulfation group content, and other chemical structure differs markedly depending on which seaweed the fucoidan derives from, as do the components possessed by each seaweed, it cannot be concluded that fucoidans from differing seaweeds possess the same activity. Determining which activities are exhibited by which seaweed's fucoidans is thus of crucial importance when using fucoidans in health food product design.

(1) Tumor Growth Suppression and Extending Lifespans in Animals with Gallbladder Cancer

When given food including 0.7% fucoidans for a period of four weeks, mice implanted with the sarcoma 180 tumor showed tumor growth suppression of 54% compared to mice given food without added fucoidans. Rats with cancer chemically induced by continuous subcutaneous administration of azoxy methane similar showed a 52-week survival rate of 79% when given water with 0.4% Kagome seaweed-derived fucoidans added to drink freely, compared to 47% for the group drinking water without added fucoidans.

Addition of Kagome seaweed-derived fucoidans together with Meth A sarcoma cell antibodies to a medium of cultured spleen lympoctyes extracted from mice inoculated with Meth A resulted in a capacity-dependent increase in production of interferon γ and interleukin 12. This means that when the cytokines are produced within the bodies of animals with gallbladder cancer, this results in activation of cytoxic T cells, natural killer cells, and macrophages that can kill cancer cells (Chevolot et al. 2001; Duarte et al. 2001; Ponce et al. 2003; Cumashi et al. 2007; Kusaykin et al. 2008; Yang et al. 2008; Ermakova et al. 2011; Vo and Kim 2013).

(2) Suppression of Allergic Reactions

The allergic reaction suppression effects of Kagome seaweed-derived fucoidans were evaluated on adrenal cortex anaphylaxis model rats with egg albumin as an antigen. Preventive effects were observed in a line of rats that began receiving a 1% fucoidan solution seven days before the albumin was administered to their abdominal cavity. Treatment effects were observed by beginning fucoidan administration under identical conditions 19 days after the albumin was administered. In both groups, marked suppression was observed in production of immunoglobin E in the blood, which is the trigger for allergic reactions.

Oral administration of Kagome seaweed-derived fucoidans may thus have both preventive and therapeutic allergic reaction suppression effects.

(3)   *Helicobacter pylori* Infection Suppression

Following administration of an *H. pylori* suspension to infection model rats, the animals were orally administered a Kagome seaweed-derived U-fucoidan fraction every eight hours (10 mg per mouse). After one day, the infection rate was suppressed by around 30% compared to the control group for the same bacteria. After the second day, the fucoidan-administrated group showed an *H. pylori* infection rate of zero. In contrast, a simultaneous experiment using the same amount of *Fucus vesiculosus*-derived fucoidans showed the same infection rate for the control and experimental groups after one and two days.

(4)   Thrombosis Formation Suppression with Fucoidan Oligosaccharides

Kagome seaweed-derived fucoidan oligosaccharides obtained through enzymatic degradation were administered to rats for one month as a 0–0.25% solution, after which a mixture of formalin and methanol was dripped on the outer wall of the jugular vein to induce thrombosis formation. Whereas the thrombosis formation for rats given only water was 100%, those given water containing 0.05–0.25% oligosaccharides showed a 30–80% reduction in their thrombosis formation rate.

In terms of the condition of thromboses formed, roughly 80% of the group not administered oligosaccharides showed firm thromboses; for the group receiving oligosaccharides, the rate was 0–50%. Thrombosis formation in the aforementioned model mice was similarly suppressed through oral administration of depolymerized heparin, but only for fucoidans with the same negative charge as heparin. For instance, the respective numbers of sulfation residues per ten sugars for fucoidans derived from Kagome seaweed, *Fucus vesiculosus*, and Okinawa mozuku were 17, 15, and 4, with differences in the main chain and chain structure. It is difficult to image that these fucoidans would have the same level of thrombosis suppression effects.

(5)   Promoting Hepatocyte Growth Factor Production

Oral administration of one gram of Kagome seaweed-derived F-fucoidans treated with F-fucoidan lyase per kilogram a day to rats with their livers partially removed resulted in a roughly threefold increase in hepatocyte growth factor (HGF) plasma concentration after 24 h compared to a group that was not administered fucoidans. HGF is known to have anti-cancer effects, and a connection may exist between this and the suppression of cancer through oral administration of fucoidans.

HGF also has hair growth-stimulating properties, and there has been proof of the hair growth properties of kelp as transmitted by folk practice.

6.9   Chapter Summary and Conclusion

Seaweeds are emerging as a future resource with various potential applications for humankind, including food, production of functional substances, and a source for energy and paper-making.

Carbohydrates accounts for more than half the composition of green, brown, and red algae, the chief forms of seaweed. Some of the carbohydrates—the polysaccharides alginic acid, carrageenan, fucoidans, and agar—have already been put to a wide range of uses in products related to food, medicine, and industry. New uses for partial hydrolysis products from these polysaccharides are emerging by the day, and their use is expected to increase going forward.

Seaweeds living in different environments differ in the structure and properties of their physiologically active substances, which results in difficulties acquiring consistent sources. This fact has been the biggest stumbling block so far in the large-scale industrial use of seaweeds.

The paramount concern for factory production of seaweed-based polysaccharides is the acquisition of sources with consistent content and structure and their functional components. Achieving this will require mass production through farming in defined zones.

Also essential is the development of source seaweeds with high target component content. Increasing the cost-effectiveness of bioenergy production from seaweed, for example, will require increasing the content of source seaweed components (such as starch) that can be converted into energy.

Farming is a primary industry that has not been the subject of development investment. Without farming, however, it will be impossible to develop and use high value-added products from seaweed. Hopefully the future will bring the development of new, targeted strains through seaweed breeding development with biotechnology such as cell fusion and genetic manipulation.

References

Aguirre-Lipperheide, M., Estrada-Rodríyuez, F. J., & Evans, L. V. (1995). Facts, problems, and needs in seaweed tissue culture: An appraisal. *Journal of Phycology, 31*(5), 677–688.

Amano, H. (1990). Algae Biotechnology. In K. Yamaguchi (Ed.), *Marine Biochmistry* (pp. 200–209). Tokyo, Japan: Tokyo University Pub.

Aruga, U. (1990). Seaweed biotechnology. In F. Takashima (Ed.), *Marine Biotechnology and High Technology* (pp. 98–118). Tokyo, Japan: Seizando-Shoten Publishing, Co.

Augst, A. D., Kong, H. J., & Mooney, D. J. (2006). Alginate hydrogels as biomaterials. *Macromolecular Bioscience, 6*(8), 623–633.

Bierhalz, A. C. K., da Silva, M. A., & Kieckbusch, T. G. (2012). Natamycin release from alginate/pectin films for food packaging applications. *Journal of Food Engineering, 110*(1), 18–25.

Borowitzka, M. A., Critchley, A. T., Kraan, S., Peters, A., Sjøtun, K., & Notoya, M. (2009). In *Proceedings of the 19th International Seaweed Symposium, held in Kobe*, Japan, March 26–31, 2007, Springer Science & Business Media.

Brownlee, I., Allen, A., Pearson, J., Dettmar, P., Havler, M., Atherton, M., et al. (2005). Alginate as a source of dietary fiber. *Critical Reviews in Food Science and Nutrition, 45*(6), 497–510.

Chaki, T. (2005). Function of alginic acid oligosaccharides and food application. In K. Inuuye (Ed.), *Functional glyco-materials: Their development and application to food* (pp. 151–159). Tokyo, Japan: CMC Publishing, Co.

Chen, L. C.-M., & Taylor, A. R. A. (1978). Medullary tissue culture of the red alga. *Canadian Journal of Botany, 56*(7), 883–886.

Cheney, D. P., Mar, E., Saga, N., & van der Meer, J. (1986). Protoplast isolation and cell division in the agar-producing seaweed Gracilaria (rhodophyta) 1. *Journal of Phycology, 22*(2), 238–243.

Chevolot, L., Mulloy, B., Ratiskol, J., Foucault, A., & Colliec-Jouault, S. (2001). A disaccharide repeat unit is the major structure in fucoidans from two species of brown algae. *Carbohydrate Research, 330*(4), 529–535.

Costa, L., Fidelis, G., Cordeiro, S. L., Oliveira, R., Sabry, D. D. A., Câmara, R., et al. (2010). Biological activities of sulfated polysaccharides from tropical seaweeds. *Biomedicine & Pharmacotherapy, 64*(1), 21–28.

Cumashi, A., Ushakova, N. A., Preobrazhenskaya, M. E., D'incecco, A., Piccoli, A., Totani, L., et al. (2007). A comparative study of the anti-inflammatory, anticoagulant, antiangiogenic, and antiadhesive activities of nine different fucoidans from brown seaweeds. *Glycobiology, 17*(5), 541–552.

Davis, T. A., Volesky, B., & Mucci, A. (2003). A review of the biochemistry of heavy metal biosorption by brown algae. *Water Research, 37*(18), 4311–4330.

de Souza, M. C. R., Marques, C. T., Dore, C. M. G., da Silva, F. R. F., Rocha, H. A. O., & Leite, E. L. (2007). Antioxidant activities of sulfated polysaccharides from brown and red seaweeds. *Journal of Applied Phycology, 19*(2), 153–160.

Draget, K. I., Smidsrød, O., & Skjåk-Bræk, G. (2005). Alginates from algae. Biopolymers Online: Biology• Chemistry• Biotechnology• Applications **6**.

Duarte, M. E., Cardoso, M. A., Noseda, M. D., & Cerezo, A. S. (2001). Structural studies on fucoidans from the brown seaweed Sargassum stenophyllum. *Carbohydrate Research, 333*(4), 281–293.

Ermakova, S., Sokolova, R., Kim, S.-M., Um, B.-H., Isakov, V., & Zvyagintseva, T. (2011). Fucoidans from brown seaweeds Sargassum hornery, Eclonia cava, Costaria costata: Structural characteristics and anticancer activity. *Applied Biochemistry and Biotechnology, 164*(6), 841–850.

Garcia-Reina, G., Gomez-Pinchetti, J., Robledo, D., & Sosa, P. (1991). Actual, potential and speculative applications of seaweed cellular biotechnology: Some specific comments on Gelidium. *Hydrobiologia, 221*(1), 181–194.

George, M., & Abraham, T. E. (2006). Polyionic hydrocolloids for the intestinal delivery of protein drugs: Alginate and chitosan—A review. *Journal of Controlled Release, 114*(1), 1–14.

Gosch, B. J., Magnusson, M., Paul, N. A., & de Nys, R. (2012). Total lipid and fatty acid composition of seaweeds for the selection of species for oil-based biofuel and bioproducts. *Gcb Bioenergy, 4*(6), 919–930.

Hiura, N., Chaki, T., & Ogawa, H. (2001). Antihypertensive effects of sodium alginate oligosaccharides. Journal of the Agricultural Chemical Society of Japan (Japan).

Hulata, G. (2001). Genetic manipulations in aquaculture: A review of stock improvement by classical and modern technologies. *Genetica, 111*(1–3), 155–173.

Ito, K., & Hori, K. (1989). Seaweed: Chemical composition and potential food uses. *Food reviews international, 5*(1), 101–144.

Iwamoto, M., Kurachi, M., Nakashima, T., Kim, D., Yamaguchi, K., Oda, T., et al. (2005). Structure–activity relationship of alginate oligosaccharides in the induction of cytokine production from RAW264. 7 cells. *FEBS Letters, 579*(20), 4423–4429.

Jiménez-Escrig, A., & Sánchez-Muniz, F. (2000). Dietary fibre from edible seaweeds: Chemical structure, physicochemical properties and effects on cholesterol metabolism. *Nutrition Research, 20*(4), 585–598.

Kawada, A., Hiura, N., Tajima, S., & Takahara, H. (1999). Alginate oligosaccharides stimulate VEGF-mediated growth and migration of human endothelial cells. *Archives of Dermatological Research, 291*(10), 542–547.

Kim, S., Moon, S., & Popkin, B. M. (2000). The nutrition transition in South Korea. *The American journal of clinical nutrition, 71*(1), 44–53.

Klemmer, K., Waldner, L., Stone, A., Low, N., & Nickerson, M. (2012). Complex coacervation of pea protein isolate and alginate polysaccharides. *Food Chemistry, 130*(3), 710–715.

Kraan, S. (2013). Mass-cultivation of carbohydrate rich macroalgae, a possible solution for sustainable biofuel production. *Mitigation and Adaptation Strategies for Global Change, 18*(1), 27–46.

Kumar, G. R., Reddy, C., Ganesan, M., Thiruppathi, S., Dipakkore, S., Eswaran, K., et al. (2004). Tissue culture and regeneration of thallus from callus of Gelidiella acerosa (Gelidiaies, Rhodophyta). *Phycologia, 43*(5), 596–602.

Kusaykin, M., Bakunina, I., Sova, V., Ermakova, S., Kuznetsova, T., Besednova, N., et al. (2008). Structure, biological activity, and enzymatic transformation of fucoidans from the brown seaweeds. *Biotechnology Journal: Healthcare Nutrition Technology, 3*(7), 904–915.

Lahaye, M., & Robic, A. (2007). Structure and functional properties of ulvan, a polysaccharide from green seaweeds. *Biomacromolecules, 8*(6), 1765–1774.

Le Gall, Y., Braud, J., & Kloareg, B. (1990). Protoplast production in Chondrus crispus gametophytes (Gigartinales, Rhodophyta). *Plant Cell Reports, 8*(10), 582–585.

Lee, K. Y., & Mooney, D. J. (2012). Alginate: Properties and biomedical applications. *Progress in Polymer Science, 37*(1), 106–126.

Lüning, K., & Pang, S. (2003). Mass cultivation of seaweeds: Current aspects and approaches. *Journal of Applied Phycology, 15*(2–3), 115–119.

Manivannan, K., Thirumaran, G., Devi, G. K., Hemalatha, A., & Anantharaman, P. (2008). Biochemical composition of seaweeds from Mandapam coastal regions along Southeast Coast of India. *American-Eurasian Journal of Botany, 1*(2), 32–37.

Marinho-Soriano, E., Fonseca, P., Carneiro, M., & Moreira, W. (2006). Seasonal variation in the chemical composition of two tropical seaweeds. *Bioresource Technology, 97*(18), 2402–2406.

Mazarrasa, I., Olsen, Y. S., Mayol, E., Marbà, N., & Duarte, C. M. (2013). Rapid growth of seaweed biotechnology provides opportunities for developing nations. *Nature Biotechnology, 31*(7), 591.

McCandless, E., & Craigie, J. (1979). Sulfated polysaccharides in red and brown algae. *Annual review of plant physiology, 30*(1), 41–53.

Millner, P. A., Maureen, E., Callow, L., & Evans, V. (1979). Preparation of protoplasts from the green alga Enteromorpha intestinalis (L.) Link. *Planta, 147*(2), 174–177.

Miyachi, S., Sagar, N., & Matsunaga, T. Labo-manual Marine Biotechnology (pp. 29–42). Tokyo, Japan: Shokabo Pub. Co.

Moriya, C., Shida, Y., Yamane, Y., Miyamoto, Y., Kimura, M., Huse, N., et al. (2013). Subcutaneous administration of sodium alginate oligosaccharides prevents salt-induced hypertension in Dahl salt-sensitive rats. *Clinical and Experimental Hypertension, 35*(8), 607–613.

Nickerson, M., & Paulson, A. (2004). Rheological properties of gellan, κ-carrageenan and alginate polysaccharides: Effect of potassium and calcium ions on macrostructure assemblages. *Carbohydrate Polymers, 58*(1), 15–24.

Notoya, M. (2000). Algae farming and Biotechnology. In F. Takashita (Ed.), *The next generation of marine biotechnology* (pp. 101–113) Tokyo, Japan: Seizando-Shoton Publishing, Co.

Ocio, M., Fernandez, P., Rodrigo, F., & Martinez, A. (1996). Heat resistance of Bacillus stearothermophilus spores in alginate-mushroom puree mixture. *International Journal of Food Microbiology, 29*(2–3), 391–395.

Ocio, M., Fiszman, S., Gasque, F., Rodrigo, M., & Martinez, A. (1997). Development of a restructured alginate food particle suitable for high temperature-short time process validation. *Food Hydrocolloids, 11*(4), 423–427.

Percival, E. (1979). The polysaccharides of green, red and brown seaweeds: Their basic structure, biosynthesis and function. *British Phycological Journal, 14*(2), 103–117.

Pereira, R., & Yarish, C. (2008). Mass production of marine macroalgae.

Polne-Fuller, M., Giborm, A. (1987). Calluses and callus-like growth in seaweeds: Induction and culture. In M. A. Ragan & C. J. Bird (Eds.), *Twelfth International Seaweed Symposium. Developments in Hydrobiology*, vol 41. Dordrecht: Springer.

Polne- Fuller, M. (1988). In T. J. Stadler, M. Mollion, C. Verdus, Y. Karamanos, H. Morvan, & D. Christiaen (Eds.), *Algal Biotechnology* (pp. 17–31). London and New York: Elsevier Applied Science.

Ponce, N. M., Pujol, C. A., Damonte, E. B., Flores, M. L., & Stortz, C. A. (2003). Fucoidans from the brown seaweed Adenocystis utricularis: Extraction methods, antiviral activity and structural studies. *Carbohydrate Research, 338*(2), 153–165.

Pranoto, Y., Salokhe, V. M., & Rakshit, S. K. (2005). Physical and antibacterial properties of alginate-based edible film incorporated with garlic oil. *Food Research International, 38*(3), 267–272.

Reddy, C., Gupta, M. K., Mantri, V. A., & Jha, B. (2008). Seaweed protoplasts: Status, biotechnological perspectives and needs. *Journal of Applied Phycology, 20*(5), 619–632.

Renn, D. (1997). Biotechnology and the red seaweed polysaccharide industry: Status, needs and prospects. *Trends in Biotechnology, 15*(1), 9–14.

Rowley, J. A., Madlambayan, G., & Mooney, D. J. (1999). Alginate hydrogels as synthetic extracellular matrix materials. *Biomaterials, 20*(1), 45–53.

Saga, N. (1982). A new method for pure culture of macroscopic algae, the one step selection method. *Japanese Journal of Phycology, 30*(1), 40–45.

Saga, N., & Sakai, Y. (1983). Axenic tissue culture and callus formation of the marine brown alga Laminaria angustata. *Nippon Suisan Gakkaishi, 49*(10), 1561–1563.

Sakai, T., & Kato, I. (2005). Functionality of fucoidan derived from kelp and their application to heath foods. In K. Inouye (Ed.), *Functional glyco-materials: Their development and application to food* (pp. 401–410). Tokyo, Japan: CMC Publishing, Co.

Sega, N., Uchida, T., & Sakai, Y. (1978). *Bulletin Japanese Society Science Fisheries, 44*, 87–?.

Smidsrod, O., & Skjakbrk, G. (1990). Alginate as immobilization matrix for cells. *Trends in Biotechnology, 8*, 71–78.

Stevenson, T. T., & Furneaux, R. H. (1991). Chemical methods for the analysis of sulphated galactans from red algae. *Carbohydrate Research, 210*, 277–298.

Terakado, S., Ueno, M., Tamura, Y., Toda, N., Yoshinaga, M., Otsuka, K., et al. (2012). Sodium alginate oligosaccharides attenuate hypertension and associated kidney damage in Dahl salt-sensitive rats fed a high-salt diet. *Clinical and Experimental Hypertension, 34*(2), 99–106.

Tseng, C. (2001). Algal biotechnology industries and research activities in China. *Journal of Applied Phycology, 13*(4), 375–380.

Ueno, M., Tamura, Y., Toda, N., Yoshinaga, M., Terakado, S., Otsuka, K., et al. (2012). Sodium alginate oligosaccharides attenuate hypertension in spontaneously hypertensive rats fed a low-salt diet. *Clinical and Experimental Hypertension, 34*(5), 305–310.

Ugwu, C., Aoyagi, H., & Uchiyama, H. (2008). Photobioreactors for mass cultivation of algae. *Bioresource Technology, 99*(10), 4021–4028.

Uno, T., Hattori, M., & Yoshida, T. (2006). Oral administration of alginic acid oligosaccharide suppresses IgE production and inhibits the induction of oral tolerance. *Bioscience, Biotechnology, and Biochemistry, 70*(12), 3054–3057.

Usov, A. I. (2011). Polysaccharides of the red algae. *Advances in carbohydrate chemistry and biochemistry, 65*, 115–217.

Vo, T.-S., & Kim, S.-K. (2013). Fucoidans as a natural bioactive ingredient for functional foods. *Journal of Functional Foods, 5*(1), 16–27.

Waaland, S. D. (1975). Evidence for a species-specific cell fusion hormone in red algae. *Protoplasma, 86*(1–3), 253–261.

Wang, L., Wang, X., Wu, H., & Liu, R. (2014). Overview on biological activities and molecular characteristics of sulfated polysaccharides from marine green algae in recent years. *Marine drugs, 12*(9), 4984–5020.

Wheeler, W., Neushul, M., & Woessner, J. (1979). Marine agriculture: Progress and problems. *Experientia, 35*(4), 433–435.

Yamada, N. (2001). Carbohydrates and polysaccharides of seaweeds, Science of Seaweed Utilization, (pp. 85–104). Tokyo, Japan: Seizando-Shoten Publishing, Co.

Yang, C., Chung, D., Shin, I.-S., Lee, H., Kim, J., Lee, Y., et al. (2008). Effects of molecular weight and hydrolysis conditions on anticancer activity of fucoidans from sporophyll of Undaria pinnatifida. *International Journal of Biological Macromolecules, 43*(5), 433–437.

Yoshida, T., Hirano, A., Wada, H., Takahashi, K., & Hattori, M. (2004). Alginic acid oligosaccharide suppresses Th2 development and IgE production by inducing IL-12 production. *International Archives of Allergy and Immunology, 133*(3), 239–247.

Microalgae, a Biological Resource for the Future

7

Contents

7.1   What Are Microalgae?

7.1.1   Definition of Microalgae

People often speak of the seas as being blue, but the blue color of the oceans is said to be the result of the water's depth and various physical factors. A beautiful sea is a deep shade of indigo, but not all seas are blue. We see different colors ranging from green to brown, which are the result of phytoplankton in the water. The term plankton refers to floating organisms that drift with the water's currents, encompassing the two categories of phytoplankton and zooplankton. Phytoplankton consists of single-cell algae that belong, like green plants on land, to the category of primary producers, generating cells and energy through photosynthesis. These are also referred to as microalgae (Spolaore et al. 2006).

The term algae is typically used to refer to lower plants without systematically differentiated roots, stalks, and leaves which perform photosynthesis with chlorophyll. The category includes not only familiar seaweeds like sea mustard, kelp, and laver but also microalgae like *Chlorella* that consist of a single cell. As a category,

© Springer Nature Switzerland AG 2019
S.-K. Kim, *Essentials of Marine Biotechnology*,
https://doi.org/10.1007/978-3-030-20944-5_7

microalgae includes only photosynthesizing organisms of microscopic size and does not include seaweeds.

Microalgae are single-celled algae measuring less than 50 μm in size. They occasionally exist as colonies of many such cells. The term picoplankton refers to especially small organisms measuring 2 μm in size or less. Jellyfish has sometimes been included under the definition of plankton, but the category of zooplankton typically refers to organisms of sizes between 10 μm and 1 mm. Tiny microalgae thus serve as a food source for larger zooplankton, just as small fish serve as a food source for larger fish.

The entire natural world exists in this form of food chain, with organisms consuming other organisms and being consumed in turn. Many other varieties of organisms besides microalgae, zooplankton, and fish inhabit the seas; in some instances, fish consume microalgae directly. The food chain is thus not simple, but quite complex. It is through this food chain relationship, however, that carbon is circulated through nature.

The microalgae at the base of the food chain are a form of plant and are thus capable of surviving on their own through photosynthesis. This means that they can produce the organic materials they need from light energy and carbon dioxide (CO_2); despite their small size, the organisms are capable of living somewhat independently.

The question of how much organic matter microalgae are capable of creating in all the world's seas has long been the topic of ecological research. According to one 1975 report, net annual primary production of organic matter on Earth was calculated at 115×10^9 ton in dry weight for land and 55×10^9 ton for the seas, which would mean that the seas account for the equivalent of roughly one-third of land-based production. Net annual primary production refers to total outward production minus the amount consumed through respiration.

Reported figures on marine production volumes vary significant from one researcher to the next. Several factors account for these studies' failure as yet to calculate a precise value for total marine primary production, despite their having taken place well before that production. Perhaps the biggest factor has been the fact that microalgae, the producers of the seas, are not evenly distributed across all ocean waters (Pulz and Gross 2004).

Twenty-five percent of total marine production comes from a portion accounting for less than 10% of total marine area, while 50% of production occurs within a portion representing 30% of total area. This means that some ocean areas are highly productive while others are not. Although it may be possible to calculate an actual volume by measuring at as many sites as possible, there are limits to how many boats can be sent out onto the waters to investigate production volumes there. Technology for launching artificial satellites to conduct periodic monitoring has been continually developed to address this issue. Indeed, an artificial satellite was launched in the U.S. in the 1970s to measure the amount of chlorophyll at the ocean surface. Chlorophyll, a green pigment that absorbs light energy and is essential for photosynthesis, provides a reflection of the concentration of microalgae in a given area. Under sufficient illuminance, a correlation emerges between microalgae production rate and chlorophyll quantity, as well as a correlation between this illuminance and the rate of photosynthesis.

It is thus possible to calculate the amount of photosynthesis by measuring chlorophyll quantity and illuminance. At the time, the lack of technology needed for continuous chlorophyll measurement prevented precise calculation of its quantity, but later measurements applied this method to measure the chlorophyll quantity at the ocean's surface and found it to be quite significant. Beginning in 1995, artificial satellites launched by the U.S., Japan, and the countries of Europe allowed ongoing temporal and spatial measurement of chlorophyll amounts at the ocean surface (Borowitzka 1999).

7.1.2   The World of Microalgae

The countless populations of the marine ecosystem form a complex food chain, an interrelationship through which energy is transmitted and the ecosystem's circulation and functions can be maintained. As primary producers, microalgae occupy a very important role in this marine ecosystem, and the composition of these populations may transform with changes in factors of their physical, chemical, and biological environment and the geography and oceanographic characteristics of the waters that they inhabit. Moreover, because microalgae possess photosynthetic pigments, they are capable of synthesizing high-energy organic compounds from inorganic materials in environments where light is present. Chlorophyll-a in particular, as the pigment present at the highest rate among microalgae, represents a basic nutrient stage for the marine ecosystem, as well as an important indicator in measuring microalgae's primary production capacity. The primary products thus synthesized provide the foundation for the flow of energy through the seas and fresh waters, while the oxygen produced through photosynthesis supplies most of the respiration needs of underwater organisms. While they share the common quality of generating oxygen and producing organic matter by fixing CO_2 through photosynthesis, these microalgae also exhibit many differences among species (Spolaore et al. 2006).

Microalgae can be classified into many types: those that lack flagella and live by floating, those that possess a few flagella and are capable of actively swimming, those that possess scale-like objects around their cell, those that are encased in a calcified shell, those that encase in a helmet-like shell, and those that despite their photosynthetic capabilities consume bacteria and organic matter with broom-like bearing hairs (like the strands of a broom). In addition to photosynthesis pigment type, microalgae can be categorized into species by chloroplast shape, storage polysaccharides, reproductive cycle, form, ecology, molecular biology line, and cell system characteristics resulting from the presence or absence of a cell wall and pyrenoids. Besides differences in shape, cells also exhibit great outward differences in coloration, including deep blue, green, yellow, brown, and red varieties (Muller-Feuga 2000).

The microalgae of the seas thus exhibit greater diversity than land-based plants in terms of several aspects, including photosynthetic pigments, cell shape, and motility. With all of this diversity and its vast production volumes, how might we put microalgae to work for us?

7.1.3   Uses of Microalgae

Microalgae algae are known to produce many different useful substances as primary metabolite, including proteins, fats, sugars, and extractives. Most current studies on microalgae are focused on industrial applications for food, cosmetics, energy production, biofertilizer, bioactive substances, and wastewater treatment (Fig. 7.1).

Microalgae grow much faster than other plants, and because they possess an outstanding balance of nutrients as food products (a single cell contains large amounts of proteins and carbohydrates, along with other nutrients such as vitamins, minerals, and dietary fiber), they are capable of being used in their entirety (Table 7.1). *Chlorella* in particular was first used as a health supplement in Japan and Taiwan in the 1960s, and has been reported to have outstanding effects in terms of promoting the expulsion of harmful materials such as cadmium and dioxin from the body, encouraging *lactobacillus* growth in the intestines, and preventing hardening of the arteries. Reports from testing of macrophage phagocytic capabilities in the abdominal cavities of white mice have shown an increase of fully

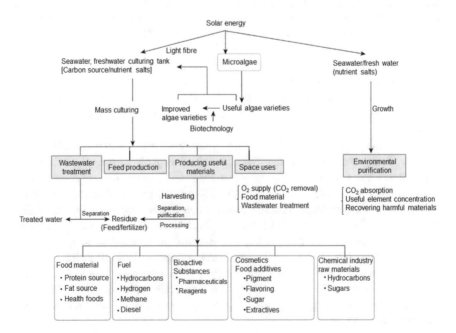

Fig. 7.1 Methods of microalgae culturing and utilization

Table 7.1 Comparison of microalgae and ordinary food components (% dry weight)

Food and microalgae	Protein	Carbohydrates	Fats	Ash
Rice	8	77	2	1
Eggs	47	4	41	1
Milk	26	38	28	0.7
Chlorella sp.	48.4	19.8	11.8	15.6
Dunaliella salina	57.0	31.6	6.4	7.6
Spirulatina platensis	58.5	13.5	6.5	9.0

Table 7.2 Microalgae subject to mass culturing

	Major varieties	Growth rate	Countries produced	Major uses
		(g (dry leaf bodies)/ m² day)		(Useful ingredients)
Chlorella (Green alga)	*Chlorella pyrenoidosa* C. vulgaris C. ellipsoidea	2–20	Japan, Taiwan	Health food (Proteins, fats, vitamins, minerals) please keep (protein, fats, vitamins, minerals) in hanging with "health food" and follow the same with the below columns
Spirulina (Blue-green alga)	*Spirulina platensis* S. maxima	8–14	Mexico, North America, Thailand, Taiwan	Health food (Proteins, vitamins, minerals, phycobilins, carotenoids)
Dunaliella (Green alga)	*Dunaliella bardawil* D. salina Teodoresco	5–25	Israel, North America, Australia	Health food (β-carotene, glycerol)
Porphyridium (Red alga)	*Porphyridium cruentum* P. aerugineum	10–20	France	Phycoerythrin, phycobiline (pigment), polysaccharides, arachidonic acid

220% in microbial phagocytic functions for macrophages in the *Chlorella* dosage group (Wilde and Benemann 1993; Lee 1997; Muller-Feuga 2000; Becker 2007; Raja et al. 2008; Costa and De Morais 2011).

Spirulina, which has recently become the focus of some attention, has long been consumed in Africa as a foodstuff, with records of its cultivation as food by the indigenous population in Mexico. *Spirulina* is roughly 100 times larger than *Chlorella* and is easy to culture. In addition to its low production cost, its digestibility is reported to be quite high at around 80%. Use as a foodstuff was permitted in Mexico in 1973 on this basis. As shown in Table 7.2, foodstuffs made with microalgae are already on the market in several countries (Ötleş and Pire 2001; Capelli and Cysewski 2010; Hoseini et al. 2013).

Another variety, *Dunaliella*, is a brown alga that lacks a cell wall. The vitamin A precursor β-carotene accounts for 10–15% of its total mass, raising hopes for future

development as a new functional material. Production of extracellular polysaccharides from *Porphyridium* has almost reached the commercialization stage.
Development is also under way on marine microalgae containing large amounts of
polyunsaturated fatty acids such as EPA, DHA, and γ-linoleic acid.

7.2   Screening and Culturing Marine Microalgae

Marine ecosystems can be broadly categorized into coastal and oceanic regions.
Each of these environments forms its own characteristic ecosystem. Marine environments also vary significantly by geographic location and site conditions. In the
case of coastal regions, brackish water basin form through the mixture of freshwater
and seawater at estuaries where inland water flows in, or at inland lagoons. Environments vary especially greatly by their linkage to the open sea (i.e., the inner or
outer portion of bays, warm or cold currents), the type or structure of seabed (reef,
pebble, sand, or mud), and the quantity and quality of industrial and living drainage
entering into them, among other factors. Microalgae distribution is subject to
influences from these environmental conditions and exhibits temporal and spatial
changes in the standing crop and component species due to variable environmental
factors on a diurnal and seasonal basis (Kellam and Walker 1989; Gerwick et al.
1994; Matsunaga et al. 1999; Doan et al. 2011).

Planktonic (floating) microalgae reported in seawater and brackish water basin
belong to categories such as dinoflagellates, diatoms, haptophytes, raphidophytes,
chrysophytes, xantophytes, eustigmatophytes, cryptophotes, euglenophytes,
chlorophytes, and cyanobacteria. Some varieties are benthic and survive by attaching
to organic or inorganic substrates, while others exist in symbiosis with other
organisms. While their distribution is thought to be confined to the euphotic zone,
numerous microalgae are distributed over a wide variety of marine environments
spanning both coastal and open sea regions. When these microalgae are used as
research material, they may be acquired pre-separated and pre-preserved from algae
banks. Screening of flora present in the natural world is another important and
effective means of acquiring and culturing strains suited to one's research purpose.

This section will focus on explaining about the screening and culturing of marine
microalgae and examining their application to other marine microalgae.

Figure 7.2 shows the full manipulation stage and areas of consideration.
Microalgae handling is basically the same way as with other microbes (such as
bacterial) and requires manipulation under aseptic conditions (Borowitzka 1995;
Kumazawa 1991; Yamaguchi 1992).

7.2.1   Collecting Marine Blue-Green Algae

The choice of a collection site is based on a comparison and examination of the
physiological characteristics of the target blue-green alga (or other microalgae) and

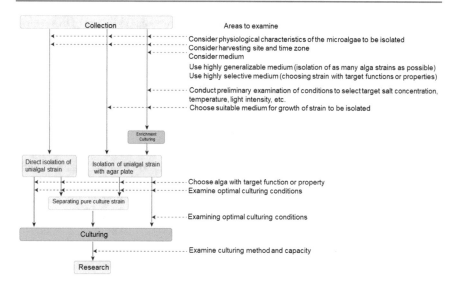

Fig. 7.2 Flow chart of marine microalgae screening and culturing

the features of the collection site. At the same time, one should also bear in mind that the environment conditions of the screening site will not necessary conform to the optimal culturing conditions for the blue-green alga (or other microalgae) isolated from that site.

Sample collection in open-ocean regions requires a suitably equipped research vessel. For microalgae collection, a plankton net or water sampler (e.g., Nansen or vertical sampler) is used. Water sampling is used to collect algae with fine structure or smaller nanoplankton (2–20 μm) and phytoplankton (under 2 μm) (Saino and Hattori 1978).

7.2.2 Separating Marine Blue-Green Algae

The basic methods used to isolate blue-green algae from a sample involves first ① using a stereoscopic or biological microscope to examine the sample and extract a sample from a region with a high density of the target algae, after which it is ② cultured on an agar plate to isolate a unialgal strain species or ③ subjected to enrichment culturing to increase the density of the target algae, which is then cultured on an agar plate to isolate a unialgal strain. Other approaches include ④ the capillary pipet approach to large microalgae or algae that do not grow on an agar plate or ⑤ the micromanipulator approach, which are used to isolate a unialgal strain directly by microscope to be ⑥ cultured in a liquid medium (Saino and Hattori 1978).

Versatile media are used to obtain a maximum of species through enrichment culturing, while highly selective media are used to obtain a specific group. In selecting enrichment culturing conditions, the environment conditions of the sample collection site must be considered alongside the characteristics of the target alga (e.g., temperate or light intensity and saline concentration).

While pure (axenic) strains may be obtained through isolation manipulation, the result is often a unialgal strain coexisting with bacteria.

7.2.3 Examining Culturing Conditions

To ensure an adequate supply of stable research material (blue-green algae or other microalgae), consideration must be given to the cultural conditions, including light, temperature, pH, saline concentration, and type or composition of the culturing solution. Indicators that can be used to compare culturing conditions include number of cells per volume of culturing solution, weight (wet or dry), packed cell volume, and biogenic substances such as proteins or chlorophyll, and sequential rises in the solution's absorbance. An explanation on how to find the exponential phase generation time (the time needed for a single division) and specific growth rate will be provided later. Because specific cell function activity and amounts of particular components often vary by propagation time, consideration should also be given to obtaining cells with high levels of activity or content in the biogenic substance under examination.

7.2.4 Culturing Methods

Culturing methods including batch culture, which involves culturing in a specific volume of medium and harvesting at a suitable time for the research purposes, and continuous culture, in which cells are supplied continuously. The latter includes approaches of semi-continuous culturing, in which part of the culturing solution is regularly replace in a new medium; the turbidostat approach, which maintains a fixed cell density; and the Chemostat approach, which adjust the addition of specific nutrient salts to ensure a constant proliferation rate.

With the typically batch culturing approach, observation of the solution at any given time during culturing will show cells at various different stages of the cell cycle (although this state cannot be described as fully irregular). A contrast approach is synchronous culturing, in which overall cell proliferation in the solution is made to conform to the same cell cycle. Because this approach supplies cells that are physiologically and biochemically uniform, it is used for studies of changing physiological functions within the cell cycle.

7.3   Mass-Production of Microalgae

The single-celled green alga *Chlorella* has drawn attention as a food source because of its high photosynthetic performance and protein content. Studies on its bulk culturing have been carried out around the world since World War II. Established in 1957, the Japan Chlorella Research Institute has conducted research on bulk culturing that has led to the development the open and closed circulation methods that today serve as the prototypes for bulk culturing of microalgae (Aaronson and Dubinsky 1982).

Research on *Chlorella* production as a source of protein was halted after its costs were found to be prohibitive in temperate climates such as Japan's. Experiments with the application of *Chlorella* algal bodies at the time, however, found physiological effects for both humans and animals, and their use for purposes other than protein began drawing attention. Commercial *Chlorella* production began in 1964, and the alga has been developed into a foodstuff with both nutritional and pharmacological potential (Pires 2015).

Commercial production of *Chlorella* is currently under way in countries such as Japan, Taiwan, and Bulgarian. Most of the algal bodies produced are consumed as health food. Applications span a diverse range of fields besides food, including agriculture, marine products, and bioenergy. Because of the great diversity of microalgae, this section will first explain about the bulk culturing methods used for *Chlorella* production, their characteristics, and current uses of *Chlorella*.

7.3.1   Chlorella Production

A. Bulk Culturing

(1) Forms of Culturing

Large-scale culturing of *Chlorella* may be classified into three types according to the method used to supply carbon and energy sources: photoautotrophic culture, which uses inorganic carbon to multiply through photosynthesis; mixotrophic culture, which uses involves photosynthesis and use of organic as well as inorganic carbon; and heterotrophic culture, which uses organic carbon (Liu and Hu 2013).

Photoautotrophic culture typically takes place in outdoor settings and has a yield of roughly 2.0–23.3 g dry leaf bodies per square meter per day. In principle, the use of solar energy should allow for the most economical production of leaf bodies. Because the upper limit of production is determined by amount of sun exposure and the process requires a large culturing area and large amounts of medium, however, culturing temperatures cannot be adjusted in the winter, and productivity from photoautotrophic culturing in a temperate region such as Korea is low. Moreover, the process also requires a supply of carbon dioxide.

Because concentrations of carbon dioxide in the air are low, maximum photosynthesis activity cannot be obtained, and the rate of dissolution into the culture medium is low. Under typical agitation conditions, a rate of carbon dioxide dissolution greater than 2.4–9.6 g/m^2 day cannot be expected in the air. Production of 24 g of leaf body dry weight per square meter-day typically requires a supply of carbon dioxide. For this reason, the culture ground must be maintained in an enclosed system to increase the usage rate when carbon dioxide is artificially ventilated (Liu and Hu 2013).

The mixotrophic culture approach attempts growth simultaneously through organic carbon such as acetic acid and through photosynthesis. It is general performed with outdoor media. Mixotrophic multiplication of *Chlorella* includes both multiplication through organic carbon and multiplication through photosynthesis. Because productivity per unit area and unit medium is higher than with photoautotrophic culture, some degree of culturing temperature control is possible. The absence of light as a limiting factor results in higher leaf body concentrations and easier harvesting. Another great advantage is that large quantities of CO_2 are emitted into the culture medium as the *Chlorella* cells consume one mole of acetic acid, which can then be used for photosynthesis.

Heterotrophic culture uses organic carbon such as glucose or acetic acid to multiply *Chlorella* in a tank with no light present. Conditions within the tank can be fully controlled for culturing. High-concentration culturing of 50 g/L or more in cell dry weight can be achieved, and productivity per unit area is very high (Fig. 7.3) (Liu and Hu 2013; Maruyama and Ando, 1992).

While *Chlorella* cells can synthesize the pigments and proteins needed for photosynthesis even when cultured under light-free conditions, the amounts are slightly lower than in photosynthesizing *Chlorella* cells.

Chlorella production is performed through selection of one of these culturing methods. In many cases, two approaches are used simultaneously, such as pure-breed culturing through heterotrophic or photoautotrophic culture and productive culturing through mixotrophic culture.

(2) Strains

Because proliferation properties differ between strains (Table 7.3), the productivity of mass culturing depends on the strain used. Natural lines are typically separated into many strains, and those with outstanding proliferation are used with the culture method suited to the purpose. Varieties typically used include *Chlorella vulgaris*, *C. ellipsoidea*, *C. pyrenoidosa*, and *C. regularis*.

For photoautotrophic culture using sunlight, it is beneficial to use strains with how solar light usage efficiency (energy transformation efficiency). At around 150 klx, sunlight is more intense than the photosynthesis saturation value (2.7–28 klx) for the *Chlorella* genus. Light that is more intense than the saturation value cannot be used for photosynthesis. Accordingly, it is beneficially to select and use strains with high photosynthesis saturation values to increase energy transformation efficiency.

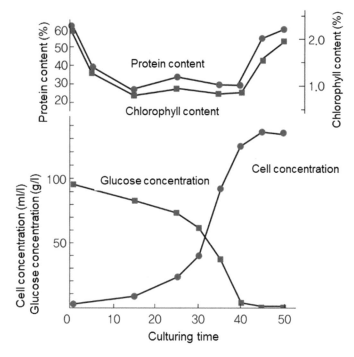

Fig. 7.3 Example of heterotrophic culture (Jar fermenter used to culture *C. vulgaris* K-155 under dark conditions with 200 L of culture medium at 30 °C and glucose used as a carbon source)

Table 7.3 Growth properties of *Chlorella* strains

Strain	Specific growth rate (μC/h)	Culturing temperature (°C)	Light exposure	Light saturation value (klx)	Organic carbon
Chlorella pyrenoidosa	0.080	25	+	5.4	
Emerson strain	0.041	25	–		Glucose
	0.026	25	–		Sodium acetate
C. pyrenoidosa 7-11-05	0.086	25	+	5.4	
	0.264	39	+	15	
C. ellipsoidea	0.128	25	+		
Tamiya strain	0.045	25	–		Glucose
	0.033	25	–		Sodium acetate
C. regularis S-50	0.28	36	+	28	
	0.28	36	–		Sodium acetate

(continued)

Table 7.3 (continued)

Strain	Specific growth rate (µC/h)	Culturing temperature (°C)	Light exposure	Light saturation value (klx)	Organic carbon
C. vulgaris 211/8 k	0.110	30	+		
	0.098	30	–		Glucose
C. vulgaris 211/8b	0.075	25	+	2.5	
C. vulgaris K-73107	0.25	38	+		
	0.26	38	–		Glucose
	0.22	38	–		Acetic acid
	0.11	28	+	15	

For culturing that uses organic carbon, it is advantageous to use strains with a high rate of proliferation when using organic carbon. The rate of microbe proliferation can be expressed as $dx/dt = \mu x$ for cell concentration x (g/L), time t, and specific growth rate µC/h, which allows for comparison between the proliferation rate and the specific growth rate.

The specific growth rate for different forms of culturing varies greatly between strains. Some strains proliferate better through photosynthesis than through heterotrophic multiplication, while others show no difference between the two.

The specific growth rate is also strongly influenced by the culturing temperature, and high-temperature strains with optimal growth temperatures of 35–40 °C may exhibit high specific growth rates (Table 7.3, Fig. 7.4). Specific growth rates for fast-growing strains are 0.2 µC/h or greater, with a time of 2.5 to 3 h required for doubling (Maruyama and Ando 1992).

(3) Culture Medium

The medium in mass culturing is the raw material for *Chlorella* leaf body production. Potassium nitrate (KNO_3) or urea is used for a nitrogen source, with other important components including monopotassium phosphate (KH_2PO_4), magnesium sulfate ($MgSO_4 \cdot 7H_2O$), and iron(ii) sulfate ($FeSO_4 \cdot 7H_2O$) as well as A_5 solution as a trace ingredient (Table 7.4). With nitrogen accounting for 7–10% of the leaf body, respective dry weights of 1.5 and 5.2 g can be produced from 1 g of KNO_3 and elemental nitrogen. Because photosynthetic culture typically involves a supply of carbon dioxide, no carbon source is added to the medium.

For purposes of mixotrophic or heterotrophic culturing, a carbon source such as glucose or acetic acid is added to the medium. In the case of acetic acid, a small

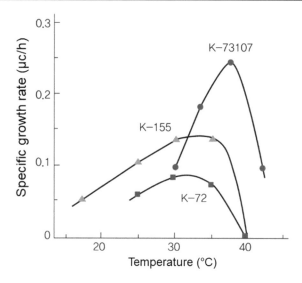

Fig. 7.4 Relationship between culturing rate and specific growth rate for multiple strains

Table 7.4   Medium used in mass culturing

Medium Ingredients (g/L)	A	B	C	D	E
Potassium nitrate (KNO_3)	5.0		1.25		
Urea		3.0		6.0	1.5
Monopotassium phosphate (KH_2PO_4)	1.25	1.25	1.25	1.0	1.2
Magnesium sulfate ($MgSO_4 \cdot 7H_2O$)	2.5	2.5	1.25	1.0	0.6
Calcium chloride ($CaCl_2$)				0.03	
Ferrous sulfate ($FeSO_4 \cdot 7H_2O$)	0.003	0.003	0.02	0.015	
EDTA-Na–Fe					0.015
EDTA-Na_2		0.037			
A_5 solution (mL)[a]			1.0	3.0	2.0
Trace elements	$A_4 \cdot B_7^{b}$				
Citric acid				0.25	
Glucose					20
Acetic acid (mL)[c]				100	

[a]Dissolve 2.86 g H_3BO_3, 1.81 g $MgCl_2 \cdot 4H_2O$, 0.22 g $ZnSO_4 \cdot 7H_2O$, 0.08 g $CuSO_4 \cdot 5H_2O$, and 0.021 Na_2MoO_4 in 1L of pure water and add one drop of H_2SO_4
[b]Add 1.0 mL solution containing 0.6 ppm Fe, 0.5 ppm B and Mn, 0.05 ppm Zn, 0.02 ppm Cu, and 0.1 ppm Mo, Co, Ni, Cr, V, W, and Ti
[c]Add small quantities according to amount consumed

about of acetic acid or sodium acetate is added at the beginning of the culture process, and small additional amounts are supplied using the medium's rising pH as an indicator. Carbon is a major component accounting for 51.4–72.6% of the leaf body, and the carbon source growth yield (dry weight of leaf bodies produced divided by grams of carbon source consumed) is an important indicator in leaf body production (Kumar et al. 2010).

In heterotrophic culture, glucose and acetic acid have respective growth yields of 0.4 and 0.35, which means that 40 and 35 g of leaf bodies can be produced with 100 g of carbon source. Carbon dioxide has a growth yield of 0.5, and in principle it should be possible to produce around 1 kg dry weight from 1 L of CO_2. In culturing with open aquaria, however, the surface yield drops because of the low rate of CO_2 usage (Maruyama and Ando 1992).

(4) Harvesting and drying

Once culturing is completed, *Chlorella* cells are separated from the culturing solution and undergo processing and drying to become dry leaf bodies.

Separation and enrichment of *Chlorella* cells involves use of a centrifuge. Enrichment of around five times can be achieved through a single passage of a cell suspension. *Chlorella* cells are first separated from the culturing medium through enrichment, after which water is added to enrich them and remove any remaining medium. The end result is a cell suspension with a leaf body concentration of 100 g dry weight per liter or greater. For the production of seedlings for seafood, a *Trochelminthes* generation product has been developed in this cell suspension state.

Next, the cell suspension is subjected to three minutes of heat treatment with a plate heater at 100 °C. Heating is performed to deactivate enzymes within the cell, increasing preservability of the dry leaf bodies and the digestive absorption rate for humans and animals (Taub and Dollar 1964; Watanabe et al. 2005; Maruyama and Ando 1992) (Table 7.5).

Table 7.5 Components of dry *Chlorella* leaf bodies

Energy (kcal/100 g)	408	Vitamins (mg/100 g)	
Proximate composition (g/100 g)		Vitamin B_1	1.85
Protein	63.5	Vitamin B_2	5.70
Lipids	11.5	Vitamin B_6	0.912
Carbohydrates	12.5	Vitamin B_{12}	0.077
Crude fiber	2.3	Vitamin C	58
Ash	6.2	Ergosterol	86
Water	4.0	Vitamin E	12
Vegetable fiber (g/100 g)	20	Linoleic acid	1800
Pigments (g/100 g)		Linolenic acid	2000
Chlorophyll	2.0	Niacin	22.3
Carotene	0.041	Pantothenic acid	2.85
Inorganic element (mg/100 g)		Folate	0.049
Calcium	98.5	Biotin	0.21
Iron	160	Choline	380
Magnesium	221	Inositol	221
Potassium	962	Vitamin K_1	1.0

For drying, a <u>spray dryer</u> is used. An atomizer is used to spray a mist of *Chlorella* concentrate in the dryer <u>tower</u>, and moisture is removed instantaneously with a wind of around 180 °C. The short heating time results in a low level of change in the drying process. Leaf bodies obtained in this way serve as raw materials for various products using *Chlorella* (Maruyama and Ando 1992).

7.4 Biotechnology of Microalgae

Because microalgae have long been used for biochemistry research and as feed for parent and offspring fish and shellfish, culturing techniques have been established for many types. Their relatively fast growth and ease of genetic manipulation are raising hopes for their application in production of useful materials and addressing global environment issues. Research to date, however, has been confined to production of mutant strains, genetic manipulation, and fixation for a relative few microalgae (Ng, Daphe H. P. et al., 2015).

7.4.1 Mutant Strain Production

Mutants are sometimes used to obtain strains containing particular useful substances. Mutations are produced through contact with mutagens or exposure to ultraviolet rays, X-rays, and gamma rays and subjected to selective treatment to create mutant strains. Examples obtained to date include resistant mutants; auxotrophic mutants; temperature- and ultraviolet-sensitive mutants; mutants with photosynthesis, nitrogen fixing, or carbon metabolism defects; and morphological mutants (Kao et al. 2012).

Among blue-green algae, single-celled *Anacystis nidulans* (*Synechococcus*) and multi-celled *Anabaena variabilis* are the most commonly used. Mutant production has also been attempted with *Agmenelum quadruplicatum, Synecocystis, Synechococcus cedrorum, Aphanocapsa, Gloeocapsa alpicola, Aphanothece halophytica, Nostoc muscorum,* and *Spirulina platensis.*

Chlamyodomonas species—*C. reinhardtii* in particular—are most commonly used among green algae. Some research has also been conducted with *Chlorella, Scenedesmus,* and *Euglena.*

7.4.2 Genetic Manipulation

Prokaryotes such as brown algae and eukaryotes such as green algae have a number of basic differences that affect genetic manipulation. Prokaryotes are haploid, whereas eukaryotes have haploid and diploid generations. In prokaryotes, there is no interaction with the organelle genome within individual genes, while mitochondrion and chloroplast genomes in eukaryotes interact with the nuclear genome.

Prokaryote genes do not include introns, while most eukaryote genes do. It is for this reason that brown algae, which are relatively easy to manipulate, have chiefly been used to date.

Many brown algae possess plasmids (extra-chromosomal genes capable of autoreproduction). Because of uncertainties concerning their physiological functions, however, a marker must be introduced for use as a vector (DNA used to transport foreign DNA into a host).

For example, the R2 strain of *Anacystis nidulans*, which possesses the integral plasmid pUH24 (7.8 kbp), has been transformed with the *E. coli.* plasmid pRI 46, which contains Tn901, a transposon (a sequence that moves chromosomal genes or plasmids) coding for ampicillin resistance (Ap^r). While the frequency was very low, manipulated pUH 24::Tn 901 plasmids (that is, where Tn 901 has been shifted to pUH24) have been found to occur in analysis of five Ap^r transformation strains. Because Tn 901 shifting is an issue in the use of this recombinant plasmid as a vector, it was truncated with the restriction enzyme BamHI and the transposon fixed. The resulting plasmid (called pUC1) possessed one section each of BamHI and XhoI and two of Bgl II and coded for Ap^r. Cloning of several blue-green algae vectors in this way allowed the production of mutant strains with drug-resistant properties.

One example involved the use of the *A. nidulans* R2's pUH24 and the colon bacillus vector pACYC to produce a shuttle vector (a vector capable of independent growth on cells from either of two organisms) and pUC 303 (tolerance to Cm^r, Sm^r, chloramphenicol and streptomycin), which were combined with an *A. nidulans* R2 · pUH24 delation strain. The result was a high-efficiency transformation in which drug tolerance remained stable for a long period of time.

Cloning of acetate reductase in *A. nidulans* R2 has also been achieved with the comid vector pPUC29, which includes a phage DNA *cos* region in the shuttle vector.

Because cosmids have properties intermediate between those of phage vectors and plasmid vectors, DNA fragments of up to 50 kbp can be introduced and a gene library (including recombinants of all chromosomal DNA factors with vectors and all chromosome pair regions) can be produced.

The above is an example of genetic manipulation through transformation. Other forms of manipulation have attempted through electroporation, or the use of an electric pulse to produce a tiny hole in the cell membrane through which genes can be physically introduced, and, in the case of filamentous blue-green algae such as *Anabaena* sp. or *Nostoc* sp., through conjugative transfer using colon bacilli.

DNA introduced in transformation may be used under fully in vitro conditions. When a restriction enzyme line exists in the host, transformation may be achieved by altering the DNA beforehand with the corresponding methylase. In the conjugative transfer method, however, the DNA introduced is transformed and cloned within the *E. coli*, and lines transformed in vitro cannot be used.

One approach used to address this issue has been to clone methyl transferase DNA as a helper plasmid for conjugation. Transformation is performed in vivo and conjugated with the blue-green alga. This method has been used for conjugative transfer in *Fremellia diplosiphon*, a blue-green alga with complementary adaptation capabilities. Factors have been cloned to control *F. diplosiphon*'s complementary adaptation capabilities at the transcription level.

To date, most conjugative transfer lines have used shuttle vectors with filamentous blue-green algae and plasmids derived from those algae. Recently, conjugative transfer with a wide range of host vector plasmids has been reported for the single-celled blue-green alga *Synechocystis* PCC6803. In this line, the plasmid used for conjugative transfer was pKT210, a wide-ranging vector plasmid from an Inc Q population with replication units from RSF1010. This use of various vectors to introduce genes through conjugation is expected to be a fruitful area in the future (Dunahay et al. 1996; Radakovits et al. 2010; Larkum et al. 2012; Takeyama and Matsunaga 1989).

7.4.3 Genetic Manipulation in Marine Blue-Green Algae

Genetically manipulation of marine blue-green algae is still in its early stages, including the conjugative transfer method described below. The only examples reported to date involve two species with natural transformation capabilities: *Agmenellum quadruplicatum* PR6 and *Synechococcus* sp. PCC 7002. As with freshwater blue-green algae, the gene to be introduced needs only to be mixed with the cell to allow the cell's absorption of the DNA and its combination with the genome or replication of plasmids. Intrinsic plasmids have been discovered for *A. quadruplicatum* PR6 and used with a shuttle vector line with an *E. coli* in an attempt to express foreign proteins.

One example is an attempt to express pesticidal proteins produced by *Bacillus thuringiensis* in *A. quadruplicatum*. To date, genetic manipulation in marine blue-green algae has been restricted to the main strain. As marine biotechnology develops in the future, replacement genetic manipulation techniques will have to be developed quickly for screened marine blue-green algae with the ability to produce useful materials or other special capabilities.

In Japan, Matsunaga et al. attempted to develop a genetically manipulated variety of the marine blue-green alga *Synechococcus* sp. NKBG042902. A preliminary examination of plasmids that could be used for transformation line shuttle vectors for this strain showed four types: pSY08 (>10 kb), pSY09 (around 10 kb), pSY10 (2.7 kb), and pSY11 (2.3 kb). Among these, pSY11 was used to create a shuttle vector with *E. coli*, as it has a relatively high copy number and low molecular weight. pSY's restriction enzyme sites are shown in Fig. 7.5. As a plasmid, its copy number is high at 30–50 per genome. A 1.4 kb Hind III fragment from pSY18 has been introduced at a pUC18 multicloning site to form the hybrid plasmid pUSY02 (Moore et al. 1988; Takeyama and Matsunaga 1989).

As a host, *Synechococcus* sp. NKBG042902 possesses pSY_{11}, raising concerns that the shuttle vector may be excluded due to incompatibility. To address this, the curing strain 042902-YG1116 was created for the same strain. This curing strain is obtained by treating the 042902 strain with acridine orange for random selection and cloning, and does not include pSY_{10} or pSY_{11} (Matsunaga 1992).

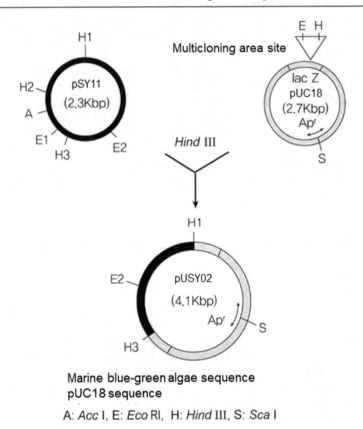

A: *Acc* I, E: *Eco* RI, H: *Hind* III, S: *Sca* I

Fig. 7.5 Map of restriction enzyme pSY11 and construction of the *E. coli* shuttle vector pUSY₀₂ with the same plasmid

Because this strain does not possess *A. nidulans*'s marked DNA absorption capabilities, transformation competence must be established in the cell at the time of *E. coli* transformation. The resulting shuttle vector pUSY02 has been found to be capable of transforming the marine blue-green alga *Synechococcus* sp. NKBG042902. This shuttle vector can also likewise transform the freshwater blue-green alga *A. nidulans*.

To date, marine blue-green alga transformation has not been achieved with the shuttle vector pECAN8 used with *A. nidulans* R2. This means that although this shuttle vector has shown potential for a broad range of uses in blue-green algae from different environments (freshwater and marine), independent genetic manipulation lines must be developed for marine blue-green algae.

The microprojectile method has also been noted recently as a transformation technique. This method involves using highly pressurized air or gunpowder to launch a metal particle coated with the desired gene or plasmid (DNA carrier) at high speed to embed it in the target cell for transformation. This method has allowed transformation to take place without plant cells being subjected to protoplast formation, and DNA can also be introduced into organelles within the cell.

Matsunaga et al. have attempted to apply the microprojectile approach to increase transformation efficiency and manipulability in blue-green algae. While particles of tungsten and gold have been used thus far as DNA carriers in the microprojectile method, they are relatively large, with a diameter of around 0.1–0.5 μm that has rendered them unsuitable for transformation in prokaryotes. Application research has been carried out on magnetic particles synthesized with the cells of magnetotactic bacteria. These magnetic bacterial particles are very small (around 50 nm) and even and are covered with an evenly surfaced organic membrane that allows for easy chemical modification. This has resulted in explorations of the applicability of magnetic bacterial particles as DNA carriers.

Modification of magnetic bacterial particles with γ-APTES, glutaraldehyde, and triamine allows for adsorption and fixing of over twice as much DNA as the carriers previously used with particles with a DNA fixation weight of 2 μg/mg. Transformation with magnetic bacterial particle DNA carriers through the microprojectile approach using highly pressurized air in the marine blue-green alga *Synechococcus* sp. and the freshwater blue-green alga *Anabaena* sp. PCC 7120 showed a cellular magnetic response of up to 1.5% in marine blue-green algae and 7.5% in freshwater blue-green algae when sprayed at a pressure of 100 kgf/cm^2, indicate a very high efficiency of DNA carrier introduction within the cell. Although the respective shuttle vectors pUSY02 and pRL5 were mixed to the DNA carrier in the marine and freshwater blue-green algae in this experiment, transformation could be achieved under these conditions. Observation of these magnetic responsive cells with a transmission electron microscope showed DNA carriers to be present near the membrane in both.

In the future, transformation is expected to be made possible through increases in the DNA carrier spraying speed. Many marine blue-green algae have surfaces sheathed in polysaccharides or cells that are rendered incapable of agglutination or growth when washed. For this reason, the microprojectile approach appears to be a promising method for application with marine blue-green algae that cannot be simply transformed.

An interesting property was recently reported in pSY10, one of the plasmids for the marine blue-green alga *Synechococcus* sp. As mentioned, the shuttle vector PUSY02 was developed from the same strain's high-replication plasmid pSY11. While a defined copy number has been identified for pSY11 regardless of culturing conditions, pSY10 shows the same copy number as pSY11 under normal culturing conditions, with a copy number that increased sharply according to those conditions. While pSY10's copy number is 30–50 per genome (similar to pSY11) when the strain is cultured beforehand in a medium with low saline concentration, it rises sharply when transferred to a medium with a high saline concentration, reaching 150 under a concentration of 0.5–2% and around 250 under a concentration of 3% within 20 h (Fig. 7.6).

This phenomenon has not been observed at all in three other plasmids from the same strain, including pSY11. The rise in pSY10's copy number was also not

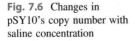 **Fig. 7.6** Changes in
pSY10's copy number with
saline concentration

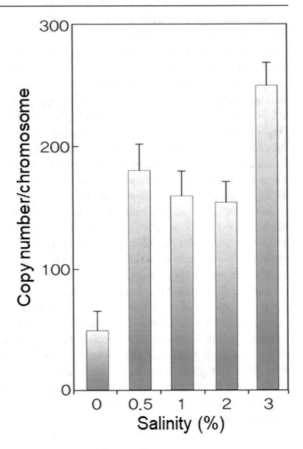

observed when potassium chloride (KCl) or sorbitol was added due to osmotic
pressure and ionic strength stresses associated with salts other than sodium chloride
(NaCl). From this, it was determined that pSY10 selectively responds to NaCl
concentration changes, resulting in a rapid rise in the copy number. It is therefore
expected that this can be used to create shuttle vectors for which copy numbers can
be adjusted through saline concentration. This, along with controllable promoter
genes, offers a potentially promising avenue to the development of foreign gene
expression systems (Van Baalen 1962).

As a donor bacterium, the *E. coli* S17-1 strain has been used with the plasmid
RP4 inserted in the genome. With this strain, high-efficiency conjugative transfer
can be achieved without the use of other helper or cargo plasmids when transformation is performed with plasmids carrying a Mob site specifically recognized by
RP4 at the time of transfer or plasmids with mob or bom. This process is illustrated
in Fig. 7.7. Initially, insertion in marine blue-green algae through conjugative
transfer was attempted with the suicide vector plasmid pSUP1021, which is combined with Tn5. The pSUP1021 replication unit comes from pACYC184 and
undergoes little to no replication outside of *E. coli*.

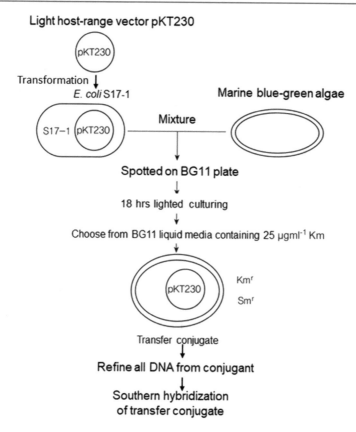

Fig. 7.7 Introducing genes into a marine blue-green alga through conjugative transfer with the *E. coli* S17-1

The resulting conjugant was confirmed through the expression of kanamycin resistance, resulting from Tn5's entry into the bacteriophage genome to code for the same transposon. In other words, replication occurs in this line at the time of conjugation, and the plasmid is introduced into the recipient cell. Once it has entered the genome, the effectiveness of the conjugative transfer method can be assessed without taking into account differences in replication units due to plasmid instability or variations between types.

In this experiment, seven strains of marine blue-green algae were used. Five strains of single-celled *Synechococcus* sp., one strain of *Synechocystis*, and one strain of the filamentous *Pseudanabaena* were used as recipient cells, while *E. coli* S17-1 was used as a donor cell for conjugative transfer of the plasmid pSUP1021. The result showed a high frequency of strains with kanamycin resistance for cells of all seven strains after conjugation. This was the first example of the use of conjugative transfer to introduce genes with marine blue-green algae. As this method can be used to produce various deactivated mutant strains containing Tn5, it may be broadly applicable in the future in analysis of marine blue-green alga genes.

Conjugative transfer with the wide-ranging host vector pKT230 has also been used for introduction with *Synechococcus* sp., which has the fastest growth rate of the marine blue-green algae used in research to date. pKT230 is a wide-ranging host vector of the IncQ population with an RSF1010 replication unit, and codes for resistance to kanamycin and streptomycin.

Similarly, conjugation with *E. coli* S17-1 resulted in strains with phenotypes for kanamycin and streptomycin resistance appearing at a frequency of 10^{-5} to 10^{-4} conjugants per recipients. This wide-ranging host vector can be consistently recovered and <u>rotated</u> with *E. coli*, suggesting ample potential for use as a marine blue-green alga and *E. coli* shuttle vector. Successful introduction of pKT230 has also been achieved with conjugation in the marine blue-green algae *Synechococcus* sp. and *Synechocystis* sp., and in particular in *Synechococcus* ATCC29403 (PCC7335), a marine blue-green alga with complementary adaptation capabilities. These findings testify to the broad applicability of the conjugative transfer approach with marine blue-green algae.

To date, no blue-green algae (including freshwater varieties) with conjugation capabilities or transfer plasmids have been found. Gene exchange has been found in blue-green algae as a result of natural DNA absorption and cyanophage infection, but gene donation through conjugation from other Gram-negative bacteria is believed to have taken place. Hopefully, strains with conjugation capabilities can be isolated in the future for marine blue-green algae as well (Matsunaga 1992).

7.5   Microalgae Use in CO$_2$ Fixation and Production of Beneficial Substances

7.5.1   Biosolar Reactor CO$_2$ Fixation Using Marine Blue-Green Algae

Culturing of blue-green algae to date has involved the issue of artificial ponds. This method is also used today to culture *Spirulina* and other varieties in regions with strong sunlight. When blue-green algae reach high concentrations, however, sunlight penetrates only 10 cm or so below the water's surface, which presents natural limitations. The use of artificial ponds is thus only suitable for production in sites where land or staffing is inexpensive (Aresta et al. 2005).

When culturing microalgae, the area of light per unit volume must be greatly increased to raise efficiency. Glass tube and foil reactors have been developed to achieve this, but the area of light per unit volume never exceeded around 100 m^{-1} in either case. The area can be increased to around 500 m^{-1} with the use of optical fibers, but uniform scattering is difficult to achieve. For example, scattering through etching on a quartz fiber surface results in some surfaces that do not scatter light at all and others that scatter it strongly.

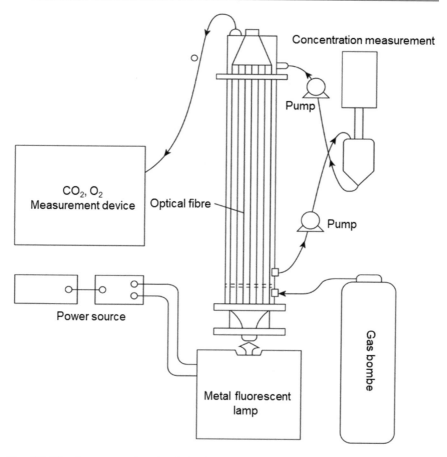

Fig. 7.8 Biosolar reactor using photolytic fibers

Matsunaga et al. used a new biosolar reactor with plastic and dispersion fibers as shown in Fig. 7.8 for high-density culturing of marine blue-green algae. This biosolar reactor has a light-absorbing area per unit volume of around 700 m^{-1} and 100% light emission efficiency (scattered light/light consumption). It has been used to grow the marine blue-green alga *Synechococcus* sp. at concentrations 10–50 times higher than those achieved before.

The same biosolar reactor has also been used with *Synechococcus* sp. (a marine blue-green alga that produces glutamic acid <u>outside</u> the leaf body) for continuous transforming the production of useful substances using carbon dioxide. If 5 mol CO_2 is presumed to result in 1 mol glutamic acid, the efficiency reaches as high as 28%. Continuous conversion of CO_2 to glumatic acid with this biosolar reactor using *Synechococcus* sp. resulted in a glutamic acid production rate per unit area around 13.6 times higher than with existing batch reactors and around 11.5 times higher than with reactors using immobilized leaf bodies. Efficient of light usage was also 5.4 times and 6.8 times higher, respectively. This light dispersion fiber biosolar reaction can allow for more efficient usage of marine blue-green algae (Matsunaga et al. 1991).

Biosolar reactors have thus been applied for CO_2 fixation. *Synechococcus* sp. has been cultured within the reactor, and CO_2 concentrations in the supply and emission gas have been measured to calculate the amount of CO_2 fixed. If the luminous intensity on the optical fiber surface is 20 μE^{-1} m^{-2} S^{-1} and the initial blue-green algae concentration is 6.8, 2.22 gL^{-1} of CO_2 will have been fixed within 12 h, resulting in a 0.97 gL^{-1} increase in *Synechococcus* sp. biomass.

Development of these biosolar reactors for high-density blue-green algae culturing has also necessitated development of monitoring and control systems. Matsunaga et al. developed sensors for one-line monitoring of blue-green algae volume within a bioreactor using phycocyanin fluorescence. Such monitoring systems are expected to be used in the future in new engineering for marine blue-green algae culturing (Matsunaga 1992).

7.5.2 Bioactive Substances in Blue-Green Algae

Blue-green algae are believed to contain various bioactive substances. From a molecular evolution perspective, production of vegetable bioactive substances is an area of particular interest. Matsunaga et al. developed a hot-water extract from various marine blue-green algae and compared the number of plant bodies produced from a carrot culturing cell, which resulted in the discovery of plant body regeneration promotion substances from *Synechococcus* sp. and other blue-green algae extracts. Induction of adventitious embryony and plant body regeneration was found in culture cells from *Cnidium* and from carrots, in which adventitious embryony and plant body redifferentiation is difficult to achieve.

The plant body regeneration promotion effects achieved with blue-green algae extract could not be induced with natural organic substances such as the amino acids used in plant tissue culturing. The quantity of known natural plant hormones in blue-green algae extract was also below the effective concentration in plant physiology terms. This indicated a possibility that a plant body regeneration promotion substance exists in blue-green algae extract (Matsunaga et al. 1991).

Major low and high molecular weight active ingredients have been found to exist in extracts from marine blue-green algae. One example is a low molecular weight alkaline substance that promotes somatic embryo maturation. High molecular weight fractioning has been found to promote young plant body and chlorophyll production in somatic embryos. Additional fractionation of this high molecular weight fraction resulted in isolation of GPF9.4, a substance that promotes plant body formation. GPF9.4 is a polysaccharide consisting of hexosamine, xylose, glucose, galactose, and an unidentified deoxysugar. GPF9.4 at a concentration of 10 mgL^{-1} has been found to promote the young plant body formation rate to 3.8 times when it is not added, and the amount of chlorophyll by 3.9 times.

Examination of blue-green algae extract's germination promotion effects in artificial seeds with carrot somatic embryos showed a germination rate of over 90% for artificial seeds containing the high molecular weight fraction (100 mgL^{-1}), indicating improvements in the artificial seed's germination defective conditions.

All marine blue-green algae also have very high protein content of nearly 50%. In some cases, the content is as high as 78.4%. Growth is also very fast, with a doubling time of nine hours, suggesting possibility for use of a new source of single-cell protein (SCP). The fatty acids contained in marine blue-green algae also include unsaturated fatty acids with 16 and 18 carbon atoms. Unsaturation occurs as a result of strong light and low temperatures during culturing, suggesting possible applications as a new oil source in addition to its pharmacological actions (Matsunaga et al. 1991; Fish and Codd 1994).

Additionally, marine blue-green algae have also been found to produce substances that can be used for cosmetics, including UV ray absorption substances, tyrosinase inhibitor, and superoxide dismutase. Sulfated polysaccharides produced by blue-green algae have also exhibited blood clotting and antiviral activity. Blue-green algae have been used commercial in pigment production as well, including phicocyanine, allophycocyanin, phycoerythrin, and red pigments (Vo et al. 2015).

7.6 Industrial Applications of Microalgae

7.6.1 Using Microalgae in Industry

The first research on microalgae in Korea took place with phytoplankton as part of a marine survey in 1915. Before Korea regained independence in 1945, research was primarily conducted by the Japanese. The first reports on phytoplankton research by Korean scholars came in the mid-1950s. Research on microalgae became more active in the late 1960s, and numerous academic achievements began to be realized. Research during this time focused chiefly on microalgae typology and ecological index organisms, but by the late 1980s it had expanded into physiological ecology, biochemistry, genetic engineering, and bioengineering. Recently, the focus has shifted to research and development for use of microalgae as a key biological resource in applications for the environment and bioindustry. Current research centers on increasing biomass and bioenergy capabilities by promoting photosynthesis mechanisms to absorb carbon dioxide, which is a cause of global warming (Pires et al. 2012).

Rapid advancements in biotechnology have resulted in microalgae gaining greatly in industrial usage value. Each microalgae has its own genetic properties, serving not only as a potentially useful food source for humans but also a source of various functional bioactive substances (such as antioxidants, anticancer agents, immunomodulators, and agents to combat skin aging), functional food and cosmetic ingredients, environmentally friendly ingredients, and bioenergy materials.

Biotechnology research with microalgae includes analysis of natural compounds for the development of alternative energy sources, health and functional foods, health supplements, skin cosmetics, skin ointments, moisturizers, aromatic plant activators, feed for marine farming, and pharmaceuticals. Additionally, there has been a wide range of biotechnological research with photobioreactors and fermenter culturing methods for the mass production of microalgae to produce specific pigments with high value-added, carbohydrates, amino acids, and unsaturated fats and as materials for wastewater treatment, fertilizer, and energy. Because microalgae have a relatively low cell density compared to other microbes, research and development has been conducted in connection with environmental industries to use waste resources (such as industrial waste and wastewater from livestock raising) to ensure economic efficiency in the culturing process by increasing productivity with reductions in the costs of mass culturing (Hong and Lee 2015).

7.6.2 Seeking Potential Bioactive Substances in Microalgae

Microalgae are photosynthesizing microbes that ① are easy to apply for industrial purposes, ② contain relatively large amounts of functional bioactive substances, and ③ possess high value-added. Another characteristic is that mass production can be achieved by optimizing value-added products through culturing conditions without the need for genetic recombination technology (Michalak and Chojnacka 2015).

In the advanced economies, research and development is already under way on health foods using microalgae such as *Chlorella*, *Spirulina*, *Dunaliella*, and *Haematococcus*, as well as microalgae-derived antioxidants such as lutein, astaxanthin, and cantaxanthin. Clinical reports indicate superior antioxidant effects to vitamin E (tocopherol) or beta carotene. Astaxanthin is a high value-added item that sells for US$2500 per kg, with sales reaching the billions of dollars in the U.S. alone (Fig. 7.12, photo 7.4).

The recent publication of research reporting various bioactive substances in microalgae has made them the focus of new attention. Among the substances reported are antimicrobial agents, enzyme inhibitors, cytotoxins, anti-inflammatories, superoxide dismutase, activators, anticancer agents, and anti-HIV agents.

The author's own laboratory has confirmed liver cell protection and liver fibrosis suppression effects in peptides isolated from enzymatic hydrolysates of the diatom *Navicula incerta*. The results of treatment of liver cells damaged by alcohol use with the microalgae-derived peptides are shown in Fig. 7.9. While liver cells treated with alcohol showed disruption to normal cell formation due to expression of inflammation-inducing factors as a result of cell surface damage, liver cells were found to be protected when treated with microalgae-derived peptides, due to suppression of alcohol-induced liver cell damage.

Peptides isolated from *N. incerta* enzymatic hydrolysates were also shown to be effect in suppressing liver fibrosis. As seen in Fig. 7.9, examination of liver cell

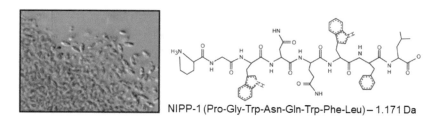

NIPP-1 (Pro-Gly-Trp-Asn-Gln-Trp-Phe-Leu) – 1.171 Da

Peptides isolated from the microalga *Navicula incerta*

| Normal cells | Alcohol-treated cells | Peptide-treated cells |

Fig. 7.9 Liver cell protection effects of microalgae-derived peptides

Vacuole

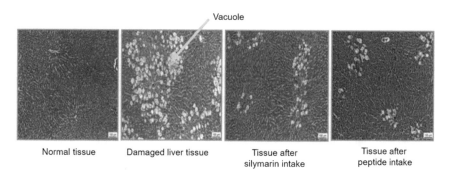

| Normal tissue | Damaged liver tissue | Tissue after silymarin intake | Tissue after peptide intake |

Fig. 7.10 Liver fiber tissue reduction effects of microalgae-derived peptides in mice with carbon tetrachloride-induced liver damage

tissues after feeding samples containing microalgae-derived peptides to animal models with liver damage induced by the hepatotoxin carbon tetrachloride showed marked reductions in the many vacuoles formed within the liver cells as a result of the toxin. The effect was found to be comparable to that of silymarin, which is currently used in the treatment of liver disease (Fig. 7.10) (Kang et al. 2012).

Additionally, bioactive peptides have also been isolated from a microalgae ferment produced by *Chlamydomonas,* one of the most prominent photosynthetic flagellates, or single cell organisms with flagella, after it is fermented by microbes such as *Candidata utilis, Bacillus subtilis* (Fig. 7.11).

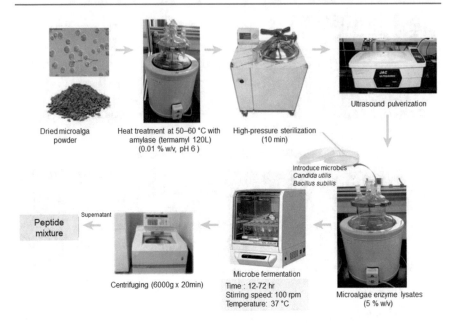

Fig. 7.11 Peptide isolation from microbial culturing of the microalgae *Chlamydomonas*

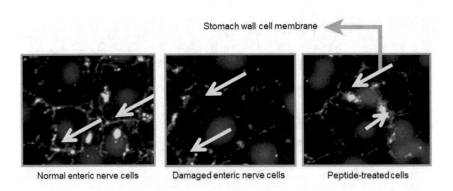

Fig. 7.12 Enteric nerve cell protection with immuno-staining

Immuno-staining was also used to examine the effects of peptides isolated from microalgae ferment on enteric nerve cells in the stomach wall following damage with *Helicobacter pylori*. As shown in Fig. 7.12, stomach wall nerve cells damaged by *H. pylori* exhibited damage due to collapse in the cell membrane boundaries. In contrast, those treated with peptides exhibited similar cell wall formation to normal cells and gradual recovery (Cheung et al. 2015).

Many other functional foods and pharmaceuticals have been obtained from marine resources as well. Marine microalgae are thus expected to continue contributing to human lives in various ways as a new resource for the future, particularly in high value-added functional food and Chinese medicine production.

7.7 Chapter Summary and Conclusion

As recently as the mid-1980s, microalgae research in Korea was limited to basic typological studies. Microalgae themselves remained utterly unfamiliar to the general public. Today, they are at the heart of marine bioindustry, with potential to drive global clean energy trends in the future and open a veritable Pandora's box of future marine bioindustry possibilities. They are without a doubt a biological research with vast untapped potential.

Primary producers in aquatic environments, microalgae are a vast biological resource in terms of both volume and variety, with tens of thousands of species producing more than 20 billion ton of organic matter per year. Microalgae typically grow far more quickly than land-based plants and can be easily cultured in both freshwater and seawater, or in any environment with light energy. They possess great potential as a material in bioindustry, allowing for low-cost production of industrially useful high molecular weight substances such as proteins, fats, sugars, and pigments as well as substances with specific physiological functions.

Taking advantage of microalgae's usefulness will first require mass culturing through artificial purification and isolation. Microalgae culturing is essential not only for phycological research but for basic science and understanding of aquatic ecosystems. Mass-cultured microalgae are also the focus of active use and development in a variety of industry areas as a source of food for marine and livestock forming and a material in fertilizer, functional health supplements, food additives, pharmaceuticals, industry, wastewater treatment, atmospheric purification, and bioenergy.

An invisible biological resource war is now under way around the world. Through active biotechnology research, microalgae may yet become not only a subject of academic interest but a high value-added bioindustry resource for Korea's future through development of high-functioning antioxidants, pharmaceuticals, health foods, functional cosmetics, and bioenergy.

References

Aaronson, S., & Dubinsky, Z. (1982). Mass production of microalgae. *Experientia, 38*(1), 36–40.

Aresta, M., Dibenedetto, A., & Barberio, G. (2005). Utilization of macro-algae for enhanced CO_2 fixation and biofuels production: Development of a computing software for an LCA study. *Fuel Processing Technology, 86*(14–15), 1679–1693.

Becker, E. (2007). Micro-algae as a source of protein. *Biotechnology Advances, 25*(2), 207–210.

Borowitzka, M. A. (1995). Microalgae as sources of pharmaceuticals and other biologically active compounds. *Journal of Applied Phycology, 7*(1), 3–15.

Borowitzka, M. A. (1999). Commercial production of microalgae: Ponds, tanks, and fermenters. *Progress in Industrial Microbiology, 35*, 313–321.

Capelli, B., & Cysewski, G. R. (2010). Potential health benefits of spirulina microalgae. *Nutrafoods, 9*(2), 19–26.

Cheung, R. C. F., Ng, T. B., & Wong, J. H. (2015). Marine peptides: Bioactivities and applications. *Marine Drugs, 13*(7), 4006–4043.

Costa, J. A. V., & De Morais, M. G. (2011). The role of biochemical engineering in the production of biofuels from microalgae. *Bioresource Technology, 102*(1), 2–9.

Doan, T. T. Y., Sivaloganathan, B., & Obbard, J. P. (2011). Screening of marine microalgae for biodiesel feedstock. *Biomass and Bioenergy, 35*(7), 2534–2544.

Dunahay, T. G., Jarvis, E. E., Dais, S. S., & Roessler, P. G. (1996). Manipulation of microalgae lipid production using genetic engineering. *Applied Biochemistry and Biotechnology, 57*(1), 223.

Fish, S. A., & Codd, G. (1994). Bioactive compound production by thermophilic and thermotolerant cyanobacteria (blue-green algae). *World Journal of Microbiology and Biotechnology, 10*(3), 338–341.

Gerwick, W. H., Roberts, M. A., Proteau, P. J., & Chen, J.-L. (1994). Screening cultured marine microalgae for anticancer-type activity. *Journal of Applied Phycology, 6*(2), 143–149.

Hong, S. J., & Lee, C. G. (2015). Microalgal systems biology through genome-scale metabolic reconstructions for industrial applications. In S. K. Kim (Ed.), *Handbook of marine microalgae: Biotechnology advances* (pp. 353–370). Oxford, UK: AP.

Hoseini, S., Khosravi-Darani, K., & Mozafari, M. (2013). Nutritional and medical applications of spirulina microalgae. *Mini Reviews in Medicinal Chemistry, 13*(8), 1231–1237.

Kang, K. H., Qian, Z. J., Ryu, B., Karadeniz, F., Kim, D., & Kim, S. K. (2012). Antioxidant peptides from protein hydrolysate of microalgae Navicula incerta and their protective effects in HepG2/CYP2E1 cells induced by ethanol. *Phytotherapy Research, 26*(10), 1555–1563.

Kao, C.-Y., Chiu, S.-Y., Huang, T.-T., Dai, L., Hsu, L.-K., & Lin, C.-S. (2012). Ability of a mutant strain of the microalgae *Chlorella* sp. to capture carbon dioxide for biogas upgrading. *Applied Energy, 93,* 176–183.

Kellam, S. J., & Walker, J. M. (1989). Antibacterial activity from marine microalgae in laboratory culture. *British Phycological Journal, 24*(2), 191–194.

Kumar, M. S., Miao, Z. H., & Wyatt, S. K. (2010). Influence of nutrient loads, feeding frequency and inoculum source on growth of *Chlorella vulgaris* in digested piggery effluent culture medium. *Bioresource Technology, 101*(15), 6012–6018.

Kumazawa, S. (1991). Screening, cultivation of marine microalgae, In S. Miyachi, N. Saga, & T. Matsunaga (Eds.), *Labo-manual marine biotechnology* (pp. 18–28). Tokyo, Japan: Shokabo Publishing Co.

Larkum, A. W., Ross, I. L., Kruse, O., & Hankamer, B. (2012). Selection, breeding and engineering of microalgae for bioenergy and biofuel production. *Trends in Biotechnology, 30*(4), 198–205.

Lee, Y.-K. (1997). Commercial production of microalgae in the Asia-Pacific rim. *Journal of Applied Phycology, 9*(5), 403–411.

Liu, J., & Hu, Q. (2013). Chlorella: Industrial production of cell mass and chemicals. In *Handbook of microalgae culture: Applied phycology and biotechnology* (pp. 327–338).

Maruyama, I., & Ando, Y. (1992). Mass culturing microalgae: Chlorella. In K. Yamaguchi (Ed.), *Utilization of microalgae* (pp. 18–30). Yokyo, Japan: Kouseisha Kouseikaku Publishing Co.

Matsunaga, T. (1992). Development of biotechnology, In K. Yamaguchi (Ed.), *Utilization of microalgae* (pp. 81–101). Tokyo, Japan: Kouseisha Kouseikaku Publishing Co.

Matsunaga, T., Takeyama, H., Sudo, H., Oyama, N., Ariura, S., Takano, H., et al. (1991). Glutamate production from CO_2 by Marine Cyanobacterium *Synechococcus* sp. *Applied Biochemistry and Biotechnology, 28*(1), 157.

Matsunaga, T., Takeyama, H., Nakao, T., & Yamazawa, A. (1999). Screening of marine microalgae for bioremediation of cadmium-polluted seawater. *Journal of Biotechnology, 70*(1–3), 33–38.

Michalak, I., & Chojnacka, K. (2015). Algae as production systems of bioactive compounds. *Engineering in Life Sciences, 15*(2), 160–176.

Moore, R. E., Patterson, G. M., & Carmichael, W. W. (1988). New pharmaceuticals from cultured blue-green algae. *Biomedical Importance of Marine Organisms, 13*(1988), 143–150.

Muller-Feuga, A. (2000). The role of microalgae in aquaculture: Situation and trends. *Journal of Applied Phycology, 12*(3–5), 527–534.

Ng, D. H. P., Ng, Y. K., Shen, H., & Lee, Y. K. (2015). Microalgal biotechnology: The way forward, In S. K. Kim (Ed.), *Handbook of marine microalgae: Biotechnology Advances* (pp. 69–77). London, UK: Academic Press.

Ötleş, S., & Pire, R. (2001). Fatty acid composition of Chlorella and Spirulina microalgae species. *Journal of AOAC International, 84*(6), 1708–1714.

Pires, J. C. (2015). Mass production of microalgae. In *Handbook of marine microalgae* (pp. 55–68). Elsevier.

Pires, J., Alvim-Ferraz, M., Martins, F., & Simões, M. (2012). Carbon dioxide capture from flue gases using microalgae: engineering aspects and biorefinery concept. *Renewable and Sustainable Energy Reviews, 16*(5), 3043–3053.

Pulz, O., & Gross, W. (2004). Valuable products from biotechnology of microalgae. *Applied Microbiology and Biotechnology, 65*(6), 635–648.

Radakovits, R., Jinkerson, R. E., Darzins, A., & Posewitz, M. C. (2010). Genetic engineering of algae for enhanced biofuel production. *Eukaryotic Cell, 9*(4), 486–501.

Raja, R., Hemaiswarya, S., Kumar, N. A., Sridhar, S., & Rengasamy, R. (2008). A perspective on the biotechnological potential of microalgae. *Critical Reviews in Microbiology, 34*(2), 77–88.

Saino, T., & Hattori, A. (1978). Diel variation in nitrogen fixation by a marine blue-green alga, *Trichodesmium thiebautii*. *Deep Sea Research, 25*(12), 1259–1263.

Spolaore, P., Joannis-Cassan, C., Duran, E., & Isambert, A. (2006). Commercial applications of microalgae. *Journal of Bioscience and Bioengineering, 101*(2), 87–96.

Takeyama, H., & Matsunaga, T. (1989). Genetic recombination of photosynthetic bacterium, microalgae, In T. Matsunaga (Ed.), *Marine bio marine new materials and new substances* (pp. 163–167). Tokyo, Japan: CMC, Publishing, Co.

Taub, F. B., & Dollar, A. M. (1964). A Chlorella-Daphnia food-chain study: The design of a compatible chemically defined culture medium 1, 2. *Limnology and Oceanography, 9*(1), 61–74.

Van Baalen, C. (1962). Studies on marine blue-green algae. *Botanica Marina, 4*(1–2), 129–139.

Vo, T. S., Ngo, D. H., & Kim, S. K. (2015). Nutritional and pharmaceutical properties of microalgal spirulina, In S. K. Kim (Ed.), *Handbook of marine microalgae: Biotechnology advances* (pp. 299–308). Oxford, UK: AP.

Watanabe, K., Takihana, N., Aoyagi, H., Hanada, S., Watanabe, Y., Ohmura, N., et al. (2005). Symbiotic association in Chlorella culture. *FEMS Microbiology Ecology, 51*(2), 187–196.

Wilde, E. W., & Benemann, J. R. (1993). Bioremoval of heavy metals by the use of microalgae. *Biotechnology Advances, 11*(4), 781–812.

Yamaguchi, K. (1992). Present situation and future direction, In K. Yamaguchi (Ed.), *Utilization of microalgae* (pp. 9–17). Tokyo, Japan: Kouseisha Kouseikaku Publishing Co.

Developing Functional Materials with Marine Organisms

8

Contents

8.1   Introduction

Efforts to explore and develop drugs and other bioactive substances from natural resources have a history dating back more than three centuries. These efforts have been particularly active over the past few decades. In the past, the chief targets of these explorations were land-based resources, and plants in particular. This was due to humankind's long history of using land-based plant resources for medicinal purposes, which resulted in a large base of knowledge, and to the ease with which resources could be mass-produced. Research into land-based resources to date has been so wide-ranging that virtually no species remain unstudied (Newman et al. 2003).

For this reason, the world's advanced economies have begun turning their research investment attention to marine ecosystems, which have remained an unexplored area. Recent advancements in collection, aquaculture, and analysis techniques and academic understanding of related areas have resulted in even more active research in this area. In terms of diversity of species, the oceans are an almost unlimited repository of resources, with over 80% of all organisms on the planet (around 500,000 species) known to inhabit marine environments. In the East,

© Springer Nature Switzerland AG 2019
S.-K. Kim, *Essentials of Marine Biotechnology*,
https://doi.org/10.1007/978-3-030-20944-5_8

natural marine products also have a very long history as frequent ingredients in Oriental medicine and folk remedies.

Marine environments form distinctive ecosystems that differ from those on land. The secondary metabolites developed by marine organisms to stay alive in the competition for survival of the fittest—particularly those lacking physical defense capabilities—are often quite different chemically from those of land-based organisms. While these secondary metabolites are understood to have been developed as a means of chemical defense, the substances also exhibit powerful bioactivity when introduced into humans and other mammals. For this reason, research into the development of new leading bioactive materials from marine organisms and their use toward human health has been the focus of recent attention.

Early studies into marine natural products consisted mainly of simple phytochemical studies motivated by academic curiosity. As the final goal has shifted to the development of marine-based pharmaceuticals, the trend has moved toward multidisciplinary research into bioactive ingredients involving a number of different areas, including pharmacology, ecology, biochemistry, and medicine. Pharmacological research into marine natural products has also expanded from early studies of toxicity (including tetrodotoxin and saxitoxin) toward various forms of pharmacological activity, including anti-cancer (cell toxicity), antiviral, and antiinflammatory properties (Fanning et al. 2011).

Pharmacological research groups investigating the bioactivity of marine natural products have not been especially diverse, their primary focus being on a few forms of activity such as anticancer, antiviral, antiinflammatory, and antibacterial effects. While it is difficult to reach any definite conclusions, research reports to date have uncovered bioactivity in terms of anticancer (cell toxicity) and antiviral effects in marine natural products with great frequency.

A number of barriers must be cleared before a pioneering bioactive substance discovered in or chemically synthesized from natural products can be developed into a commercial item. In sheer probabilistic terms, only a few of the tens of thousands of candidate substances ever reach the final development stage. In the case of marine natural products (as opposed to land-based products), there is the additional obstacle of securing resources. In addition to the quantities needed for the pre-clinical and clinical stages, consideration in developing items for pharmaceutical production must also be given to whether 100% of the necessary resources for pharmaceutical development can be obtained.

In terms of the ability to obtain resources, pioneering bioactive substances from marine sources can be classified into two categories: those for which source organisms can be cultured or farmed, and those for which they cannot. Numerous other practical barriers also exist, although large quantities of source materials can theoretically be obtained through mass culturing.

Culturing marine microorganisms entails a number of difficulties. Examples included whether strains only grow at low cell densities, whether culturing requires high saline concentrations that can lead to corrosion in production facilities, or whether suitable growth can only be maintained at low temperatures, resulting in longer culturing times (and, by extension, longer production times) and a greater risk of contamination by other microorganisms. The most fundamental concern, however, is a lack of adequate knowledge to date on the culturing conditions for marine microorganisms. With many pharmaceutical companies continuing to research solutions to problems with marine microorganism fermentation, methods for addressing these issues are likely to emerge in the near future. Indeed, the U.S. company Martek Bioscience is currently mass-culturing the microalga *Cryptheco-dinium cohnii* to produce docosahexaenoic acid (DHA).

This chapter will examine the marine organism-derived antibacterial, anti-inflammatory, and anticancer substances that have received attention to date as pharmaceutical materials. It will also share about active research efforts currently under way on research reagents derived from these materials, as well as ingredients in cosmetics and functional foods and new biomaterials.

8.2 Pharmaceutical Materials

8.2.1 Anti-cancer Agents

One of the most active areas of bioactive substance development using marine organisms is the development of anticancer agents. Collaborative research with industry and academia to develop anticancer agents from marine organisms has been taking place under the leadership of the National Cancer Institute (NCI) in the U.S. The NCI's search for anticancer effects in various sponges have shown anti-cancer activity appearing frequently in the sponges (phylum Porifera), tunicates, bony fish (superclass Osteichthyes), and comb jellies (phylum Ctenophora). The chief methods currently used to search for anticancer effects are *in vitro* and *in vivo* approaches using cancer cells from humans and other animals; a variety of simpler *in vitro* approaches are also used for mechanism-based searches, including DNA cleavage assays, topoisomerase assays, protein kinase C assays, collagenase assays, and angiopoiesis inhibitor assays (Singh et al. 2008).

One fast, simple, and highly sensitive in vitro anticancer effect assay used with cancer cells employs KB human nasopharyngeal carcinoma cells. This approach involves observing the cancer cell toxicity of search specimens. Typically, an ED_{50} value (50% effective dose) value of 20 µg/ml or less is viewed as a significant anticancer effect for crude extract, and 10 µg/ml or less for the pure substance.

One search method using live animals as the P388 murine leukemia approach. In P388 searches, an increase of lifespan (ILS) of 20% or more in mice (T/G 120%) is seen as indicating lack of significance as an anticancer agent. Other approaches using L1210 lymphoid leukemia, B16 melanoma, M5076 sarcoma, and Mx-1 human mammary tumors are used for more sophisticated measurement of anti-cancer effects. For these searches, a T/G of 150% or more is seen as indicating that a substance is worth being used for clinical testing.

NCI is currently working to develop *in vitro* search approaches using around 60 types of human cancer cells, including leukemia, lung cancer, colon cancer, central nervous system tumors, melanoma, ovarian cancer, renal cancer, prostate cancer, and breast cancer. As a first step, searches are performed for the 60 types of cancer cells for specific prescriptions. If selective activity is found in a particular cancer cell, the human cell is transplanted to an athymic mouse for animal testing. This system allows for logical and efficient searches for anticancer effects. Ideal though it may be, however, it entails tremendous maintenance costs and is beyond the capabilities of most ordinary institutes.

For this reason, several simpler search methods have been developed. One of these is the sea urchin egg assay. Newly fertilized sea urchin eggs rapidly undergo cell division, and the observed effects of search samples in the division process may be used as an indicator of anticancer effects. This method can be useful not only in tracking substances to suppress cell division, but also in studying the mechanisms of prospective anti-cancer agents. Another advantage is that the results can be observed quickly without the need for particular equipment or technology. Time and effort can be saved by using this method to perform an initial search and selection of materials, which are then subjected to a later examination for biological applications.

In addition to the sea urchin egg assay method, the crown-gall potato disc assay and brine shrimp bioassay methods are currently being used to examine anti-cancer effects. A crown gall is a type of tumor that forms on plants due to the plant bacterium *Agrobacterium tumefaciens*. Examination of the presence and extent of crown gall formation when a potato section is inoculated with *A. tumefaciens* and a search sample can serve as an indicator for anti-cancer effects. While the inability to distinguish germicidal effects on *A. tumefaciens* and tumor suppression effects remains a problem, experiments have shown a relatively strong correlation with the P388 search method for biological application, which means that this can serve as a simple laboratory-based approach.

The brine shrimp assay is a simple toxicity test, in which eggs from the brine shrimp (*Artemia salina*) are incubated and the toxicity of a larva search sample is examined to serve as an indicator of cell toxicity. While this approach cannot strictly be called specific to examination for anti-cancer effects, it may be used as a fast and convention method when tracking active substances.

8.2.2 Antiviral and Antibacterial Substances

While viral diseases can typically be prevented with vaccines, many of them, including acquired immune deficiency syndrome (AIDS), cannot be easily eradicated with the vaccine approach. The need for viral disease medications is growing, yet the number of antiviral agents developed to date is not large, and those that do exist are chiefly nucleosides (Ehresmann et al. 1977; Sastry and Rao 1994).

Sesquiterpene hydroquinone, a substance isolated from the Mediterranean sponge *Dysidea* sp., possesses substantial antiviral effects, while avarone and avarol have HIV-1 reverse transcriptase inhibition functions that suggest potential applications in AIDS treatment (Fig. 8.1).

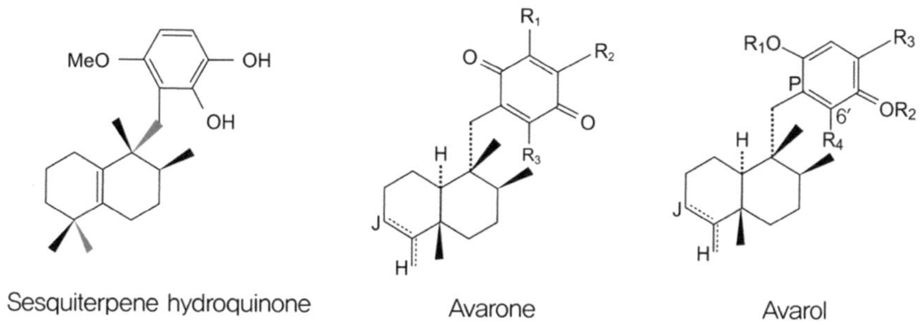

Fig. 8.1 Substances isolated from the Mediterranean sponge *Dysidea* sp.

Adenine arabinoside (Ara A, Vidarabine), a nucleoside isolated from the Mediterranean marine sponge *Eunicella cavolini*, has been used in antiviral medication due to its strong germicidal properties against viruses with DNA. Eudistomin from the colonial sea squirt *Eudistoma olivaceum* has been shown to have germicidal effects in addition to its antiviral properties against the herpes simplex virus HSV-1 (Fig. 8.2). Ptilomycalin A from the Caribbean sponge *Ptilocaulis spiculifer* and the Red Sea sponge *Hemimycale* sp. is effective in stopping HSV growth and has anticancer and antifungal effects (Fig. 8.3). Patellazole B from the sea squirt *Lissoclinum patella* (Fig. 8.4) has shown powerful antiviral effects against HSV, while glycolipids from the blue-green algae *Lyngbya lagerheimii* and *Phornidium tenue* has been reported to suppress growth in HIV-1.

Sulfated polysaccharides from algae, including fucoidan metabolites from the brown alga *Laminaria japonica*, are known to impede various viruses, including HSV and HIV (Vo et al. 2012).

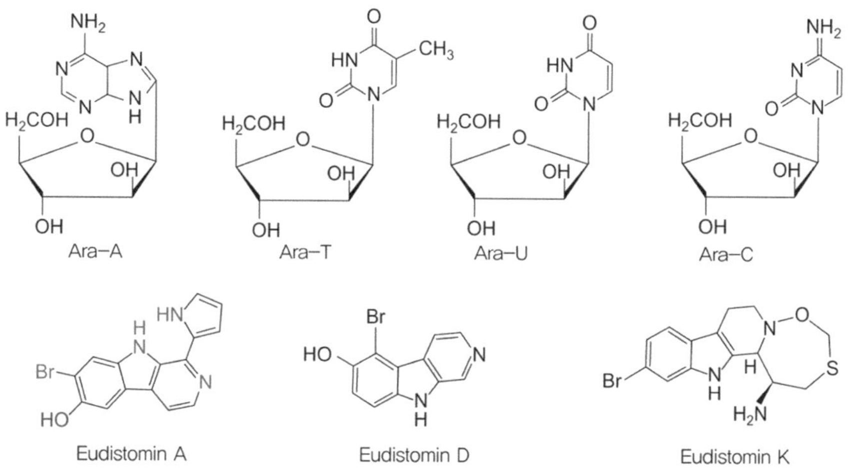

Fig. 8.2 Substances isolated from sponges

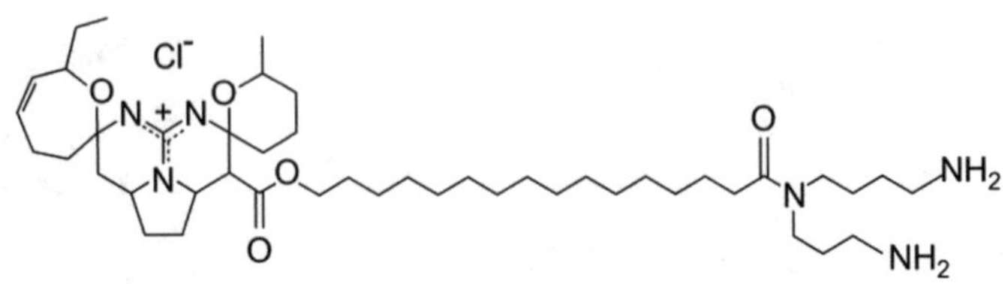

Ptilomycalin A

Fig. 8.3 Substance isolated from the marine sponges *Ptilocaulis spiculifer* and *Hemimycale* sp.

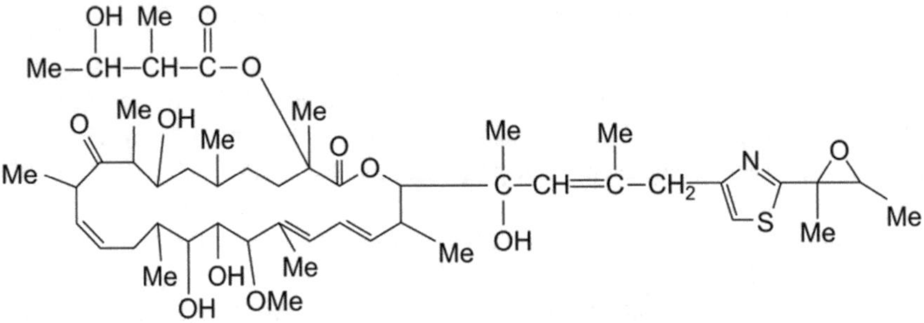

Patellazole B

Fig. 8.4 Substance isolated from the sea squirt *Lissoclinum patella*

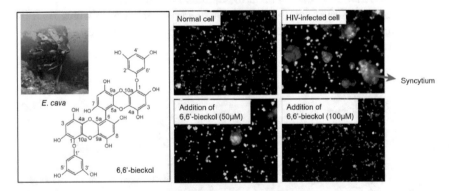

Fig. 8.5 Disruption of syncytium formation in HIV-infected cells with 6,6'-bieckol from *E. cava* (Artan et al. 2008; Karadeniz et al. 2011)

When a cell is infected with HIV, cell membrane fusion results in the formation of a syncytium to produce a large multinuclear cell. Recently, the author's laboratory found that 6,6'-bieckol from the brown alga *Ecklonia cava* notably suppressed syncytium formation in HIV-infected cells, suggesting possible application in treatments for AIDS (Fig. 8.5).

Several types of antibiotics have been isolated from a marine microorganism discovered near a wastewater treatment site outlet on the Mediterranean island of

Sardinia. One of them, cephalosporin C, possesses strong antibacterial properties and is effective against penicillin-resistant bacteria. Its discovery led to the development of several different synthetic cephalosporins, which are widely used today in treating illnesses due to bacterial infection (Fig. 8.6).

A saponin mixture from the edible sea cucumber *Sticophus japonicus* is also known to have marked growth suppression properties against athlete's foot fungus, and its chief ingredient holotoxin A has been used in athlete's foot medication (Fig. 8.7) (Kim 2018).

Derivatives of pseudomonic acid from the marine microorganism *Alteromonas* sp. have recently shown powerful antibacterial effects against *Staphylococcus aureus*, while new caprolactams isolated from as-yet unknown Gram-positive marine microorganisms from deep-sea sediments have shown antiviral activities against HSV-II at concentrations of 100 µg/ml (Fig. 8.8). New ring-shaped

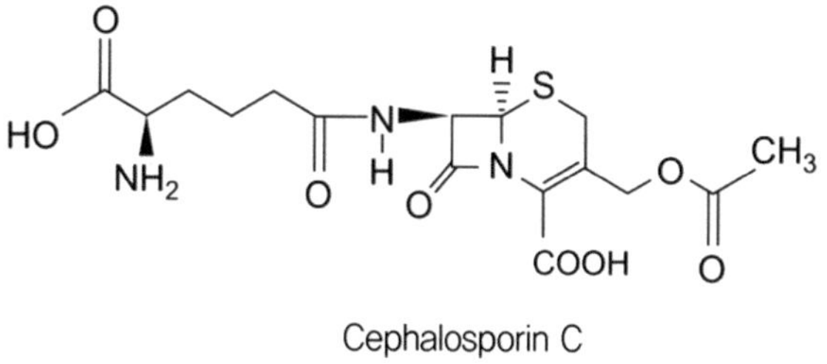

Cephalosporin C

Fig. 8.6 Cephalosporin C derived from marine microorganisms

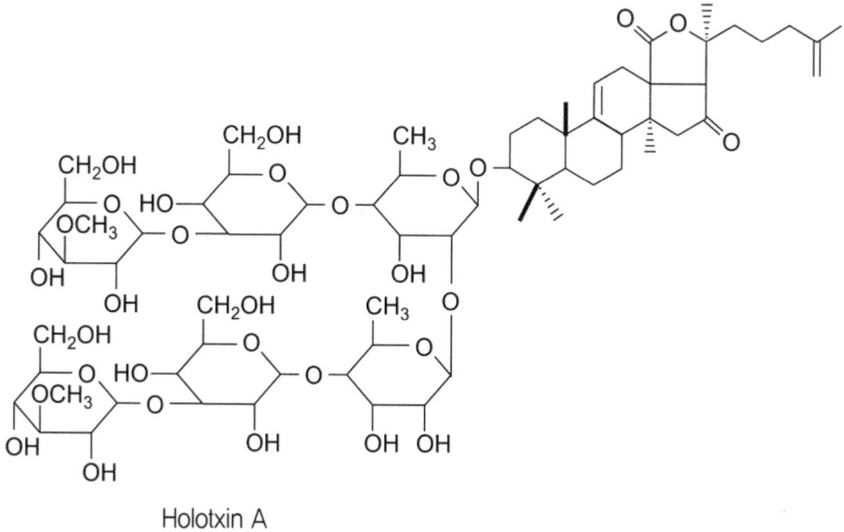

Holotxin A

Fig. 8.7 Holotxin A from the sea cucumber *S. japonicas*

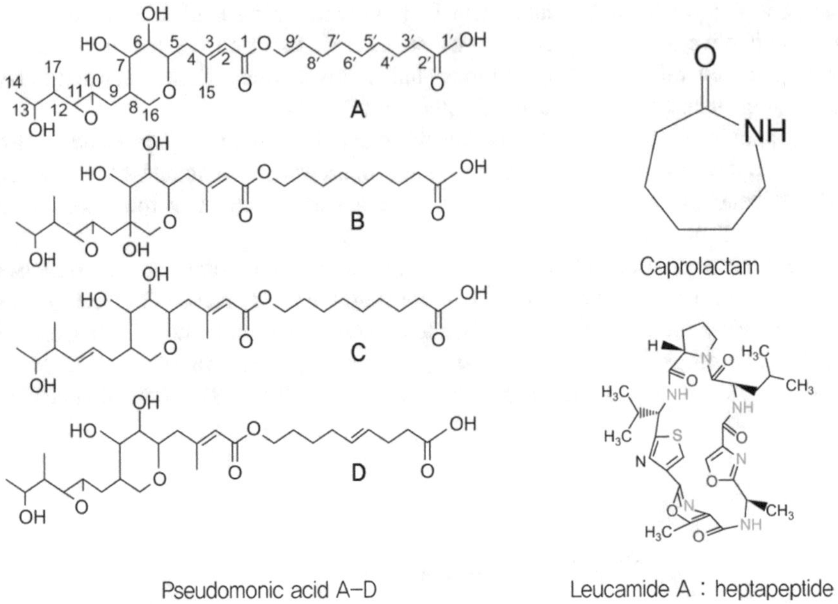

Pseudomonic acid A–D

Caprolactam

Leucamide A : heptapeptide

Fig. 8.8 Substances isolated from marine microorganisms

heptapeptides isolated from the sponge *Hymeniacidon* sp. have been shown to have powerful antifungal properties against the fungus *Cryptococcus neoformans*.

8.2.3 Pharmacologically Active Substances

The red alga *Digenea simplex*, better known as Corsican weed, has been used over the years as an insectifuge. Its effective ingredient is L-α-kainic acid, a kind of amino acid, and it has been used together with santonin as an antihelminthic that paralyzes parasites (roundworms, whipworms, and tapeworms) by blocking tissue respiration through suppression of dehydrogenase activity within their bodies (Fig. 8.9) (Endo et al. 1986).

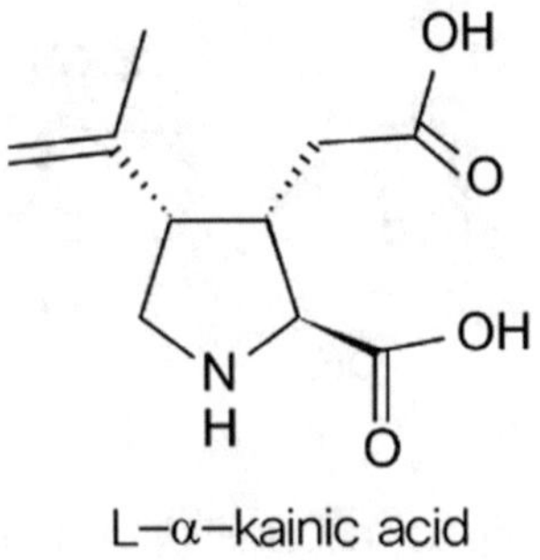

Fig. 8.9 L-α-kainic acid
from the red alga *D. simplex*

L–α–kainic acid

Inuits, who consume more fish and shellfish than meat-favoring Westerners, are noted for their very low morbidity rate from conditions such as hypertension and heart disease. This has been shown to result from the activities of eicosapantaenoic acid (EPA) and docosahexaenoic acid (DHA), which are unsaturated fatty acids found in fish oil. EPA, which has been shown to have effects in reducing platelet aggregation suppression, neutral fats, and cholesterol, was marketed in Japan in 1990 as a treatment for arteriosclerosis obliterans (Childs et al. 1990).

DHA is another nutrient that is expected to have a wide range of pharmacological properties. Research on DHA's pharmacological functions to date has shown effects in improving senile dementia and calcium functions, along with anticancer, antiallergy, and lipid reduction properties. Because EPA and DHA are nutrients present in large quantities in fish, they are expected to potentially have pharmacological properties second only to pharmaceuticals (Fig. 8.10).

From the fact that flies settle and die on the carcasses of bristle worm *Neanthes japonica* (which is used as bait in ocean fishing), nereistoxin has been isolated from the annelid (Fig. 8.11). Not only has it shown insecticidal properties against many other injurious pests besides flies, but its acute toxicity toward mammals is relatively low. Many kinds of related compounds have been synthesized, one of which —cartap, known commercially as Padan—is nontoxic to warm-blooded animals, degrades within the body or in nature, and is particularly effective against insects resistant to organophosphorus and chlorinated pesticides. This has led to its development into a widely used pesticide for farming.

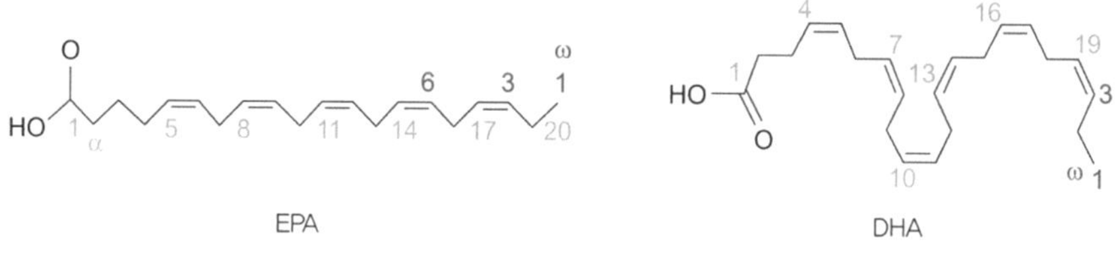

Fig. 8.10 Structures of EPA (eicosapentaenoic acid) and DHA (docosahexaenoic acid)

Fig. 8.11 Nereistoxin from the annelid *Neanthes japonica*

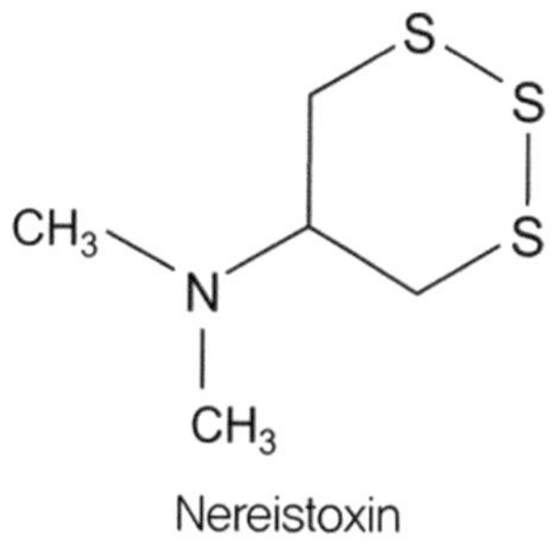

Nereistoxin

8.2.4 Anti-inflammatory Substances

Marine natural substances with strong anti-inflammatory and pain-relieving proper-ties have made many contributions to studies on arachidonic acid metabolism and calcium ion (Ca^{2+}) movement in inflammation reactions. Inflammation reactions are ultimately the result of calcium ion emission and movement within the cell. The calcium ion movement mechanism begins when an agonist combines with a receptor. The receptor communicates a signal by messenger of G-proteins (guanine nucleotide-binding proteins), activating phospholipases such as PLA_2 or PLC. The phospholipases hydrolyze phospholipids in the cell membrane, producing secondary signal transmitters such as inositoltriphosphate (IP_3) and arachidonic acid (Fig. 8.12). Calcium ions within the cell separate when IP3 binds to receptors on the rough endoplasmic reticulum. Calcium ion emission outside the cell has been found to be determined by the separation of arachidonic acid (Lewis and Lewis 1989).

Arachidonic acid passes through the cyclooxygenase pathway and is metabo-lized into prostaglandins, prostacyclin, or thromboxane. Another pathway is the lipoxygenase pathway, through which it is metabolized in tetraenoic acid, leuko-triene, or lipoxin. When these arachidonic acid metabolites bind to receptors in the calcium channel, movement of calcium ions outside the cell occurs. Many of the drugs identified so far as functioning on inflammation reactions produce their anti-inflammatory effects through regulation of phospholipid metabolism or cal-cium ion movement. Substances that are found to hinder phospholipase or calcium ion movement may thus be used to develop anti-inflammatory analgesics.

Substances with powerful anti-inflammatory and pain-relieving properties have been discovered in marine natural materials. Sesterterpenoid manoalide, which is found in the sponge *Luffariella variabilis*, has been found to have powerful anti-inflammatory properties and is currently undergoing clinic testing. Pseu-doterosine, a diterpene ribiside isolated from the Caribbean gorgonians *Pseu-dopterogorgia bipinata* and *P. elisabethae*, has powerful anti-inflammatory and pain-relieving properties and inhibits eicosanoid biosynthesis by reversibly coun-teracting lipoxygenase and phospholipase A_2 (Fig. 8.13).

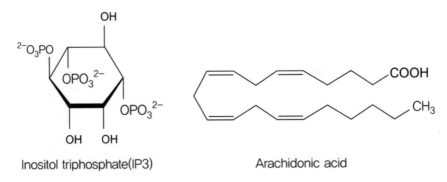

Inositol triphosphate(IP3)   Arachidonic acid

Fig. 8.12 Structures of inositol triphosphate (IP_3) and arachidonic acid

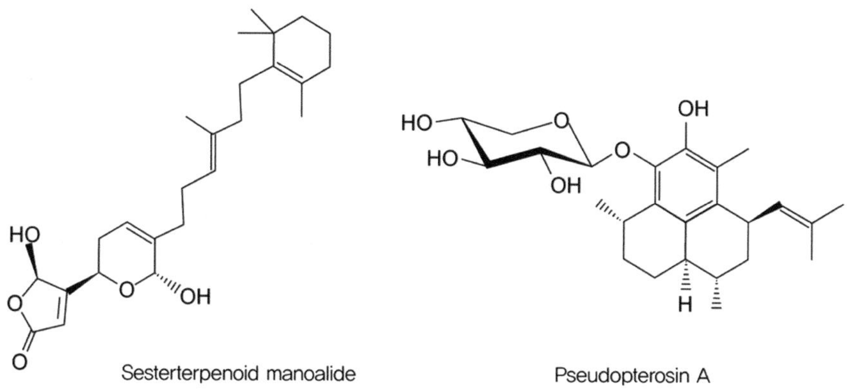

Fig. 8.13 Substances isolated from sponges

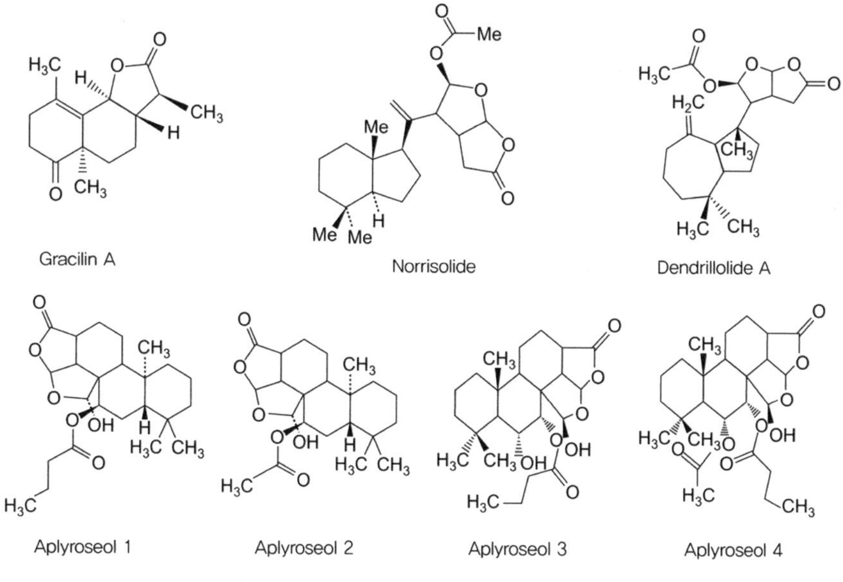

Fig. 8.14 Substances isolated from marine organisms

Other anti-inflammatory substances found in marine organisms include gracilin A, norrisolide, dendrillolide A, and aplyroseols (Fig. 8.14) (Kim et al. 2013).

8.2.5 Research Reagents

Plant and animal toxins have long been in wide use as research reagents in neurophysiology. In the mid-1960s, the poison of the puffer fish was found to suppress the flow of sodium ions into the cell by specifically blocking the membrane, which consists of proteins that selectively allow sodium ions in the ion-selective channel to pass. The poison was ultimately used in various forms of research, including studies on nerve transmission functions. Great developments were achieved in the field of neurophysiology as a result.

The discovery of these specific reagents has contributed greatly to academic development, and their importance as research reagents has only grown as recent research in the life sciences progresses to the molecular level. The acquisition of so much information about biological phenomena involving specific receptors has naturally led to a large increase in demand for these specific reagents (Fusetani 2005).

As an example, one may point to the discoveries of okadaic acid and calyculin A, specific inhibitors for protein phosphatases 1 and 2A. These compounds were first isolated from sponges as cell toxins, and their mechanism was found to take place through enzyme inhibition, leading to their widespread use in analyzing the biochemical phenomena involved in protein phosphorylation and dephosphorylation. Great advancements were thus achieved in our understanding of information transmission within the cell, muscle contraction, and carcinogenesis. A few years ago, British companies in Canada were found to have focused on increased production in response to a global shortage of the research reagent kainic acid, a revelation that underscored the industry importance of research reagents.

8.3 Bio Materials

8.3.1 Underwater Adhesives (Anticipated Uses in Dentistry and Surgery)

Recently, adhesives have been used in place of bolts in car part assembly and in place of needlework in textiles. Adhesively hold a particularly important position in the shoe industry. Growing uses in recent years have led to the development and marketing of a great variety of adhesives. Despite their varied development, however, there is one environment today where adhesives are not yet put to adequate use: underwater.

Water disrupts adhesion in various ways, as when it inserts itself between the adhesive and the attached substance to weaken or degrade the adhesive. The problem is severe enough for some to simply argue that "water and adhesives don't mix." While adhesives have been developed that are capable of maintaining sufficient hardness underwater once solidified, no adhesive yet exists that can achieve adhesion with sufficient hardness underwater or in damp environments. For example, adhesives that can be used in damp environments are urgently needed not only for the building of underwater structures, but also for dental treatment and surgery (Dove and Sheridan 1986).

A. Underwater Adhesion by Attached Organisms

Various forms of attached organisms inhabit the seas, including shellfish and sea squirts. They have earned a negative reputation as "fouling organisms" because of the economic damage they can cause with their powerful adhesion to underwater rocks, the bottoms of boats, fishing nets, and artificial structures such as generators, which can diminish the speed of boats and block mesh knots and waters. So strong is the adhesive force of sessile organisms that they are difficult to remove once attached to a surface. Such are the mysterious principles of nature: if their adhesion were weaker, they would be vulnerable to being carried away on incoming waves and would be unable to survive (Kamino 2008).

Attached organisms thus represent a powerful form of underwater adhesion that humans have been unable to achieve. Their adhesive substances have been to found to consist of proteins, and active research is currently under way to identify the genes for those proteins for mass-production of underwater adhesives through genetic engineering.

The most representative of attached organisms may be the sea mussel (*Mytilus edulis*). Sea mussels not only affix their oysterlike shell to surfaces such as rocks, but also secrete anywhere from a few to occasionally as many as 100 threads called byssi to anchor their bodies in place. A byssus consists of a thread portion and a velum, which resembles a sucker. This velum plays a role in adhering to rock, metal, concrete, and plastic surfaces (Fig. 8.15).

So strong is the sea mussel's adhesion that removal requires clipping of the thread, as the velum cannot be removed. Byssi are synthesized in an organ called a "foot." At the back of the foot is a crack where the byssi form; here is where the necessary ingredients are secreted for byssus formation. When a sea mussel is removed and placed in an aquarium filled with seawater, the individual strands of its byssi can easily be seen extending outward as it stretch out its foot (Silverman and Roberto 2007).

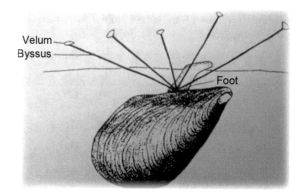

Fig. 8.15 Adhesion by a sea mussel (The byssi have been rendered large for ease of understanding.)

Velum
Byssus
Foot

B. Three Proteins Created by Byssi

The main constituents of the byssus are proteins. This was recognized relatively early on, but closer research has not been a simple matter. Studying the ingredients requires dissolving of the byssi and the refinement of their ingredients through various methods. Once formed, however, a byssus is almost completely resistant to dissolution with water, enzymes, or solvents, which prevents refinement through simple methods. Around 1980, one of the main ingredients in the byssus was successfully extracted from byssus secretion tissue in the U.S.

Byssus ingredients secreted from the foot could be eluted because they had not yet hardened. The ingredient was a protein with a molecular weight of around 130,000 daltons and a structure consisting mostly of ten repeating ten-amino acid units known as decapeptides (Fig. 8.16) and hexapeptides (with the centermost four amino acids removed). Decapeptides have a fascinating structure, containing rare amino acids that are almost never found normally in proteins, including hydroxyproline, dihydroxyproline, and 3,4-dihydroxyphenylalanine (DOPA). Hydroxyproline and dihydroxyproline are amino acids consisting of proline with one or two hydroxyl groups (–OH). Hydroxylproline had been previously reported chiefly as an ingredient in collagen, but it was the first time that dihydroxylproline had been discovered as a protein ingredient.

DOPA consists of tyrosine with one additional hydroxyl group and has been reported present in proteins in spider webs and the outer surface of insect eggs. As serine and threonine also have hydroxyl groups in their side chains, these proteins have a relatively large number of hydroxyl groups. The proteins have led in turn to the discovery of secondary and tertiary byssus constituents. The first of the proteins to be discovered has become known as "byssus protein 1." Protein 1 is thought to exist in the byssus thread and velum, surrounding the left side of the byssus.

While the amino acid sequence for protein 1 has been identified, its insolubilization mechanism is not yet fully understood. So far, it has been speculated that a DOPA-based bridging reaction is involved to some extent in this insolubilization. In other words, insolubilization is believe to occur due to protein 1 precursors synthesized in the cells of the foot: tyrosine is first hydrated through catechol oxidase (tyrosinase) to become DOPA, which in turn becomes dopaquinone,

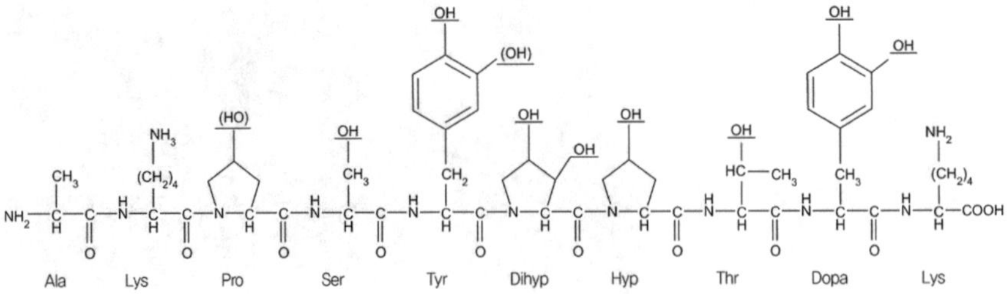

Fig. 8.16 Structure of the decapeptide (a constituent peptide in byssus proteins)

creating what is known as a quinone bridge with lysine and binding within or between molecules.

In the 1990s, a "protein 2" with a molecular weight of around 70,000 daltons was found in the velum portion. Like protein 1, protein 2 also contained DOPA, but differed in not containing either hydroxyproline or dihydroxyproline. Another characteristic was an abundance of cysteine, which is not found in protein 1. Japan's Inou et al. identified protein 2 as having a structure with 11 repeating sequences that are very similar to the cell growth factor known as epidermal growth factor (EGF).

Structures like EGF are known to exist in proteins in the extracellular cell matrix integrating communications between cells, but byssus protein 2 marked the first case of it being discovered as a component in *in vitro* tissues such as byssi. Protein 2 is thought to create the basic structure of the velum, but the presence of sequences shared with protein 1 at the end of the molecule raises a strong possibility that it insolubilizes through a similar mechanism. It may also bind together with protein 1.

A third protein measuring around 6000 daltons in molecular weight has also been discovered recently. This protein similarly contains DOPA in its amino acid sequence, but its sequence also includes an abundance of a rare amino acid known as hydroxyarginine (arginine with an additional hydroxyl group). Many quite similar proteins are present in protein 2, suggesting that they form a protein family. These proteins are also present in the velum and are thought to play an important role in adhesion, but the details are not yet understood (Hwang 2013).

C. Creating Adhesive Proteins with Genetic Engineering

Understanding proteins' properties and conformation will first and foremost require refinement and purification of the proteins themselves. Purification and analysis are rather time-consuming processes for adhesion proteins that have hardened and resist dissolving. Recent developments in genetic analysis technology, however, have enabled genetic cloning such that a full protein sequence can be identified quickly if even a portion of it is known. Indeed, identification of the full sequence of protein 1 in the sea mussel is a process that takes approximate half a year; for proteins 2 and 3, genes can be isolated and the sequence determined in roughly two to three months each. Genetic cloning has already become an essential technique for protein research. Its advantages extend beyond the ability to efficiently determine sequences; cloned genes can also be inserted into microorganisms and culture cells to produce proteins.

A research group at one U.S. genetics company achieved expression from the introduction of part of the genes from protein 1 in yeast. The resulting recombinant protein itself could not be used to obtain adhesive proteins (as noted before, DOPA production from tyrosine could not be generated through a hydroxyl group addition reaction), but the researchers reported that they achieved adhesive activity by applying catechol oxidase isolated and refined from button mushrooms and bacteria to the recombinant protein. While it was not determined whether the protein that they obtained had the same form as natural protein 1, the research at least demonstrated the potential for mass production of adhesive proteins, as it was able to produce proteins with adhesive properties through recombination.

Once all of the reactions that occur in byssus synthesis within the foot are identified, as well as the enzymes responsible for catalyzing those reactions, it should be possible to product adhesive proteins identical to those found in nature. Enhancement of adhesive protein gene sequences by recombinant DNA methods to create more powerful adhesive proteins may become more than wishful thinking.

The research team of Dr. Cha Y. J. of the Pohang University of Science and Technology, South Korea, has used molecular bioengineering production technology to commercialize bio adhesives from the sea mussel. The team has isolated genes for adhesive proteins from the mussels and used them to develop new forms of hybrid bio adhesives that can be used as underwater adhesives (Choi et al. 2015).

As seen in Fig. 8.17, the production of sea mussel adhesive proteins through genetic engineering first requires gene cloning to produce recombinant microorganisms capable of expressing the proteins. To do this, they first located genes that code for adhesive proteins in the sea mussel's foot, using the polymerase chain reaction (PCR) method to amplify the genes. Specific portions of DNA were recognized and clipped with restriction endonuclease for placement in plasmid gene carriers to create recombinant plasmids. These recombinant plasmids were then transformed in a *E. coli* production host to produce recombinant microorganisms. To produce the sea mussel proteins from the recombinant microorganism, protein

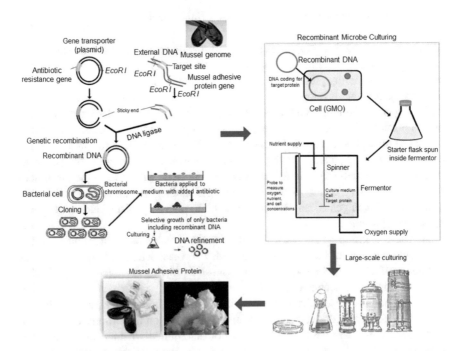

Fig. 8.17 Production process for sea mussel adhesive proteins

expressed was induced by culturing in a fermentor with supplies of oxygen and nutrients. Recombinant microorganisms expressing sea mussel adhesive proteins in their cytoplasm were disrupted and the proteins isolated, refined, and finally freeze-dried to produce a mussel adhesive protein powder. In the case of actual production, a large fermentor may be used sequentially for large-scale culturing to mass-produce recombinant sea mussel adhesive proteins.

In addition to the use of microorganisms and recombination technology, another form of protein production that is under consideration involves the use of culture cells. As with recombinant production using microorganisms, expression may also be achieved through introduction of byssus protein genes into cells. Natural byssus proteins may be produced as well by culturing the sea mussel foot cells that are originally responsible for byssus protein production (Wool and Sun 2011).

The future creation and commercialization of adhesives capable of working underwater are expected to allow them to be used not only in underwater construction but also in dental and surgical treatment settings.

8.3.2 Algae Use in Functional Paper Manufacturing

The chief roles of paper and cardboard are for recording and preserving writing and information, packaging and protecting objects, and absorbing fluids for disposal. Recent years, however, have seen growing demand for papers with additional special functions besides these three. To answer this demand for diverse and functional paper types, the sources used to make them have broadened from the natural fibers typically used to date to include synthetic, metal, and inorganic fibers. Paper processed to perform specific functions is drawing particular notice as a new material in such high-tech industry fields as electronics and bioengineering (Balcıoğlu et al. 2007).

The material used for the majority of paper and pulp to date consists of fibers obtained from land-based plants (trees) through mechanical or chemical processing. These plants have served to sustain human environments and been of great service in our lives, but the situation to date is one in which the global environment is facing serious crisis as a result of the various kinds of paper and fuel that we use. One of the biggest issues we will have to address in the 21st century is that of the global environment—an area closely linked to the very survival of humanity.

Human activities to date have burgeoned to a level qualitatively and quantitatively beyond the load capacities of the earth's circulation of matter. Rises in atmospheric carbon dioxide, methane, and carbon monoxide have given rise to global warming, and human activities have exacerbated the emission of chlorofluorocarbons (CFCs) that destroy the ozone layer in the stratosphere, the burning of chemical fuels that cause acid rain, and desertification of the earth due to loss of tropical forest.

Preservation of the earth's environment is thus crucial for the continued flourishing of humankind, and use of the purifying functions of the seas—which cover 70% of the earth's surface—have emerged as one means of achieving this. The recent shift toward paperless practices in information-related industries has resulted in declining demand for paper. At the same time, however, the ease of printing has led to explosive increases in the demand for different forms of paper (including information and printing paper, hygienic paper, and cardboard), which has accelerated the devastation of foreign resources.

It was in this context that the Earth Summit in Brazil in June 1992 saw the declaration of plans to continue pursuing sustainable development by keeping the load on the global environment within the earth's environmental allowance. Civilizations may have grown by consuming forests, but humankind's development has fed upon paper. In that sense, the continued rise in demand for paper, chiefly in developing economies, has necessarily led to an increase in demand for timber. One way of addressing this burgeoning demand for timber is through the use of marine polysaccharides—products of the active purification activities of marine organisms —as a raw material for paper.

A. Paper from Algae

If our goal is make paper from marine cellulose, our first thought may go to algae, which possess cellulose in their cell walls. Among the different alga types, cellulose is contained in green algae; the cell walls of brown algae are chiefly alginic acid, while those of red algae consist of carrageenan, amylose, and/or amylopectin.

Almost none of these have a cell wall structure consisting of polysaccharides that can be isolated in fiber form. One possible approach is artificial spinning, or extraction of polysaccharides from the cell wall for reconstitution in fibrous form. For a saccharide to be reprocessed into fiber form, its molecular structure must consist of a lineament with few branches. In that sense, alginic acid offers the most typical example.

Alginic acid is the chief polysaccharide in brown algae (Fig. 8.18), characteristically assuming a soluble sol form in sodium salt. From this, a water-insoluble gel form can be obtained from the salt by removing a certain quantity of cationic metal ions such as magnesium and mercury. Wet spinning can be achieved by applying these properties to a spinning method using a nozzle. The ability to spin alginic acids was discovered some time ago; during World War I, alginic acid fibers were used in Britain to make military tents (Rehm 2009).

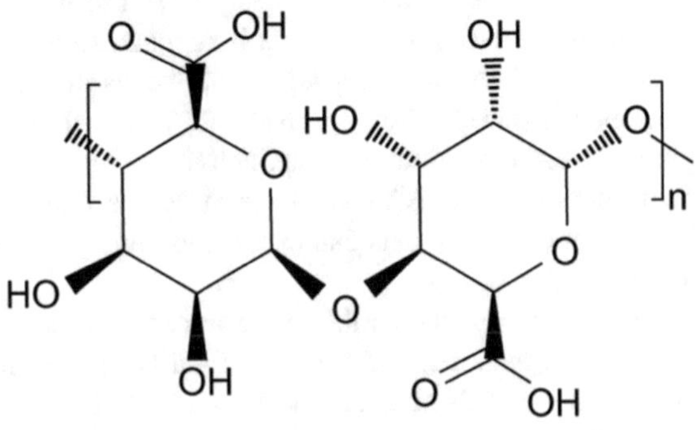

Fig. 8.18 Alginic acid structure

Paper made from alginic acid fibers is outwardly similar to artificially woven paper types such as conventional paper, and has been used in the making of speaker cones. Paper from cellulose fiber was originally used for speakers, but suffered problems with friction noises from the fibers; the use of paper made from alginic acid fiber was a solution to this issue. Alginic acid paper is also edible and can be used for internal food packaging. More interestingly, reagents such as barium acetate or acetic acid can be added to alginic acid fibers to produce thin film superconducting paper.

In Korea, paper production technology has undergone rapid growth since it was first introduced around 1900 for the making of currency. Today, the country ranks eighth worldwide in paper production and 21st in consumption, but remains almost entirely dependent on imports for both the raw materials and special paper types. If the abundance of natural algae resources can be used to develop new functional papers, perhaps there will come a day when South Koreans favor paper made from domestic ingredients.

8.3.3 Natural Liquid Crystals: A Cutting-Edge New Marine Material

A. Liquid Crystals: A Fourth Non-liquid State of Matter

Liquid crystals, which are frequently encountered in electronic calculators and digital watches, are a material with very unique properties. Typically, matter exists in three states (gas, liquid, or solid), depending on its temperature. Some forms of matter, however, have special properties that do not correspond to any one of these states. Liquid crystals represent a fourth state of matter with both liquid and solid properties. For example, the well-known cholesterol derivative cholesterol myristate exists in crystalline solid form at room temperature (20 °C), but dissolved into liquid form when heated to 71 °C. The resulting liquid is highly turbid compared to water or alcohol. When heated further to 86 °C, the turbid liquid enters a clear liquid state that does not change with additional heating. The resulting turbid liquid is neither solid nor liquid, but a different state that we refer to as a "liquid crystal."

Substances that form liquid crystals typically consist of stiff and inflexible molecules shaped like long rods. In the liquid crystal phase, the molecular configuration is not rigid and can be influenced by outside conditions (temperature, magnetic field) to undergo phase transitions through changes of parameter and orientation order. Liquid crystal displays take advantage of the electro-optical effects of the liquid crystal obtained through this phase transition in the liquid crystal's molecular configuration.

The first liquid crystal-related substance to be discovered was myelin (a lyotropic liquid crystal in the broad sense) by Virchow in 1854. The first person to discover the liquid crystal phenomenon was the Austrian biologist Reinitzer.

In 1888, Reinitzer was studying the melting behavior of organic matter related to cholesterols in plants when he found that cholesteric benzoate had two melting points. He observed that when cholesteric benzoate crystals were heated to 145.5 ° C, they dissolved to form a white, turbid liquid, but at 178.5 °C they transformed into a clear liquid. Reinitzer shared his discovery with the German physicist Lehmann. The following year in 1889, Lehmann used a polarizing microscope he had devised with the latest heating equipment and showed the two melting points possessed by cholesteric benzoate. He discovered that the substance exhibited birefringence in the liquid state; when cooled, it displayed various beautiful colors like a pearl before reaching crystal form. Because it possessed the flowing properties of a liquid and the optical properties of a solid, Lehmann called it *fliebende Krystalle*, or "flowing crystal" in German.

In 1922, the French mineralogist Friedel proposed that liquid crystals represented an intermediate phase, which he termed "mesophase." Through optical observation of liquid crystals, he categorized them into three structures: nematic, smectic, and cholesteric (Fig. 8.19) (Palffy-Muhoray 2007).

Nematic liquid crystals consist of configuration of rod-shaped molecules in parallel. Each molecule is relatively free to move around the major axis, and no lattice structure exists. This accounts for the crystals' high fluidity and low viscosity.

In smectic liquid crystals, rod-shaped molecules form a stratified structure, where the constituent molecules are arrayed in parallel such that they are nearly vertical at the surface of each layer. Bonds between molecule layers are relatively weak, allowing for relative ease of slipping. As a result, smectic liquid crystals show two-dimensional fluid properties. At the same time, they are also far more viscous than ordinary liquid crystals.

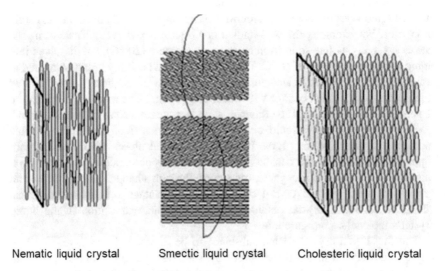

Nematic liquid crystal   Smectic liquid crystal   Cholesteric liquid crystal

Fig. 8.19 Molecular structures of liquid crystals

Like smectic liquid crystals, cholesteric liquid crystals form a stratified structure, but the major axis molecules form a parallel configuration similar to those in nematic liquid crystals within the surface of each layer. The molecular axis configuration also deviates slightly in direction between contiguous layers, and the overall liquid crystal has a helical structure. The optical properties of cholesteric liquid crystals—including their optical rotation, selective light scattering, circular polarization, and dichromism—are produced as a result of this helical structure.

B. Use of Liquid Crystals in Liquid Crystal Displays

The smallest particle producing an image on a liquid crystal display is called a cell; images are displayed as individual cells reflect (polarize) or block light. The basic process of using liquid crystals to produce a liquid crystal display comes in three main types: base material production, cell production, and module production.

In simple terms, the base material production process involves removing impurities from a transparent glass electrode board and creating a cast base material that can be used for the display electrodes. The next process, cell production, involves applying an organic polymer resin on two cast base material from the previous process, conferring orientation layer and insulation layer functions. Sealing adhesive and spacers are applied to produce a bowl-shaped cell with adjustable thickness; the liquid crystal material injected here, and the polarizing plate is attached to create the liquid crystal display shape. The final module production process involves assembling a base material arrange with the cell, a driver direct circuit to drive the cell, and a power circuit, and a framework to support electronic contact between the cell's electrodes and the driver circuit. When an electrical signal arrives in the resulting liquid crystal display, liquid crystal materials in each cell react to it and either reflect or block light, producing the desired images.

C. Liquid Crystals from Marine Organisms

Cholesterol, the starting substance for cholesteric liquid crystals, is one of a group of compounds known as sterols that have a cyclopentaphenanthrene carbon framework. The sterols that make up its basic structure are found among a broad range of animals, plants, and microorganisms; cholesterol in particular is found in many animal cells. It is also abundant in the membranes of cells in higher-order animals, including red blood cells and myelin sheaths, but scarce in the internal membranes of mitochondria and bacterial cell membranes.

Recently, researchers have tried to use these abundant cholesterol compounds in animal cells to produce liquid crystals, which are drawing attention as a next-generation display device. This section will examine liquid crystals that can be made from substances present in the seas' abundant marine life.

For some years, cholesteric liquid crystals have been made with cholesterol extracted from squid, sardines, and other marine animals. The reason for this is the

Table 8.1 Cholesterol content of various animals

Animal	Cholesterol (mg/100 g)	Animal	Cholesterol (mg/100 g)
Cod	37	Ayu (muscle)	53
Mackerel	80	Ayu (skin)	397
Sea bream	104	Ayu (viscera)	827
Sea smelt	178	Common eel (muscle)	132
Herring	70–80	Common eel (skin)	306
Tuna	112	Common eel (liver)	290
Herring roe (dry)	242	Egg	630
Flounder roe	275	Beef	80–125
Salmon roe (pickled)	370	Cow liver	260–320

large amount of cholesterol found in the viscera and skin of these marine organisms (Table 8.1).

For example, the longfin inshore squid (*Loligo pealei*) has a cholesterol content of around 170–460 mg per 100 g. Over 30% of membrane lipids in its nerve cells are cholesterol, making it a valuable source of cholesterol. In contrast to animals, plants contain large amounts of sterol compounds such as sitosterols and stigmasterols instead of cholesterols.

Sitosterols are also abundant in marine organisms such as red algae, sea urchins, and starfish, while cholestanols (where the double bond in cholesterols has been reduced) are found in large quantities in sponges and other marine life. Cholesteric liquid crystal formation has also been found in sitosterol acidic ester and cholestanol benzoic acid esters (Fig. 8.20).

D. Liquid Crystals from Fatty Acid Chain of Lipid Bilayer

Typically, cells are separated from the outside world by a cell membrane. The organelles within the cell are also surrounded by biomembranes. These membranes separating cells from the outside world consist basically of a lipid bilayer. A lipid bilayer is a two-dimensional film created as lipid molecules such as lecithin associated due to hydrophobic interactions in water; it is called a bilayer because it consists of a two-molecule layer measuring around 5 nm in thickness.

Biomembranes contain various molecules in their lipid bilayers, including proteins, glyco proteins, and cholesterols. Lipid molecules composition also take various forms, as shown in Table 8.2. The bilayer's molecular structure is a lyotropic liquid crystal in that it forms a regular association structure in water, but is also thermotopic in that phase changes also occur due to temperature.

Differential thermal analysis for a bilayer made from a single pure lipid such as dipalmitoyllecithin shows a heat absorption peak at 41 °C. Endothermic temperature varies according to fatty acid molecule chain length: 58 °C for disteroyl

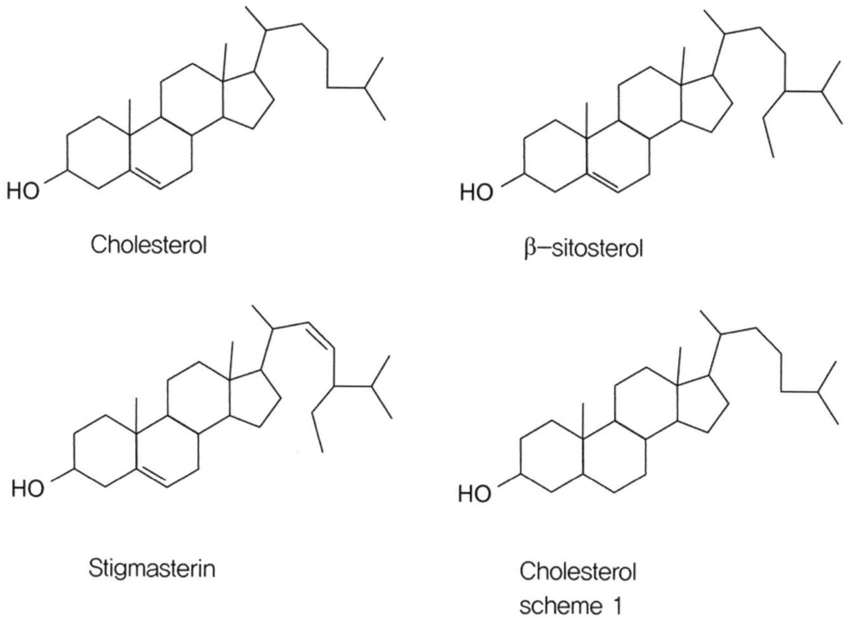

Cholesterol

β–sitosterol

Stigmasterin

Cholesterol
scheme 1

Fig. 8.20 Sterol compounds

Table 8.2 Fatty acid composition in squid nerve cell membranes

Tissue and lipids	Brain	Fin nerves
Tissue weight (mg)	39.2	41.8
Total lipid weight (mg)	7.48	5.35
Cholesterols (%)	33	39
Cardiolipin (%)	2	0
Free fatty acids (%)	3	0
Phosphatidylethanolamine (%)	32	20
Phosphatidycholine (%)	19	29
Phosphatidylserine and phosphatidylinositol (%)	8	5
Sphingomyelin (%)	1	5

lecithin and 23 °C for dimyristoyl lecithin. This heat absorption is the result of changes in the form of molecules in the fatty acid chain; from a solid, trans zigzag stage, it changes to a highly fluid liquid crystal state with a gauche conformation (alternating rotated isomers; see Fig. 8.21).

The presence of unsaturated fatty acids lowers the phase transition temperature. In the case of dioleyl lecithin, the temperature is around −22 °C. To preserve life activities, the biomembrane must always be kept in a state to allow activity, and it is entirely appropriate that the biomembrane lipids largely contain unsaturated fatty

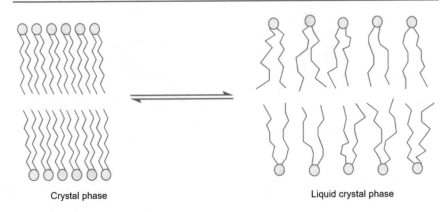

Crystal phase   Liquid crystal phase

Fig. 8.21 Solid-liquid crystal transition in the lipid bilayer

Table 8.3 Fatty acid composition in cod muscle lipids

Fatty acids	Triglycerides (%)	Phosphatidylcholine (%)	Phosphatidylethanolamine (%)
Stearic acid (18:0)	3.3	0.7	3.8
Oleic acid (18:1)	19.5	9.7	11.3
Linoleic acid (18:2)	3.5	0.7	0.8
Linolenic acid (18:3)	0.4	0.5	0.5
EPA (20:5)	6.8	21.5	20.6
DHA (22:6)	9.8	30	46.6

acids. Table 8.3 shows the fatty acid composition of cod muscle lipids. Phospholipids contain more unsaturated fatty acids than triglycerides, while the high amounts of eicosapentaenoic acid (EPA), which is known to be effective against thrombosis, and docosahexaenoic acid (DHA), which is known to be effective in improving memory, indicate that marine organisms are sources of a wide variety of fatty acids.

E. Liquid Crystal Materials from Marine Archaebacteria

Marine organisms inhabit a more diverse range of environments than land-based organisms in terms of temperature, pressure, hydrogen ion concentrations, and saline concentrations. Marine organisms are extremely adaptable to different environments; some life forms have even been found inhabiting areas near thermal vents at depths of 2600 m under the sea. These bacteria that have adapted to high-temperature, high-pressure, and highly saline environments are known as thermophiles. Comparison in terms of homology in their RNA base sequences shows them to be archaebacterial like methane-producing bacteria. The lipids in archaebacterial cell membranes have a different chemical structure from those found in eukaryotes and eubacteria (Zillig et al. 1987).

While the framework in ordinary lipids is a fatty acid chain and glycerine esters, in archaebacteria it is an ester bond with saturated isoprenoids. Thermophile and methane bacteria cell membranes also contain a ring-shaped tetraether structure and closed diether tail. Tetraether lipids have a single-layer rather than a bilayer structure. The characteristic chemical structure of archaebacterial cell membrane limits is believed to be a specific membrane structure that has evolved to respond to extreme conditions (such as high temperatures, high pressures, and high saline concentrations). Research to understand liquid crystal behavior and protein inter-actions is still lacking, however, and a physiochemical approach is needed.

In 1977, the Japanese researcher Kunidake and others found dialkyl ammonium chloride (which has a simplified phospholipid chemical structure) formed a similar bilayer structure to a biomembrane. This lipid was a purely synthetic compound that is not present in any organism, and the synthesis led to a worldwide trend of biomembrane engineering studies using synthesized bilayers. The model chosen by chemists from among the different synthetic candidate bilayers was a compound from the cholesterol family, a double chain lipid typically seen in animal and plant cell membranes.

Fuhrhop et al. succeeded in synthesizing a ring-shaped lipid similar to those found in archaebacterial cell membranes and successfully formed a single-molecule membrane vesicle in water. Additional modeling resulted in synthesis of a single-chain compound with two hydrophilic groups, resulting in formation of a single-molecule membrane. While a phase transition is believed to exist at 40 °C in black membranes produced with terpenoid lipids from archaebacterial, no clear phase transition temperature was observed in the single-molecule membrane produced by Fuhrhop et al. Phase transitions typically appear in single-chain compounds, with transition temperatures depending on chemical structure, and in particular on the structure of aromatic portions. It still appears rather premature to confer synthetic molecular membrane research the status of biomembrane engineering on a par with protein or gene engineering.

In contrast, it is not at all premature to draw inspiration from the various archaebacteria that sustain life activities in extreme conditions, and to attempt to develop new functional membrane and liquid crystal materials. With researchers in synthetic molecular membranes and marine biotechnology currently joining forces, the day when marine organism-derived liquid crystals are produced and put to use may not be far away (Shimomura 1989).

8.3.4 Developing New High-Temperature Superconductors with Alginic Acid

Humans have long fantasized about what might lie in unknown worlds across the sea. The medieval Age of Navigation marked a new beginning in linking that land to the Old World; after that, the world became one. Boats were companions in achieving this dream: from primitive dugout canoes to the sailboats of the Medieval Era, the steam ships of the modern era, and today's high-speed passenger ships, boats have been improved in shape and function over time to suit the purposes of long-term travel. Early boats moved purely by human-generated force; after that came boats that harnessed the power of the wind. Today, most boats are driven by motors and propellers. Recently, however, a historic form of boat has been under development—one with propeller-free, electronic propulsion (Crow et al. 1991).

The electronic propulsion system on this propeller-free boat is based on the Fleming Left Hand Rule, one of the basic principles of electromagnetics. It became the subject of much research in the U.S. after first being proposed by Rice in the 1960s, but numerous technological difficulties arise, and no satisfactory results were achieved. In 1976, the first real development of a propeller-free boat came when a Kobe University team led by Sachi devised a model that used a superconductor for an electrical propulsion device. In 1991, Japan's Shipbuilding Industry Foundation and Mitsubishi Heavy Industries held a launch ceremony for the world's first superconductor-powered electrical ship, the Yamato-1. The ship was just a test vessel measuring 30 m and 280 tons, with a design speed of 8 knots (around 15 km) and a capacity of 10 people. Once a superconducting coil and superconductor magnetic shielding technology are developed and issues are resolved with alternative refrigerants for superconducting materials, the future may bring development of a superconducting ship capable of traveling at over 100 knots (about 180 km) without a propeller. (Konishi 1989).

A. What Is Superconductivity?

What is the superconductivity that has made these things possible? Typically, the term "conductor" is used to refer to materials through which an electric current travels easily, and "insulator" to those in which it does not. As ordinary metals like copper increase in temperature, they lose resistance as their atoms experience lattice vibration. When a current flows, resistance arises as a result of this lattice vibration, and electricity is lost. (In other words, less than 100% of electricity is received.) When the metal's temperature is lowered, its electrical resistance decreases. As it reaches a certain point close to an absolute temperature of 0 K (−273 °C), some inherent electrical resistance remains in the metal despite any further cooling. In some materials, however, electrical resistance suddenly falls to zero at a certain temperature. This phenomenon is known as superconductivity (Wu et al. 1987).

Superconductivity was first discovered in 1911 by the Dutch physicist Kamerlingh Onnes. In 1908, he succeeded in compressing helium gas to produce liquid helium at an absolute temperature of 4 K (−269 °C). He then used this to cool substances to an absolute temperature close to 0 K (−273 °C). As he was studying the relationship between temperature and resistance, he found that when he cooled mercury to low temperatures, its resistance suddenly disappeared near 4.2 K, the evaporating temperature for liquid helium. It was the first discovery of a superconductor.

Another discovery in superconductivity was made in 1933 by Germany's Meissner and Oschenfeld, who found that superconductors not only lacked resistance but demonstrated a self-repulsion effect in which a magnetic field inside the superconductor was expelled outward. Now called the Meissner effect, it is recognized alongside the absence of resistance as one of the most basic characteristics of superconductivity.

The first to successfully ascertain the cause of superconductivity were the U.S. scientists Bardeen, Cooper, and Schrieffer. Their theory, which was announced in 1957, came to be known as the BCS Theory after their initials. It was a somewhat complex theory, holding that lattice vibration occurred as electrons of opposite spin joined into "Cooper pairs," and that while electrons always moved with the same energy at opposing speeds at the relatively macro-level scale, they flowed within the lattice with a general lack of resistance as it vibrated. The three were awarded the 1972 Nobel Prize in Physics. While lattice vibration in a material typically produces resistance to electrical current, in a superconducting state it produces Cooper pairs, and electrons in this state slip out of the atom's lattice as though surfing waves.

B. Types of Superconductors

Superconductors come in two main types, known as Type I and Type II. Type I is the kind of superconducting material initially discovered by Kamerlingh Onnes; the most representative example may be mercury. Because these materials have a relatively low threshold, they pose numerous difficulties for application. Type II is divided in turn into low-temperature and high-temperature superconductors. Because Type II materials have a relative higher threshold than Type I materials, it was the discovery of Type II superconductors that truly paved the way for superconductor applications.

The biggest difference between Type I and II concerns the intermediate stage when undergoing a phase transition in a superconducting state. Type I superconductors have this intermediate stage. It exists momentarily, so that if any part of the superconductor undergoes a phase transition, the rest of it immediately does so as well. Type II superconductors, in contrast, have a mixed state in which superconductor and normal conductor coexist. The mixed state is physically stable and can continue as long as the threshold is not exceeded (Schnyder et al. 2012).

Low-temperature superconductors—so called because of their low threshold temperature—become superconductors like Type I superconductors in the presence of liquid helium. Low-temperature superconductors have a high critical current density, but a relatively low critical field or temperature. High-temperature superconductors use liquid nitrogen as a refrigerant, which makes them far more economical than low-temperature superconductors.

C. High-Temperature Superconductors

The key issue in superconductor application is temperature. More than 1000 types of superconducting materials have been discovered, including metals, organic substances, and ceramics. Only around five to six have been commercialized, including niobium-titanium (Nb–Ti) and niobium-tin (Nb–Sn) alloys. This owes to fact that superconductivity occurs only at extremely low temperatures, which requires the use of expensive liquid helium (4 K/−276 °C) for refrigeration. Also, because the helium gas needed to make liquid helium is light and tends not to remain in the atmosphere, cooling costs are prohibitively expensive, and the technology cannot be used except for purposes other than advanced precision machinery.

After superconductivity was discovered, few believed in the existence of a substance that would exhibit it when cooled to 77 K (−200 °C), a temperature achievable with highly inexpensive liquid nitrogen. In 1986, however, lanthanum-based LBCO developed by Bednorz and Müller was reported to become a superconductor at the relatively high temperature of 30 K. In 1987, the discovery of oxide-based superconductors developed by Paul Chu of the University of Houston that exhibited superconductivity at 77 K drew attention to the area of high-temperature superconductors. Examples of high-temperature superconductors that are currently the focus of attention include rare earth oxides that are titanium-based (critical temperature 30 K) and yttrium-based (»90 K), bismuth oxides, and mercury-based forms (134 K) (Burns 1992).

D. Commercializing High-Temperature Superconductors

For high-temperature superconductors to be used commercially, they must be fashioned into thin, rodlike wires that allow for easy passage of current. Recent rapid strides in thin film technology have resulted in the production of outstanding forms of film.

High-temperature superconducting ceramic are not only difficult to make but also fragile. New forms of wire production are currently being explored to address these issues. The most typical way of fashioning a wire from high-temperature superconducting ceramic is powder sintering. In this method, ceramic powder is joined to a pipe made of copper, silver, or another metal and pulled until it is slender. Another approach currently being researched involves mixed an organic binding agent into the ceramic powder to produce a wire form, which is then burned into a wire.

Wires produced in this way have roughly the same critical temperature as a regular block of superconducting ceramic. They differ, however, in having a critical current density (the maximum current capable of flowing in the superconductor) of 10^7 A/cm^2 and over twice the utility level. This can be attributed to the sintered wire materials containing remaining particles of the ceramic powder used as an ingredient; the presence of many gaps between particles prevents a current from flowing easily. When the crystalline particles in the material are arranged in a disordered way, critical current density is constrained by the particles' direction relative to the current flow, preventing the superconducting ceramics from achieving maximum performance. This problem is not something that can be solved simply by shaping powder into wire form and sintering it.

The most widely recognized method for producing ceramic wire is the sol-gel method. This method involves gelling an organic sol containing a metal ion into wire form, then heating it into ceramic form. Since this method involves mixing the constituent atoms in ceramic at the atomic level without a powdered process, it is less uniform than the powder sintering method, and the ability to produce precise wire forms is uncertain. For this reason, research is under way on an approach involving use of the sol-gel method on several organic metal salt gels to produce superconducting ceramic wire. The resulting wires are not precise and long like the wires currently produced; they are typically hollow like tubes and contain numerous gas bubbles (McGinn 1998).

E. Producing High-Temperature Superconducting Wires with Alginic Acid

As sodium alginate, alginic acid $[(C_5H_7O_4COOH)X \cdot (YH_2O)]$ becomes a viscous liquid when dissolved in water. The resulting solution acquires gelling properties when the sodium ions are replaced with hydrogen or multivalent metal ions. For this reason, research is currently focusing on the fashioning of high-temperature superconducting materials into wire through the sol-gel method using alginic acid, a polysaccharide present in algae such as sea mustard and kelp. The principle involves using taking advantage of the gelling properties of alginic acid by binding the metal ions needed for high-temperature superconducting ceramics to an alginic acid gel in wire form and then heating the surface to produce a high-temperature superconducting ceramic wire (Crow et al. 1991).

An example of the alginic acid method of producing high-temperature ceramic wire is the use of $YBa_2Cu_3O_x$ as a high-temperature superconducting ceramic and creation of an alginic acid precursor which is then shaped. For the alginic acid precursor, a 5% sodium alginate solution is carefully directed from a nozzle toward 1 N hydrochloric acid. The sodium ions in the sodium alginate solution are replaced with hydrogen ions from the HCl and gel in the same form produced as they emerged from the nozzle. The gelled alginic acid precursor is washed with distilled water and placed in a sodium acetate/barium acetate/copper (I/II) acetate solution. As the hydrogen ions in the alginic acid precursor are replaced by Y · Ba · Cu ions, an alginic acid wire is produced. The stoichiometric ratio of Y · Ba · Cu bonding with hydrogen ions in the alginic acid wire is 1:2:3; more copper than yttrium or barium is also found when an acetate solution is used. This is a result of selectivity in alginic acid's ion exchange capacity.

This alginic acid wire is then washed again with distilled water and subjected to a load as it dries at room temperature to produce the final wire. Magnification with a scanning electron microscope showed a smooth, evenly surface wire with no gas bubbles or evidence of metal salt deposition.

The alginic acid wire exhibits a tensile strength of 146 MPa and 5.7% expansion. A firing temperature of 900 °C is inadequate, but firing proceeds when the temperature is raised to 950 °C. The empty spaces are removed, and the cross-section assumes a continuous, smooth, nearly cylindrical surface.

The length of alginic acid wire is constrained by the size of the furnace, but is typically around 150 mm. Theoretically, there is no limit to the possible length, which means that longer wire could be produced. The wire shrinks in diameter by around one-third when fired at 950 °C. The minimum diameter of alginic acid wire currently being produced is around 70 μm, although even smaller wires could be made.

F. Characteristics of Wires Produced with the Alginic Acid Method

$YBa_2Cu_3O_x$ wire produced with the alginic acid method has a maximum tensile strength of 192 MPa, or five times higher than $YBa_2Cu_3O_x$ wire produced with the powder sintering method. The alginic acid method's $YBa_2Cu_3O_x$ precursor is also more precisely designed. The cross-section of a wire fired at 950 °C after hardening a $YBa_2Cu_3O_x$ powder with the powder sintering method contains remaining particles from the source powder, resulting in many gaps between particles. This is a significant difference from wires produced with the alginic acid method (Poncelet et al. 1992).

In $YBa_2Cu_3O_x$ wires produced with the alginic acid method, electrical resistance depends on temperature: as the temperature falls, it slowly declines until the temperature reaches around 90 K, at which point it begins dropping sharply from 85 K all the way to zero. 85 K is the maximum critical temperature for $YBa_2Cu_3O_x$ wires made with the sol-gel method. Currently, the critical current density at a critical temperature of 77 K is 10^5 A/m^2, which decreases as the wire's critical temperature is increased above 90 K.

Critical current density does not readily decline when crystalline particles within the wire are directionally disordered and not arrayed in a particular direction. This phenomenon also appears in wires produced through other methods besides the alginic acid approach and is a fundamental issue encountered in the making of high-temperature superconducting ceramics. Attempts are currently being made to solve this problem by raising critical current density through increased orientation of crystalline particles in wines made with the powder sintering approach. Research should also be conducted to increase the alignment of crystalline particles in high-temperature superconducting ceramic wires made with the alginic acid method.

G. What Happens to the Alginic Acid When the Wires Are Fired?

When high-temperature superconducting ceramic wires made with the alginic acid method are fired, the alginic acid is ultimately converted into water and carbon dioxide that evaporate into the air. This fact can be ascertained from thermogravimetric analysis (TGA) and differential thermal analysis (DTA) of the alginic acid wire firing process. Weight decreases slowly as the temperature rises from room temperature to 180 °C, and the DTA curve similarly slows a slow rate of heat absorption before reaching this temperature, at which point the water in the alginic acid evaporates. As it continues heating from 180 to 340 °C, most of the water and organic matter in the alginic wire is broken down, and the weight falls to roughly half its initial level. This can be confirmed through observation of changes in a sample's infrared absorption spectrum as the heat treatment temperature changes: no major change to the spectrum is absorbed through a heat treatment temperature of 210 °C, but once it reaches 340 °C, nearly all absorption disappears except for the 3450, 1600, and 1420 cm^{-1} wavelengths. In particular, absorption disappears for the 1200–800 cm^{-1} range of absorptions indicating the C–O–C and C–O–H in the alginic acid framework, suggesting that the main framework of the alginic acid is broken down almost completely at this stage (Fig. 8.22).

The pyranose ring that makes up alginic acid contains two types of monosugars with differing three-dimensional structures: malonic acid (M) and glucuronic acid (G). The blocks that contribute to gelling through bonding with metal ion due to alginic acid are called "G-G blocks" because G and G are used. It is thus better to have more G-G blocks to arrange the metal ions at as high a density as possible. Measuring the ratios of G-G blocks in alginic acid is not a simple task; it is typically viewed as the ratio of M to G (M/G). For commercial-grade alginic acid, the M/G ratio is in the range of 0.9–1.3. The form discussed here is obtained from commercial-grade alginic acid; precision in superconducting ceramic wires can be increased through the use of alginic acid with a higher G content.

The M/G ratio for domestically produced alginic acid extracted from kelp is typically greater than 1.0, making it unsuitable as a material for producing high-temperature superconducting ceramic wires. In the future, extracts with higher G content will need to be obtained from kelp. As described above, $YBa_2Cu_3O_x$ high-temperature superconducting ceramic wires can be produced with sodium alginate, which exhibits gelling properties as it bonds with multivalent metal ions. The resulting cross-sections are more precise than those produced with the powder sintering approach. With further research to optimize the composition and firing of alginic acid wires, higher critical temperatures and critical current destiny will be achieved. Since it is applicable to all water soluble multivalent metal ions, the

Fig. 8.22 Infrared absorption spectrum for an alginic acid [Y·Ba·Cu] precursor

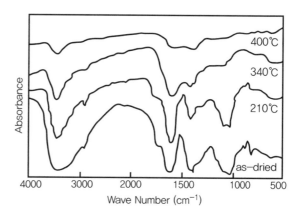

alginic acid method can also be applied with other high-temperature supercon-ducting ceramics besides $YBa_2Cu_3O_x$, through the format of complex with different metals.

H. Applications and Future of High-Temperature Superconductors

In the 21st century, high-temperature superconductors have a very broad range of applications and could represent the start of a new Industrial Revolution. We can expect to see them starting to emerge in the future. In the field of electricity and power generation, development of high-temperature superconducting wires could substantially improve the low efficiency of current copper wires, allowing the practical usage of the power that is lost to heat generation and the stockpiling of relatively large amounts of power. In electronics, the use of superconductor-based chips in place of semiconductors will enable the manufacturing of computers far faster than those in use today.

Prospects for applications in medical equipment are also bright. Development of superconductor-based magnetic resonance imaging (MRI) could allow for more precise diagnosis of human illnesses, prolonging lifespans through the elimination of misdiagnosis and the early diagnosis of disease. Perhaps most notable is the development of high-temperature superconductor magnetic levitation-based trains currently under way in advanced economies. This transportation advancement allows the generation of much faster speeds than current high-speed rail. Once it is established as a means of public transportation, a person could conceivably travel between Seoul and Busan in just 40 min.

High-temperature superconductors can also be used with ships to allow for travel at very fast speeds. In terms of potential applications, high-temperature supercon-ductor manufacturing could fairly be called a new Industrial Revolution. Given the fact that even more outstanding high-temperature superconductors can be produced by using alginic acid—one of the main ingredients in the sea mustard and kelp Koreans eat every day—research into marine bioresources will be needed not only in some relevant fields of application, but in a diverse range of other areas as well.

8.3.5 Chitin in Artificial Skin

A. Components and Function of Skin

In addition to wrinkles, the surface of human skin also pores for hair and sweat. Pores in turn abound with different types of hair. Downy hair in particular is rounded and raised to form a horny layer, which is known to be related to the skin's sense of touch. Skin color is determined not only by the pigments and quantity of blood in the skin, but also by the light reflected from the skin's surface and its level of transparency. In addition to race, gender, age, and part of the body, skin color

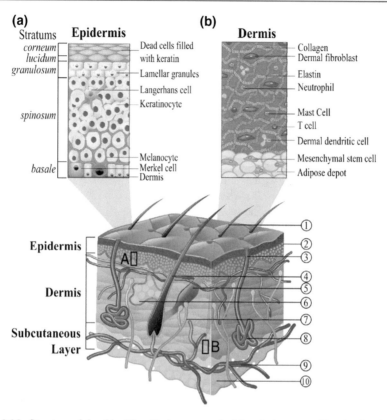

Fig. 8.23 Structure of the skin. The skin is composed of 3 main layers—epidermis, dermis and subcutaneous layers containing (1) hair shaft; (2) stratum corneum; (3) sweat-pore; (4) hair follicle; (5) arrector pili muscle; (6) sebaceous gland; (7) nerve; (8) eccrine sweat gland; (9) cutaneous vascular plexes; (10) adipose depot. Detailed structure of **a** epidermis is shown with the stratum layers and **b** cellular composition and dermis. Figure adopted from Gaur et al. (2017)

also differs as a result of nutritional condition, irregularities in endocrine, and intestinal or skin disease. The skin consists of an outermost epidermis, an internal layer dermis, and subcutaneous tissue (Fig. 8.23) (Dutta et al. 2004).

All cases are the result of numerous cells coming together to form tissue. Skin thickness differs according to the part of the body, but measures 0.03–1.00 mm for the epidermis and around ten times that for the dermis. Subcutaneous tissue consists largely of lipid, resulting in large differences not only between parts of the body but also between individuals. Numerous blood vessels form a network through the dermis and subcutaneous tissue (but not the epidermis), supplying nutrients as the blood circulates. Sensory and autonomous nerves are also distributed in the skin, where they participate in perception, smooth muscle movements in the skin and blood vessels, and the secretions of sebaceous and sweat glands.

In addition to showing human characteristics, skin also betrays our emotions. Our faces redden when we are embarrassed or angry, and turn pale when we are shocked. Skin also responds to outside stimuli, perceiving sensations of touch, coldness, warmth, and pain. These sensations are each transmitted from the tips of sensory nerves and are immediately translated into actions.

Skin is tough and elastic. Because of the abundance of fatty tissue at its lowest layer and tough corneous tissue at its topmost layer, it protects the body with its resistance to mechanical stimuli such as bruising, pressure, and friction. The presence of a fatty membrane on the skin's surface blocks permeation by water, while its high acidity prevents germs from growing. When the skin is exposed to sunlight, it turns red and pigments increase. This is the result of the ultraviolet rays in sunlight, which are harmful when they penetrate the body excessively. As the skin absorbs UV rays, the body is protected by the formation of the pigment melanin. The skin is further the place where our immune system is built. This is why a rash developed on the skin when we contract chicken pox or smallpox confers a lifetime of immunity, and why inoculation prevents us from contracting smallpox at all.

Skin includes sweat glands, which produce sweat, and sebaceous glands, which produce oils. These are constantly being secreted onto the skin's surface, giving it a shiny appearance. Soluble substances typically are not absorbed by the skin, but can be absorbed when dissolved in fat or alcohol. Permeation is particularly easy when a substance has been emulsified with the addition of an emulsifying agent. The skin is also a poor conductor of warmth, preventing external overheating and regulating the radiation of warmth outside the body. Eighty percent of the heat radiated outside the body comes from the skin, a process that occurs through heat conduction and dissipation and through evaporation. Skin thus performs many important roles: covering the body and protecting important internal organs, while expelling moisture and salts to regulate and maintain a specific body temperature.

B. Wound Healing Effects

The skin completely loses its regeneration capabilities when the epidermis is destroyed—for example, in the case of a burn severe enough to result in destruction to the dermis. The results can sometimes be fatal. Even a less severe wound, if left untreated, can form a keloid scar (a form of benign tumor caused by abnormal skin growth resulting from a burn). More distressingly for some, the scars that result from burning may last for a lifetime. Normally, the best way to treat these is by autodermic grafting (removing skin from another part of the body for grafting). Skin will not engraft unless it comes from the patient him or herself. No examples of engrafting with non self skin have been found in the past except between identical twins. Because of these limits on graftable skin, it has been the dream of medical practitioners to develop bona fide artificial skin that is capable of skin regeneration (Dai et al. 2011).

That dream has now become a reality thanks to chitin extracted from crab shells. Typically, the following effects are required for wound dressings near artificial skin:

① pain reduction, ② drying prevention, ③ prevention of bacterial growth, ④ protein and red blood cell loss prevention, ⑤ protection of exposed tendons, blood vessels, and nerves, and ⑥ promotion of wound healing. Another requirement is cladding materials of the wound's surface should be simple, affordable, and amenable to mass-production. These wound dressings exist in synthetic and living types. Synthetic dressings react strongly to foreign matter and subject to issues with biocompatibility and usage limitations. The skins of pigs, sheep, and dogs have long been used as live dressings.

Live pig skin was used in extremely rare cases, but its practicality was low. Chicken egg membranes and the human amnion (the thin membrane encasing a fetus in the womb) have proven similarly unusable. Other options included processed freeze-dried pig skin and collagen, which are currently in limited use. In contrast, artificial skin made with chitin taken from crab shells may be seen as compensating amply for these deficiencies as a wound protection agent (Kim 2018).

C. Making Artificial Skin

The first study to suggest the potential for chitin as a medical material may be potentially useful in making artificial skin. Examining the role of ground shark cartilage ointment in fast wound recovery, the researchers were examining which components of the cartilage contributed to the effect when they implicated an ingredient called N-acetylglucosamine. Based on the fact that chitin was a con-stituent of N-acetylglucosamine, they examined its relationship with the healing of cuts in animal experiments using a powdered version. Wounds to which chitin powder were applied were found to close faster than a control group to which it was not (Burke et al. 1981).

The researchers went on to report findings from an examination of effects with thread and film forms. Due to shaping method inadequacies, they did not succeed in practical application, and the research was halted midway through. The central institute at Japan's Unitika subsequently began researching methods of producing chitin molds and succeeded in creating high-quality products that could be used for medical purposes.

At the same time, animal studies and clinic research were being conducted in various fields by researchers at healthcare research institutions around the country who were interested in the material. Some fascinating findings were produced as a result, one of which was the development of a product called Beschitin W, which could be used as artificial skin in the broad sense of wound surface protection. Beschitin W is manufacturing according to the following steps. Chitin is a mucopolysaccharide formed by β-1,4 bonding of N-acetyl-D-glucosamine, only with the hydroxyl group of the fibrin's glucose residuereplaced by an aminoacetyl group. For manufacturing, the outer framework from a crustacean (such as a crab or shrimp) is used. Calcium and proteins are removed to obtain a highly refined chitin powder with less than 0.2% ash content (Kumar 2000). Following treatment to increase solvency, the powder is dissolved with amide solvent to produce a chitin solution with a concentration of around 10%.

After accounting of the solute and extrusion of aqueous solution from a micropore nozzle, a roller at constant speed is used to produce a multifilament with an outer diameter of several microns. Remaining solvent is removed with water, and the chitin is cut to lengths of around 5 mm for treatment to increase protein adsorption, after which a paper is formed with polyvinyl alcohol as a binder. In other words, the process is the same as creating a non-woven fabric with a thickness of 100–120 μm (Kibo 1994).

D. Chitin-Based Artificial Skin: Astonishing Results

Precise growth was observed when mouse-derived fibroblasts were grown on a transparent chitin film. Adhesion conditions were also superior to those of cellulose films, showing that Beschitin W has positive effects on epidermal tissue formation without negatively influencing the growth of normal cells at the wound's surface.

Artificial skin made with chitin can be used in most wound protection agents. Excellent effects in terms of pain relief, resistance to adhesive solvents, drying, and epidermis formation have been observed with use in thermal burns and skin ulcers. Halfside testing (a method that involves dividing the epithelium in two halves and treating with several degrees of material) has been conducted at 22 university and general hospitals nationwide for donor sites (averaging 15 thousandths of an inch) with similar dressing protection agents, freeze-dried pig skin (LPS), freeze-dried dermal skin (LDPS), collagen membranes, and bioplan (silicon applied to nylon fibers and treated with peptides). The method involves observing pain relief effects, adhesion, resistance to solvents, dryness, epidermis formation, and extraction quantities while treatment was being carried out, and comparing differences through the treatment's completion (Kumar 2000).

No side effects were observed during the treatment period. Negative results were observed in all forms of safety testing, including acute toxicity, sub-acute toxicity, epidermal reaction, febrility, physical property, grafting, cell toxicity, hemolytic, mutagenicity, and antigenicity testing. High degrees of safety were observed in all cases.

When epidermis cells from chitin are grown beforehand as a porous body and grafted onto a deep wound, a perfect epidermis is formed from the chitin's breakdown as the cells grow. This real-life example of artificial skin has long been marketed in Japan under the dame Beschitin W, and is currently being exported to South Korea. As it is used more and more in clinical settings, its characteristics will be better understood, and its applications may extend to include recovery from injuries to other parts of the body besides skin.

The functionality of chitin also holds great potential for improvement through chemical modification. If this comes to pass, its possibilities as a biomaterial may expand beyond what has been described here. The time may soon arrive when chitin from the shrimp and crab shells that are currently waste products from marine processing are used to heal wounds from burns or traffic accidents without leaving a scar.

8.4 Functional Food Materials

8.4.1 Fish Fats and Disease Prevention

Unlike other island countries, Japan is a place where cold and warm currents come together. Its waters are some of the most hospitable to formation of plankton, which are primary producers in the ocean food chain. It may thus be described as a country ringed by a sea with abundant stores of marine living resources. The Japanese people themselves have unwittingly enjoyed many benefits from this sea and its rich marine resources—a fact attested to by Japan's lifespans, which are the longest in the world. A 17-year cancer prevention institute survey of the nutritional habits of 265,000 Japanese people found their lifespans to increase with greater frequency of fish consumption. It was also found that increases in the morbidity rate and death rate from lifestyle disease could be suppressed through fish and shellfish consumption (Table 8.4) (Burr et al. 1989; Kim 2018).

At the same time, Michael Crawford of the United Kingdom's Institute of Brain Chemistry and Human Nutrition has observed that the reason for Japanese children having the highest IQs in the world is due to their having eaten mainly fish and other seafood from a young age. In Japan, the consumption of fish has been promoted in disease prevention campaigns with the slogan "Eat fish and your brain improves."

What is it about fish that makes it so good for the body? To be sure, it contains proteins, vitamins, and other nutrients that are healthy for the human body. But it has been shown to be particularly good because of the presence of two key ingredients in fish fats: eicosapentaenoic acid (EPA) and docosahezaenoic acid (DHA) (Erdman et al. 2011).

Studies on EPA date back 40 years. Ninety percent purified EPA from sardine oil was found to have platelet clotting suppression affects, leading its marketing for the first time in Japan as a treatment of arteriosclerosis obliterans in 1990. Clinical

Table 8.4 Mortality rates ranked by fish and shellfish consumption frequency (relative risk)

Cause of death	Fish/shellfish consumption frequency				Relative risk
	Daily	Frequent	Occasional	None	
All deaths	1.00	1.07	1.12	1.32	9.134
Cerebrovascular disease	1.00	1.08	1.10	1.10	4.541
Heart disease	1.00	1.09	1.13	1.24	3.919
Hypertension	1.00	1.55	1.89	1.79	4.143
Liver cirrhosis	1.00	1.21	1.30	1.74	3.768
Stomach cancer	1.00	1.04	1.04	1.44	2.144
Liver cancer	1.00	1.03	1.16	2.62	2.109
Uterine cancer	1.00	1.28	1.71	2.37	4.142
No. examined	1,412,710	2,186,368	203,945	28,943	

testing has shown the medication to be easy to use with few side effects. In 1994, EPA was also shown to be effective in reducing neutral fats and cholesterol. Additional approval was granted by the Japanese Health Ministry for adaptation as an antihyperlipidemic agent, suggesting that its market as a medical treatment will only grow. Research and development on DNA, another chief ingredient in fish fat alongside EPA, has been hindered in the past by the inability to obtain highly purified forms for research. Since the discovery of sources in fish such as tuna and bonito, related products have been developed (Gillies and Schaefer 2011).

A. EPA Production from Marine Microorganisms and Genetic Engineering

Mass production of EPA has been enabled by the discovery of microorganisms that produce it. Although EPA is a main ingredient in fish lipids, fish cannot synthesize it on their own. Rather, EPA is present in fish muscles and lipids as a result of a food chain in which the primary producers are algae and phytoplankton. In the past, almost no cases of its production by microorganisms had been reported.

In 1986, EPA-producing bacteria were successfully identified among the intestinal microorganisms in blue fish such as mackerel, Japanese horse mackerel, and sardines, all of which have large amounts of EPA in their lipid (Fig. 8.24). Attempts were made to mass-culture the bacteria to produce EPA for use in medical treatment; unfortunately, the approach proved to be less cost-effective than sardine oil. Because EPA-producing bacteria are bacterial, however, it should be possible to isolate the genes from the biosynthesis system behind EPA production. Currently, genetic engineering methods are being used in efforts at EPA production. It is a process that involves producing EPA by manipulating EPA biosynthesis system genes isolated from the stomach in microorganisms, algae, and higher plants that do not have the capability to produce EPA themselves (Certik and Shimizu 1999).

Productivity constraints may exist on inexpensive mass production of EPA from microorganisms. However, if EPA can be produced by organisms capable of making large amounts of oil (such as yeast, mold, or algae) or by higher plants from which oil can be extracted (such as soybeans or colza), this will usher in the development of new functional food products.

It may seem like a dream now, but successful EPA production has already been reported from the isolation of EPA biosynthesis genes and their introduction through genetic manipulation into E. coli, which have no capabilities for producing EPA on their own (Fig. 8.25).

Success has also been reported in conferring EPA production capabilities on blue-green algae through introduction of EPA genes. The time may soon come when vegetables, grains, and fruit are enabled to produce EPA, allowing us to easily consume EPA that was previously only available through fish and shellfish.

B. Hidden Treasures in Fisheries Processing By-Products

Tuna orbit lipid contains as much as 30% DHA (Table 8.5), but only around 6–7% EPA. Whereas previously identified fatty acids in fish consisted of large amounts of

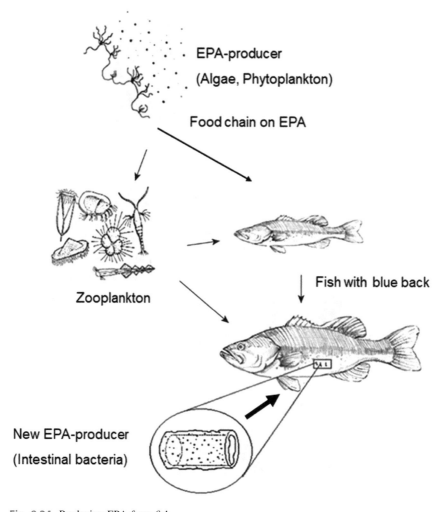

Fig. 8.24 Producing EPA from fish

EPA and only a small percentage of DHA, here the situation is the exact reverse. Moreover, fish heads have traditionally been treated as seafood waste products, with only a small portion used for feed or fertilizer. Fatty oils have been used as boiler fuel in place of heavy oils. For this reason, orbital fat from tuna and skipjack tuna appears to offer a rather good potential source of DHA (Shahidi 2006).

Highly pure (over 99%) DHA, which is used as a reagent, is extremely expensive, costing around US$220 for 100 mg. This has prevented its use in animal experiments. DHA has had to be extracted from orbital lipid, and its purification has allowed access to large quantities of inexpensive, highly pure DHA, ushering in rapid advancements in DHA research. Because this orbital lipid is fresh and

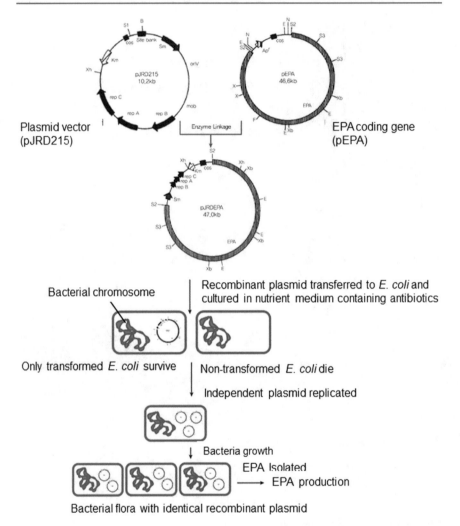

Fig. 8.25 Producing EPA from *E. coli* through genetic manipulation

presents no hygiene concerns, it may also be used in food. The discovery of large concentrations of DHA in orbital lipid has provided the impetus for rapid developments of foods, and health supplements in particular (Yazawa 1989).

C. DHA Pharmacological Activity and Development of Functional Health Foods

DHA is a nutrient that is believed to have a wide range of pharmacological functions, including nervous system development, learning improvement, retinal reflex improvement, anticancer properties, antiallergy properties, and lipid reduction properties. Many findings have already been reported on these functions. DHA may

Table 8.5 DHA and EPA content of orbital fat in marine fish species

Marine fish (orbital lipid)	DHA (%)	EPA (%)
Bigeye tuna	30.6	7.8
Black tuna	28.5	6.1
Yellow tuna	28.9	4.5
Bonito	42.5	9.5
Striped marlin	28.4	3.9
Swordfish	9.6	3.4
Yellowtail amberjack	10.8	3.3
Greater amberjack	20.5	6.5
Japanese horse mackerel	15.3	15.3
Sardine	12.1	22.6
Japanese bullhead shark	29.0	3.0
Cloudy catshark	12.5	13.4

be viewed as an effective nutrient for preventing various diseases and a functional health food with preventive medical characteristics (so-called "third functions"). A fundamental distinction between DHA and EPA is that the former is able to pass through the blood-brain barrier. It is thus DHA and not EPA that contributes to nervous functioning and memory and learning effects.

Over the years, a large base of data has been assembled from *in vitro* (test tube) and *in vivo* (animal) studies on the pharmacological functioning of DHA and its ability to "improve intelligence" in a broad sense. As yet, however, research on its mechanism and studies with human subjects remain insufficient. Much has been learned about DHA's mechanism, and improvements in patients with senile dementia have been reported in clinic studies with human subjects. In addition to proven improvements in atopic dermatitis, other important pharmacological functions include prevention of cancer and mutation, lowering of blood cholesterol, and suppression of increases in blood pressure. Rapid advancements have recently been made in the development of health food containing oils with large amounts of DHA, which have grown to become a very large market (Jun et al. 2004).

That market is expected to grow further in the future as advancements are made in DHA's antioxidant properties, stability, and techniques for pulverization and emulsification.

South Korea's total tuna production stood at 254,000 tons in 2011. Of that amount, 54,000 tons were used for processing of canned tuna, accounting for around 65% of the 82,000 tons in total canned production. Tens of thousands of tons of tuna parts are discarded during processing, including heads, viscera, and bones. Hopefully, functional materials can be developed in the near future to use these parts to prevent various diseases.

D. Improvements in Senile Dementia

The term "dementia" comes from the Latin meaning "to be out of one's mind." In the past, it has been regarded as a natural physiological phenomenon of aging.

However, dementia is a nervous system condition that is result of definite degenerative lesions to the brain. Particularly common among senior citizens aged 65 and older, it is a clinic syndrome that leads to impairments in professional, social, and interpersonal activities as a result of disorders in memory, judgment, and thinking functions. It can also occur before the age of 65 as a result of traumatic head injuries, cerebrovascular disorders, and various metabolic disorders. Many were shocked to learn that after developing Alzheimer's disease, Ronald Reagan reportedly forgot that he had even been President of the United States. In South Korea, family discord as a result of dementia has emerged as a serious social issue (Summers et al. 1986).

Improvements in senile dementia have been reported from the administration of DHA (an unsaturated fatty acid abundant in fish) as a health food. Previously, animal tests involving the administration of DHA showed improvements in brain functioning; recent reports also indicate outstanding results in improving human brain functions. In Japan, clinical testing was performed on the brain functioning improvement effects of DHA in 13 individuals aged 57–94 who had been hospitalized for dementia and another five with Alzheimer's disease. For half a year, patients received daily oral administration of 10–20 DHA capsules (70 mg) of DHA produced by a Japanese pharmaceutical company. Ten out of thirteen patients with cerebrovascular dementia were reported in the Japanese press as showing increased will to live and decreased fancies, while slight improvements were also reported in the willpower, interpersonal relations, and composure of the five Alzheimer's patients.

Indirect testing of intellectual functioning has shown gradual declines in brain functioning for patients who were not given DHA, whereas patients administered DHA showed improvements in mathematical ability and judgment over the course of half a year. No side effects were reported except in the case of one patient who suffered stomach pain due to excess fat consumption.

According to the research team, the activity of living nerve cells increases as DHA enters the brains of dementia patients whose neurons are partially dying off. For this reason, DHA may be viewed as an effective agent in improvement and prevention, but not as a basic treatment for dementia. The increase in senile dementia in Japan is seen by some as related to changes in eating habits, as fish-based diets are replaced with meat-centered ones. South Korea is experiencing a similar shift, and the prevalence of dementia-related symptoms appears likely to increase. In addition to dementia improvement effects, clinical testing has also shown a 50% improvement rate among pediatric atopic dermatitis patients, and various other forms of clinic tests are being performed as well. More reliable research findings in the future appear likely to testify even more to the usefulness and effectiveness of DHA as a health food (Kim 2018).

E. Importance in Preventive Medicine

As a food nutrient, DHA may also be expected to exhibit pharmacological activity as a medicine. While its effects may be seen as similar to other nutrients that have

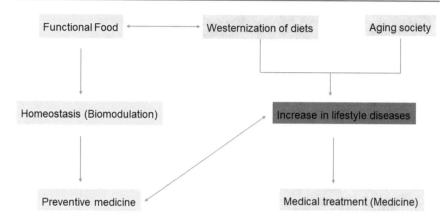

Fig. 8.26 Functional foods and disease prevention

proven useful in preventing various diseases, closer consideration will need to be extended to these preventive effectives.

Prevention may be thought of as falling into two categories. One is prevention in the sense of preventing a disease from causing death. In other words, if a patient with a particular disease is administered DHA or another functional substance and that patient's condition improves, this may be regarded as "prevention of death." Another form of prevention involves preventing disease from occurring in the first place. This prevention of the very contraction of a disease can be understood ultimately as the same thing as delaying the onset of disease past the patient's lifespan.

The aforementioned epidemiological studies have predicted that people who eat large amounts of fish live longer and have a lower morbidity rate from disease. If animal testing shows DHA consumption to result in cancer not occurring prior to the animal's death, that animal cannot be said to have contracted cancer. Ultimately, it is an instance of prevention of disease. This is an approach that should be considered not only with cancer but also with arteriosclerosis, allergies, senile dementia, and range of other diseases (Fig. 8.26) (Rose et al. 2008).

As shown in Fig. 8.27, health declines to "semi-health" status as aging processes. If left neglected, this results in the occurrence of disease and ultimately death (a health level of zero). By administering biofunctional substances to a patient in a semi-healthy state, it is possible to prevent disease from occurring and restore health. For this reason, consumption of biofunctional substances is necessary as a means of suppressing aging. (Kim 2018).

8.4.2 Functionary Dietary Fiber from Algae

Nutrition is a process by which an organism receives substances that it needs to survive and reproduce from outside the body. In the case of human beings, five

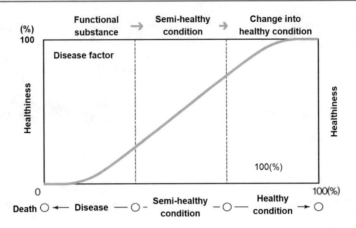

Fig. 8.27 Functional foods and health

nutrients have been reported as essential: carbohydrates, lipids, proteins, vitamins, and minerals. In advanced economies where food is abundant, however, obesity and lifestyle diseases have emerged as serious social issues. The choice of foods that are rich in nutrients has become a major area of interest for long and healthy lives (Ortiz et al. 2006).

Dietary fiber is not easily digestible, and has often been viewed in food terms as a non-nutrient that does not serve as an energy source or ingredient in body tissues. Recently, however, it has been recognized as a useful food ingredient in terms of human health. Dietary fiber is a category that encompasses all food ingredients not digested by human digestive enzymes, and is contained most commonly in vegetables.

The importance of dietary fiber was first recognized as the possibility was raised that the steady increase in the proportion of carbohydrates in Western diets was etiologically related to lifestyle diseases. Indigenous Africans show extremely low incidence of conditions that are common among Westerners, including heart disease, diabetes, and digestive conditions such as constipation, diverticulitis, hemorrhoids, and colorectal cancer; obesity is also extremely rare. The reason for these differences in disease occurrence between Westerners and Africans stems from the amount of fiber consumed: the Africans examined showed fecal content many times higher than for Westerners (Kim 2018).

A. Algae Components

Algae grow in markedly different environments from land-placed plants, and their components are correspondingly different. Forms of dietary fiber include the carbohydrate cellulose, which makes up the cell well in land plants; the hemicelluloses xylan, mannan, and galactan; pectin; the aromatic hydrocarbon polymer lignin; and the wax polyphenol-based cutin. Non-structural carbohydrates include pectin and konjak mannan. Sap gums such as gum Arabic, gum karaya, and tragacanth gum and seed gums such as locastobin gum, guar gum, and tamarind gum are carbohydrates used as stabilizers and emulsifying agents in food, cosmetics, and medicine (Wei et al. 2013).

Algae typically have thick cell walls. Cellulose is found in most types, but not in large quantities; mannan and xylan are present in red and green algae. The chief ingredients of the dietary fiber contained in algae are the mucilaginous polysaccharides present between cells. These include agar, carrageenan, funoran, and porphyran in red algae; alginic acid and fucoidans in brown algae; water soluble uronic acid sulfated polysaccharides and water soluble neutral sulfated polysaccharides in green algae. Laminaran, a storage polysaccharide in brown algae, is also a dietary fiber that is not broken down by human digestive enzymes. Characteristics of these forms of dietary fiber in algae include the presence of ingredients not found in land plants, such as agar, carrageenan, and alginic acid. Many carboxyl groups and ester-linked sulfate groups are present, although this is believed to be the result of structural differentiation to regulate the passage of ions for growth in highly saline environments.

All of the above are derived from plant life. Animal-based forms of dietary fiber also exist, including the chitin that is the chief ingredient in shrimp and crab shells, and the hyaluronic acid present in the connective tissues of mammals. Derivatives resulting from chemical modifications to polysaccharides, including methyl cellulose and polydextrose, are also treated as dietary fiber.

B. Algae Consumption

A national nutritional study on the consumption of dietary fiber by South Koreans calculated an average of 20 g per day per person—a far higher level than in either Japan (16 g) or the U.S. (13 g).

Dietary fiber contribution rates calculated for different types of food include 38% for vegetables, mushrooms, and algae; 29% for rice/grains, potatoes, sweet potatoes, and seeds; 9% for tofu; and 14% for fruit. By themselves, algae accounted for around 7%. In Japan, dietary fiber consumption was down to 16 g from 21 g thirty years before, a decline that has reportedly become even sharper in recent years. Different regions also show considerable variation in consumption. Among Americans, survey reports in the U.S. have shown daily dietary fiber consumption of 13 g. Contribution rates included 28% from vegetables, 19% from breads, 17% from fruits, and 14% from beans. The numbers show considerable differences in dietary fiber consumption from South Korea, which may be attributable to differences in diet (Lee et al. 2010).

C. Roles of Algae Within the Body

Physicochemical properties of dietary fiber vary according to its components, but major physiological roles include water adsorption, ion exchange, and gel formation. Table 8.6 shows the roles of dietary fiber in different parts of digestive organs. Because people naturally chew more when consuming food containing large

Table 8.6 Interactions of digestive organs and dietary fiber

Part of the digestive system	Effects of dietary fiber
Mouth	Increased chewing, increased feeling of satiation
Stomach	Absorption, swelling, continued feeling of satiation, suppression of over intake
Small intestine	
– Duodenum	Prolonging time in stomach, improving glucose tolerance, reducing insulin secretion
– Jejunum	Decreasing cholesterol micelle formation, decreasing cholesterol absorption, normalizing cholesterol concentration in body
– Ileum	Reducing bile acid reabsorption
– Jejunum, ileum	Changing rate of food movement, changing secretion of digestive hormones, normalizing digestive functions, preventing disorders in nutrient movement due to toxic substances
Large intestine	
– Colon	Altering intestinal flora, bile acid metabolism, and cholesterol metabolism; reducing carcinogen production; binding or diluting carcinogens
– Rectum	Increasing excretion of cholesterol, bile acid, and metabolites; increasing number of excretions; smoother excretions

amounts of dietary fiber, much of the saliva secreted in the mouth is absorbed, resulting in swelling of the food. This swelling in the food provides a sense of satiation and suppresses overconsumption, which may represent a means of preventing obesity.

The stomach contains a pylorus, which serves as a gateway regulating the rate at which nutrients are received. As food that is swollen from saliva absorption enters the stomach, it remains there for a longer period of time. At this time, the pylorus functions to prevent food from rapidly entering the duodenum and suppress spreading, allowing for more gradual absorption in the small intestine and improved glucose tolerance. Soluble dietary fiber has been shown to be more effective in suppressing rises in glucose than the insoluble kind (Barsanti and Gualtieri 2014).

A cycle of enterohepatic circulation occurs as cholesterol is observed in the jejunum (inside the small intestine) and bile acid in the ileum (lower small intestine). Blood cholesterol content has been shown to be particularly strongly suppressed by soluble dietary fiber, which has strong gel formation capabilities. This too is a result of dietary fiber blocking the movement of cholesterol and bile acid within the intestinal tract; decreasing absorption through bonding; and suppressing the formation of micelles (colloidal aggregates of numerous small particles; one example is starch solution) consisting of fat-soluble components due to hydrophobic properties from gel formation. When dietary fiber is absent, volume declines rapidly in the small intestine as nutrients are digested and absorbed. When dietary fiber is abundant, volume does not change significantly, but movement time within the small intestine increases, resulting in repeated impairment of nutrient absorption due to toxic substances. Part of the dietary fiber that enters the large intestine is also broken down by intestinal bacteria. The bacterial bodies increase due to bacterial nutrients, but around 50% of fecal volume consists of bacterial bodies. The feces are thus ultimately a different form of dietary fiber.

Lack of dietary fiber also results in reduced water retention capacity in the large intestine. As hardening results from the absorption of water in the large intestine, volume within the rectum decreases, and one does not sense the need to evacuate the bowels. As a result, the movement period within the large intestine is prolonged and the excretion time is reduced, causing constipation. During constipation, the formation of bile acid and carcinogenic substances increases, prolonging contact between the intestine's mucous membrane and carcinogenic substances and increasing the incidence of cancer.

As South Korean diets become increasingly westernized, childhood obesity has recently been on the rise. Abnormal accumulation of fat is detrimental to quality of life (QOL) and may result in serious risks in terms of prognosis. Visceral obesity in particular is seen in medical terms as requiring treatment due to its perceived close relationship with arteriosclerosis and other conditions. Treatment of obesity must be geared to reversing the balance of energy intake and usage and increasing the burning of calories. Exercise and other supplementary therapies should be implemented with a focus on diet; in cases where such treatment cannot be offered or its effects are inadequate, drug therapy should be provided. With any therapy, however, sustaining reductions over a long period of time is highly difficult, and progressive treatment through improvements in dietary habits should be combined with therapies aimed at behaviors.

South Koreans mothers have long made sure to eat seaweed soup after childbirth. Even without any scientific understanding of why algae should be good for recent mothers, the ancestors of modern-day Koreans are reported to have eaten seaweed soup because life experience had taught that it clears the blood. Recently, some headway has been made in solving this riddle.

In addition to being abundant sources of minerals and vitamins, all algae have been reported to include ingredients with antibacterial and antiviral property and effects in controlling blood pressure and blood cholesterol and assisting in antitumor activities and the clotting of red blood cells and lymphocytes.

D. Algal Polysaccharides and the Removal of Heavy Metals in the Body

Alginic acid and other algal polysaccharides are now being considered as a means of absorbing and expelling radioactive metal contaminants within the body. This approach is based in algae's high rate of metal uptake. For example, algae such as *Ascophyollum nodosum* (K = 5700) and *Fucus vesiculosus* (K = 1200) show high rates of iron uptake. Green algae typically exhibit concentrations of cesium (Cs), while concentrations of strontium (Sr) and plutonium (Pu) are respectively seen in brown and red algae.

Effective elimination of contaminating metals in the body requires that ① the toxins are removed, ② the toxins are not absorbed into the body, ③ the metals are

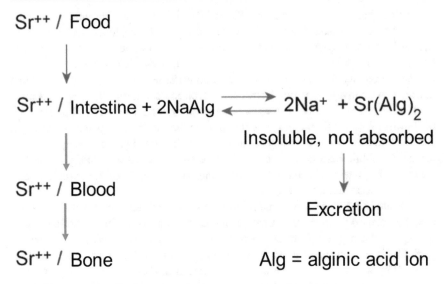

Fig. 8.28   Removal of fixed strontium from the body with sodium alginate

reacted to with relative specificity, ④ stabilization is achieved with stomach acidity and enzymes in the stomach region, and ⑤ prices remain affordable.

Alginic acid, an important polysaccharide in brown algae, has been examined as an optimal compound to achieve this. As shown in Fig. 8.28, it has proven effective in the elimination of radioactive strontium (Sr).

Similarly, sodium alginate also works to eliminate heavy metals such as cadmium (Cd) and strontium (Sr).

8.4.3   Physiological Functions of Chitin and Chitosan

Recent discoveries about the physiological functions of the chitin and chitosan in crab and shrimp shells have resulted in many people wanted to learn more about them. While these functions are startlingly diverse and the effects are large, the molecular structure of the compound is actually quite simple (Jeon et al. 2000).

Chitin is a polysaccharide consisting of a linked sequence of the polysaccharide N-acetyl-D-glucosamine. It is found in the shells of crustaceans such as shrimp and crabs, the exoskeletons of insects, and the cell walls of mushrooms and other fungi. Like cellulose in plants, it is a natural polymer that functions to support and protect organisms. Chitosan is not as abundant as chitin, but can be found in the cell walls of some molds. The term chitosan refers to a form of compound in which the acetyl component has been removed from the chitin polysaccharide structure (deacetylated; see Fig. 8.29). Crustaceans such as crabs and shrimp generate upwards of 144 million tons of waste around the world, while the amount of chitin discarded in seafood manufacturing has increased by 120,000 tons per year.

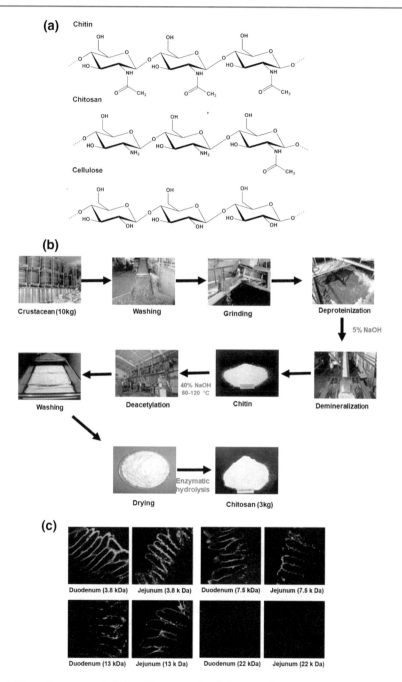

Fig. 8.29 **a** Structures of chitin, chitosan, and cellulose and **b** schematic procedure for the isolation of chitosan from chitin. **c** Absorption of chitosan oligosaccharides in animals

Today, chitin and chitosan are being used in a wide range of areas. In medicine, they are used as absorbent stitching fibers, ingredients in artificial skin, and wound treatments. They are used in cosmetics for their outstanding hydration efforts; in wastewater treatment as wastewater coagulants and heavy metal absorbents; in health food to boost immunity and combat cancer; and in paper-making, textiles, agriculture, and other industries.

The first industrial production of chitin and chitosan occurred in Japan in 1970. In South Korea, an estimated 4 million tons are produced each year. On its own, chitin has few uses; in most cases, it is used as an ingredient in the manufacturing of chitosan. As seen in Fig. 8.29, many different chitin and chitosan lysates can be produced, each of them with different physiological functions. Recent years have seen the discovery of new functions in low-molecular-weight monosaccharides and chitin and chitosan oligosaccharides obtained from chitin and chitosan hydrolysis. These are performing a crucial role in maximizing use of as yet unexploited chitin and chitosan (Prashanth and Tharanathan 2007).

Similarly, active research is also being conducted to use chitin-degrade enzymes such as chitinase and chitosanase to produce molecules with low molecular weight. This section will consider the chitin and chitosan obtained from the shells of crabs, shrimp, and other crustaceans and the production and uses of their oligosaccharides (Kim 2018).

A.   How Are Chitin and Chitosan Produced?

Chitin and chitosan are extracted from the shells of crabs and shrimp. These shells, however, consist of proteins, lipids (pigments), carbohydrates (chitin), and minerals (calcium carbonate), and carbohydrates extracted from them may be either chitin or chitosan. Chitin and chitosan both consist solely of carbon (C), hydrogen (H), oxygen (O), and nitrogen (O)—elements that bond with each other in the thousands or tens of thousands according to specific rules.

To produce chitin alone from a crab shell, one must first remove the mineral calcium carbonate ($CaCO_3$) component. Hydrochloric acid (HCl) is used to remove it according to the following principle.

When combined with an acid such as hydrochloric acid, calcium carbonate (also known as limestone) produces calcium chloride ($CaCl_2$) and carbonic acid (H_2CO_3). In this case, the calcium chloride and carbonic acid can be easily removed, the former because it dissolves in water and latter because it is broken down by the powerful hydrochloric acid and water and carbon dioxide. Once the calcium carbonate is removed, the material can be immerse in thin sodium hydroxide (NaOH) solution to break down the proteins and pigments; once these are removed, chitin is produced (Fig. 8.29 (b)) (Lee et al. 2003).

After cellulose, chitin is one of the most quantitatively abundant materials in the natural world. The fundamental reasons for our failure to take advantage of it concerns the lack of solvents to readily dissolve it and its weak reactivity. Its use has begun to increase with the discovery of numerous functions as chitosan is produced from chitin and the substance is dissolved in weak organic acid obtains greater reactivity. Chitosan is a form in which the N-acetyl group (CH_3CO^-) is removed from the chitin structure through heat treatment with a solution of 40% sodium hydroxide. Creating 100% pure chitosan is no easy feat: all chemical reactions entail side reactions, and what is typically referred to as chitosan is a form in which around 70% or more of acetyl groups have been removed from chitin (Kim and Rajapakse 2005).

To produce chitosan in its purest form, the concentration of sodium hydroxide solution is increased and the temperature is raised to around 100 °C with a slightly increased reaction time. This may result in the chitosan's viscosity decreasing, ultimately reducing the size of molecules. For this reason, the production process may differ slightly depending on the purpose of chitosan use. When producing chitin, one should take care to apply treatment as soon as the shrimp or crab shell emerges, as it tends to decompose readily. Decomposition sets in when the shell is left for even one day, and the shell should be dried immediately unless treatment is being applied immediately. The use of decomposed raw materials will degrade the quality of the chitin or chitosan (Kim 2010).

In addition to increasing production costs, drying results in slight reduction in molecular weight. It is therefore advisable in both quality and production terms to apply immediate and continuous treatment to a fresh shell. Since chitin is stable with respect to bases, a good way to prevent decomposition is to treat fresh crab shell as quickly as possible with a basic solution.

B. Producing Oligosaccharides with Physiological Functions

As mentioned previously, physiological functions are often exhibited more in the products of chitosan once degraded into suitable sizes than in chitosan itself. Such products include oligosaccharides, series of ten or so monosaccharides produced from the breakdown of chitin and chitosan. Oligosaccharides have been shown to be highly effective in fighting cancer, while their slightly sweet flavor makes them more palatable than chitin or chitosan on their own, and they exhibit strong absorption rates in the body (Fig. 8.29 (c)). Approaches to oligosaccharide production include the chemical method, or the use of strong hydrochloric acid to break down molecules, and the biological method, which uses enzymes (Kim and Rajapakse 2005).

In South Korea, the marketing of oligosaccharides produced from chitosan broken down with hydrochloric acid has been the source of some controversy. The use of hydrochloric acid to break down chitin and chitosan results in the production of some harmful byproducts. Also, failure to completely remove the hydrochloric acid can result in it entering the body and having serious effects on sugar metabolism. This can reportedly result in diabetes occurring due to problems in insulin functioning.

For this reason, the South Korean Ministry of Health and Welfare's Food Standards Codex permits only the use of chitin and chitosan oligosaccharides produced with enzymes. Because the enzymes capable of degrading chitin and chitosan are so expensive and show such low rates of activity, their batch use in industrial production of chitin- and chitosan-based oligosaccharides presents a large number of issues. For this reason, the author is currently working on domestic development and production of the world's first technology for using a membrane enzyme reaction device for continuous production of chitosan oligosaccharides from chitosan hydrolysate (Fig. 8.30).

C. Uses of Chitin, Chitosan, and Oligosaccharides

Crab and shrimp have long been used as ingredients in meals and snacks. As this shows, chitin is non-toxic, and the chitosan produced through deacetylation of chitin is believed to be non-toxic as well. It should also be noted that chitosan is observed to have physiological functions that include reduction of blood cholesterol levels (Rinaudo 2006).

Because chitosan has antibacterial and antifungal properties, it may be considered for use as a food preservative. Also, because of its bacteriostatic properties with regard to plant pathogens, it may be considered as a potential soil improvement agent or natural pesticide. It also undergoes various forms of polycation formation when dissolved in acidic solution, giving it outstanding coagulant properties for wastewater treatment, and it is used as a cosmetic ingredient because of its excellent moisturizing properties. Its outstanding absorption by the body and affinity at the cellular level have resulted in its drawing attention as a medical material in the production of sutures and artificial skin.

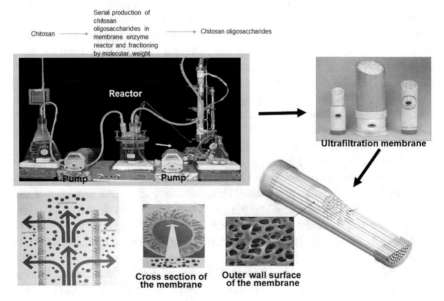

Fig. 8.30 Automated membrane enzyme reactor device for production of chitosan oligosaccharides

(1) Dietary Fiber

Recent recognition of the importance of dietary fiber has led to active research in the field. Since chitin and chitosan are not hydrolyzed by human digestive enzymes, they may be used as sources of dietary fiber. Mice that were fed high-cholesterol feed (0.5% cholesterol) with 5% added chitosan powder for 20 days showed no effects on growth, but marked reductions in blood and liver cholesterol.

Chitosan has also been shown to have functions in terms of processing harmful toxins present in food. If has been found to exhibit toxicity prevention effects with food dyes Red No. 2 and Red No. 105 that were not present in dietary fiber. This is the result of pigments binding strongly with the chitosan, resulting in reduced absorption.

(2) Antibacterial and Antifungal Agents

Chitosan and its lysates have already been shown to inhibit the growth of fungi that are pathogenic to plants. They have also demonstrated marked growth suppression effects on *E. coli*, *B. subius*, *S. aureus*, and other bacteria. At the same time, bacterial and fungal growth suppression efforts have been shown to differ according to chitosan's molecular weight (Table 8.7 and Photo 8.1).

In the case of *Streptococcus mutans* (which has been implicated in cavity formation), up to 99% growth suppression is achievable even with the low molecular weight chitosan oligosaccharide III, indicating that it can be used in the development of cavity prevention products. Storage periods for products such as kimchi and tofu may also be increased with the additional of chitosan, resulting in improvements in resilience and other properties. This suggests that chitosan may be ideal as a natural food preservative.

(3) Medicines

Similar anti-inflammatory and anticancer effects to polymeric chitin have recently been found in chitin oligosaccharides. The N-acetyl chitohexose in particular has

Table 8.7 Fungal and bacterial activity reduction effects of chitosan and chitosan oligosaccharides

Gram positive (+) bacterium	Antibacterial activity (%)			
	Chitosan	COS I (polymer)	COS II (mesomolecule)	COS III (small molecule)
Streptococcus mutans	100	100	99	99
Microporus luteus	>99	70	67	63
Staphylococcus aureus	100	97	95	93
Staphylococcus epidermidis	0.99	82	57	23
Bacillus subtilis	98	63	60	63

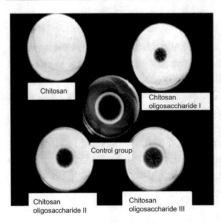
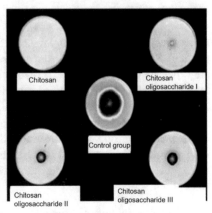

(a) Effects on *Aspergillus niger*   (b) Effects on *Alteraria mali*

Photo 8.1 Fungal growth suppression effects of chitosan and chitosan oligosaccharides (Jeon et al. 2001)

been shown to exhibit the strongest activity. With the use of chitin in the recent commercialization of absorbent sutures, wound treatment accelerators, and artificial skin, the potential for additional uses in future medicines appears great.

(4) Diagnostics

Chitin oligosaccharides have long been known to degrade in the presence of lysozymes. Attempts are currently being made to create the ρ-nitrophenyl penta-N-acetyl-β-chitopentaoxide derivative (ρ-nitrophenol bonded to the hexose lysate produced by lysozymes) for use as a lysozyme substrate. In addition to being widely used in medicine as an antiphlogistic agent, lysozymes have been used as a food preservative. They are also present in the blood and urine, where they are reported to show higher than normal values when a person has contracted a disease. Chitin oligosaccharide derivatives may therefore have potential applications in clinic diagnosis of lysozyme activity, medicine, and food quality management.

(5) Cholesterol Improvement

When people consume lipids such as cholesterols, the lipids are typically not absorbed in their regular state, but broken down by lipase (a fatty acid hydrolase secreted by the pancreas), with the resulting products absorbed by the viscera. Because lipids does not dissolve in solutions such as water, however, most lipidss cluster together within the body (similar in principle to the way that vegetable oils and water do not mix). Since these cannot be broken down by lipase, they tend not to be absorbed by the body. Here, bile acid secreted by the duodenum plays a role in helping to loosen the clusters of lipid so they can be broken down by lipase (Fig. 8.31).

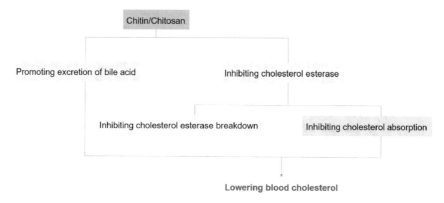

Fig. 8.31 Anti-cholesterol properties of chitin and chitosan

In terms of its chemical structure, however, this bile acid an anion framework. Chitosan has cation ions, which combine with the anion in bile acid to suppress lipase activity and inhibit absorption of cholesterol, thus allowing for the prevention of various lifestyle diseases (Kang et al. 2012).

(6) Cosmetics

When applied to wigs, chitosan has shown moisturizing, anti-static, and lorcia formation properties. In terms of pharmaceutical preparation, it has demonstrated thickening, protective colloid formation, and chelating properties.

The chitosan molecule includes amino (NH_2) and hydroxyl (OH^-) groups, rendering it fairly hydrophilic. Because of its high water holding capability, its strong moisturizing capabilities may keep to maintain the moisturizing property of hair, where the maintenance of a specific moisture level is typically recognized as important for health. Hair also accumulate static electricity from the wind when brushed in low-moisture environments. This can cause hair to stick up or dust to attach and make combing difficult, resulting in a messy hair. Chitosan adheres to form a film on the hair, but the resulting planning of the hair's surface reduces friction resistance. Because of chitosan's strong water holding capacity, it also increases conductivity on the hair's surface, preventing the accumulation of static electricity. Carboxyl chitin, a chitin derivative, dissolves more easily in water than chitosan. Because of its strong moisturizing properties, it is used in cosmetics as a moisturizer. Low molecular weight chitosan and phosphorylated chitin also inhibit adsorption of the cavity-causing bacterium *S. mutans* on saliva-treated hydroxya-patite. Phosphorylated chitin is also being considered for use as a cavity prevention agent due to its strong effects in suppressing adsorption of other bacteria (*Strep-tococcus*) within the oral cavity (Xinyi 1998).

(7) Anticancer Properties of Chitosan Oligosaccharides

Victory over cancer remains a distant dream for modern medicine. Many anticancer agents have been developed, but problems with their side effects have resulted in demand for more effective treatments. While the spread of cancer may be checked with early discovery and removal of the area where it has developed, failure to do so necessitates the use of anticancer agents to prolong life. The anticancer agents developed to date are mostly chemical substances with serious side effects. The reason for these side effects stems from the fact that they do not selectively damage only cancer cells, but destroy normal cells as well. For this reason, immunotherapy-based treatments are preferable to chemical-based ones whenever possible (Huang et al. 2006).

Immunity is a system in which white blood cells, macrophages, T-cells, and B-cells consume and eliminate foreign materials (antigens) when they enter the body. White blood cells are present in the body and functioning to detect and consume relatively small quantities of bacteria and other foreign materials. The name macrophage (meaning "big eaters") comes from the fact that these cells consume rather larger foreign materials than white blood cells. The fragments produced after macrophage consumption are expelled outside of the macrophage's body, at which point T-cells respond by making a determination on whether the material is foreign. When necessary, they order B-cells to attack; once a B-cell receives this order, is produces large quantities of antibodies designed to react specifically to and remove that material.

Chiton, chitosan, and their oligosaccharides are known to boost these immune capabilities and suppress cancer or prevent its spread. Their anticancer effects are reported to be the result of increased macrophage activity or strengthening of T-cells so that B-cells produce more antibodies.

Testing with mice to examine the anticancer effects of chitin and chitosan has all been done through abdominal or subcutaneous injection rather than oral administration. Currently, the only way to consume chitin or chitosan is by eating it, as it is a health supplement rather than a medical product. Because chitin and chitosan are insoluble polymers, their effects will be clearly manifested only when they are consumed in soluble chitin and chitosan oligosaccharide form.

Ascites tumor and cervical cancer suppression effects were examined in mice administered samples of chitosan oligosaccharides of differing molecular weights for a one-month period. As shown in Table 8.8, the highest cancer suppression rate was seen in mice administered samples of chitosan oligosaccharides with molecular weights of 5–10 kDa. Cancer suppression rates were very low for both lower and higher molecular weights, indicating that oligosaccharide size is very important in anticancer effects (Kim 2019).

(8) Controlling High Blood Pressure

High blood pressure occurs as a result of both genetic factors and excessive salt consumption. One of the many factors discovered to cause high blood pressure is

Table 8.8 Cancer suppression effects of chitosan oligosaccharides by molecular weight

Sample	Dose (mg/Kg/day)	Number of mice	Sarcoma 180 tumor cells		Uterine cervix carcinoma tumor cells	
			Tumor weight (mg)	Inhibition rate (%)	Tumor weight (mg)	Inhibition rate (%)
Control		12	1032.5 ± 839.5		912.2 ± 612.1	
COS I (5~10kDa)	50	12	1147.0 ± 933.9	-	965.2 ± 839.7	-
	20	12	901.0 ± 741.7	12.7	803.3 ± 641.8	11.9
	10	12	795.5 ± 384.8*	22.9	772.7 ± 592.2	15.3
COS II (1~5kDa)	50	12	345.2 ± 218.6*	66.6	240.5 ± 202.5*	73.6
	20	12	665.6 ± 340.1	35.5	352.3 ± 331.1	61.4
	10	12	739.5 ± 351.8	28.4	669.5 ± 562.3	26.6
COS III (< 1kDa)	50	12	904.0 ± 510.3	12.4	665.0 ± 477.9	27.1
	20	12	874.8 ± 516.6	15.3	841.1 ± 602.1	7.8
	10	12	973.5 ± 417.1	5.7	879.9 ± 650.3	3.5

$^*P < 0.05$ versus the control group

angiotensin I converting enzyme (ACE), which is present in the body. This enzyme causes blood pressure to rise as angiotensin I is converted into angiotensin II, a substances that increases blood pressure. Normal blood pressure can be maintained by inhibiting this enzyme. Eating less salt is another way to reduce high blood pressure.

Crucially, while sodium ions (Na^+) from the intake of salt (NaCl) into the body have traditionally been viewed as responsible for rising blood pressure, recent experiments with chitosan have shown the chloride ions (Cl^-) to be the substance causing the blood pressure increase (Xia et al. 2011).

Chitosan is a large molecule and a fiber that is partially excreted from the body without being absorbed into the intestines when consumed. The chitosan structure also contains an amino group (positively charged) that bonds with the chloride ions in salt (negatively charged) or bile acid and allows it to be excreted from the body. Bile acid assists the absorption of lipid as it is synthesized into cholesterol and secreted into digestive organs when lipid is consumed. By blocking the reabsorption of lipid into the intestines, chitosan reduces cholesterol in the body and helps expel chloride ions from the body.

As shown in Fig. 8.32, actual testing of blood pressure reduction in mice fed chitosan and alginic acid has shown blood pressure to be reduced through chitosan consumption (Je and Ahn 2011).

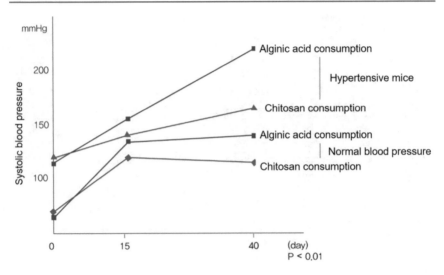

Fig. 8.32 Systolic blood pressure changes in mice fed high-sodium feed and chitosan

8.5 Natural Cosmetic Ingredients

South Korea has traditionally been home to numerous traditional and folk remedies, most notably Korean medicine. Most of these, however, have depended upon land-based plant ingredients, while ingredients from marine sources have been used only rarely.

The same situation can be seen with cosmetics, which handles many ingredients that are used in traditional remedies. Europe, by contrast, has traditional marine-based remedies using marine-derived ingredients such as seaweed, seawater, and sea mud—an approach collectively known as thalassotherapy. Thalassotherapy has recently found its way to South Korea as well, and women in particular have shown a growing interest in marine-derived ingredients. The bovine spongiform encephalopathy (BSE) issue as well has led many to avoid mammal-derived ingredients, resulting in increased interest in marine-based ingredients as substitutes (Khan and Abourashed 2011).

Cosmetics are substances applied to the human body by rubbing, spraying, and similar means for the purposes of cleansing and beautifying, increasing attractiveness, altering facial appearance, or maintaining soft skin and hair. In these cases, actions on the body are moderated. In other words, the aim of cosmetics is to use mild actions to improve the state of skin or hair through continued, long-term use; the effects are not as strong as they are in pharmaceuticals.

In addition to being mild, it is important that ingredients must also be well suited to cosmetic purposes and enjoy a favorable image to be attractive for cosmetic purposes. Because cosmetics are applied directly to the skin, they must demonstrate appropriate stability with regard to the skin.

Marine ingredients from algae and other sources satisfy the above criteria in terms of cosmetic effects and relationships and appear to offer outstanding materials for cosmetic use.

In the past, those wishing to known what ingredients were used in cosmetics could consult official compendiums of materials in the pharmaceutical approval precedents, including the "Cosmetic Ingredient Standards" and "Standards for Ingredient Mixtures by Cosmetic Type." Today, regulations on cosmetics are becoming looser, and there is no need for registration in the compendiums described above. The rules listed in them are useful, however, as a standard for judging quality.

Table 8.9 shows some of the marine-derived ingredients found in the compendiums. Because this table is based on ingredients in the pharmaceutical approval precedents, it also includes ingredients such as whale oil that are no longer in use.

The listings show basic ingredients of cosmetics to include oils such as squalane and squalene; sodium alginate, which is used in moisturizers and thickeners; and fish scales, which are used for pearl pigment agent (Hineno 2005).

Table 8.9 Marine-derived cosmetic ingredients in the pharmaceutical approval precedents

Name	Source	Name	Source
Pottasium alginate	Brown algae	Agar powder	Agar (*Gelidium amansii*)
Sodium alginate	Brown algae	Desulfated aluminum silicate	Sea mud
Calcium alginate	Brown algae	Chitin	Snow crab and red crab
Propylene glycol alginate	Brown algae	Chitosan	Crustaceans
Sulfated alginate	Brown algae	Fish scale membrane	Hairtail (*Trichiurus lepturus*)
Squid ink	Squid	Chlorella extractives	*Chlorella*
Black-banded sea krait lipid	*Laticauda semifasciata*	Succinyl chitosan	Crustaceans
Dried seawater	Seawater	Squalene	Sharks
Seawater extractives	Brown, red and green algae	Squalane	Deep-water sharks
Hydrolyzed extractives	Squid	Soluble collagen solution	Fish
Oyster extractives	Oysters	Conchiolin powder	Pearl oysters
Carrageenan	Red algae	Pearl powder	Shellfish
Carboxylmethyl chitin solution	Crabs	Hydroxyethyl chitosan solution	Crustaceans
Dried chlorella	*Chlorella*	Hydroxypropyl chitosan solution	Crustaceans

Natural extracts also contain skin-friendly biomolecules that can be obtained through bioprocessing. These can be described as functional biocosmetics that emphasize prevention of skin aging and whitening effects. In contrast with the chemical cosmetics used in beauty treatments, these functional products are known as "cosmeceuticals," a portmanteau of "cosmetics" and "pharmaceuticals."

The seven main functions of cosmetics are skin moisturizing, antioxidation, UV protection, wrinkle improvement, whitening, acne prevention, and hair growth and scent effects. The recent emergence of edible nutricosmetics has broken down the boundary between cosmetics and food. This shows that biocosmetics are expanding into the domains of pharmaceuticals and food thanks to advancements in gene manipulation, bioprocessing technology, and technology to explore useful natural bioresources.

Figure 8.33 shows the number of instances in which cosmetic ingredients including the word "sea" (such as "seaweed" or "dried seawater") were included in cosmetic patent requests in Japan over the past 10 years. This testifies to the increasing number of cosmetic ingredient patents listed as being related to the marine field. Since applications are submitted one to two years before publication, this means that attempts to use marine ingredients in cosmetics have been increasing since the late 1990s. It was during this time period that many began increasingly avoiding ingredients derived from mammals and birds, turning new attention to the new resource represented by marine ingredients. As the figure shows, algae accounted for the largest number of marine ingredient types, while dried seawater (sea salt) and sea mud were also used (Fig. 8.33).

Table 8.10 shows the effects and properties of different marine organism-based ingredients. Thanks to continued discovery of new functions, seaweed-derived polysaccharides have a broad range of applications, as seen with agar, alginic acid, carrageenan, chitin, and chitosan.

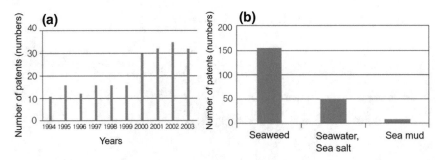

Fig. 8.33 Number of patents registered for marine-derived cosmetic ingredients in Japan (**a**); number of registrations by marine-derived material patent content (**b**)

Table 8.10 Effects of marine ingredients in public patent listings

Algae	Skin care cosmetics (Moisturizing, cracking prevention, promoting hyaluronic acid production, promoting collagen production, promoting fibroblasts, inhibiting elastase, antioxidation etc.)
Seawater/sea salt	Bathing (Skin moisturizing, cracking prevention and treatment etc.) Skin care cosmetics (Whitening, moisturizing, cracking prevention etc.)
Sea mud	Cleansers Packs (moisturizing, eliminating skin wastes etc.) Skin care cosmetics (moisturizing, whitening etc.) Hair care products (sleek feel, moisture etc.)

Among these is alginic acid, a polymer of D-mannuronic acid and L-guluronic acid that generates viscosity in brown algae such as Kelp (*Saccharina japonica*), Sea mustard (*Undaria pinnatifida*), Gulf weed (*Sargassum* sp.), and fusiforme (*Sargassum fusiforme*). These substances are currently used as natural thickening agents in cosmetics, while carrageenan, an acidic polysaccharide consisting of galactose and anhydrogalactose, is used as a cosmetic stabilizer and dispersant.

Compounds of sargachromenol, sargachromenol E, and sargachromenol D isolated from the brown alga *Sargassum horneri* have also proven effective in preventing UV ray damage (Fig. 8.34).

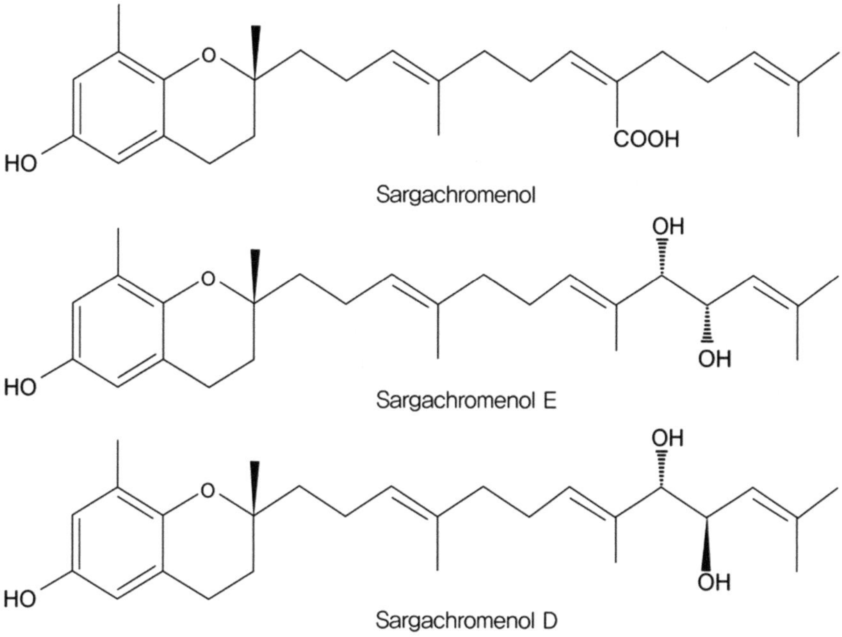

Fig. 8.34 Sargachromenol compounds isolated from *S. horneri*

As seen in Fig. 8.35, degradation of elastin through UV damage has been shown to be suppressed by sargachromenol compounds.

The author has found the phlorotannin compound 7-phloroeckol (isolated from *E. cava*) to strongly inhibit synthesis of tyrosinase and melanin. In addition to using it to develop functional whitening ingredients, the Jeon et al. also found dieckol to suppress wrinkles by blocking DNA damage from UV rays (Figs. 8.35–8.37).

Dieckol has also been shown to be effective against the inflammatory skin condition atopy. In animal models with induced atopy, treatment with *E. cava* extract containing dieckol showed outstanding effects in relieving atopy, as shown in Fig. 8.37.

Ingredients in red algae extracts have been shown to have skin moisturizing, antioxidant, UV protection, and hair growth promotion effects. As shown in Fig. 8.38, extract from the Indonesian red alga *Eucheuma cottonii* was found to have hair growth effects comparable to those of the hair loss medication Minoxidil.

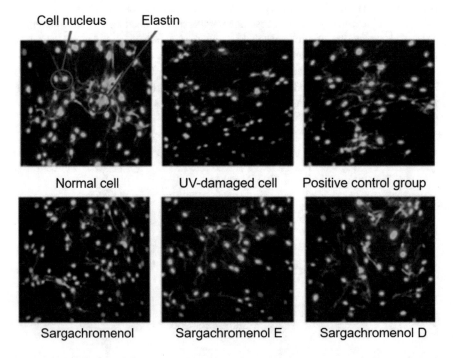

Fig. 8.35 Elastin degradation suppression effects of sargachromenol compounds from *S. horneri*

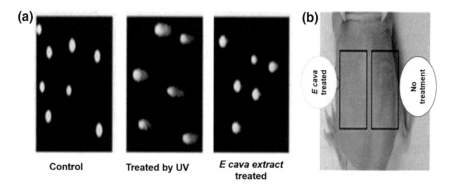

Fig. 8.36 DNA damage prevention effects of dieckol in animal models with induced wrinkles (**a**) and wrinkle improvement effects (**b**)

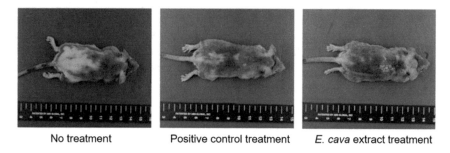

Fig. 8.37 Atopy suppression effects of *E. cotonii* extract in animal models with induced atopy. Positive control group: Betamethasone 17.21.-dipropiana

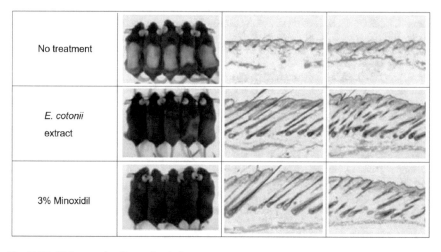

Fig. 8.38 Hair growth effects of red algae extract

8.6 Chapter Summary and Conclusion

Great strides have been made over the past 30 years in research into functional materials obtained from marine organisms, particularly in the world's advanced economies. The isolation of more than 10,000 new materials already and identification of their structures and characteristics have changed the very concept of organic matter in hitherto land organism-based natural product chemistry. The powerful bioactivity and distinctive reaction mechanisms found in many functional materials obtained from marine organisms have contributed greatly not only to medicine and pharmacology, but also basic and applied research in fields related to biology, ecology, and biochemistry. In industry terms, the recent registration of numerous patents for bioactive substances and functional materials derived from marine organisms has led to development of pharmaceuticals, health supplements, functional cosmetics, and industrial products.

The development of new and useful materials from marine organisms is a very important field in both academic and industrial terms. In South Korea, however, this area of research has long been inactive due to a failure to fully recognize its importance. Hopefully, recent increases in government support for the field will result in active development of high value-added products from the unexploited marine organism resources being produced in South Korea's seas.

References

Artan, M., Li, Y., Karadeniz, F., Lee, S.-H., Kim, M.-M., & Kim, S.-K. (2008). Anti-HIV-1 activity of phloroglucinol derivative, 6, 6'-bieckol, from *Ecklonia cava. Bioorganic & Medicinal Chemistry, 16*(17), 7921–7926.

Balcıoğlu, I. A., Tarlan, E., Kıvılcımdan, C., & Saçan, M. T. (2007). Merits of ozonation and catalytic ozonation pre-treatment in the algal treatment of pulp and paper mill effluents. *Journal of Environmental Management, 85*(4), 918–926.

Barsanti, L., & Gualtieri, P. (2014). *Algae: Anatomy, biochemistry, and biotechnology*. CRC Press.

Burke, J. F., Yannas, I. V., Quinby, W. C., Jr., Bondoc, C. C., & Jung, W. K. (1981). Successful use of a physiologically acceptable artificial skin in the treatment of extensive burn injury. *Annals of Surgery, 194*(4), 413.

Burns, G. (1992). *High-temperature superconductivity*.

Burr, M. L., Gilbert, J., Holliday, R. A., Elwood, P., Fehily, A., Rogers, S., et al. (1989). Effects of changes in fat, fish, and fibre intakes on death and myocardial reinfarction: Diet and reinfarction trial (DART). *The Lancet, 334*(8666), 757–761.

Certik, M., & Shimizu, S. (1999). Biosynthesis and regulation of microbial polyunsaturated fatty acid production. *Journal of Bioscience and Bioengineering, 87*(1), 1–14.

Childs, M. T., King, I. B., & Knopp, R. H. (1990). Divergent lipoprotein responses to fish oils with various ratios of eicosapentaenoic acid and docosahexaenoic acid. *The American Journal of Clinical Nutrition, 52*(4), 632–639.

Choi, B. H., Kim, B. J., Kim, C. S., Lim, S. H., Yang, B. S., Seo, J. H., Cheong, H. Y., & Cha, H. J. (2015). Mussel-derived bioadhesives. In S. K. Kim (Ed.), *Springer handbook of marine biotechnology* (pp. 1321–1330). London, New York: Springer.

Crow, J. E., Stein, S., & Wahlers, R. (1991). *The development of high transition temperature superconducting ceramic thick films and wire composites*. Department of Materials Research and Technology, Florida State University, Tallahassee.

Dai, T., Tanaka, M., Huang, Y.-Y., & Hamblin, M. R. (2011). Chitosan preparations for wounds and burns: Antimicrobial and wound-healing effects. *Expert Review of Anti-infective Therapy, 9*(7), 857–879.

Dove, J., & Sheridan, P. (1986). Adhesive protein from mussels: Possibilities for dentistry, medicine, and industry. *Journal of the American Dental Association, 112*(6), 879.

Dutta, P. K., Dutta, J., & Tripathi, V. (2004). *Chitin and chitosan: Chemistry, properties and applications.*

Ehresmann, D., Deig, E., Hatch, M., DiSalvo, L., & Vedros, N. (1977). Antiviral substances from California marine algae 1. *Journal of Phycology, 13*(1), 37–40.

Endo, M., Nakagawa, M., Hamamoto, Y., & Ishihama, M. (1986). Pharmacologically active substances from southern Pacific marine invertebrates. *Pure and Applied Chemistry, 58*(3), 387–394.

Erdman, J., Oria, M., & Pillsbury, L. (2011). *Eicosapentaenoic acid (EPA) and docosahexaenoic acid (DHA).*

Fanning, L., Mahon, R., & McConney, P. (2011). *Towards marine ecosystem-based management in the wider Caribbean.* Amsterdam: Amsterdam University Press.

Fusetani, N. (2005). Research development of marine natural substances. In N. Fusetani (Ed.), *Biotechnological applications of marine natural substances* (pp. 1–5). Tokyo, Japan: CMC Pub. Co.

Gaur, M., Dobke, M., & Lunyak, V. V. (2017). Mesenchymal stem cells from adipose tissue in clinical applications for dermatological indications and skin aging. *International Journal of Molecular Sciences, 18*(1), 208.

Gillies, P. J., & Schaefer, E. J. (2011). *Clinical benefits of eicosapentaenoic acid in humans.* Google Patents.

Hineno, T. (2005). Cosmetics. In N. Fusetani (Ed.), *Biotechnological application of marine natural substance* (pp. 178–192). CMC.

Huang, R., Mendis, E., Rajapakse, N., & Kim, S.-K. (2006). Strong electronic charge as an important factor for anticancer activity of chitooligosaccharides (COS). *Life Sciences, 78*(20), 2399–2408.

Hwang, D. S., Choi, Y. S., & Cha, H. J. (2013). Mussel-derived adhesive biomaterials. In S. K. Kim (Ed.), *Marine biomaterials* (pp. 289–309). NW: CRC Press.

Je, J. Y., & Ahn, C. B. (2011). Antihypertensive actions of chitosan and its derivatives. In S. K. Kim (Ed.), *Chitin, chitosan oligosaccharides and their derivatives; biological activities and applications* (pp. 263–269). CRC Press.

Jeon, Y.-J., Shahidi, F., & Kim, S.-K. (2000). Preparation of chitin and chitosan oligomers and their applications in physiological functional foods. *Food Reviews International, 16*(2), 159–176.

Jeon, Y.-J., Park, P.-J., & Kim, S.-K. (2001). Antimicrobial effect of chitooligosaccharides produced by bioreactor. *Carbohydrate Polymers, 44*(1), 71–76.

Jun, S.-Y., Park, P.-J., Jung, W.-K., & Kim, S.-K. (2004). Purification and characterization of an antioxidative peptide from enzymatic hydrolysate of yellowfin sole (*Limanda aspera*) frame protein. *European Food Research and Technology, 219*(1), 20–26.

Kamino, K. (2008). Underwater adhesive of marine organisms as the vital link between biological science and material science. *Marine Biotechnology, 10*(2), 111–121.

Kang, N.-H., Lee, W. K., Yi, B.-R., Park, M.-A., Lee, H.-R., Park, S.-K., et al. (2012). Modulation of lipid metabolism by mixtures of protamine and chitooligosaccharide through pancreatic lipase inhibitory activity in a rat model. *Laboratory Animal Research, 28*(1), 31–38.

Karadeniz, F., Zafer Karagozlu, M., Kong, C.-S., & Kim, S.-K. (2011). In vitro anti-HIV-1 activity of the aqueous extract of *Asterina pectinifera. Current HIV Research, 9*(2), 95–102.

Khan, I. A., & Abourashed, E. A. (2011). *Leung's encyclopedia of common natural ingredients: Used in food, drugs and cosmetics.* Wiley.

Kibo, H. (1994). Application of artificial skin (pp. 71–104). Tokyo, Japan: Gihodo Pub. Co.

Kim, S.-K. (2010). *Chitin, chitosan, oligosaccharides and their derivatives: Biological activities and applications.* CRC Press.

Kim, S. K. (2018). Utilization of chitin artificial skin, Healthcare Using Marine Organisms (pp. 47–51). N.Y. US.: CRC Press.

Kim, S. K. (2018). Fish oil is an adult disease preventive medicine, Healthcare Using Marine Organisms (pp. 185–188). CRC Press.

Kim, S. K. (2018). How to make chitin, chitosan, and chitosan oligosaccharides?. In *Healthcare using marine organisms* (pp. 13–17). CRC Press.

Kim, S. K. (2019). Physiological properties of chitin, chitosan, and chitosan oligosaccharides. In *Healthcare using marine organisms* (pp. 19–35). CRC Press.

Kim, S. K. (2018). Development of an ointment from sea cucumber extract for treatment of athlete's foot (pp. 191–194). Healthcare Using Marine Organisms, CRC Press.

Kim, S. K. (2018). Functional food and disease prevention, Healthcare Using Marine Organisms (pp. 1–6). N.W.: CRC Press.

Kim, S.-K., & Rajapakse, N. (2005). Enzymatic production and biological activities of chitosan oligosaccharides (COS): A review. *Carbohydrate Polymers, 62*(4), 357–368.

Kim, S. K., Vo, T. S., & Ngo, D. H. (2013). Marine algae: Pharmacological values and anti-inflammatory effects. In S. K. Kim (Ed.), *Marine pharmacognosy: trends and applications* (pp. 273–280). N.Y. US.: CRC Press.

Konishi, H. (1989). Developing new high-temperature superconductors with alginic acid. In T. Matunaga (Ed.), *Marine bio: Marine bio new materials and new substances* (pp. 243–244). Tokyo, Japan: CMC Pub. Co.

Kumar, M. N. R. (2000). A review of chitin and chitosan applications. *Reactive & Functional Polymers, 46*(1), 1–27.

Lee, H.-W., Park, Y.-S., Choi, J.-W., Yi, S.-Y., & Shin, W.-S. (2003). Antidiabetic effects of chitosan oligosaccharides in neonatal streptozotocin-induced noninsulin-dependent diabetes mellitus in rats. *Biological and Pharmaceutical Bulletin, 26*(8), 1100–1103.

Lee, H. J., Kim, H. C., Vitek, L., & Nam, C. M. (2010). Algae consumption and risk of type 2 diabetes: Korean National Health and Nutrition Examination Survey in 2005. *Journal of Nutritional Science and Vitaminology, 56*(1), 13–18.

Lewis, D. A., & Lewis, D. (1989). *Anti-inflammatory drugs from plant and marine sources.* Birkhäuser.

McGinn, P. J. (1998). Commercializing high-temperature superconductors. *JOM Journal of the Minerals Metals and Materials Society, 50*(10), 15.

Newman, D. J., Cragg, G. M., & Snader, K. M. (2003). Natural products as sources of new drugs over the period 1981–2002. *Journal of Natural Products, 66*(7), 1022–1037.

Ortiz, J., Romero, N., Robert, P., Araya, J., Lopez-Hernández, J., Bozzo, C., et al. (2006). Dietary fiber, amino acid, fatty acid and tocopherol contents of the edible seaweeds *Ulva lactuca* and *Durvillaea antarctica. Food Chemistry, 99*(1), 98–104.

Palffy-Muhoray, P. (2007). Orientationally ordered soft matter: The diverse world of liquid crystals. *Electronic-Liquid Crystal Communications (e-LC).*

Poncelet, D., Lencki, R., Beaulieu, C., Halle, J., Neufeld, R., & Fournier, A. (1992). Production of alginate beads by emulsification/internal gelation. I. Methodology. *Applied Microbiology and Biotechnology, 38*(1), 39–45.

Prashanth, K. H., & Tharanathan, R. (2007). Chitin/chitosan: Modifications and their unlimited application potential—An overview. *Trends in Food Science & Technology, 18*(3), 117–131.

Rehm, B. H. (2009). *Alginates: Biology and applications.* Springer.

Rinaudo, M. (2006). Chitin and chitosan: Properties and applications. *Progress in Polymer Science, 31*(7), 603–632.

Rose, G., Khaw, K.-T., & Marmot, M. (2008). *Rose's strategy of preventive medicine: The complete original text.* New York, USA: Oxford University Press.

Sastry, V., & Rao, G. (1994). Antibacterial substances from marine algae: Successive extraction using benzene, chloroform and methanol. *Botanica Marina, 37*(4), 357–360.

Schnyder, A. P., Brydon, P., & Timm, C. (2012). Types of topological surface states in nodal noncentrosymmetric superconductors. *Physical Review B, 85*(2), 024522.

Shahidi, F. (2006). *Maximising the value of marine by-products.* Woodhead Publishing.

Shimomura, M. (1989). Nature crystal material. In T. Matunaga (Ed.), *Marine bio: Marine bio new materials and new substances* (pp. 181–187). Tokyo, Japan: CMC Pub. Co.

Silverman, H. G., & Roberto, F. F. (2007). Understanding marine mussel adhesion. *Marine Biotechnology, 9*(6), 661–681.

Singh, R., Sharma, M., Joshi, P., & Rawat, D. S. (2008). Clinical status of anti-cancer agents derived from marine sources. *Anti-Cancer Agents in Medicinal Chemistry (Formerly Current Medicinal Chemistry-Anti-Cancer Agents), 8*(6), 603–617.

Summers, W. K., Majovski, L. V., Marsh, G. M., Tachiki, K., & Kling, A. (1986). Oral tetrahydroaminoacridine in long-term treatment of senile dementia, Alzheimer type. *New England Journal of Medicine, 315*(20), 1241–1245.

Vo, T. S., Ngo, D. H., & Kim, S. K. (2012). Anti-HIV activities of marine macroalgae. In S. K. Kim (Ed.) *Handbook of marine macroalgae: biotechnology and applied phycology* (pp. 417–421). UK: Wiley-Blackwell, Oxford.

Wei, N., Quarterman, J., & Jin, Y.-S. (2013). Marine macroalgae: An untapped resource for producing fuels and chemicals. *Trends in Biotechnology, 31*(2), 70–77.

Wool, R., & Sun, X. S. (2011). *Bio-based polymers and composites.* Elsevier.

Wu, M.-K., Ashburn, J. R., Torng, C., Hor, P. H., Meng, R. L., Gao, L., et al. (1987). Superconductivity at 93 K in a new mixed-phase Y-Ba-Cu-O compound system at ambient pressure. *Physical Review Letters, 58*(9), 908.

Xia, W., Liu, P., Zhang, J., & Chen, J. (2011). Biological activities of chitosan and chitooligosaccharides. *Food Hydrocolloids, 25*(2), 170–179.

Xinyi, X. W. Z. F. H. (1998). Antimicrobial of chitooligosaccharides and its application to food preservation. *Journal of Wuxi University of Light Industry, 4.*

Yazawa, K. (1989). EPA derived from fish intestine bacteria. In T. Matunaga (Ed.), *Marine bio.: Marine bio. new materials and new substances* (pp. 99–107). Tokyo, Japan: CMC Pub. Co.

Zillig, W., Holz, I., Klenk, H.-P., Trent, J., Wunderl, S., Janekovic, D., et al. (1987). Pyrococcus woesei, sp. nov., an ultra-thermophilic marine archaebacterium, representing a novel order, Thermococcales. *Systematic and Applied Microbiology, 9*(1–2), 62–70.

Marine Bioenergy Production

9

Contents

9.1 Marine Biomass

Biomass is the total of organisms (including plants, animals, and microbes) within a given space as converted into a numerical quantity. In South Korean ecology, corresponding terms include "existing organism quantity" and "organism quantity." In a contemporary environment where people are obliged to use solar energy in addition to chemical fuel and nuclear energy, the word "biomass" has been used in applied science fields beyond its original ecological sense. Today, biomass has been positioned as the specific focus in attempts to obtain useful materials and fuels by taking advantage of the ways in which organisms accumulate solar energy (Gao and McKinley 1994).

Biomass is currently being considered alongside nuclear power and solar energy as an energy source for the future. Particularly high hopes are being pinned on the use of theoretically unlimited solar energy and biomass—an aggregate of that solar energy—as a new energy resource.

The total biomass of the Earth is estimated to be 1.0×10^{12} ton of carbon equivalent, a total that corresponds to 100 times the current annual energy consumption and five times the amount of petroleum deposits. One-tenth of total biomass is reproduced each year through photosynthesis, a process that uses solar

© Springer Nature Switzerland AG 2019
S.-K. Kim, *Essentials of Marine Biotechnology*,
https://doi.org/10.1007/978-3-030-20944-5_9

energy. To date, scarcely any of the unlimited resources offered by biomass have been used. Once it becomes possible to produce energy with higher efficiency, this biomass may well become the main energy production system of the 21st century. Energy produced with marine biomass—including algae, microalgae, and marine microorganisms—could be used as an energy resource for the future (Kim and Manivasagan 2015).

Algae have long been valued as a food source in Korea: sea mustard, kelp, laver, and many other forms of algae have become established parts of Korean culinary culture. In addition to food, algae have a wide range of other uses, including livestock feed, crop fertilizers, pharmaceuticals, cosmetics, food additives, and industrial materials. Recently, efficient use of the solar energy that falls on the world's oceans—which account for two-thirds of the Earth's surface area—has been discussed as a way of dealing with the predicted energy crisis from the rapid increase in the world's population and the exhaustion of fossil energy resources such as petroleum and coal. Algae in particular have drawn great attention as a form of marine biomass resource to achieve this (Lee et al. 1971).

As a peninsula surrounded on three sides by water, Korea possesses relatively abundant algae resources. Because of its relatively small size and dearth of other resources, it has focused its efforts on alternative energy development since the oil shocks of the 1970s, though without the anticipated results.

While Korea uses roughly 15 different types of algae for food, there are known to be over 200 types worldwide. Examples of food algae include laver, sea mustard, kelp, and green laver. The 2010 seafood annual put the total of natural and farmed production of algae at roughly 10,000 ton. Other examples with potential uses as food or marine biomass resources include two species of blue-green algae, 35 of green algae, 106 of brown algae, and 254 of red algae, for a total of 397 species. As ways of boosting production to use these as marine biomass resources, growing and farming technologies can be improved and superior varieties can be obtained through seed technology development. It is against this backdrop that research on algae biotechnology has advanced rapidly in the U.S., Europe, and Japan, with the aim of promoting the currently somewhat underperforming field of applied phycology (Lee et al. 1992).

Because microalgae are single-celled organisms inhabiting the water, they do not require fields or soil. With sufficient water and light, it is possible to use them to produce and stockpile fuel (hydrocarbons) from their photosynthetic process of carbon dioxide in the atmosphere. Under stable immune conditions, energy production efficiency is reportedly around ten times higher than for land plants. Another advantage in comparison with land plants is that microalgae can be harvested throughout the year and are well suited to automation due to their ease of farming and harvesting. Today, microalgae are drawing global attention as a final means of producing next-generation biofuels (Brennan and Owende 2010).

This chapter will examine the use of algae, microalgae, and marine microorganisms to produce bioenergy.

9.2   Producing Ethanol with Algae

As a fossil fuel, petroleum exists in limited deposits, and the carbon dioxide produced with its combustion has been identified as responsible for global warming. More and more attention is focusing on the development of alternative energy sources to take its place.

Nuclear power had been regarded as one major alternative source of energy. After a 2011 mega-earthquake and tsunami in Japan resulted in an explosion at the Fukushima Nuclear Power Plant with catastrophic consequences, many are now calling for changes or reconsideration not only of the future construction of additional nuclear powerful plants but also the operation of existing ones (Visschers and Siegrist 2013).

In addition to requiring members of the public to conserve energy more to prepare for a future energy shortage, there have also been calls around the world to make greater use of natural energy sources such as solar, water, and wind power and increase the use of sustainable energy sources such as bioethanol from biomass.

In addition to farming crops and food resources such as soybeans, corn, sugarcane, and sugar beets, other resources used for ethanol production to date have included timber, hay, leaves, barley straws, and rice straws (Pimentel and Patzek 2005).

The use of corn, soybeans, sugarcane, and other crops as biofuels in Brazil, the U.S., and Europe has led to problems with a global food shortage amid skyrocketing grain prices. This in turn has led to the recent identification of non-edible biomass in the form of algae and lignocellulose biomass from weeds, rice straw, and waste lumber. In the case of lignocellulose biomass, the lignin surrounding the cellulose prevents extraction and must be removed first. Cellulose is not readily degradable, making saccharification difficult. Because of these factors and difficulties related to chemical treatment, progress has not been made in the development of production technology (Pimentel and Patzek 2005).

In contrast, algae may be seen as a more or less unlimited source of ethanol biomass. Though a portion of seaweeds are currently used for food, they can also be mass-produced in the oceans (Bush and Hall 2006).

Here, "seaweeds" is understood to refer to the larger varieties of green, brown, and red algae. While chiefly distributed in coastal regions, they are produced in seas around the world in quantities that exceed the yields of tropical rain forest zones. Other great advantages of algae biomass compared to the farming of land biomass for ethanol production include the fact that it grows naturally in the vast environment of the seas, which means that it does not compete with edible crops for farmland or water and presents no environmental load in terms of pesticides (Percival 1979).

As an ethanol source, algae contain large amounts of sugar, including cellulose polysaccharides, starch polysaccharides, high-molecular weight sulfate polysaccharides, alginic acid (a uronic acid compound), and sugar alcohols such as mannitol. Because algae either do not possess or have only very small quantities of the lignin that prevents cellulose polysaccharide extraction in land biomass, they may be seen as amenable to usage as a fuel for ethanol fermentation (Wei et al. 2013).

Among green, brown, and red algae types, green algae have the lowest resource value, being restricted to estuaries and freshwater-influenced environments where seawater and freshwater mix. Red algae exist in numerous varieties, but most of these are small, and while there is strong demand for mucilaginous polysaccharides such as agar and carrageenan as thickening agents in foods and daily essentials, there is little hope for their use as an ethanol production source. In that sense, brown algae are expected to serve as a source for ethanol, existing in numerous large varieties such as kelp, sea mustard, sargassum, and giant kelp and having the largest production yields of the three algae types. At the same time, they also contain a complex mixture of sugars, with large quantities of mannitol and other sugar alcohols in addition to fucoidans, alginic acid, laminaran, and cellulose polysaccharides, which poses a large obstacle to their use in whole form.

Also, algae have softer tissues and higher moisture content than land biomass, making them prone to rotting and foul odors when removed from the ocean. Measures to prevent this are therefore essential.

9.2.1 Algae-Based Ethanol Production Process

The process of producing ethanol from marine biomass follows a sequence of algae drying, crushing, pulverization, liquefaction, saccharification, and other preliminary processing before ethanol fermentation and refinement (concentration and isolation) (Sato 2011).

A. Liquefaction

Liquefaction is necessary for simple enzyme treatment or microbial fermentation once the algae's ingredients have been extracted. Methods of algae liquefaction include ① extracting sugars from dried powder, ② using enzymes on live algae to break down the replenishing polysaccharides in the cell wall or between cells, and ③ liquefying live algae under conditions of intense heat and pressure.

Method ① is used with land biomass, but reduces the energy balance because of the large energy amount needed for drying and pulverization.

The enzyme treatment in ② involves liquefaction through treatment of the algae's structural polysaccharides with cellulose or the digestive enzymes of algivorous mollusks. Liquefaction under mild conditions is possible with enzymes that are capable of breaking down structural polysaccharide bonds, but the acquisition and costs of suitable enzymes can be an issue.

Liquefaction into a kind of paste form is achieved by treating the dried, powdered form of kelp or other brown algae with enzymes that break down the fibrin in the cell wall or alginic acid-degrading enzymes that reduce the molecular weight of the mucopolysaccharides (Alginic acid) between cells and reduce their viscosity. The red algae *Gelidium amansii* can be treated with sodium chlorite to remove its lignin, after which β-galactosidase and xylanase are used for liquefaction and saccharification.

The intense temperatures and pressures in ③ can be used for liquefaction or molecular weight reduction of proteins and other polymer compounds. For algae, polysaccharide saccharification can reportedly be achieved as well. This approach, however, poses difficulties in terms of treatment of large volumes of algae due to capacity issues with the internal pressure vessels (Huang et al. 2011).

B. Saccharification

Saccharification includes processes of acid hydrolysis, degradation under high temperature and pressure, and enzymolysis. Acid hydrolysis may be achieved through treatment with 3% sulfuric acid for 60 min at 120 °C. Most polysaccharides can be broken down to monosugars through acid hydrolysis, but monosugar retention may also be reduced by excessive degradation. Too much breakdown reduces the recovery rate for monosugars, which in turns lowers the recovery rate for ethanol. It is also necessary to remove the sulfuric acid following hydrolysis. Alkali neutralization offers a simple means of removing the sulfuric acid, but this results in the formation of basis that can affect the growth and fermentation of yeast and other microorganisms that participate in ethanol fermentation.

An alternative approach is to treat the decomposition solution with ion-exchange resin to isolate the monosugars and sulfuric acid. Once recovered, the sulfuric acid can be reused, but measures must also be taken to prevent corrosion of the device by the acid, which increases the investment cost.

When treated at various pressures and temperatures, algae transform into algal tissues or components, making monosugar extraction easier or causing decomposition. Land biomass such as lumber is broken down to emit its components under supercritical conditions (temperature of 374 °C or greater, pressure of 22 MPa or greater) or subcritical conditions (close to supercritical; see Fig. 9.1). Treatment of algae with high pressures at supercritical or subcritical conditions (e.g., 500–1000 Mpa at 60–80 °C for 30 min) results in liquefaction and saccharification taking place simultaneously. This method cannot be used for the processing of large amounts of algae, as it requires a pressure vessel capable of withstanding ultra-high pressure.

With enzyme-based saccharification, the type of enzyme used depends on the components and bonding methods of the algal sugar. Algae polysaccharides contain various constituent sugars, and the component monosugars and bonding methods are varied (Table 9.1).

In the case of polysaccharides consisting of D-glucose, such as fibrin with is D-glucose beta-1,4 bond, fibrinogenase may be used for saccharification. For laminaran and other polysaccharides with a D-glucose β-1,3 bond, β-1,3 glucanase can be used. The fibrinogenase XP-425 has been developed as an enzyme capable of breaking down various kinds of bonds, including glucose β-1,4 and β-1,3 bonds and bonds between xylose, and can be used for a broad range of liquefaction and saccharification.

Just as substances have freezing (melting) points and evaporation (condensation) points, so most also have a triple point and critical point. The critical point is a characteristic of the substance; each substance has a different one. "Critical point"

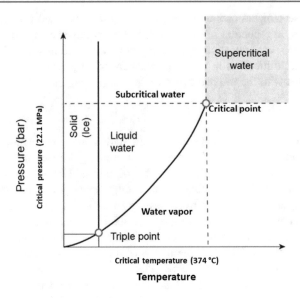

Fig. 9.1 Phase equilibrium gradients for substances

Table 9.1 Types and components of algal glycans

Glycan	Type	Major components	Major linkage types
Brown algae polysaccharides	Cellulose group polysaccharides	Glucose	β-1,4-linkage
	Laminaran	Glucose	β-1,3-linkage
	Fucoidan	Fucose, galactose, sulfate groups	α-1,2-linkage, α-1,3-linkage
	Alginic acid	Mannuronic acid, gulonic acid	α-1,4-linkage, β-1,4-linkage
Sugar alcohol	Mannitol		
Green algae polysaccharides	Cellulose group polysaccharides	Glucose	β-1,4-linkage
	Green algae starch	Glucose	α-1,4-linkage
Red algae polysaccharides	Cellulose group polysaccharides	Glucose	β-1,4-linkage
	Agar	Galactose, glucuronic acid Anhydrogalactose, sulfate groups	α-1,3-linkage, β-1,4-linkage
	Carrageenan	Galactose, anhydrogalactose Sulfate groups	α-1,3-linkage, β-1,4-linkage

refers to conditions that satisfy the critical temperature and critical pressure simultaneously. In the case of water, the critical point is a critical temperature of 374 °C and a critical pressure of 221 bar (218.3 atm). "Critical temperature" refers to a state where a substance's temperature is high enough that no amount of pressure causes liquefaction, while "critical pressure" refers to a state where a substance's pressure is high enough that it does not evaporate no matter how high the temperature (Matsumoto et al. 2003).

C. Ethanol Fermentation

Fermentation is the transformation of monosugars into ethanol once complex components have been reduced to low molecular weight. The sugars that serve as the main materials in ethanol fermentation include glucose and mannitol in brown algae, and glucose, galactose, and xylose in green and red algae. Microorganisms such as yeasts and bacterial play a role in ethanol fermentation. The optimal approach in this case would be to allow single microorganisms to convert all components into ethanol. The characteristics of the enzymes made available to the microorganisms in fermentation, however, are such that this cannot be relied upon. Research has indicated the possibility of using gene improvement to produce yeasts with a wider range of substrate specificity, which raises hopes for the future.

Glucose can be converted to ethanol by a range of microorganisms. Yeasts that can produce ethanol from glucose include *Saccharomyces cerevisiae*, *Pachysolen tannophilus*, and *Pichia angophorae*; bacteria include *Zymomonas mobilis*. Ethanol can also be produced from galactose, which is present in large volumes in red algae, after its conversion to glucose-6-phosphate. The sugar alcohol mannitol is converted into ethanol through the actions of the yeast *Pichia angophorae* and the bacteria *Zymobacter palmae*. Various microorganisms contained in *Aspergillus oryzae*, a fungus used in traditional forms of Korean liquor, have been found to produce ethanol as they break down the alginic acid in kelp; mannitol was found to be the source of this ethanol.

Takeda et al. were the first to succeed in using genetically manipulated bacteria from *Sphingomonas* sp. to produce ethanol for alginic acid, which is present in large quantities in brown algae. As an example of a major brown algae sugar being converted entirely through ethanol fermentation, this marked a great advancement in ethanol production from algae biomass, which is one of the most abundant of resources (Takeda 2011; Wang et al. 2011; Lee and Lee 2012).

9.2.2 Future Tasks

In bioethanol production, the ethanol produced from a single glucose molecule is limited to two molecules. This requires a overall manufacturing process of efficient ethanol fermentation with low energy consumption. More specifically, algae containing large amounts of sugars suitable for ethanol fermentation must be sought and developed, and farming and harvesting methods must be established for these useful algae. Also in need of development of liquefaction and saccharification methods with high sugar recovery rates, the identification and culturing of efficient fermentation microbes, and low-energy input ethanol refinement approaches.

The most important thing is to acquire algae as a source, which will require large-scale farming of promising algae species. Farming methods for the algae used in food are quite costly, and new farming methods with lower costs will need to be developed. Also needed are methods for harvesting algae and dedicated harvesting vessels that reflect the characteristics of specific sea area and the structure of farming facilities.

Because the concentration of ethanol produced from fermentation with algae is relatively low, low-concentration alcohol solution must be isolated and enriched. New isolation methods need to be developed that apply membrane separation to the existing distillation approach, allowing for the separation and enrichment of low-concentration alcohol.

Once these problems are solved, perhaps ethanol can be produced from algae for use as a renewable energy source.

9.3   Methane Production from Algae

9.3.1   Methane Production from Algae

Several methods are currently being applied to use biomass for energy. Methods for conversion into biofuel consist chiefly of thermochemical and biochemical conversion. For thermochemical conversion, methods such as pyrolytic gasification, fast pyrolysis, and carbonization are used. For biochemical conversion, microorganism reactions (fermentation) are used to convert biomass into methane gas, alcohol, or other fuels.

The fuels obtained through these conversion methods can be converted by combustion into heat, electrical, or kinetic energy. For heat energy conversion, boilers are sometimes used. For conversion to electrical energy, engines or turbines with a generator are used. Because engines and turbines release high-temperature waste gases, cogeneration systems are often used to produce electrical and heat energy simultaneously through heat exchange with these gases. The fuels obtained can also be used for kinetic energy in the form of vehicle fuels (Park and Li 2012).

In terms of fuel conversion technology, thermochemical conversion is typically used when moisture content is low, and biochemical conversion is used when it is high. Because thermochemical conversion is accompanied by high-temperature reactions, the temperature is reduced as a result of evaporation of moisture contained in the biomass. As a consequence, the necessary reaction temperature for conversion may not be obtained in cases where moisture content is high. Biochemical conversion, in contrast, uses the reactions of microorganisms, and conversion takes place at a temperature range where moisture does not evaporate. An advantage of the thermochemical conversion approach is the low quantity of non-reacting residue, but for materials with a high moisture content, biochemical conversion can be used for more efficient fuel conversion.

Algae inhabit the sea and have high moisture content (around 90%), which means that the biochemical approach is better for fuel conversion. With biochemical conversion, biodegradability is often an issue. Algae do not contain lignocellulose, which is difficult to degrade; microorganism decomposition is achieved relatively easily, and algae provide one of the best-suited materials for fermentation treatment.

One of the methods commonly used for biochemical conversion today is methane fermentation. In methane fermentation, organic components of biomass are broken down through various microorganism actions, resulting in its ultimate conversion to biogas consisting of methane and carbon dioxide. Biogas (methane gas) production is achieved through various reactions. The carbohydrates (sugars), proteins, and lipids that make up the organic components of biomass are broken down into low-molecular weight monosugars, amino acids, and fatty acids through the actions of hydrolyzing bacteria. Next, low-molecular-weight organic acids such as acetic acid are produced through the actions of acid fermentation bacteria. Through the use of methane-producing bacteria, the resulting acetic acid can be used to produce biogas containing methane and carbon dioxide. Methane gas is also formed from the carbon dioxide and hydrogen released in the acid production process.

The biogas produced from methane fermentation consists of roughly 60% methane and 40% carbon dioxide and may be used as a fuel for boilers, gas engines, and other gas devices. Biogas also contains trace quantities of hydrogen sulfide, and iron oxides and activated carbon are used as desulfurizing agents for gas refinement.

In methane fermentation, it is impossible to convert all of the organic components of the biomass into biogas, and some are emitted as residues. These residues contain undegraded organic components and the microbes responsible for methane fermentation. Typically, the waste liquid produced in methane fermentation tanks is used as liquid fertilizer or dehydrated and drained of moisture to produce a solid mass that can also be used for fertilizer.

Other forms of biochemical conversion besides methane fermentation include alcohol and hydrogen fermentation. Whereas only the carbohydrate (sugar) organic components are converted in hydrogen and alcohol fermentation, lipids and proteins can also be used in methane fermentation, resulting in greater fuel conversion efficiency. Another advantage is that the fuels obtained from methane and hydrogen fermentation are gases and do not require isolation from the fermentation solution, which means there is no energy loss due to distillation. Conversely, methane fermentation requires longer retention periods, which translates into larger fermentation tanks and equipment areas than the other approaches. Because methane and hydrogen are gases with low energy density, they are more difficult to store and transport than liquid alcohol. Each method has its pros and cons, and the choice of conversion methods must take into account not only conversion efficiency but also the installation site and the form of energy usage (Hossain et al. 2008).

In the case of methane fermentation of algae, one approach long considered in the U.S. has involved the use of fermentation from the farming stages to produce fuel from giant kelp (Nakashimada and Nishio 2011).

9.3.2   The Role of Methane Fermentation in Algae Systems Using Cascades

The idea of using marine biomass for energy was first proposed in 1968 by the U.S. scientist Howard Wilcox. By 1990, government agencies, universities, and private enterprises were collaborating on a marine biomass energy program. For this program, giant kelp (*Macrocystis pyrifera*) was proposed as a cultivar. A species of brown algae, giant kelp is fast-growing and reaches lengths of up to 43 m.

The algae biomass conversion process proposed for this program is illustrated in Fig. 9.2. Following preliminary treatment, the kelp is separated into compressed cake and juice. The cake is recovered as energy through anaerobic digestion (methane fermentation), but alginic acid is extracted as a high value-added material. One approach considered for anaerobic digestion was codigestion of the cake with organic waste materials such as livestock excreta as a way of providing the inorganic nutrients needed for microorganism growth and increasing the biogas quantity. The fermentation residue following anaerobic decomposition may be used as a feed additive or composted for fertilizer, while a portion may be hydrolyzed for additional methane fermentation (Nakashimada and Nishio 2011).

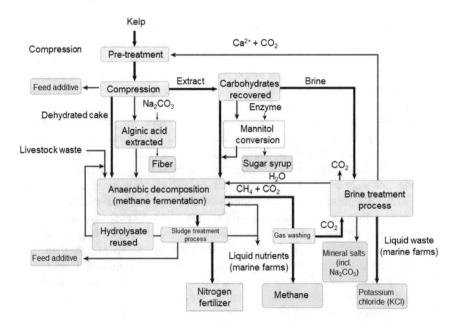

Fig. 9.2   Overview of the production of energy and chemical productions from algae

The juice, in turn, is separated into water-soluble sugars and salts. In the case of sugars, mannitol is abundant and can be recovered for energy use through methane fermentation, but another approach that has been considered involves enzymatic conversion to produce syrup and other higher value-added products. Other products that are expected to prove useful include fucoidan, chlorophyll, polyphenols, vitamins, and carotenes. The remaining brine contains inorganic salts concentrated in the seaweed body. For example, dried kelp contains 5.3 g for every 100 g, while dried sea mustard contains 5.2 g. Potassium recovery from brine is assumed to take place in the production process. Algae also have large amounts of ionic groups such as carbonyl and sulfate groups in the high-molecular weight polysaccharides of their cell wells. Since these adsorb heavy metals, they can be used for the recovery of rare and other heavy metals. The algae usage process can thus be divided into the extraction and use of high value-added materials from the seaweed bodies and energy recovery from the extract residue through methane fermentation (Chen and Oswald 1998).

Figure 9.3 shows an overview of more detailed considerations of energy production from marine biomass in Japan. Tangle weed (*Laminaria japonica*), one of the large algae in Japan's coastal regions, has been proposed as a cultivar. *L. japonica* seedlings are first mass-cultured in land-based tanks, after which they are transported a system for sea cultivation (seaweed body farming). Once grown, the *L. japonica* is harvested and useful high value-added materials are recovered, while fuel gas is produced through methane fermentation. This, along with the use of methane fermentation as a key technique for energy recovery, is the same approach attempted in the U.S (Yokoyama et al. 2007).

For this project, a million tons of kelp per year was produced at a farm (1 km^2) 60 m under the sea. In cases of energy recovery through methane fermentation alone, the amount of methane gas energy produced is 2.3×10^{11} kcal, while energy used amounts to 0.85×10^{11} kcal for kelp cultivation and harvesting and 0.6×10^{11} kcal for fermentation. Although the energy balance is positive, this approach is not economically feasible without the extraction of high value-added

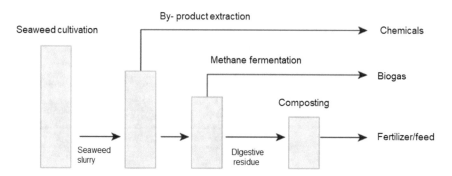

Fig. 9.3 Overview of the production of energy and chemical products from algae

byproducts. The prices of crude oil and other energy sources have risen sharply since then, but this has not translated into large improvements in economic feasibility (Ni et al. 2006).

That feasibility improved if byproducts can be extracted and sold, but generation of high value-added requires large energy consumption amounting to 3.7×10^{11} kcal for the recovery and refinement process. This results in a minus value for actual energy production, indicating that this cannot practically be used as an energy production system. When one considers that energy issues will only grow more severe in the future, however, the importance of building systems for exploiting methane fermentation of algae and its positive energy recovery rate becomes great. An important part of this is optimizing the methane fermentation approach in terms of algae biomass.

9.3.3   Process of Methane Fermentation of Algae

The process of methane fermentation has long been a subject of research, and the ultimate conversion of organic matter in algae into methane and carbon dioxide is understood to take place through the combined efforts of groups of microorganisms. The methane production process is commonly divided into steps of hydrolysis, organic acid generation, and methane generation. The hydrolysis and organic acid generation components can in turn be divided into three steps (Fig. 9.4) (Grala et al. 2012).

Organic matter in wastewater—especially high-molecular weight substances such as fiber, proteins, lipids, and starches—is first hydrolyzed into low-molecular weight substances by fermenting microorganisms, resulting in the production of alcohols such as ethanol and short-chain fatty acids (VFA, e.g., propionic acid and butyric acid). In the ecosystem, a diverse range of bacteria function to perform hydrolysis and produce acids. In particular, a range of obligate anaerobic bacteria (such as *Clostridium*, Bacteroides, and *Butyrivibrio*) and facultative anaerobic bacteria (such as *Bacillus*, *Lactobacillus*, and *Micrococcus*) are known to be involved (Nagai and Nishio 1989; Lamed and Bayer 1988).

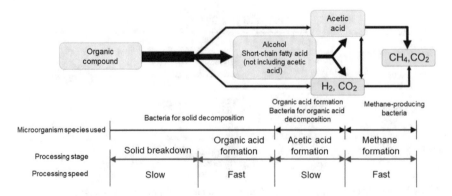

Fig. 9.4 Overview of the methane fermentation process

Metabolite types and ratios vary greatly by microorganism and fermentation substrate. For example, ethanol is produced along with short-chain fatty acids from glucose (a product of hydrolysis of the fibrin contain in algae fiber) under anaerobic conditions.

In contrast, ethanol production is unlikely to be achieved under anaerobic conditions with alginic acid, a major component of brown algae. This is due to the fact that the alginic acid monomer β-D-mannuronic acid and its C-5 epimer α-L-gluronic acid are acidic sugars lacking the reduction capability needed for ethanol production. As per the following reaction formula, the chief product from alginic acid under anaerobic conditions is acetic acid:

$$C_6H_8O_6 \rightarrow 2CH_3COOH + 2CO_2$$

In contrast, mannitol ($C_6H_{14}O_6$), which is another main component of brown algae, may provide an excellent substrate for ethanol production. Because the enzymes normally used in ethanol production do not degrade mannitol, other bacteria besides those enzymes must be used to degrade mannitol unless genetic manipulation is applied. Seaweed proteins, which are also present in large quantities in brown algae in addition to sugars, lack the necessary reduction to produce ethanol, and are typically converted to short-chain fatty acids by anaerobic microorganisms. The process of brown algae hydrolysis and organic acid generation by anaerobic microorganisms thus results in short-chain fatty acids as a major metabolite, without much alcohol production.

Short-chain fatty acid such as propionic acid and butyric acid that are created during the organic acid formation process as broken down by the hydrogen-producing acetic acid bacteria into acetic acid, hydrogen, and carbon dioxide. The organic matter that methane-producing bacteria are capable of using is very limited; acetic acid and hydrogen serve as direct substrates for methane formation in artificial methane fermentation. Depending on the methane-producing bacteria, this can be divided into production of hydrogen and carbon dioxide and production of methane and carbon dioxide from alcohol and short-chain fatty acids. The growth substrate for methane-producing bacteria is also supplied by different microorganisms.

In the case of hydrogen- and acetic acid-forming bacteria, hydrogen-decomposition methane bacteria typically break down the hydrogen, which is removed to sustain growth. As this indicates, a large number of microorganisms are involved in the methane fermentation process. It is chiefly the methane production and hydrolysis processes, however, that serve to limit the rate.

9.3.4 Yield of Methane Fermentation of Algae

Algae contain sugars that are easily hydrolyzed and have lower lignin content than land-based plants. For example, while corn has lignin content of 15.1%w/w relative

Table 9.2 Methane production yields of algae, plants, and food waste

	VS decomposition rate (%)	Methane yield (m^3/kg) added VS^{-1}
Kelp (*Laminaria*)	46–60	0.23–0.30
Sea moss (*Gracilaria*)	50–85	0.28–0.40
	18–39	0.05–0.19
Giant kelp (*Macrocystis*)	34–80	0.14–0.40
Sea lattuce (*Ulva*)	62	0.31
	41–56	0.14–0.23
Guld weed (Sargassum)	–	0.12–0.20
	20–40	0.08–0.14
Poplar	–	0.08–0.14
Food waste	–	0.54

to dry weight, the lignin content of sea lettuce (*Ulva* sp.) is just 2.7%w/w. As a result, algae can be fermented through simply pulverization treatment (Ventura and Castanon 1998).

Table 9.2 shows methane yield values for various algae, plants, and food residue. Kelp (*Macrocystis*) and *Gracilaria verrucosa* have shown methane yields of up to 0.40 m^3/kg-VS with the breakdown of 80% of volatile solids (VS). Compared to food residue, the methane yields are rather low.

For example, the theoretical methane yield for *Gracilaria* based on its composition is 0.46 m^3/kg-VS. Nearly ideal methane fermentation has been achieved, with a yield close to the theoretical maximum. Methane yields for other algae are reported to be lower than for *Macrocystis* or *Graciliara*. Since this is due to a low rate of VS decomposition, methane yields should be improved in the future with the establishment of pre-treatment approaches such as pulverization or hydrolysis (Grala et al. 2012).

9.3.5 Algae Methane Fermentation Conditions

The process of biogas production through anaerobic decomposition is chiefly influenced by factors such as hydraulic retention time (including seaweed body biomass), sludge retention time, organic matter load, pH, and temperature. In the case of algae biomass, sludge retention time has the greatest impact on methane production: yields increase with longer sludge retention times.

Habig et al. examined the effects on methane fermentation from sludge retention time for Gulf seaweed (*Sargassum*), sea moss (*Gracilaria*), and sea lattuce (*Ulva*) crushed to around 2–3 cm at moderate temperature, with semi-continuous spinning only at the time of substrate additions. In the case of *Ulva*, a 30-day retention time reportedly resulted in a methane yield of 0.14 m^3/kg-added VS^{-1} and a VS decomposition rate of 41%, while a 50-day retention time resulted in these values respectively improving to 0.23 m^3/kg-added VS^{-1} and 56%. This means that the hydrolysis of seaweed body solids is the rate limiting step of methane fermentation.

This also means that with mixed spinning culturing equipment and the same added substrate concentration, reduced solid retention time is associated with increased organic matter load.

Chynoweth reported that while increased organic matter load lowered methane yields, it also increased concentrations of acetic acid and short-chain fatty acids such as propionic and butyric acid. This means that along with hydrolysis of algal body biomass, the methane production process can also easily become a rate-limiting step in the methane fermentation of algae biomass.

With methane fermentation of algae, high concentrations of salt are present in the biomass. Reduced methane fermentation due to these salts can also be a problem with non-algae biomass. The major factor causing the reduced fermentation is the low salt tolerance of the methane-producing bacteria that decompose acetic acid. Accordingly, it is also important to increase the organic matter load by raising methane-producing bacteria activity through desalinization.

Anaerobic decomposition is typically performed at two temperature ranges, medium (35 °C) and high (55 °C). The types of microorganisms that function at each temperature range vary greatly, but both can be used for processing of algae biomass.

Otsuka et al. obtained a methane yield of 180 ml/g-VS through anaerobic decomposition of *Ulva* at medium temperature. Comparison of methane yields in VS terms for anaerobic decomposition of a *Scenedesmus* spp. and *Chlorella* spp. mixture under medium- and high-temperature conductions in terms showed high-temperature fermentation to result in increased organic acid decomposition and improved methane yields compared to medium-temperature fermentation (Otsuka et al. 2004).

Hansson used a mixture of sea lattus (*Ulva*), sea lattus (*Cladophora*), and *Chaetomorpha* collected from the coastal Baltic Sea to compare characteristics from methane fermentation at medium and high temperatures. Medium-temperature fermentation resulted in methane yields of 250–350 ml/g-added VS-1 and VS decomposition rates of 50–55%, indicating superior performance relative to high-temperature fermentation (Hansson 1983; Prabandono and Amin 2015).

9.3.6 Equipment for Methane Fermentation of Algae

The design of biogas plants, which are key to the methane fermentation process, is determined to some extent by physical characteristics such as the viscosity and settling of organic waste and flora. Traditionally, methane fermentation has involved the use of continuously supplied water and a spinner to mix the methane-producing bacteria in the form of a continuous stirred tank reactor (CSTR) or a completely mixed tank reactor using internal circulation of gas or culturing solution (Fig. 9.5b).

This approach is often used today with surplus sewage sludge and livestock excreta processing, as it is structurally simple to remain and mainly and allows for simultaneous processing of solids. Obviously, it can also be used as a tank for

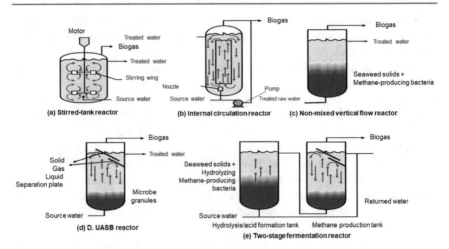

Fig. 9.5 Various tank reactors for anaerobic decomposition of algae

anaerobic decomposition of algae biomass. As mentioned previously, a long sludge retention time is necessary to obtain high methane yields from algae. In a completely mixed fermentation tank, an identical hydraulic retention time is also necessary for a longer sludge retention time. This necessitates the use of a large fermentation tank, which is not practical when recovering energy from the seas. Accordingly, Fannin et al. devised the non-mixed vertical flow reactor (NMVFR) as a reactor in which the sludge retention time alone can be increased (Fig. 9.5c).

In this reactor, solid-containing pulverized algae are sent below the reaction tank. High-settling solids become concentrated within the reactor, while only the seeping water is removed from the reactor's upper portion, resulting in a longer sludge retention time compared to hydraulic retention time. High methane yields and culture stability for the same sludge load were obtained through the use of this reactor compared to a completely mixed reactor.

As mentioned in the explanation of the algae methane fermentation process, methane fermentation can be divided broadly into hydrolysis, organic acid formation, and methane production processes, each of which involves completely different classes of microorganisms in its reactions.

For anaerobic decomposition of algae, increased load rate resulted in marked accumulation of short-chain fatty acids for both the completely mixed and NMVFR types. This is due to the fact that the hydrolysis and organic acid formation rates under high-load conditions are higher than the methane production rate; if manipulation continues in this way, pH becomes more acidic and methane fermentation ultimately stops. Because organic acid and methane formation are carried out by completely different microorganisms, however, a two-stage system was developed in which a high-speed fermentation tank was used for acid production alone and connected to a tank for methane fermentation with solution containing large concentrations of short-chain fatty acids. For drainage with low solid concentrations (5% or less), an upflow anaerobic sludge blanket (USAB) was developed to allow treatment with retention times of one to two days (Fig. 9.5d). Also developed and commercialized were high-speed methane fermentation processes such as the expanded granular sludge bed (EGSB) approach and the upflow anaerobic filter process (UAFP) approach. These have been widely used since the 1990s, especially in the domestic and overseas food processing industries.

A UASB takes advantage of the way that the microorganisms involved in methane fermentation naturally come together to form microorganism granules. By maintaining a high density of bacterial bodies in the reactor, substantially larger treatment speeds can be achieved than with the previous methane fermentation approach. Hopefully, high-speed processing will become available in which a high-speed methane fermentation tank is connected to a methane production tank, so that organic acids accumulate at high concentrations and can be swiftly converted to methane by methane fermentation bacteria even when the organic matter load is increased.

Fannin et al. used a NMVFR at high load conditions (11.2 kg-VS/m^3/day) as a hydrolysis and organic acid formation tank for the giant kelp *Macrocystis*. The supernatant was then placed in a methane production task to produce methane with a yield rate of 0.29 m^3/kg-added VS. While lower than the maximum methane yield for giant kelp as shown in Table 9.2, this was the result of a low solid hydrolysis rate in terms of hydrolysis and organic acid production; achieving higher methane yields would require a different pre-treatment process instead of crushing (Nakashimada and Nishio 2011).

9.3.7 Issues with Algae Methane Fermentation

A rather troublesome problem area remains with anaerobic decomposition of algae, namely the fact that seaweed bodies contain high conditions of sulfate ions, salt (NaCl), and heavy metals. Among these, heavy metals are a particularly serious issue (Cecchi et al. 1996).

The high molecular weight polysaccharides in algae cell walls contain large amounts of ionic groups such as carbonyl and sulfate groups, which adsorb heavy metals. As a result, algae that grow in back bay regions with little seawater movement accumulate high concentrations of heavy metals such as cadmium. (In Sweden, algae are classified as toxic waste.) This limits the ability to use residue from anaerobic decomposition as a biofertilizer.

But anaerobic decomposition can also be helpful in removing heavy metals. As previously noted, it is best to apply a two-stage process for anaerobic composition of algae (Omil et al. 1995). In a two-stage process, solid organic matter is solubilized and hydrolyzed in the first reactor, after which chiefly organic acids are produced by anaerobic fermentation using microorganisms. The resulting organic acids can be subjected to high-speed methane fermentation with the UASB method. The heavy metal ions contained in the seaweed extract become separated during hydrolysis under low pH conditions, and an absorbent can be used at this time to remove them. Improved metal solubilization has been reported in non-algae energy crops such as willows, sugar beets, and grass with a two-stage process at a pH 4. Once heavy metals are removed, organic acid drainage can undergo methane fermentation in the second fermentation tank.

Nkemka and Murto subjected algae to organic acid fermentation and removed heavy metals using a porous cryogel from a polyacrylamide base to which imin-iodiacetic acid had been introduced as a ligand. The results reported respective reductions of 79, 59, 70, and 41% for cadmium, copper, nickel, and zinc, with no problems for methane fermentation (Singh and Olsen 2011).

9.4   Biohydrogen Production

In addition to polluting the atmosphere, the use of fossil fuels also leads to global warming through the accumulation of carbon dioxide emissions. It is a situation that urgently calls for development of new energy materials that do not cause atmospheric pollution (Levin et al. 2004).

By unit weight, hydrogen's exothermic energy is three times higher than that of petroleum. In addition to potentially offering a superior energy source, it does not carry a risk of atmospheric pollution from its combustion. This, along with the possibility of conversion to electrical power and use as both a solid and liquid fuel, has resulted in it drawing attention as a key fuel source for the future. Because hydrogen is not produced as a simple substance anywhere on Earth, however, other energy sources must be used to produce it artificially.

Approaches considered for the production of hydrogen include such physical and chemical methods as water electrolysis and pyrolysis. All of these methods require large amounts of energy for hydrogen production. For this reason, solar energy is a beneficial source for hydrogen production. The use of bioprocesses involving photosynthetic bacteria capable of efficiently exploiting solar energy is now drawing attention as one of the most promising potential approaches.

The use of photosynthetic bacteria for hydrogen production offers many advantages: the production system is quite simple, the resulting gases can easily be separated into hydrogen and carbon dioxide, and renewable biomass or waste can be used as a source. Once efficient hydrogen production systems using photosynthetic bacteria become established, they may well become the chief energy production systems of the future.

Photosynthetic bacteria are typically prokaryotes that survive under anaerobic conditions through exposure to sunlight. They exist throughout the soil and hydrosphere and perform photosynthesis and nitrogen fixing through the use of solar energy and suitable electron donors. They possess chlorophores and bacterial chlorophyll to perform photosynthesis, as well as large amounts of carotenoids as supplementary pigments. Their photosynthesis differs significantly from that of plants in that they only have one photochemical reaction center, and water decomposition through photosynthesis does not occur. Organic matter (typically organic acids) and sulfur compounds are used as electron donors instead of water.

The ability of photosynthetic bacteria to produce hydrogen when exposed to sunlight was first discovered in 1949. Since then, much research has been conducted on the hydrogen formation mechanism, the relevant enzymes, and hydrogen production.

To produce hydrogen, photosynthetic bacteria must be in anaerobic conditions with sunlight and a nitrogen source. For sunlight exposure, intense light with illuminance of 5000–20,000 lx (at the container surface) is used. The resulting gas is over 90% hydrogen, with the remainder consisting of carbon dioxide.

To produce hydrogen from unexploited marine resources and marine biomass with the highest efficiency using photosynthetic bacteria, it is better to use the bacteria to convert sea-grown biomass into hydrogen while still in seawater. Freshwater photosynthetic bacteria cannot grow or produce hydrogen in seawater. Marine photosynthetic bacteria are reported to have significantly higher hydrogen production capabilities than traditional forms of freshwater bacteria. With farther-ranging explorations in the future, marine bacteria with even better hydrogen production capabilities may be found.

Stability with respect to oxygen can be improved through a technique that involves fixing the bacterial body with a high molecular weight gel supporter. This allows for greater ease of isolation and enables reuse and serial usage of bacteria bodies, increasing the efficiency of hydrogen production over the long term. In terms of applied research, hydrogen has reportedly been produced from the products of marine blue-green algae decomposed with acids or bases after being grown in a biosolar reactor. This research has drawn attention for showing the potential to establish hydrogen production systems using marine biomass and photosynthetic bacteria.

Hydrogen production from marine biomass using photosynthetic bacteria has not yet reached the commercialization stage. In light of future energy supply needs and environmental issues, however, such systems are the focus of high hopes, and commercialization technology will certainly be developed before long. The chief goal in commercializing hydrogen production using photosynthetic bacteria lies in increasing the bacteria's hydrogen production capabilities.

In the future, as new high-performing photosynthetic bacteria are explored and metabolically controlled, and research advancements in the area of marine biotechnology (such as molecular breeding and cell engineering techniques) allow for the development of photosynthetic bacteria with superior hydrogen production capabilities, systems will hopefully be commercialized for the production of hydrogen from marine biomass (Beer et al. 2009).

9.4.1   Fermentation Hydrogen Generation Pathways and Metabolites

Organisms oxidize sugars such as glucose to produce adenosine triphosphate (ATP) as an energy source. Three chief sugar metabolism pathways have been identified in microorganism: the Embden-Meyerhof (EM) pathway, the Entrner-Doudoroff (ED) pathway, and the pentose phosphate (PP) pathway. Each of them has been acquired by microorganisms for the production of ATP (Wrighton et al. 2012).

As an example, bacteria from the genus *Enterobacter* chiefly use the EM pathway to oxidize glucose to produce 2 mol each of the reduction product nicotinamide adenine dinucleotide (NADH, reduced), pyruvic acid, and ATP through the following reaction:

$$C_6H_{12}O_6 + 2NAD^+ + 2ADP + 2Pi \rightarrow 2CH_3COCOOH + 2NADH + 2H^+ + 2ATP$$
$$(9.1)$$

This alone does not generate sufficient ATP. To decompose new glucose, metabolites are produced through a reaction between reduced NADH and pyruvic acid, and NADH is oxidized to NAD^+ for reuse. It is at this point that hydrogen is generated, although this generation does not occur in the production process for all metabolites. As shown in Fig. 9.6, it chiefly occurs with the selection of a pathway where metabolites are generation through production of acetyl-CoA.

For example, not only does a lactic acid production reaction not use the acetyl-CoA production pathway, but, as shown in Formula (9.2), pyruvic acid and NADH react at a 1:1 ratio for lactic acid production and do not contribute to hydrogen generation. In the production of acetic acid, butyric acid, acetone, or butanol, however, the acetyl-CoA production pathway is used and hydrogen is generated (Tanisho 2011).

Lactic acid production reaction

$$CH_3COCOOH + NADH + H^+ \rightarrow CH_3CHOHCOOH + NAD^+ \qquad (9.2)$$

Acetic acid production reaction

$$CH_3COCOOH + H_2O + Ferredoxine(Fd) \rightarrow CH_3COOH + CO_2 + FdH_2 \quad (9.3)$$

$$FdH_2 \rightarrow Fd + H_2 \qquad (9.4)$$

$$NADH + H^+ \rightarrow NAD^+ + H_2 \qquad (9.5)$$

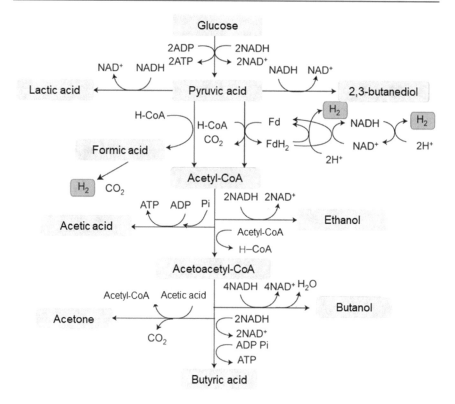

Fig. 9.6 Major metabolic pathway for bacteria under anaerobic conditions

Table 9.3 shows the hydrogen yields and chief metabolites for major hydrogen fermentation bacteria. Because many bacteria metabolically produce butyric acid, acetic acid, and lactic acid, and because these organic acids are present in fermentation waste, the processing or use of this waste poses major issues for hydrogen fermentation.

9.4.2   Bacterial Hydrogen Production Rate and Yields

Many hydrogen-producing bacteria have been reported to date. Table 9.4 shows leading bacteria that have been found to produce hydrogen at a fast rate or with high substrate yields. Because yields and rates differ for batch culturing and continuous culturing, the two are shown separately. Continuous culturing of the same bacteria results in higher bacterial density in the fermentation tank, which can increase the hydrogen production rate.

Table 9.3 Hydrogen yields and metabolites for major bacteria (Wang and Wan 2008)

Bacteria	Yield (mol-H_2/mol)	Substrate	Major metabolites
Clostridium			
C. butyricum	2.35	Glucose	Butyric acid, acetic acid
C. acetophilius	1.82	Glucose	Butyric acid, acetic acid
C. perfringens	2.14	Glucose	Butyric acid, acetic acid, lactic acid, ethanol
C. acetobutylicum	1.35	Glucose	Acetic acid, butanol, acetone
C. butylicum	0.78	Glucose	Butyric acid, acetic acid, butanol, isopropanol
Escherichia coli	0.75	Glucose	Acetic acid, formic acid, succinic acid, lactic acid, ethanol
Serratia kielensis	0.91	Glucose	Acetic acid, lactic acid, ethanol
Aerobacillus polymyxa	0.82	D-xylose	Ethanol, butanol
	1.70	Mannitol	Acetic acid, lactic acid, ethanol, butanol
Enterobacter aerogenes st. E.82005	1.0	Glucose	Butyric acid, acetic acid, lactic acid, ethanol, butanediol
	1.6	Mannitol	
	2.5	Sucrose	

Anaerobic bacteria generally have faster production rates and higher yields than facultative anaerobic bacteria, but their breeding and growth are reduced in the presence of oxygen, and the production rate and yields are significantly impacted. In contrast, facultative anaerobic bacteria exhibit low rates and yields, but can breed with oxygen when it is present or produce hydrogen through fermentation when it is absent, offering the advantage of easier handling when using bacteria for industrial hydrogen production.

Production rates are relatively low for bacteria capable of growing at high temperatures of 60 °C or more, but these are theoretically capable of generating the highest hydrogen yields (Tanisho and Ishiwata 1994).

As seen in Table 9.4, the mesophilic bacterium HN001 strain currently has the highest hydrogen production rate among bacteria. *E. aerogenes* is capable of producing hydrogen from mannitol (a chief component of kelp) with a yield of 1.6 mol-H_2/mol-mannitol, although starch cannot be used. Mannitol cannot be used with HN001 strain, but starch can, producing a hydrogen yield of 2.5 mol-H_2/mol-glucose. In continuous culturing with a supply of materials at an average hydraulic retention time of 1 h, it can be generated at a rate of 4L-H_2L^{-1} h^{-1}.

As a bacterium, *E. aerogenes* is desirable for hydrogen production from algae, but a new bacterium with a higher hydrogen yield and generation rate than *E. aerogenes* will need to be discovered for its commercialization.

Table 9.4 Hydrogen production rates and yields for major bacteria (Tanishio 2011)

Batch culture	Temp (°C)	Substrate	Yield (mol-H$_2$/mol)	Production rate (NL/L.h)	(NL/g.h)
Obligate anaerobes bacteria					
Clostridium sp. No. 2	36	Glucose	2	0.54	–
C. paraputrificum M-21	37	*N*-acetylglucosamine	2.5	0.69	–
Mesophilic bacterium HN001	47	Glucose	2.4	3.58	0.99
Facultative anaerobes bacteria					
Enterobacter aerogenes E.82005	38	Glucose	1	0.47	0.38
E. cloacae IIT-BT 08	36	Sucrose	3	0.78	0.65
High-temperature bacteria					
Thermotoga maritima	80	Glucose	4	0.22	–
Thermotoga elfii	65	Glucose	3.3	0.07	0.11
Caldicellulosiruptor saccharolyticus	70	Sucrose	3.3	0.18	0.27
Clostridium thermocellum	60	Cellobiose	1	0.16	0.31
Continuous culture	Temp (°C)	Substrate	Yield (mol-H$_2$/mol)	Production rate (NL/L.h)	(NL/g.h)
Obligate anaerobes bacteria					
C. butyricum LMG1213tl	36	Glucose	1.5	0.49	–
Clostridium sp. No. 2	36	Glucose	2.4	0.47	–
C. pasteurianum	40	Sucrose	1.6	13.71	0.38
Facultative anaerobes bacteria					
E. aerogenes E.82005	38	Molasses	1.3	0.81	0.38
E. aerogenes E.82005	38	Glucose	1	2.73	0.38
E. aerogenes HU-101 m AY-2	37	Glucose	1.1	1.3	–
High-temperature bacteria					
Thermococcus kodakaraensis KOD1	85	Pyruvate	2.2	0.2	1.32

9.4.3 Hydrogen Production Using Photosynthetic Microorganisms

A. **Photosynthetic Microorganism Hydrogen Production Mechanism and Its Principles**

Microorganisms that produce hydrogen through photosynthesis can be broadly divided into photosynthetic bacteria and seaweeds such as green and blue-green algae according to the photosynthesis mechanism, the enzymes catalyzing hydrogen production, and available electron donors (Table 9.5).

Table 9.5 Types of hydrogen-producing photosynthetic microorganisms (Wakayama 2011)

Classification of photosynthetic microorganism		Major microbe generic name	Hydrogen-producing enzyme	Hydrogen production source
Green algae		*Botryococcus*	H₂ase	Water
		Chlamydomonas	↑	↑
		Chlorella	↑	↑
		Dunaliella	↑	↑
		Scenedesmus	↑	↑
Blue-green algae	No heterocyst	*Mycrocystis*	↑	↑
		Oscillatoria	↑	↑
		Spirulina	↑	↑
		Synecococcus	↑	↑
	Heterocyst	*Anabaena*	N₂ase	↑
		Nostoc	↑	↑
Photosynthetic bacteria	Purple non-sulfur bacteria	*Rhodobacter*	↑	Organic material
		Rhodobium	↑	↑
		Rhodopseudomonas	↑	↑
		Rhodovulum	↑	↑
	Purple sulfur bacteria	*Chromatium*	↑	Sulfides
		Ectothiorhodospira	↑	↑
		Thiocapsa	↑	↑

H₂ase hydrgenase; *N₂ase* nitrogenase

Algae have two forms of photosystems, known as Photosystem I (PS I) and Photosystem II (PS II), and perform oxygen-producing photosynthesis under conditions of light exposure. Algae that are capable of producing hydrogen directly underwater through photosynthesis are able to produce energy and fix CO_2 simultaneously.

Photosynthetic bacteria possess either PS I or PS II as a photosynthesis and perform non-oxygen-producing photosynthesis under anaerobic conditions when exposed to light radiation. Since they are able to use short-chain fatty acids and other organic matter in wastewater as electron donors, they are capable of energy production and wastewater treatment simultaneously (Boichenko et al. 2004).

The main reason that photosynthetic microorganisms produce hydrogen is to moderate the reducing power that results from photosynthesis. Not only photosynthetic microorganisms but all microorganisms must oxide organic or inorganic matter to obtain energy for growth, at which point handling of the remaining reducing power (H^+ or e^-) becomes an issue. When glucose is oxidized into CO_2, some reducing power remains as carbon is oxidized ($C^0 \rightarrow C^{4+}$), which is consumed when producing hydrogen through H^+ reduction.

$$C_6H_{12}O_6 + 6H_2O \rightarrow 12H_2 + 6CO_2 \Delta G = -34 \text{ kJ}$$

B. Enzymes and Processing System for Hydrogen Production by Photosynthetic Microorganisms

Photosynthetic microorganisms typically have either nitrogenase (N_2ase), hydrogenase (H_2asee), or both as enzymes for hydrogen production.

The main enzyme involved in hydrogen production by blue-green or photosynthetic bacteria with heterocysts is N_2ase, which ordinarily catalyzes nitrogen-fixing reactions in the air.

$$N_2 + 6H^+ + 6e^- + 12ATP \rightarrow 2NH_3 + 12ADP + 12Pi$$

N_2ase has relatively low substrate specificity and catalyzes an irreversible H^+ reduction reaction in the presence of a ferredoxine (Fd) electron donor.

$$2H^+ + 2Fd_{red} + 4ATP \rightarrow H_2 + 4ADP + 4Pi + 2Fd_{ox}$$

Since NH_4^+ serves to hinder hydrogen production in biomass, they must be eliminated during the pre-treatment process when applied toward wastewater treatment.

The major enzyme involved in hydrogen production with blue-green algae without heterocysts is H_2ase, which reversibly catalyzes an H^+ reduction reaction in the presence of an electron donor (ED).

$$2H^+ + 2ED_{red} \rightleftharpoons H_2 + 2ED_{ox}$$

Since the structure and metabolic role of H_2ase differs among microorganisms (Fe, NiFe, or FeS is absent depending on the central metal), various substances are known to serve as electron donors (e.g., Fd, $CytC_3$, or $CytC_6$).

When producing hydrogen with photosynthetic microorganisms, a cascade approach can be used with the large masses of seaweed body biomass that are generated in addition to hydrogen to generate gaseous, solid, and liquid fuel (Fig. 9.7).

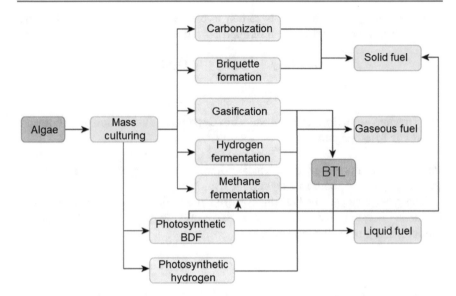

Fig. 9.7 Fuel production system using photosynthetic microorganisms (BDF: biodiesel fuel, BTL: biomass to liquid) (Wakayama 2011)

C. Gaseous, Solid, and Liquid Fuel

(1) Gaseous Fuel

Photosynthetic microorganisms are capable of producing hydrogen directly under conditions of light radiation. Production of hydrogen requires a process of separating and refining hydrogen and CO_2 through pressure swing adsorption (PSA; see Fig. 9.8).

When the large amounts of seaweed body biomass resulting from hydrogen production are supplied to the hydrogen production process, methane is obtained as a gaseous fuel with supplies of hydrogen to the methane fermentation process, while hydrogen and CO_2 has is obtained with supplies to the gasification process (Jones and Mayfield 2012).

A closed photobioreactor is used to collect hydrogen when generated directly by photosynthetic microorganisms, while an open photobioreactor is used when seaweed body biomass is used as a production source for gaseous fuel.

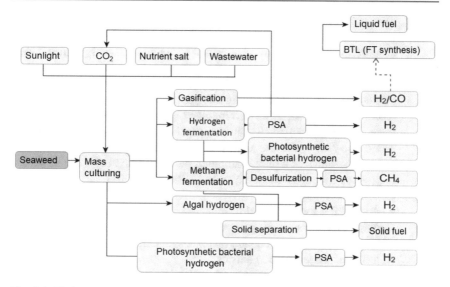

Fig. 9.8 Hydrogen production system using photosynthetic microorganisms (PSA: pressure swing adsorption, BTL: biomass to liquid)

(2) Solid Fuel

Large amounts of seaweed body biomass are formed when photosynthetic microorganisms are used for hydrogen production. These can potentially serve as alternatives to coal as a power plant fuel if they can be cheaply dehydrated, dried, and converted to briquette form. Seaweed body biomass can also be carbonized and converted to solid fuel in the presence of a large and affordable high-temperature heat source.

(3) Liquid Fuel

The seaweed biomass produced in large volumes with hydrogen production through photosynthetic microorganisms contains relatively large concentrations of lipids. Biodiesel fuel may be produced through the supply of seaweed body biomass to the solid-liquid separation (dehydration), lipid extraction, methyl esterification, and refinement process. Since algae accumulate sugars (glycogen) in bacterial bodies under dark conditions, bioethanol can also be produced through their supply to the ethanol fermentation, distillation, dehydration, and refinement process.

(4) Photobioreactors

A photobioreactor (PBR) is a system used when producing hydrogen through photosynthetic microorganisms. Since hydrogen production through photosynthetic microorganisms inextricably entails the use of solar energy, various forms of PBR have been developed to efficiently receive, transmit, and disperse solar light in ways that reflect solar radiation characteristics.

The biggest issue when producing hydrogen through photosynthetic microorganisms is the matter of supplying light efficiently. Since light is absorbed and scattered by photosynthetic microorganisms, light intensity rapidly declines at the receiving service in culturing solution, resulting in a relatively irregular distribution of light intensity within the PBR. Hydrogen production by photosynthetic microorganisms is reliant on light intensity; reactions become saturated and inhibited under high light intensity conditions, resulting in a marked drop in light-to-hydrogen conversion efficiency. These two-issues, the irregularity of light intensity distribution in PBRs, and the existence of a photo-saturation point, will be need to be addressed to produce hydrogen efficiently with PBRs (Pulz 2001).

(5) Closed Photobioreactors

A closed PBR must be used to make efficient use of solar energies when applying factory or power plant waste gas containing large concentrations of CO_2 for hydrogen production by photosynthetic microorganisms or mass culturing of those microorganisms.

The biggest advantage of closed PBRs is that it is possible to build three-dimensional systems for efficient usage of solar energy. At the same time, consideration should also be given to cleaning of the PBR interior, as the capital expenditures (CAPEX) and operating expenses (OPEX) are greater than with an open PBR. Tube, flat, and bag types are currently used, depending on the purpose.

(6) Open Photobioreactors

An open photobioreactor can be used when there is no contamination risk during mass culturing of photosynthetic microorganisms or when mass-cultured algal body biomass is used for hydrogen production. The biggest advantage of an open PBR is the low capital expenditure. At the same time, the culturing area is directly proportional to the installation area, which means that a vast space is needed. Among the photosynthetic microorganisms that can be mass-cultured with open PBRS in raceways or open and round ponds are *Chlorella*, which exhibits a fast growth rate; *Spirulina platensis* (when the medium is strongly alkaline); *Euglena gracilis* (when the medium is strongly acidic); and *Donaliela salina* (when the medium has a high saline concentration)

Cynatonech, which mass-cultures *Donaliela* in Hawaii, uses an affordable PBR created by digging a ditch in a lava plateau and installing vinyl sheeting (Wakayama 2011).

9.4.4 Hydrogen Production from Algal Biomass

During harvesting season, kelp has a component ratio of 79% water, 4% ash, 1% protein, 1% cellulose, 7% alginic acid, and 8% mannitol. Of the 17% organic components, it is chiefly mannitol that bacteria use when generating hydrogen. The structural formula is analogous to glucose, as shown below:

$$
\begin{array}{ccc}
\text{CHO} & \text{CHO} & \text{CH}_2\text{OH} \\
| & | & | \\
\text{HCOH} & \text{HOCH} & \text{HOCH} \\
| & | & | \\
\text{HOCH} & \text{HOCH} & \text{HOCH} \\
| & | & | \\
\text{HCOH} & \text{HCOH} & \text{HCOH} \\
| & | & | \\
\text{HCOH} & \text{HCOH} & \text{HCOH} \\
| & | & | \\
\text{CH}_2\text{OH} & \text{CH}_2\text{OH} & \text{CH}_2\text{OH} \\
\text{D-Glucose} & \text{D-Mannose} & \text{D-Mannitol}
\end{array}
$$

With two more hydrogen atoms than glucose, mannitol is stoichiometrically capable of producing one more mole of hydrogen than glucose, making it a suitable hydrogen fermentation substrate. Once a bacterium is discovered that is capable of metabolically producing acetic acid alone from mannitol, the hydrogen yield will increase to 5 mol/mol, making it a superior substrate to glucose as shown below (Hallenbeck and Benemann 2002):

• Theoretical maximum of hydrogen generated from one mole of glucose

$$C_6H_{12}O_6 + H_2O \rightarrow 2CH_3COOH + 2CO_2 + 4H_2 \tag{9.6}$$

• Theoretical maximum of hydrogen generated from one mole of mannitol

$$C_6H_{14}O_6 + H_2O \rightarrow 2CH_3COOH + 2CO_2 + 5H_2 \tag{9.7}$$

The sugars contained in algae thus have not only a theoretically higher maximum hydrogen yield than the sucrose in sugarcane, but also greater productivity. Sugarcane is one of the most productive plants among land-based biomass; in Brazil, productivity (harvest yield) is around 70–100 ton per hectare.

Kelp, in contrast, has productivity of 145 ton. Moisture content is around 30% for sugarcane and 20% for kelp. In terms of solid mass weight, this equates to nearly identical yields of 30 and 29 ton, indicator that kelp, like sugarcane, has very high productivity. Also, whereas sugarcane must be farmed for one year in a field, farmed kelp and sea lettuce spend their time from seedling to surface culturing in land-based facilities and have a short ocean surface cultivation time of six to seven months.

Also substantial increasing annual harvest yields for fermentation substrates from algae compared to sugarcane is the fact that algae can be harvested twice a year, while sugarcane can only be harvested once. Algae (kelp) are thus a highly productive form of marine biomass.

Table 9.6 Potential hydrogen yield and fuel cell electricity generation for sugar substrates from kelp and sugar cane farming (fuel cell efficiency of 1.7 kWh/N m^3)

	Kelp, *Saccharina japonica*		Sugar cane		Units
	Single-cropping	Double-cropping	Japan	Brazil	
Harvest	14,500	25,000	7,000	10,000	(ton/year km^2)
Sugar volumes	1,160	2,000	980	1,400	(ton/year km^2)
H$_2$ production	356,923	615,385	320,936	458,480	(N m^3/ year km^2)
Annual power generation	606,769	1046,154	545,591	779,415	(kWh/year km^2)
Daily power generation	1,662	2,866	1,495	2,135	(kWh/day km^2)

Table 9.6 shows a comparison of kelp and sugarcane calculated according to the following assumptions.

1. Usage of the potential harvest yield when kelp alone is farmed and when two forms of algae with different harvesting periods are farmed
2. Average harvest yield data for Japan and Brazil used for sugar cane
3. Sugar quantities are based on 8% mannitol content in kelp and 14% sugar content in sugar cane
4. Hydrogen production yield calculation based on a bacterial hydrogen yield of 2.5 mol/mol-mannitol for kelp
5. Sugar cane yield of 5 mol/mol-sucrose used for sugar cane
6. In terms of power generation, fuel cell efficiency of 48% assumed (i.e., 1.7 kWh per cubic meter hydrogen in standard state)

Assuming these conditions, 2.866 kWh per day—enough to power around 280 households—can be produced through algae farming over an ocean area of 1 km^2. Farming over an area several times larger would allow for full power supplies to island and coastal regions where gasoline costs are higher than in inland regions. South Korea in particular has an Exclusive Economic Zone four times as large as its land area; once a suitably economic approach has been achieved, it will need to attempt energy self-sufficiency through the production of hydrogen energy with algal biomass farming (Duman et al. 2014; Tanisho 2011).

9.4.5 Fermentation Hydrogen Yields for Marine Algae Components

At the time of harvesting, kelp consists of approximately 8% mannitol and 7% alginic acid in wet weight, accounting for 71% of solids. To date, *Enterobacter aerogenes* has been identified as a bacterium producing hydrogen from the main component mannitol, although the yield—1.6 mol-H$_2$/mol-mannitol—is not especially high. The search for a higher-yield bacterium for hydrogen production from algal biomass has led to the discovery of a new bacterium with a yield of 2.5 mol-H_2/mol-mannitol, producing hydrogen at a rate of 1.1L-H_2L-culture^{-1} h^{-1} (Borines et al. 2011).

In addition to having a higher yield and hydrogen generation rate than *E. aerogenes*, the new bacterium has a hydrogen generation rate roughly equal to that of the HN001 strain. A simultaneous search for a bacterium to generate hydrogen from another chief algae component, alginic acid, similarly led to the discovery of a different new bacterium with a yield of 0.7 mol-H_2/mol-alginic acid. This bacterium can be used to produce around 31 N m^3 of hydrogen from 1 ton of wet kelp according to the following calculations:

Hydrogen production from mannitol

$= $ (moles of mannitol in wet kelp) $\times$ (hydrogen yield)

$= $ (1000 kg/ton wet kelp) $\times$ 8.0% $\div$ 0.182 kg/mol) $\times$ (2.5 mol-H_2/mol)

$= $ 1099 mol-H_2/ton-wet kelp

$= $ 24.6 Nm^3-H_2/ton-wet kelp

Hydrogen production from alginic acid

$= $ (1000 kg/ton-wet kelp $\times$ 7.0% $\div$ 0.176 kg/mol) $\times$ (0.7 mol-H_2/mol)

$= $ 278 mol-H_2/ton-wet kelp

$= $ 6.2 Nm^3-H_2/ton-wet kelp

Hydrogen production from kelp

$= $ (Hydrogen production from mannitol) $+$ (Hydrogen production from alginic acid)

$= $ 24.6 Nm^3-H_2/ton-wet kelp $+$ 6.2 Nm^3-H_2/ton-wet kelp

$= $ 30.8 Nm^3-H_2/ton-wet kelp

Electricity production from kelp

$= $ 30.8 Nm^3-H_2/ton-wet kelp $\times$ 1.7 kWh/m^3-H_2

$= $ 52.4 kWh/ton-wet kelp

This comes out to 52 kWh in electricity (at a fuel cell efficiency of 48% and 1.7 kWH/m^3-H_2), or the same amount of energy found in 31 L of gasoline, the fuel used in cars.

By using free sources of algae (such as kelp and sea mustard waste or drifting sea lettuce), this capability becomes economically feasible at an electricity rate of around 0.27\$ per kWh. Because large-scale energy production with farmed algae will require additional fuel costs, new bacteria with higher yields will need to be found (Tanisho 2011).

9.4.6 Comparison of Fermentation Energy Conversion Methods and Energy Conversion Efficiency

Ethanol and methane fermentation are well-known methods for the production of fermentation energy. Their generation reactions from glucose and theoretic energy conversion efficiencies are as follows (Tanisho 2011):

1. Ethanol fermentation

   $C_6H_{12}O_6 \rightarrow 2CH_3CH_2OH + 2CO_2$

   Energy conversion ratio $= (2 \times 1371.3)/2817 \times 100 = 97.4\%$

2. Methane fermentation

   $C_6H_{12}O_6 \rightarrow 3CH_4 + 3CO_2$

   Energy conversion ratio $= (3 \times 882.4)/2817 \times 100 = 94.0\%$

3. Hydrogen fermentation

   $C_6H_{12}O_6 \rightarrow 2CH_3COOH + 2CO_2 + 4H_2$

   Energy conversion ratio $= (4 \times 285.9)/2817 \times 100 = 40.6\%$

As this shown, ethanol fermentation and methane fermentation have much higher theoretical energy conversion rates than hydrogen fermentation. Since ethanol fermentation can produce concentrations as low as 8–10%, however, a treatment process is required to enrich the concentrations to 99% or greater for usage as energy. This process is more complex than the fermentation process, and the ratio is greater. Any comparison of energy conversion ratios is therefore without practical significance unless it uses the same final usage forms rather than theoretical values. For this reason, comparison will focus on the usage of the final form for electricity.

The general process of energy production from biomass raw material follows the following stages:

1. Ethanol

   Raw material $\rightarrow$ fermentation $\rightarrow$ enrichment/separation $\rightarrow$ thermal power generation $\rightarrow$ general efficiency

2. Methane

   Raw material $\rightarrow$ fermentation $\rightarrow$ desulfurization $\rightarrow$ diesel power generation $\rightarrow$ general efficiency

3. Hydrogen

   Raw material $\rightarrow$ fermentation $\rightarrow$ desulfurization $\rightarrow$ fuel cell power generation $\rightarrow$ general efficiency

   General efficiency is assessed according to the following formula:

   General efficiency = Theoretical fermentation efficiency $\times$ (1 − processing energy) $\times$ Effective power generation efficiency

   As Table 9.7 shows, the results indicate no major difference among the three approaches. At the same time, great differences exist in post-fermentation

Table 9.7 Comparison of general biomass energy conversion efficiencies

	Theoretical conversion efficiency (%)	Processing energy (%)	Power generation efficiency (%)	General efficiency (%)	Generation method
Ethanol fermentation	97.4	25	30	21.9	Thermal power generation
Methane fermentation	94.0	10	30	25.4	Diesel generation
Hydrogen fermentation	40.6	10	60	21.9	Fuel cell generation

processing for ethanol and hydrogen fermentation. Also, while ethanol fermentation requires enrichment and distillation towers or membrane separation, the same issues in hydrogen fermentation can be resolved with a small desulfurization tower, making hydrogen fermentation potentially simpler in plant terms.

Additionally, while methane fermentation involves raw material reactor retention times from several days to weeks, these times are very short (several hours) for hydrogen fermentation. Also, the devices used for hydrogen fermentation are potentially very small, ranging from several tenths to several hundredths the size of those used in methane fermentation. Construction costs for hydrogen fermentation are therefore much lower than for ethanol or methane fermentation.

Based on the above considerations, hydrogen fermentation possesses some advantages in spite of its low theoretical energy conversion efficiency, in that the general energy conversion ratio for its final usage form is almost identical to the other two forms, while the devices can be simpler and smaller.

9.4.7 Biohydrogen Production Technology Using Marine Ultra-High-Temperature Archaebacteria

The development of converting carbon monoxide (CO), which is a chief component of the gaseous by-products of South Korea's steelmaking plants, into hydrogen (H_2) has opened the door to the establishment of a new green energy resource for the future.

Recently, the research team of Dr. Kang at the Korea Institute of Ocean Science and Technology established South Korea's first-ever strain improvement and culturing technology for *Thermococcus onnurimeus* NA1, an ultra-high-temperature archaebacterium found in the deep sea, to develop a technique for mass-production of high-efficiency biohydrogen from renewable resources such as formic acid, starch, and carbon monoxide. For the use of NA1 in biohydrogen production, the team developed a biohydrogen generation fermentation tank system using 10, 30, and 300 L of ultra-high-temperature archaebacteria.

A high-temperature anaerobic bioreactor follows a fermentation process in which a fermentation tank is filled with NA1, a high-temperature anaerobic microorganism (archaebacterial) that grows at temperatures of 70–90 °C and shuns oxygen. Raw materials such as CO are then supplied, and the NA1 produce hydrogen as they consume the CO and other materials.

The Linz-Donawitz converter gas (LDG) in current iron manufacturing plant consists of around 60% CO, with most steel plants using a portion of that as a heat source for their power generation. One way of using this gas by-product to produce hydrogen would be to use the marine ultra-high-temperature archaebacterium *Thermococcus onnurimeus* NA1 to consume formic acid or CO from the gas to produce hydrogen while continuing to grow with the generation of ATP bioenergy.

Plans are currently under way to use the pressure swing adsorption (PSA) method to separate gas from the CO and formic acid consumed by NA1, efficiently separating out highly pure (CO-free) hydrogen to use in fuel cells, electricity, desulfurization, and industrial materials (Fig. 9.9).

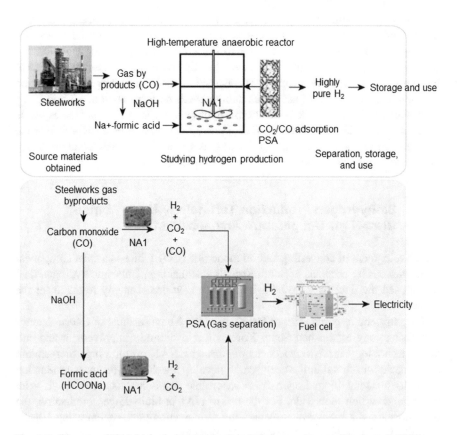

Fig. 9.9   Diagram of NA1 biohydrogen production, isolation, and usage (Singh et al. 2005)

9.5 Biodiesel Production with Microalgae

Research on the use of microalgae to make fuel intensified as a result of the two oil shocks of the 1970s. In the U.S., the National Renewable Energy Laboratory (NREL) conducted an 18-year Aquatic Species Program (ASP) between 1978 and 1996 in an effort to use microalgae for biodiesel production. This research and development was not sustained as oil prices stabilized and difficulties were encountered in the maintenance and management of outstanding species when mass-culturing microalgae in ponds, along with technical hurdles related to molecular microalgae improvements. With food, environment, and energy issues emerging at the global scale over the past years, however, the rise in crude oil prices has sparked renewed interest in biofuels such as biodiesel and ethanol (Miao and Wu 2006; Khan et al. 2009; Lardon et al. 2009; Mata et al. 2010; Campbell et al. 2011; Yang et al. 2011).

Biofuel production to date has chiefly involved the production of ethanol and biodiesel from sources such as corn, sugarcane, beans, rapeseed, and palm. A number of difficulties have surfaced, including the finite nature of these resources and the ethical issues attendant on converting grain into energy. Microalgae are not in competition with grains, however, and offer number advantages, including the ability to use idle farmland for biodiesel production, the relative ease of molecular biological improvements compared to plants, and biodiesel yields up to ten times greater or more than soybeans per unit area of farmland.

Like plants, microalgae synthesize organic matter through photosynthesis, using carbon dioxide, water, and solar energy. At the same time, they possess a number of advantages as renewable energy sources: they have faster growth rates than plants, they are more amenable to functional improvements through gene manipulation that allow the production of a greater variety of useful substances, and they are not traditional food crops. Indeed, biodiesel production (with 30% oil content) using microalgae amounts to around 58,700 L per hectare of farmland, or roughly 130 times greater than the 446 L/ha for soybeans. It is for this reason that the science journal *Nature* described microalgae biodiesel as "green gold," an analogy to the name "black gold" for crude oil. Through mass culturing of microalgae, it is possible to use carbon dioxide to produce large amounts of biomass, which can be converted into biofuel.

Mass culturing of microalgae is thus a form of green technology that is capable of preventing global warming through its absorption of atmospheric CO_2, while at the same time producing biodiesel from the resulting algae biomass. Moreover, biological conversion and processing of CO_2 is an environmentally friendly approach that takes advantage of photosynthesis, one of the basic means by which matter circulates in the natural world. Other advantages include the fact that it takes place at normal temperature and pressure, which allows for a simple process, and that the biomass produced through it can be used for useful substances.

9.5.1   Biodiesel Production with Microalgae

A. Environmental Stress Response and Lipid Accumulation

Most microalgae are autotrophs that grow by using CO_2 as a carbon source, performing photosynthesis when exposed to light in the presence of inorganic nutrients (compounds containing nitrogen, phosphorus, potassium, etc.). Reports also indicate that some of them are capable of growth under mixotrophic or heterotrophic conditions using organic nutrients, although these properties are not yet fully understood.

Some microalgae are reported to accumulate lipids in the form of oil droplets as their cells are subjected to stress from large changes in their optimal growth environment. They accumulate oil especially quickly when deprived of nitrogen, which makes them useful when selecting strains that accumulate lipids. Other microalgae accumulate large amounts of lipids on a continuous basis even under ordinary culturing conditions, although these species tend to have relatively slow growth rates.

Other potential sources of environmental stress besides nitrogen deprivation include physical factors such as light quantity, temperature, and CO_2 concentration; chemical factors owing to changes in pH, nutrients, and toxins; and biological factors related to growth, symbiosis, and bacteria.

Microalgae in the natural world are understood to accumulate lipids as a response to these environmental stresses, although the physiological relationship between lipid accumulation and the molecular response mechanism toward environment stresses has not been identified.

Like the lipids contained in the higher plant seeds used for food or as raw material for biodiesel, the lipids accumulated by known microalgae consist mainly of triacylglycerols. Triacylglycerols are neutral lipids with a three fatty acid ester linked to a single molecule of glycerol. As shown in Fig. 9.10, the length of the fatty acid carbon chain synthesized in the cell and degree of unsaturation varies by species and growth environment. Since increased efficiency of triacylglycerol production may reduce biofuel production costs, control of the fatty acid carbon number can result in fuel quality management and high value-added (Hu et al. 2008; Mata et al. 2010; Chen et al. 2011; Bulgariu 2015).

B. Microalgae Characteristics

Microalgae inhabit a broad range of environments. They can be understood as inhabiting any place in which sunlight can be exploited, including polar and snowy regions, hot springs, freshwater and seawater, soil, plant surfaces, building surfaces, and deserts. Symbiotic algae are also present in the cells and bodies of animals such as corals and jellyfish; some species have been found in caves, where they have adapted to environments with very weak light. Not all of these have led to the discovery of suitable strains for algae-based biofuel. This raises the question of what characteristics a strain should have to make it appropriate for algae-based biofuel production. At least three points must be considered:

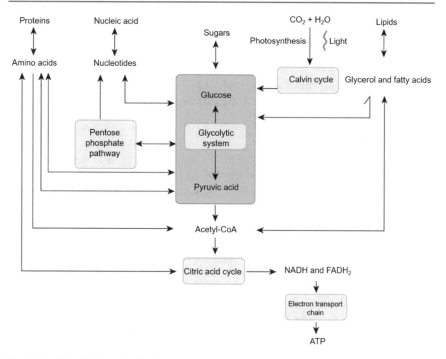

Fig. 9.10 Simplified metabolic diagram

1. Oil content must be high: The higher the oil content, the greater the oil productivity.
2. Resistance to pollution even in extensive outdoor environments must be strong: Extensive culturing is essential to suit the goal of providing low-cost fuel, but raises issues concerning contamination by similar organisms or predators. To withstand pollution, a species must have a fast growth rate and be capable of growing in extreme environments.
3. The species should be easy to harvest: For large-scale production, cells must be recovered at a concentration of at least 1 g/L. This results in massive amounts of liquid that must be handled. Self-cohesion is an extremely beneficial property for cell recovery. Harvesting is also easier when the algae are rod-shaped (such as *Spirulina*).

As was learned with the American Selling Price, it can be difficult to find a strain that satisfies all of these characteristics. For instance, *B. braunii*, which secretes straight-chain hydrocarbons outside the cell, satisfies ① with oil content of around 50% and ③ with its ability to produce masses of dozens to hundreds of cells, but falls short in terms of ② because of its slow growth rate. The cyanobacterium *Spirulina platensis* (*Arthrospira platensis*) has been cultured very successfully outdoors, is fast-growing, and thrives in alkaline environments, satisfying ② and ③, but has never accumulated much oil.

While tens of thousands of microalgae species are known to exist, only a few have been applied successfully to industrialization through outdoor culturing. Most of these are strains that satisfy condition ② and are capable of producing high value-added materials. When the goal is to use this as a basis for producing algae-based biofuels, ② is the top priority and ① is also essential, while separate technology will need to be developed for ③ (such as the use of bacterial to induce auto-flocculation).

C. Collecting Microalgae

Extreme environments are suitable for the collection of strains that satisfy ②. Unfortunately, Korea has almost no acidic or alkaline hot springs, which means that microalgae inhabiting them have to be obtained from overseas. Being able to mass-culture in acidic or alkaline conditions reduces the likelihood of culturing being impeded by contamination. Microalgae that are capable of growth under high temperatures often have fast rates of cell division (growth rates), and it may be possible to obtain strains that are easy to culture outdoors. Outdoor culturing cannot be performed during the winter, however. The ideal environment for collection is one which water flows into nature from a spring and microorganism populations are formed (Lien and Knutsen 1975; Vonshak 2017).

The following equipment must be prepared: a plankton net, a urethane sponge, a medicine spoon, a spuit, a plastic bottle, thin polyethylene bags (small, medium, and large) with zippers, a disposable tube, tweezers, and a permanent marker. The date, location, weather conditions, atmospheric temperature, water temperature, pH, and collector's name should be recorded in the collection log. A permit is needed for collection in national papers and privately owned or managed land. To avoid potential conflict, a collection permit should be obtained from the government when collecting overseas. To ensure safety while collecting in extreme environments, collection should be performed in groups of two or more people.

The following are two simple yet practical suggestions for collection. When temperature conditions for collected samples change while they are being transported to a laboratory or other place for treatment, it may not be possible to isolate a strain with the anticipated properties. Using a portable incubator with a battery-operated LED lamp inside is a convenient way of at least supplying light and preserving temperature when transporting samples. If a portable incubator is not available, an expanded styrol container or disposable pocket warmer may be effective over short periods.

Damage to the seaweed body can be minimized and survival increased through the use of a melamine sponge when rubbing or absorbing moisture from microbial mats, various surfaces, and hydrosphere samples. Depending on the aims, the use of a melamine sponge to absorb medium beforehand may also produce good results. For those wishing to apply immediate pressure to the sample, dedicated medium may be kept on hand in a disposable tube.

D. Isolating Microalgae

Typically, collected samples are placed in various media for enrichment culturing. The purpose of enrichment culturing is to allow for isolation of various forms of microalgae with various characteristics that are present in the sample.

Algae cells in natural environments live under conditions in which they become subject to stress when they do not absorb sufficient nutrients. As a result, they may become damaged and lose species diversity if subjected to immediate isolation with dilution or pipetting.

It is because of these concerns that the enrichment medium is prepared. If the aim is to produce algae-based biofuel, however, there is no need to use cells that are frail enough to be damaged by pipetting or dilution. It is better to use hardy strains that grew even when subjected to immediate isolation without enrichment culture (Andersen and Kawachi 2005).

The following are the three principal means of isolation (Kumazawa 1991):

1. Single cell isolation by micropipette

   First, a micropipette is prepared by applying a Pasteur pipette to an alcohol lamp and stretching it out. This is an advanced technique in which the micropipette is connected to valve rubber, and single-celled microalgae are drawn out as the Petri dish sample is observed under a microscope. Once drawn out, the cells are moved to the target medium for culturing. Use of a 96-well plate for culture allows for easy observation of growth. Sterilization can also be achieved easily through repeated cycles of culturing and single cell isolation.

2. Isolating algae by agar plate

   A colony can be formed by plating a suitable diluted sample (or enrichment medium solution) with agar in a laboratory dish. This technique is often used with cell isolation, although many microalgae cannot form colonies in agar. Because strains that do not grow in agar (i.e., strains that are sensitive to various stressed from the agar) are not used in algae-based biofuel production, however, this method may prove effective. Gellan gum may be used in place of agar.

3. Extinction dilution method

   With a 96-well plate, a series of five or ten dilutions may be developed to reduce the sample's cell concentration until there is only one or more cell per well. The resulting cell frequency of occurence per well follows a Poisson distribution. (When p represents the probability of an event E over n repetitions, the probability of E occurring r times follows a binominal distribution. If the value of p is very low and the number of iterations n is very high, the probability of E occurring x times becomes $p(x) = mX * e^{-n}/x!$ (where e is the natural logarithm), which is known as a Poisson distribution. m is the average of E occurring over n iterations, and the standard deviation equals m.) Thus, if the sample is diluted to an average of 0.2–0.3 cells per well, the possibility of division taking place in a single cell is a well where growth has been confirmed is high. Isolation becomes assured as the dilution method is repeated numerous times.

The following are some suggestions for avoiding difficulties with isolation. In many cases, the target organisms form clusters within a collected sample. When the clusters are clearly visible, an effective appropriate may be pulverization using surfactants or ultrasound. Because some microalgae will be killed off by even a second of treatment in an ultrasound cleaning device, one must be prepared to sacrifice some diversity.

An efficient approach for target strains with flagella is the use of phototaxis. Isolation can be achieved relatively easily by exposing only a portion of the culture to light and gathered the cells that cluster there.

If no diatoms (a form of single-celled algae) are needed, it may suffice to place approximately 5–10 mg/L germanium dioxide (GeO_2) in the medium. By competing with silicon dioxide (SiO_2), GeO_2 deters silicon absorption, resulting in an environment where diatoms cannot grow.

It takes at least one week for single cells to divide enough to form a colony that is visible to the unaided eye. To avoid potential negative effects from medium evaporation in the meantime, the plate may be stored in a thin polythene bag with a zipper. Microalgae growth can also be promoted and the isolation process hastened by introducing small amounts of CO_2 to the bag.

Examination of the growth of the marine green alga *Chlorococcum littorale* under conditions of changing CO_2 partial pressure in the liquid medium showed a specific growth rate over ten times higher when 2% CO_2 was mixed into air (CO_2 partial pressure of 0.02) than for air alone (CO_2 partial pressure of 0.00036). The same tendency has been observed in other microalgae, indicating that CO_2 addition is an effective means of promoting microalgae growth.

E. Selection of Microalgae Containing Oil

Nile Red (7-diethylamino-3,4-benzophenoxazine-2-one, $C_{20}H_{18}N_2O_2$, 318.369 gmol^{-1}) is a lipid droplet that is widely used as a stain for plants, animals, and bacteria. The following a simple screening method using this pigment for microalgae containing high concentrations of oil. It can be performed with a fluorometer and does not require a fluorescence-activated cell sorter (FACS) or other expensive equipment. The same approach can be applied with boron dipyrromethene (BODIPY), differing only in the wavelengths.

1. Prepare a culturing solution of sufficient turbidity.

   In the case of agar plates, effective dispersion can be achieved by scratching with a platinum loop and suspension in a 10 mM phosphoric acid buffer solution (pH 7.0). Culture medium may be used as is. The target cell concentration is around 0.2 at an optical turbidity of 720 nm.
2. Place 3 ml of cell suspension in a cuvette with transparent sides.
3. Introduce 10 μl of Nile red ethanol solution (0.5 mg/mL) and spin while holding entrance down with Parafilm.
4. Measure a 550–650 nm emission pattern with excitation of 488 nm.

Fig. 9.11 Emission pattern for Nile red staining of *Botryococcus braunii* (Maximum around 570 nm; emission 488 nm)

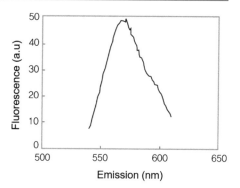

5. Peaks indicate positive (+) values. If only positive strains are found, check for autofluorescence within the same wavelength range (measure culture medium without added Nile red).

Figure 9.11 shows the NR stain emission pattern for *Botryococcus braunii*. Obviously, not all oil-containing microalgae exhibit similar patterns. A peak in the range of 570–580 nm, however, can be regarded as indicating an oil-containing microalga. While a Nile red ethanol solution was used, the cell walls may sometimes be too thick for Nile red to permeate. In such cases, the use of dimethylsulfoxide as a solvent may result in a higher staining efficiency. Because of changes in Nile red fluorescence over time, measurement should always be performed some period after staining to ensure stability.

Because Nile red achieves excitation at 488 nm, it is also well suited to devices using lasers. Some have suggested the use of a fluorescence-activated cell sorter (FACS), although in such cases, highly cytotoxic solvents such as dimethylsulfoxide are not conducive to cell growth after isolation. Conversely, the use of solvents with low cell toxicity results in insufficient staining and increased loss. This is a good illustration of how high throughput screening cannot be achieved simply through the use of FACS.

F. Which Medium to Use?

A very large variety of medium compositions exist for microalgae culturing, differing from one microalga species to the next. Which of these media is best for isolating a particular target strain? The diversity of microalgae medium competitions conversely means that there is no such thing as a one-size-fits-all medium to allow for growth of any microalgae. One medium developed in Japan has been found capable of allowing for growth by a very wide variety of marine microorganisms; it is currently marketed under the name Daigo INK. Even this medium, however, is not capable of achieving growth with all marine microorganisms.

When the aim is to collect fast-growing strains with large amounts of accumulated oil, at least three times of medium compositions should be selected. Acetic

Table 9.8 CHU13 X 4 medium components

Potassium nitrate (KNO₃)	0.2 g/L	Boron	0.5 ppm
Dipotassium phosphate (K₂HPO₄)	0.04 g/L	Manganese	0.5 ppm
Magnesium sulfate (MgSO₄·7H₂O)	0.1 g/L	Copper	0.02 ppm
Calcium chloride (CaCl₂·6H₂O)	0.08 g/L	Cobalt	0.02 ppm
Ferric citrate	0.01 g/L	Molybdenum	0.02 ppm
Citric acid	0.1 g/L		

Table 9.9 Comparison of average microalgae cell constituent elements and constituent elements of CHU13X4 (molar ratios)

	Cell	CHU13X4
Nitrogen (N)	1	1
Phosphorus (P)	0.1	0.1
Magnesium (Mg)	0.06	0.2
Calcium (Ca)	0.05	0.2
Sulfur (S)	0.05	0.2
Iron (Fe)	0.03	0.2

Cell atomic components are expressed as molar ratios to nitrogen per unit dry weight

acid or ammonia should be used as a nitrogen source, and a medium should be prepared in which the initial pH has been changed. The primary goal should be to ensure diverse conditions to produce as wide a variety of strains as possible.

Because seawater contains various kinds of dissolved inorganic salts, one would expect it to be dominated by pH-preferring microalgae in comparison with freshwater. Yet when the initial medium pH is set (for example at 6 and 10) and seawater is loaded for enrichment culturing, entirely different microflora are observed a week later. This is strong evidence that species that prefer both acidic and alkaline conditions coexist even in an environment with a pH of 8.2. At the same time, focusing too much on neutrality can result in the loss of many kinds of microorganisms. One must also be cautious of lowering pH when CO_2 is added. As noted before, growth is stimulated with the addition of gaseous CO_2. The presence of this dissolved CO_2 results in the medium pH falling. This fact must be taken into account as well when carrying out the initial pH adjustment (Rossignol et al. 1999) (Tables 9.8 and 9.9).

9.6 Developing Microalgae with High Oil Concentrations

The components of vegetable oils such as palm, soybean, and rapeseed oil that can be used to produce biodiesel fuel are fatty acid methyl esters, which are obtained from reactions involving triglycerides (the oils' chief components) in the presence of methanol and a base.

The oils produced by microalgae are similar to those obtained from plants. In that sense, biodiesel fuel can fundamentally be produced with the neutral lipids obtained from microalgae. In addition, many species of microalgae are known to be capable of accumulating neutral lipids, which can account for 50% or more of dry algal bodies. Biodiesel fuel can also be produced with hydrocarbon raw materials (Miao and Wu 2006).

The freshwater microalga *Botryococcus braunii* is considered highly promising for its ability to produce large amounts of hydrocarbons. At the same time, its growth rate is also slow, which makes its culturing (growth) difficult to control. As no other microalgae species besides *B. braunii* have be found to produce hydrocarbons accounting for double-digit percentages of dry weight, other species cannot be used in producing hydrocarbon with microalgae at the present time. For now, we must rely entirely on the success or failure of *B. braunii*.

For this reason, neutral lipids, which are accumulated universally by various species of microalgae, can be viewed as a future target material in terms of biodiesel fuel production with microalgae. We therefore need to obtain marine microalgae that are characterized by their ability to accumulate neutral lipids with the greatest efficiency.

The search among the more than 10,000 species of marine organisms for microalgae that produce oils containing neutral lipids has led to the discovery of *Navicula* sp. JPCC DA0580, a marine diatom that produces large amounts of oil. While the marine green algae strain *Scenedesmus rubescens* JPCC GA0024 was previously found to produce large amounts of oil, DA0580 has both a faster growth rate and higher oil content than GA0024 (Matsunaga et al. 2009; Takeyama and Matsumoto 2010).

(a) Optical observation
(b) Fluorescence observation (white portions are oil droplets)

Peak growth for DA0580 is achieved within one week, with oil stores reaching 40–60 wt% per dry algal body cultured over a one-week period. Oil accumulation was also found to occur from the middle to end period of culturing (Fig. 9.12).

Typically, the accumulation of oil in microalgae has required a two-stage culturing process, in which growth is induced and culturing is performed in a medium from with nitrogen nutrient salt components have been removed. In the case of DA0580, however, oils can be accumulated during growth. Since this does not entail the starvation required with other microalgae, its characteristics are superior to those of previous species: equal or greater amounts of oils can be produced with a shorter culturing period, allowing for reduced culturing costs.

The oils produced by DA0580 are meaningless if they are not suited for use in biodiesel fuel. As shown in Table 9.10, the oil components produced by DA0580 are distinctive even in comparison with *Chlorella*. While *Chlorella* possesses a broad scope of fatty acid ingredients, the oils in DA0580 consisted of 90% C_{16} fatty acids ($C_{16:0}$: palmitic acid, $C_{16:1}$ palmitoleic acid), although EPA (C_{18} fatty acid), which is typically present in diatoms, was absent.

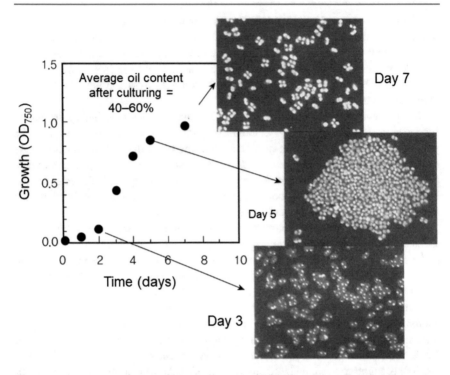

Fig. 9.12 Growth and oil accumulation in *Navicula* sp. JPCC DA 0580 (Nile red; white portions are oil droplets)

Table 9.10 Fatty acid composition for oils extracted from *Navicula* sp. JPCC DA0580

	Biodiesel fuel suitable	EU norm under for 12%							Total
	$C_{14:0}$	$C_{16:0}$	$C_{16:1}$	$C_{18:0}$	$C_{18:1}$	$C_{18:2}$	$C_{18:3}$	$C_{20:5}$	
Navicula sp. 4JPCC DA 0580	3.1	38.6	51.6	0.0	2.4	0.0	n.d.	4.3	100
Chlorella sp.	0.0	26	2.0	6.0	32	18	16	0.0	100

For the algal bodies cultured over one week, 47 wt% oil was obtained, with neutral lipid content of 32.9%, which accounted for 80% of oil stores.

Fatty acid composition has been known to vary greatly with culturing conditions. In the case of DA0580, proportions of various fatty acids changed with alterations in culturing conditions (such as temperature), with content of 0.8% C_{16}–C_{26} fatty acid hydrocarbons and 0.3% squalene. The strain showed few of the long-chain unsaturated fatty acids responsible for degradation during repurposing for biodiesel fuel, and because palmitoleic methyl esters have a low melting point of 0.5 °C, solidification does not occur at low temperatures. DA0580 oils thus have excellent fuel characteristics that have strong potential for use in biodiesel fuel. In addition, they have a high caloric value of around 37 MJ/kg, which means that oil stores can be combusted directly.

9.7   Chapter Summary and Conclusion

With the recent trend of consistently high oil prices and the Climate Change Agreement taking effect, energy has emerged as a serious societal issue. As a solution to this problem, research into developing and commercializing bioenergy-related technology for the production of clean energy and new/renewable energies using biomass is taking place actively around the world (Sims et al. 2010).

Brazil has drawn attention as a model for checking greenhouse gas emissions with its policies requiring the use of biofuels for 20–25% of automotive fuel. At the same time, the use of corn, soybeans, sugar cane, and other crops for bioenergy, has caused their prices to skyrocket, creating a global food shortage.

Numerous large-scale projects aimed at developing bioenergy with marine biomass have been pursued recently in the U.S., Europe, and Japan.

As a peninsular country surrounded on three sides by water with vast areas of ocean, Korea is in a position to take advantage of its situation by cultivating vast amounts of large seaweeds that grow by absorbing solar energy and nutrients in seawater. In addition to using these as energy resources, the country can also recover useful substances from them, including fertilizers, feed, and chemicals. Algae in particular grow biologically by fixing carbon dioxide from within seawater, which means that they can be used to prevent the CO_2 contamination that is currently becoming an issue, as well as several problems related to fossil fuels and nuclear energy.

Currently, the costs of producing energy from marine biomass such as algae and microalgae are too high for it to compete with petroleum and other fossil fuels in production cost terms. The fact remains that production costs will be significantly higher than those of petroleum even if the prices are rationalized somewhat through energy production system improvements

Because South Korea has a low rate of energy self-sufficiency compared to the oil-producing countries of the Middle East, North and South America, and Europe, however, the possibility of a fossil fuel exhaustion leading to an energy shortage is very likely to surface at some point. For the sake of its stability, it will need to develop technologies for producing energy from marine biomass.

References

Andersen, R. A., & Kawachi, M. (2005). Microalgae isolation techniques. *Algal Culturing Techniques* 83.

Beer, L. L., Boyd, E. S., Peters, J. W., & Posewitz, M. C. (2009). Engineering algae for biohydrogen and biofuel production. *Current Opinion in Biotechnology, 20*(3), 264–271.

Boichenko, V. A., Greenbaum, E., & Seibert, M. (2004). Hydrogen production by photosynthetic microorganisms. *Molecular to Global Photosynthesis, 397*–451.

Borines, M., De Leon, R., & McHenry, M. (2011). Bioethanol production from farming non-food macroalgae in Pacific island nations: Chemical constituents, bioethanol yields, and prospective species in the Philippines. *Renewable and Sustainable Energy Reviews, 15*(9), 4432–4435.

Brennan, L., & Owende, P. (2010). Biofuels from microalgae—A review of technologies for production, processing, and extractions of biofuels and co-products. *Renewable and Sustainable Energy Reviews, 14*(2), 557–577.

Bulgariu, L., & Bulgariu D. (2015). Biodiesel production from marine macroalgae, In S. K. Kim, & C. G. Lee (Eds.), *Marine bioenergy: Trends and developments* (pp. 423–458). CRC Press.

Bush, R. A., & Hall, K. M. (2006). Process for the production of ethanol from algae. Google Patents.

Campbell, P. K., Beer, T., & Batten, D. (2011). Life cycle assessment of biodiesel production from microalgae in ponds. *Bioresource Technology, 102*(1), 50–56.

Cecchi, F., Pavan, P., & Mata-Alvarez, J. (1996). Anaerobic co-digestion of sewage sludge: application to the macroalgae from the Venice lagoon. *Resources, Conservation and recycling, 17*, 57–66.

Chen, P. H., & Oswald, W. J. (1998). Thermochemical treatment for algal fermentation. *Environment International, 24*(8), 889–897.

Chen, C.-Y., Yeh, K.-L., Aisyah, R., Lee, D.-J., & Chang, J.-S. (2011). Cultivation, photobioreactor design and harvesting of microalgae for biodiesel production: a critical review. *Bioresource technology, 102*(1), 71–81

Duman, G., Uddin, M. A., & Yanik, J. (2014). Hydrogen production from algal biomass via steam gasification. *Bioresource Technology, 166*, 24–30.

Gao, K., & McKinley, K. R. (1994). Use of macroalgae for marine biomass production and CO_2 remediation: A review. *Journal of Applied Phycology, 6*(1), 45–60.

Grala, A., Zieliński, M., Dębowski, M., & Dudek, M. (2012). Effects of hydrothermal depolymerization and enzymatic hydrolysis of algae biomass on yield of methane fermentation process. *Polish Journal of Environmental Studies, 21*(2).

Hallenbeck, P. C., & Benemann, J. R. (2002). Biological hydrogen production; fundamentals and limiting processes. *International Journal of Hydrogen Energy, 27*(11–12), 1185–1193.

Hansson, G. (1983). Methane production from marine, green macro-algae. *Resources and conservation, 8*(3), 185–194.

Hossain, A. S., Salleh, A., Boyce, A. N., Chowdhury, P., & Naqiuddin, M. (2008). Biodiesel fuel production from algae as renewable energy. *American Journal of Biochemistry and Biotechnology, 4*(3), 250–254.

Huang, H., Yuan, X., Zeng, G., Wang, J., Li, H., Zhou, C., et al. (2011). Thermochemical liquefaction characteristics of microalgae in sub-and supercritical ethanol. *Fuel Processing Technology, 92*(1), 147–153.

Hu, Q., Sommerfeld, M., Jarvis, E., Ghirardi, M., Posewitz, M., Seibert, M., et al. (2008). Microalgal triacylglycerols as feedstocks for biofuel production: perspectives and advances. *The Plant Journal, 54*(4), 621–639.

Jones, C. S., & Mayfield, S. P. (2012). Algae biofuels: Versatility for the future of bioenergy. *Current Opinion in Biotechnology, 23*(3), 346–351.

Khan, S. A., Hussain, M. Z., Prasad, S., & Banerjee, U. (2009). Prospects of biodiesel production from microalgae in India. *Renewable and Sustainable Energy Reviews, 13*(9), 2361–2372.

Kim, S.-K., & Manivasagan, P. (2015). *Introduction to marine bioenergy* (pp. 20–29). Marine Bioenergy: CRC Press.

Kumazawa, S. (1991). Screening, incubation of marine microalgae, In S. Miyachi, N. Saga, & T. Matsunaga (Eds.), *Labo-manual marine biotechnology* (pp. 18–28). Tokyo, Japan: Shokabo Publishing Co.

Lamed, R., & Bayer, E. A. (1988). The cellulosome of clostridium thermocellum: *Advances in Applied Microbiology. 33*, 88–93.

Lardon, L., Helias, A., Sialve, B., Steyer, J.-P., & Bernard, O. (2009). Life-cycle assessment of biodiesel production from microalgae.

Lee, S.-M., & Lee, J.-H. (2012). Ethanol fermentation for main sugar components of brown-algae using various yeasts. *Journal of Industrial and Engineering Chemistry, 18*(1), 16–18.

Lee, I.-K., Shim, S.-C., Cho, H.-O., & Rhee, C.-O. (1971). On the components of edible marine algae in Korea-I. The components of several edible brown algae. *Applied Biological Chemistry, 14*(3), 213–220.

Lee, I., Hwang, M., & Oh, Y. (1992). Notes on marine algae from Korea (IV). *Korean Journal of Phycology, 7*(7), 257–268.

Levin, D. B., Pitt, L., & Love, M. (2004). Biohydrogen production: Prospects and limitations to practical application. *International Journal of Hydrogen Energy, 29*(2), 173–185.

Lien, T., & Knutsen, G. (1975). Inexpensive continuously operating centrifuge for rapid collection of microalgae. *Laboratory Practice.*

Mata, T. M., Martins, A. A., & Caetano, N. S. (2010). Microalgae for biodiesel production and other applications: A review. *Renewable and Sustainable Energy Reviews, 14*(1), 217–232.

Matsumoto, M., Yokouchi, H., Suzuki, N., Ohata, H., & Matsunaga, T. (2003). Saccharification of marine microalgae using marine bacteria for ethanol production. *Applied Biochemistry and Biotechnology, 105*(1–3), 247–254.

Matsunaga, T., Matsumoto, M., Maeda, Y., Sugiyama, H., Sato, R., & Tanaka, T. (2009). Characterization of marine microalgae, Scenedesmus sp. strain JPCC GA0024 toward biofuel production. *Biotechnology letters, 31*(9), 1367–1372.

Miao, X., & Wu, Q. (2006). Biodiesel production from heterotrophic microalgal oil. *Bioresource Technology, 97*(6), 841–846.

Nagai, S. & Nishio, N. (1989). Handbook of heat and mass transfer: In *Catalysis, Kinetics, and Reactor Engineering*, Vol. 3. Gulf Publishing.

Nakashimada, Y., & Nishio, N. (2011). Methane production techniques from alage. In M. Notoya (Ed.), *Seaweed bio fuel* (pp. 160–173). Japan: CMC.

Ni, M., Leung, D. Y., Leung, M. K., & Sumathy, K. (2006). An overview of hydrogen production from biomass. *Fuel Processing Technology, 87*(5), 461–472.

Nkemka, V.-N., Murto, M. (2010). Evaluation of biogas production from seaweed in batch tests and in UASB reactors combined with the removal of heavy metals. *Journal of Environmental Management, 91*(7), 1573–1579.

Omil, F., Méndez, R., Lema, J. M. (1995). Anaerobic treatment of saline wastewaters under high sulphide and ammonia content. *Bioresource technology, 54*(3), 269–278.

Otsuka, K. & Yoshino, A. (2004). A fundamental study on anaerobic digestion of sea lettuce. In *Ocean'04-MTS/IEEE techno-ocean'04: bridges across the oceans-conference proceedings*, (pp. 1770–1773).

Park, S., & Li, Y. (2012). Evaluation of methane production and macronutrient degradation in the anaerobic co-digestion of algae biomass residue and lipid waste. *Bioresource Technology, 111*, 42–48.

Percival, E. (1979). The polysaccharides of green, red and brown seaweeds: their basic structure, biosynthesis and function. *British Phycological Journal, 14*(2), 103–117.

Pimentel, D., & Patzek, T. W. (2005). Ethanol production using corn, switchgrass, and wood; biodiesel production using soybean and sunflower. *Natural Resources Research, 14*(1), 65–76.

Prabandono, K., & Amin, S. (2015). Production of biomethane from marine microalgae, In S. K. Kim & C. G. Lee (Eds.), *Marine bioenergy: trends and development*, (pp. 303–323). CRC Press.

Pulz, O. (2001). Photobioreactors: Production systems for phototrophic microorganisms. *Applied Microbiology and Biotechnology, 57*(3), 287–293.

Rossignol, N., Vandanjon, L., Jaouen, P., & Quemeneur, F. (1999). Membrane technology for the continuous separation microalgae/culture medium: Compared performances of cross-flow microfiltration and ultrafiltration. *Aquacultural Engineering, 20*(3), 191–208.

Sato, M. (2011). Ethanol production technique with high efficiency from seaweed by continuous fermentation. In M. Notoya, (Ed.), *Seaweed Bio Fuel* (pp. 129–137). Japan: CMC.

Sims, R. E., Mabee, W., Saddler, J. N., & Taylor, M. (2010). An overview of second generation biofuel technologies. *Bioresource Technology, 101*(6), 1570–1580.

Singh, A., & Olsen, S. I. (2011). A critical review of biochemical conversion, sustainability and life cycle assessment of algal biofuels. *Applied Energy, 88*(10), 3548–3555.

Singh, P., Gupta, S. K., Guldhe, A., Rawat, I., & Bux, F. (2015). Microalgae isolation and basic culturing techniques. In S. K. Kim, (Ed.), *Handbook of marine microalgae* (pp. 43–54). London, UK: Academic Press.

Takayama, H., & Matsumoto, M. (2010). Industrial applications of marine microalgae and their potential for bioenergy. *Bio industry, 27*(8), 6–12.

Takeda, H., Yoneyama, F., Kawai, S., Hashimoto, W., & Murata, K. (2011) Bioethanol production from marine biomass alginate by metabolically engineered bacteria. Energy & Environmental Science. *4*(7), 2575

Tanisho, S. (2011), Production technique of biohydrogen, In.: Masahiro Notoya (Ed.), *Seaweed Bio Fuel*, pp. 177–189. Japan: CMC.

Tanisho, S., & Ishiwata, Y. (1994). Continuous hydrogen production from molasses by the bacterium Enterobacter aerogenes. *International Journal of Hydrogen Energy, 19*(10), 807–812.

Ventura, M. R., & Castañón, J. I. R. (1998). The nutritive value of seaweed (Ulva lactuca) for goats. *Small Ruminant Research 29*(3), 325–327.

Visschers, V. H., & Siegrist, M. (2013). How a nuclear power plant accident influences acceptance of nuclear power: Results of a longitudinal study before and after the Fukushima disaster. *Risk Analysis: An International Journal, 33*(2), 333–347.

Vonshak, A. (2017). Laboratory techniques for the cultivation of microalgae. In *Handbook of microalgal mass culture (1986)* (117–146). CRC Press.

Wakayama, T. (2011). Hydrogen production using photosynthetic microorganisms, In M. Notoya (Ed.), *Seaweed Bio Fuel* (pp. 190–203). Japan: CMC

Wang, J., & Wan, W. (2008). Comparison of different pretreatment methods for enriching hydrogen-producing bacteria from digested sludge. *International Journal of Hydrogen Energy, 33*(12), 2934–2941.

Wang, X., Liu, X., & Wang, G. (2011). Two-stage hydrolysis of invasive algal feedstock for ethanol fermentation F. *Journal of Integrative Plant Biology, 53*(3), 246–252.

Wei, N., Quarterman, J., & Jin, Y.-S. (2013). Marine macroalgae: An untapped resource for producing fuels and chemicals. *Trends in Biotechnology, 31*(2), 70–77.

Wrighton, K. C., Thomas, B. C., Sharon, I., Miller, C. S., Castelle, C. J., VerBerkmoes, N. C., et al. (2012). Fermentation, hydrogen, and sulfur metabolism in multiple uncultivated bacterial phyla. *Science, 337*(6102), 1661–1665.

Yang, J., Xu, M., Zhang, X., Hu, Q., Sommerfeld, M., & Chen, Y. (2011). Life-cycle analysis on biodiesel production from microalgae: Water footprint and nutrients balance. *Bioresource Technology, 102*(1), 159–165.

Yokoyama, S., Jonouchi, K., & Imou, K. (2007). Energy production from marine biomass: fuel cell power generation driven by methane produced from seaweed. *World Academy of Science, Engineering and Technology. 28*, 320–323.

Marine Natural Substances

10

Contents

10.1 Introduction

The term "natural products" typically refers to the organic products that are produced by animals, plants, and all other lifeforms in their natural state or are present within their bodies, as opposed to those that are produced artificially. Natural products such as proteins, carbohydrates, and lipids are present in relatively large amounts within organisms and exist in many types, ranging from the component substances of the body to substances such as vitamins and hormones that are present in extremely small quantities in organisms and control their functions.

Natural products like sugars, fatty acids, and amino acids that are present in all organisms and participate directly in their metabolism are known as primary metabolites. Secondary metabolites are substances that are present only in certain organisms (such as alkaloids, terpenes, flavonoids, and antibiotics) and do not take part in metabolism, as well as compounds made from primary metabolite precursors.

The oceans cover over 70% of the earth's surface and are home to hundreds of thousands of species of animals, representing 80% of those existing on the planet. Although they are a veritable storehouse of organismal raw materials, less than 1% of them have been studied to date. While research has been done on natural products from land-based organisms, the exhaustion of research resources from those organisms has increasingly underscored the importance of natural products from marine sources, and marine organisms have drawn attention as new resources. Constraints on sample collection and other areas due to the oceans' environment have meant that research on natural marine substances has not been highly active.

© Springer Nature Switzerland AG 2019
S.-K. Kim, *Essentials of Marine Biotechnology*,
https://doi.org/10.1007/978-3-030-20944-5_10

Recent developments in different collection techniques, however, have made samples relatively easier to obtain, and natural marine substances have received growing attention (Kim 2015).

10.2   Researching Natural Marine Substances

10.2.1   Bibliographic Studies

Before any experimentation, the first step of the research process is to search the data and literature for information on what substances have been obtained from what marine organisms. Such information is easily searchable today online. Some of the leading journals providing ample information on natural marine substances include the *Journal of Natural Products* (published by the American Chemical Society) and *Natural Products Reports* (published by the Royal Society).

10.2.2   Collecting Marine Organism Samples

A. Marine Organism Samples: Types and Characteristics

Marine organisms inhabit a seawater environment with salt content of around 3%. The large amounts of salt and inorganic matter pose many difficulties in isolating bioactive substances and assessing activity. At the same time, the species diversity and abundant biomass mean that various kinds of natural substances can potentially be produced.

(1) Marine Animals

Natural classification, as opposed to artificial classification, is the classification of organisms based on structural commonalities associated with their living behaviors and habitats, as well as their physiology, ecology, genetics, and genesis. This is a form of scientific classification using taxonomic nomenclature, with the species as the basic unit. Marine organisms can be separated by natural classification into the following major categories.

1. Protozoa

Protozoa are single-celled organisms that exist as clusters of traits with one or more nuclei. Accordingly, protozoa undergo localized differentiation and may possess organs or organelles, while their bodies may be fully exposed or possess a thin membrane or shell. Protozoa are differentiated into endoplasm and exoplasm and possess pseudopods, flagella, or cilia for movement, depending on the species. Protozoa can be divided into flagellates, ciliates, rhizopods, sporozoans, and suctorians.

2. Porifera

   Sponges are among the most widely used resources for natural marine substances and are known to be some of the oldest living organisms on earth. More than 10,000 species exist, covering a wide distribution ranging from freshwater to intertidal zones and even deep-sea environments. Color and shape are not defined for those inhabiting shallow waters, and small animals often live symbiotically within their bodies. Sponges are typically classified into ordinary, calcareous, and hexactinellid varieties, based on the components or shape of their spicules (Fig. 10.1).

3. Coelenterata

   The bodies of coelenterates—as exemplified by jellyfish, sea anemones, hydras, corals, and comb jellies—possess a baglike shape with a triple cell wall and a hollow gastral cavity at the center. Around the entrance to the gastral cavity (mouth) are tentacles. In addition to digestion, the gastral cavity functions as a blood vessel supplying nutrients and oxygen. Undigested matter, waste, eggs, and sperm are discharged through the mouth. Coelenterates are carnivorous animals with mettling thread cells (poisonous threadlike organs present in coelenterates) among their integument cells. Within is a needlelike structure called a cnida. Because the organism produces mettle threads when hunting food, they are also known as cnidarians. Coelenterates possess muscles, nerves, and sensory organs, but no vascular system, respiratory organs, or excretory organs. In terms of their structural characteristics, coelenterates have one fixed side. They exist in cylindrical form like corals; hydra form like sea anemones; or disk form like free-swimming jellyfish. In addition to embryogenesis, they also propagate through budding and division. Around 9000 species exist worldwide.

4. Platyhelminthes

   Flatworms have bodies that are flattened toward the abdomen and exhibit left-right symmetry. They possess a head at one end and a tail at the other, with a mouth located on their abdomen. Flatworms include parasitic cestode, fluke trematode, and rhabdocoel flatworm types.

5. Annelida

   Annelid bodies consist of numerous metameres, exhibit zygomorphy, and possess deuterocoels. Annelids can be classified into six orders: Archiannelida, Polychaeta, Oligochaeta, Hirudinea, Clitellata, and Acanthobdellida. Among these, Polychaeta and Clitellata are considered representative types.

6. Entoprocta

   This phylum consists of a small number of organisms that live in isolation or in groups by attaching to seaweed or rock surfaces. Their bodies resemble cups with long handles attached, and they possess several tentacles around a concave portion on their upper surface.

7. Mollusca

   Mollusks account for the largest number of invertebrates after arthropods. They exhibit left-right symmetry, although changes have resulted in many exhibiting irregular forms. Their bodies consist of three sections (head, foot, and visceral sac), and most possess a mantle on their body wall. The following are the most representative types of mollusks.

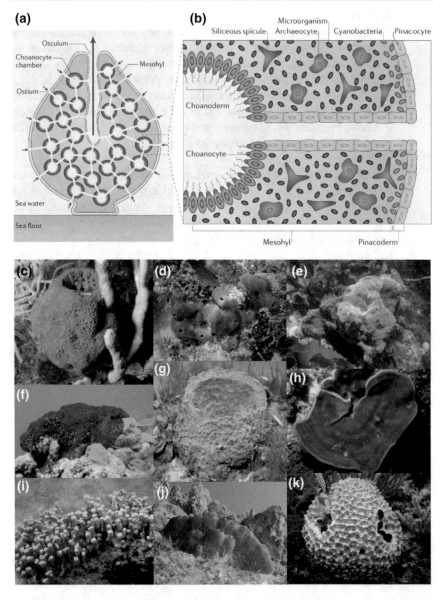

Fig. 10.1 Various sponges. Overview of characteristic demosponge (a), internal structural arrangement of demosponge (b), *Mycale laxissima* (c), *Amphimedon queenslandica* (d), *Ancorina alata* (e), *Rhopaloeides odorabile* (f), *Xestospongia muta* (g), *Cymbastela concentrica* (h), *Aplysina aerophoba* (i), *Theonella swinhoei* (j) and *Ircinia felix* (k) The figure adopted with permission from Hentschel et al. (2012)

8. Arthropoda

   Arthropods have large somite but are not regular like annelids. Their outer surface consists of tough chitin, and organismal growth is not continuous, but takes place in stages with each molting. Some 800,000 species of arthropods exist, accounting for over 75% of all animals. The most representative marine types are crustaceans such as crabs and shrimp; some species have the ability to regenerate portions of their body that have been lost. Various type of mollusks are showsn in (Fig. 10.2).

Fig. 10.2 Various mollusks. Plausible real-life inspirations for 16-bit Final Fantasy molluks. **a** A digital reconstruction of the ammonoid *Asteroceras obtusum* by Nobu Tamura. **b** A live specimen of *Nautilus pompilius*. **c** A live individual of the scallop *Argopecten irradians*. **d** Photo of a glass squid (Cranchiidae) juvenile by Uwe Kils. **e** A photo of the blue-ringed octopus *Hapalochlaena lunulata*, by Jens Petersen (All images were extracted and modified from Wikimedia Commons; images are public domain unless otherwise stated.). Figure adopted with permission from Cavallari (2015)

9. Chaetognatha

Chaetognatha is a phylum of small, transparent, floating animals. Although the most representative type is the arrow worm, around 50 species are known to exist (Fig. 10.10). Their bodies consist of three sections: a head, midsection, and tail. The head possesses a pair of eyes and stiff bristles, while the midsection and tail have fins, giving a shape similar to an arrow. Organisms typically range between one and 2 cm in size and are hermaphroditic.

10. Echinodermata

Echinodermata is a phylum that includes sea urchins, starfish, and sea cucumbers. Members exist in free-swimming and sessile forms. Echinoderms are actinomorphic coelomates with a calcareous endoskeleton and a special ambulacral system for movement. Echinoderms exist in five orders (plantlike cronoids, starfish, brittle stars, sea cucumbers, and sea urchins) and around 4000 species (Fig. 10.3).

Fig. 10.3 Various echinoderms (Arnone 2015)

11. Protochordata

Protochordates are also known as tunicates due to the thick mantle that surrounds their body. Secreted from the skin, this cuticle is similar to connective tissue and contains fibrous material; some species lack a cuticle, possessing a transparent, agar-like membrane. Organisms live independent or in colonies. Those that live in colonies have large, egg-shaped bodies with rootlike appendages at the bottom to attach to other objects. The most representative protochordates are sea squirts.

12. Vertebrata

Vertebrates are animals that possess a backbone. Fish are considered the chief example of marine vertebrates.

Fish live primarily in the water and propagate by laying eggs. They use gills to breathe, their body temperature varies with the surrounding environment, and their bodies are covered with scales. Fish have streamlined body shapes to allow them to swim with little resistance from the water. Over 2500 species of fish are known to exist around the world.

(2) Marine Plants

The term "seaweed" or "marine algae" is used to refer to plants that inhabit the sea and perform photosynthesis. Also included in this category are floating microalgae, including phytoplankton, sea grass, and blue-green algae. Typically, the term seaweed refers to sessile cryptogams. Like land-based plants, algae perform photosynthesis, but because they inhabit saline seawater, they absorb nutrients through their entire bodies. Their metabolites also differ from those of land-based plants and are important sources of natural substances. Phytoplankton in particular is a primary producer of various bioactive substances and can be easily cultured within the laboratory.

Around 25,000 species of algae exist worldwide. Over 600 inhabit the Korean coast, including 80 species of green algae, 135 of brown algae, 355 of red algae, and 48 of blue-green algae.

(3) Marine Microorganisms

Marine microorganisms possess completely different characteristics from land-based microorganisms and exist in halophile, halotolerant, psychrophile, pressure-resistant, and barophile types. These properties must be taken into account when searching for and culturing marine microorganisms. When bioactive substances are being isolated from microorganisms, the organisms must first be separated out and mass-cultured, after which the active substances are isolated and refined. The process thus requires separate microorganism isolation and culturing techniques.

The biggest advantage to microorganisms is that they possess few impurities within their bodies, which makes isolation and refinement of substances a relatively simple process. Care must be taken during culturing, however, as failure to properly culture microorganisms may result in experiments being unsuccessful. Microorganisms can be divided into bacteria, yeasts, fungi, and actinomycetes.

1. Bacteria

   Quantitatively, a large number of marine bacteria are aerobic or facultative anaerobic bacteria. They generally have low nutrient demands and belong to the category of heterotrophic bacteria. In addition to seawater, bacteria also inhabit sediment on the ocean floor, the inner and outer surfaces of other marine organisms, and the carcasses or excretions of other organisms.

   In the case of Gram-positive types, heterotrophic bacteria may be identified by sugar fermentation, oxidase activity, the presence or absence of pigments in the body, movement, the presence or absence of flagella, and salt demands. They include *Aeromonas*, *Vibrio*, *Cytophaga*, *Flavobacterium*, *Pseudomonas*, *Alteromonas*, *Chromobacterium*, *Caulobator*, *Acinetobactor*, and *Moraxella*. Gram-negative types may be classified by shape (bacillus and coccus) or by the presence or absence of motility, catalase activity, sugar fermentation, and mycelium formation. They include *Clostridium*, *Bacillus*, *Lactobacillus*, *Streptomyces*, *Nocardia*, *Athrobactor*, *Corynebacterium*, *Staphylococcus*, *Micrococcus*, and yeasts.

2. Actinomycetes

   Marine actinomycetes remain an unexplored area. While large numbers of coastal actinomycetes are believed to drift inland, some are thought to produce new secondary metabolites due to their adaptations to the marine environment.

3. Yeasts

   In addition to land environments, yeasts exist in seawater, in soil on the ocean floor, and in sediments, plankton, seaweeds, and higher plants and animals. Most of the 180 species reported to date are distributed across land and sea; around 20 species exist solely in marine environments. Physiological characteristics of marine yeasts include strong salt tolerance, a preference for low temperatures (failing to grow above 25 °C), and an optimal pH around neutral. In terms of nutrient demands, they are capable of growing in oligotrophic open-sea conditions.

10.2.3   Sampling of Marine Organism Specimens

Organisms inhabiting marine environments are very diverse in sampling sites, types, and shapes, ranging from microscopic plankton to large plants and animals, from the poles to tropical regions, and from tideland to the deep sea bottom. They are also extremely diverse in size, quantity, and specimen properties.

Because secondary metabolites exist in very small quantities as marine natural products within organisms, large samples in the dozens of kilograms are required, taking into account the recovery rate in the isolation process, the quantities used in activity measurement, and losses. Preliminary experimentation should be conducted to estimate the necessary sample amount and ensure an adequate size.

Close attention must be paid during the specimen sampling process to ensure as little mixture with other specimens as possible, and organisms must be identified through accurate classification. Because of degradation and other changes that may occur with chemical components during sampling or transportation, the specimens used in experimentation should be immediately chilled or frozen with dry or regular ice and taken to the laboratory quickly for frozen storage. Plant samples should be dried as quickly as possible for long-term preservation.

The precise sampling site and time must be recorded for samples, and appearance (color and gloss, smell, texture, and shape) should be photographed or otherwise recorded. Information should be kept in a logbook, and samples of the specimen should be separated, stored, and accurately identified.

A. Seaweeds

Seaweeds are best collected between March and June, when they are at their highest florescence. They should be collected in their entirety, including roots (adhesive organs). Shapes and colors may vary greatly even within the same type, and as many examples as possible should be collected when organisms are thought to belong to the same species.

B. Fish and Invertebrates

Depending on the type of fish, methods such as dragging, trawling, seining, jigging, and stow netting may be used; for cephalopods such as squids and octopuses, a gill net, a long line, or jigging may be used. For crustaceans, fish pots may be used with crabs and beam trawling for shrimp. Shellfish are collected by dredge or sampler.

C. Plankton

Plankton specimens may be collected by using plankton nets of different mesh sizes and filtering seawater on board the vessel or at the scene. Zooplankton typically requires 100 mesh or greater, while 20 mesh suffices for phytoplankton. When collecting large volumes, specimens may be gathered according to size through stepwise connection and seawater filtering from large to small mesh sizes.

D. Microorganisms

When collecting substances within seawater or the ocean floor, various types of water sampler or bottom sampler may be used to collect samples in sterilized canisters for transport to the laboratory. These may then be diluted and isolated on a plate containing sterilized seawater or enriched for around three days for plate medium isolation. Once separated, specimens may be obtained by mass-culturing the microorganisms in media that suit their growth conditions.

The risk of contamination may be reduced and the culturing process simplified by dividing microorganisms into various groups rather than culturing large volumes all at once. While in some instances only the bacteria or seaweed bodies are used after mass culturing and centrifuge harvesting, the culture solution should be examined for activity, as it can yield surprisingly large amounts of active substances.

10.2.4   Extracting Substances

Extraction is the first stage in the separation of substances from collected specimens. Because substance types and yields vary by extraction method, suitable solvent selection is crucial. A good approach when extracting a specific component from a sample is to finely chop or freeze-dry and pulverize it. Not only can one reduce the amount of solvent consumed in the extraction, but the extraction yield can also be improved and the enrichment time reduced.

A. Extraction Methods by Sample Type

- Frozen Sample: Thaw, place in solvent roughly triple the wet sample weight, and grind to extract.

- Dry Sample: Due to low moisture content and cell contraction, place in solvent of roughly five times the weight or higher and leave for several hours to extract; repeat extraction three times.

B. Extraction Solvent
   Various types of solvent with different polarities are used in extraction. Sequential extraction is performed, proceeding from non-polar to polar solvents. By performing the initial extraction with non-polar solvent, it is possible to separate and remove or use unneeded soluble sample components such as fats, lead, resins, and oil. It is also possible to extract substances from within the cell membrane, which is destroyed by polar solvents. Extraction solvents can provide information about which specific components enter which solvent fractions, allowing estimation of the chemical properties of those components and providing ample information for substance refinement and structure analysis.
   From low to high polarity:
   Petroleum ether < hexane < benzene < chloroform < methylene chloride < ether < ethyl acetate < acetone < butanol < isopropanol < ethanol < methanol < water.
C. Enrichment
   A rotary evaporator is used for extract enrichment. To minimize change in specimen components, solvent should be removed at as low a temperature as possible during enrichment (40 °C or less).

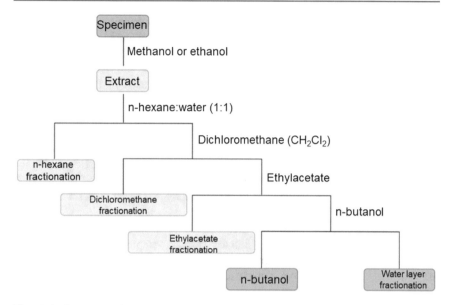

Fig. 10.4 Sequence and examples of solvents used in fractionation

10.2.5   Fractionation

Crude extract obtained from enrichment of solutions extracted through solvents from marine organism samples is strongly colored, and while some ingredients that are not dissolved by solvents are captured by filtration, other components of the organism are present, including various soluble salts, non-polar lipids, and carbo-hydrates. Fractionation must be performed to remove impurities before separating out the active ingredients in this raw extract.

With fractionation, two solvents with very different specific gravities are mixed together, resulting in the solvents splitting into two layers according to specific gravity. When extract is dissolved by these two solvents, it does so according to its affinity to each. In terms of sequence, fractionation proceeds from non-polar to polar solvents. Fractions are enriched through filtering after the removal of moisture with anhydrous sodium sulfate (Fig. 10.4).

10.2.6   Separating and Refining Substances

Careful planning of solvent conditions is required during the substance separation and refinement process. Through the aforementioned extraction and fractionation processes, a large portion of impurities are removed, leaving a mixture of sub-stances with similar properties. Separation and refinement are more difficult than extraction and fractionation and require advanced isolation techniques.

Some of the methods used for separation and refinement of natural marine substances include precipitation (using differences in solvency), fractional distillation (using differences in boiling point), sublimation (using sublimation capabilities), partitioning (using differences in partition coefficient), and chromatography. Chromatography is the most widely used of these (Heathcock 1996; Nicolaou et al. 2000).

Depending on the principle used for separation, chromatography can be separated into adsorption column chromatography, which uses differences in adsorption and desorption between a substance's stationary and mobile phase, and partition column chromatography, which uses partitioning. Various types exist, including ion exchange column chromatography (which uses an ion exchanger), thin layer chromatography (TLC), gel column chromatography, and high performance liquid chromatography (HPLC).

In adsorption column chromatography, emission begins where polarity is smallest; this is referred to as normal phase chromatography (NP). The inverse of this is partition column chromatography, in which emission begins where polarity is largest; this is known as reverse phase chromatography (RP) (IUPAC 1997).

One example of the use of column chromatography for separation and refinement of natural substances takes place in open column chromatography using silica gel. Fractions obtained with solvent are separated by solvent fraction and subjected to TLC examination and testing, after which HPLC are used for separation and refinement.

A. Adsorption Column Chromatography

Adsorption is a phenomenon is which a substance dissolved in gaseous or liquid solvent bonds physically or chemically with a solid. Solids in which adsorption occurs are known as adsorbents, while the adsorbing solute is known as adsorbate. Adsorption occurs because of interactions between molecules on the adsorbent surface and molecules in the solution. It may be classified into physical adsorption, which occurs because of physical interactions, and chemical adsorption, which results from chemical forces. Typically, physical adsorption is characterized by the relatively fast and reversible formation of adsorbents, while chemical absorption is more powerful than physical absorption and may be reversible, depending on the case.

In many cases, separation and refinement of natural substances employs physical adsorption, which is simpler than the more powerful chemical adsorption, where separation through solvent is difficult. Because adsorption and desorption in the column involves mutual action by the adsorption strength of the substance and adsorbent and the properties of the solvent separating the substance absorbed by the adsorbent, thorough knowledge of their characteristics is essential.

Typical examples of adsorbents used for separating natural substances include silica gel, active carbon, and alumina. Silica gel is the most widely used adsorbent for separation of natural substances, especially in cases of low-polarity soluble substances and non-ionic organic compounds.

Solvents with different polarities are also mixed and adjusted at appropriate ratios to adjust the polarity of solvent. The order outlined above for solvent polarity is also typically followed for the emission strength of commonly used solvents.

(1) Separation and Purification Process

Adsorbent is evenly packed in an empty tube (column) made of suitably sized glass. The specimen is adsorbed in the upper portion of the column, with separation occurring as different adsorption bands are forms sequentially with suitable solvents. Low-polarity solvents are used initially, with polarity gradually increased to isolate substances.

In the case of emissions in the author's laboratory, hexane and ethyl acetate are used as solvents with a silica open column. Polarity is altered in order of mixed solvents (hexane and ethyl acetate mixed at ratios of 100:1, 90:10, 70:30, 50:50, 30:70, 10:90, and 1:100), followed by emission with a 50:50 solvent mixture of chloroform and methanol.

B. Partition Column Chromatography

Because partition column chromatography involves separation according to solvency differences in the stationary and mobile phases adsorbed to the adsorbent, solvency must be considered rather than sample polarity when choosing stationary and mobile phases. This approach is mainly used when separating out highly polar hydrophilic substances (Whelan 2001).

For the stationary phase, a chemically bonded filler is used combining hydrophobic bases (octadecyl, ODS, C_{18}; octyl, C_8) with hydrophilic ones (aminopropyl, cyanopropyl, nitrophenyl). For the mobile phase, mobile solvents with heavy polarity such as water and methanol are often used, and two or more mixed solvents are frequently used in combination.

The emission approach used in the author's laboratory involves a reverse ODS open column in which water (tertiary distilled water) and methanol are used as solvents and polarity is altered along the sequence of the solvent mixture (water and ethanol mixture ratios of 100:1, 90:10, 70:30, 50:50, 30:70, 10:90, and 1:100).

C. Thin Layer Chromatography

In thin layer chromatography (TLC), the movement distance on the TLC plate changes according to the different partitions between the stationary phase (adsorbent surface) and mobile phase (solvent). Because each substance exhibits a different system of movement in the composition of a particular solvent, similarity or equivalence in distance serves as an indicator showing that the substance is identical (IUPAC 1997; Stoddard et al. 2007).

A typical method of chromatography involves manipulating the sample with a mobile layer of solvent on a thin membrane surface made with plaster and an

Fig. 10.5 Experimental
preparations for the rising
stage of a thin layer
chromatography (TLC) sheet

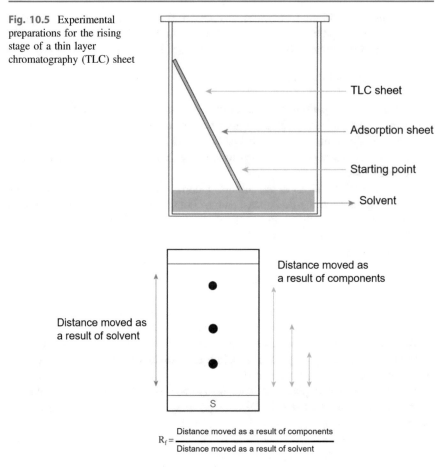

TLC sheet

Adsorption sheet

Starting point

Solvent

Distance moved as
a result of components

Distance moved as
a result of solvent

S

$$R_f = \frac{\text{Distance moved as a result of components}}{\text{Distance moved as a result of solvent}}$$

Fig. 10.6 TLC retention factor (Rf) values (IUPAC 1997)

adsorbent such as silica gel or alumina. TLC provides an important means of
analyzing or separating different types of compounds when the number of absorbents is highly diverse. As a first step, a thin membrane chromatography sheet is
prepared and a sample is dropped into it. The sheet is then placed into a chamber for
rising chromatography. The sheet is removed from the chamber and dried, and the
specimen is examined in terms of color, fluorescence, or appropriate chemical
reactions (Fig. 10.5). *Rf* is defined as the distance moved by a component divided
by the distance moved by the solvent (Fig. 10.6).

Silica gel mixed with fluorescent material may be used for a TLC sheet for
substances, which are then observed under UV light at 254 and 365 nm. This
detection method involves thoroughly evaporating the solvent on the TLC sheet and
then placing it into a vessel containing iodine crystals for a certain length of time.
A brown color is formed as the iodine vapor is adsorbed by organic matter with
double bonds or directionality. For this method, concentrated sulfuric acid is

sprinkled in the TLC and heated to around 200 °C until colors are produced. Various color-producing reagents are used, most notably including anisaldehyde, ninhydrin, and phenylmercury acetate (PMA).

D. Ion Exchange Column Chromatography

Many natural substances contain acidic, basic, or positive dissociating functional groups. Ion exchange is a method that uses the state in which compounds with dissociating groups are dissolved in solution. The ion exchange method is a form of separation that uses polymers combining various kinds of cation and anion, taking advantage of differences in the substances' ion affinity. It is typically used for protein, nucleic acid, amino acid, and peptide isolation. Exchangers can be classified broadly by polymer type into ion exchange resin, ion exchange cellulose, and dextran (Mercer 1974; Peterson and Sober 1956).

The stationary phase in ion exchange chromatography is charged ion exchange resin, which comes in a large variety of types. Ion exchange resin capacity is determined by the absolute value of the ions bonding to it. In anion exchange resin, a positively charged functional group attaches to a supporter through a covalent bond, and anion in the solute are drawn to the charged location. A positive ion exchange resin possesses a negatively charge site that bonds with cation in the solute. Different functional groups, such as carboxymethyl (CM) and diethylaminoethyl (DAEA), attach to different supporters, making them suitable ion exchange resins for chromatography.

Substances used for supporters include polystylene, polyacrylate, cellulose, Sephacel, dextran, agarose, and toyopearl. In the mobile phase, the eluant is typically an aqueous solution, and pH, salt concentration and properties, and the solvent's viscosity and dielectric constant are important considerations. Separated substances must be charged, or at least ionized.

The process of ion exchange chromatography for protein refinement is as follows (Fig. 10.7). A protein solution passes through a stationary phase column including ion exchange resin. Cation exchange resin bonds negatively charged carboxymethyl groups to a cellulose supporter, giving carboxymethyl-cellulose. Cation proteins bond with the exchange resin through electrostatic force, an adhesive force resulting from the net positive change in the pH entering the column. Once the protein has bonded, it is washed with a buffer solution that increases pH or ionic strength. These changes in the extraction solution weaken the bonds, causing the proteins to separate first from the exchange resin. Because of the significant changes to conditions in the bonding stage, solidly bonded molecules are subsequently separated out.

Protein and amino acid separation is an example of one important application of ion exchange chromatography. This chromatography is also put to effective use in separation of cation and anion. Halide ions can be separated in a Dowex-2 column with 1 M sodium nitrate ($NaNO_3$, adjust to NaOH) at pH 10.4 as an eluent, emerging in the order $F^- > Cl^- > Br^- > I^-$. Alkaline metal ions can be separated with 0.7 M hydrochloric acid (HCl) in a Dowex-50 or Amberlite IR 120 column,

Fig. 10.7 Ion exchange
chromatography for proteins

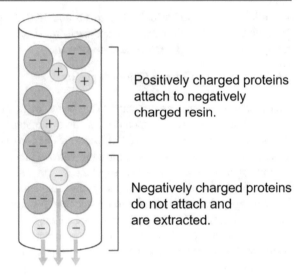

Positively charged proteins
attach to negatively
charged resin.

Negatively charged proteins
do not attach and
are extracted.

emerging in the order $Li^+ > Na^+ > K^+$. Alkaline earth metal ions can be eluted
with 1.2 M ammonium lactate in a Dowex-50 column, emerging in the order
$Ca^{2+} > Sr^{2+} > Ba^{2+}$ (Rieman 2013).

E. Gel Column Chromatography

In gel column chromatography, a molecular sieve is used for the stationary phase.
These include Sephadex, polyacrylamide, and agarose gel. Being hydrophilic, they
absorb water and may swell. When specimen molecules are larger than the maximum pore size in the expanded gel, they can pass through the gel molecules and
emerge easily from the column through spaces in the stationary phase molecules.
Smaller molecules pass through openings in the gel particles, emerging at different
rates according to their size and shape. In other words, molecules take longer to
pass and are extracted more slowly the smaller they are (Fig. 10.8) (Lathe and
Ruthven 1956).

Gels used for the separation and refinement of natural marine substances include
Sephadex G or Sephadex LH, which are made from dextran and its derivatives and
are among the most commonly used gells; biogel P, which is the chief gel made
with polyacrylamide and its derivatives; and agarose, cross linkage agarose,
polyvinyl, cellulose, and polystyrene varieties. Sephadex G, Sephadex LH-20, and
biogel P are chiefly used for the separation and refinement of natural substances.
Sephadex is mainly used for isolating proteins and tends to absorb large amounts of
water due to the strong polarity resulting from hydroxyl groups on the polysaccharides and large polymer chains. The characteristics of gels are the result of their
capacity for reabsorbing water and expanding.

Specimen injection

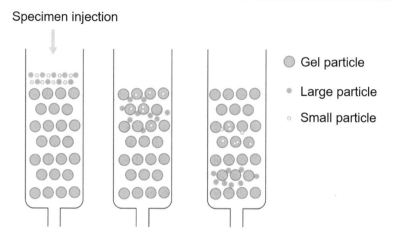

Fig. 10.8 Gel chromatography

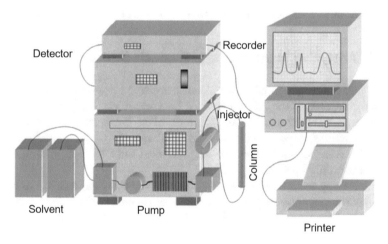

Fig. 10.9 High-performance liquid chromatography components

F. High-Performance Liquid Chromatography

In liquid chromatography, elution is not only performed at high speeds through pressure on the column. Separation tubes with filler particle sizes in the range of several millimeters and outstanding separation capabilities have been produced, resulting in significant improvements in tube efficiency. This approach is known as high-performance liquid chromatography (HPLC). HPLC boasts excellent sensitivity and allows for easy and precise analysis. Since it can be used to quickly analyze non-volatile components and substances that are unstable in the presence of heat, it is the most widely used of all separation and analysis methods. The HPLC process involves solvent, a pump, an injector, a column, a detector, and a recorder (Fig. 10.9) (Barnes 1992; Engelhardt 2012).

Once a column has been selected according to the characteristics of the substance to be analyzed (molecular weight, ionic character, solubility, etc.), a suitable mobile phase is used to separate the specimen. Once separation has been performed, a detector is selected with strong sensitivity and selectivity for the specimen, as well as a suitable band range and low detection threshold. Detectors come in many different kinds; in the case of UV/VIS detectors, specific wavelengths of light from a light source are projected through a device onto a sample within a cell known as a specimen vessel. Part of it is absorbed, while part of it passes through the specimen. For certain specimens, absorbance for light from particular wavelengths is high, and the intensity of light passing through the specimen decreases. In this case, the amount of light absorbed is related to factors such as the concentration of absorbent samples within the solution, the light wavelength, and the distance over which the light passes through the sample. This means that if a substance can be dissolved with a suitable solvent, it can be separated and analyzed with liquid chromatography, regarded of its volatility, stability with respect to light, organic and inorganic compounds, and molecular weight (Gerber et al. 2004).

10.2.7   Analyzing Marine Natural Products Structure

For the chemical structure of a natural substance to be determined, the substance used in the experiment must first be refined to at least 95% purity. Analysis cannot provide perfect information about structure when the specimen has not been thoroughly refined, which renders any structural analysis of the desired substance impossible. Analysis must therefore be performed after the specimen has been refined to maximum purity.

The following methods may typically be used for structural analysis of refined natural substances. The first group includes methods in which light and other electromagnetic waves are separated by wavelength for analysis based on molecular absorption or emission (including UV/VIS absorption, infrared absorption, and nuclear magnetic resonance), and while the second group consists of other methods (element analysis, mass analysis, X-ray diffraction analysis, etc.).

Comparison of the UV/VIS absorption spectrum with a substance's anticipated spectrum can be used to identify functional or reactive groups in a compound and gauge the presence of impurities. The IR absorption spectrum can be used to estimate the presence of functional groups in a compound and side chains such as unsaturated bonds. Nuclear magnetic resonance (NMR) can be used to determine the relative quantities of hydrogen and carbon and bonding state. Mass analysis can be used not only for molecular weight and structural units such as functional groups, but also for molecular formulas.

These methods of analysis may be used to estimate the structure of unknown specimens.

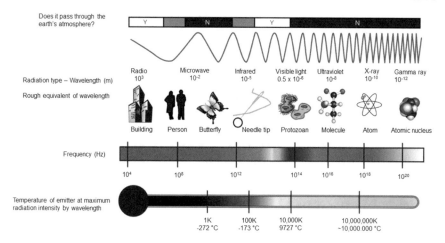

Fig. 10.10 UV and visible light regions of the electromagnetic spectrum

A. UV-Visible Spectrophotometer

The intensity of light passing through a substance is altered by factors such as reflection, refraction, scattering, and absorption. The amount of light absorbed is expressed by the ratio of the intensity of light applied to a specimen to the intensity of light emerging after passing through it. The amount of light absorbed is proportion to the concentration of absorbent compounds in the specimen and the light pass length (Skoog et al. 2017).

Beer demonstrated that reduction rate in light energy applied to a specimen was proportional to the concentration of absorbent substances in it, while Lambert showed that absorbance at particular concentrations was proportional to light pass length. These two principles were combined into the Beer-Lambert formula, which has become a basic approach to quantitative calculation for measurements using spectrophotometers. This analytical approach is referred to as UV/V is spectroscopy (Fig. 10.10).

Spectrophotometry is related to the relative ability of a compound to absorb radiant energy. Because the absorption of light by a compound is a molecular-level phenomenon resulting from the presence or absence of certain functional groups, many types of compounds do not absorb light in the UV/visible light ranges. Typically, compounds that absorb light in the UV/visible light ranges possess unsaturated double or triple bonds; functional groups containing unsaturated bonds are known as chromophores.

This light absorption is highly selective, with different chromophores exhibiting maximum absorption peaks at different wavelengths and differences in the amount of light absorbed. Saturated organic molecules do not show any absorption in the near UV or visible light range (200–800 nm), but absorption in this range does typically take place in the presence of chromophores with multiple bonds. Specific chromophores thus hold the key to confirming the presence of specific components in a specimen.

All absorbed energy is quantized, and because electron transfer is influenced by other transitions within the molecule (vibration and rotation transitions), light absorption bands in the UV/visible light range appear broad, a phenomenon that poses some constraints on observing the presence of functional groups. In the case of quantitative analysis however, this electron transfer plays an important role, as it is closely linked to molecular structure. This means that while UV/VIS photometry is not a good approach for quantitatively analyzing unknown substances, it is excellent for precise quantitative analysis of the concentrations of known substances.

B. Infrared Spectroscopy and the Structures of Natural Substances

The infrared range is located midway between the visible and microwave ranges on the electromagnetic wave spectrum. All molecules in this range have unique vibrations. When a molecule is exposed to continuously changing wavelengths in the infrared range (650–4000 cm^{-1}), rays with frequencies equal to the molecule's inherent vibration are absorbed, producing a characteristic spectrum. The use of this spectrum to analyze molecular structure is known as infrared absorption spectroscopy.

The IR spectrum can be used to identify substances by comparison with the spectra of substances that are already known. The structure can be partly ascertained from the absorption positions of multiple linkages and characteristic functional groups. Cis and trans isomers, ring positions, hydrogen linkages and other bonds, and breakdown reactions can also be confirmed (Fig. 10.11).

C. Nuclear Magnetic Resonance Analysis

The major component elements of natural substances are C, H, O, and N. The atomic nucleus spins when mass number is odd, when the atomic number is odd, or both. Nuclear magnetic resonance (NMR) focuses chiefly on hydrogen and carbon. Hydrogen may spin when the atomic and mass numbers are both one. Carbon has an atomic number and mass number of 12, so it typically cannot spin. Carbon isotopes with a mass number of 13, however, are present at a rate of 1.08% in ^{12}C and can be used for this analysis (Paudler 1987) (Fig. 10.12).

Resonance occurs in the nucleus when a specific magnetic field is applied to these elements. For example, hydrogen absorbs light at a wavelength of 100 MHz under a magnetic field intensity of 23,500 gauss, while carbon has been found to absorb 10.7 MHz of energy at 10,000 gauss. In these cases, it is only the hydrogen and carbon nuclei generating resonance, which does not occur in other nuclei. This resonance phenomenon is sometimes likened to a top, where the nuclear undergoes precession with respect to the earth's gravitational field. When an outside magnetic field is applied, the nucleus undergoes precession around its own spin according to its own oscillation frequency; the greater the magnetic field strength, the faster the

Chromophore	Compound	λ_{max} = nm
$>C=C<$	$H_2C = H_2C$	193
$-C \equiv C-$	$HC \equiv CH$	173
$>C=N-$	$(CH_3)_2 = NOH$	190,300
$-C \equiv N$	$CH_3C \equiv N$	167
$-COOH$	CH_3COOH	204
$>C=S$	CH_3CSCH_3	400
$-N=N-$	$CH_3N = NCH_3$	338
$-N=C$	$CH_3(CH_2)_3 - NO$	300,665
$C=C-C=C$	$H_2C = C - C = CH_2$ $\quad\quad H \quad\; H$	217

Fig. 10.11 Chromophores and absorption wavelengths

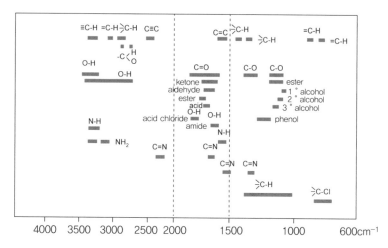

Fig. 10.12 Functional group absorption bands (cm^{-1}) (Larkin 2017)

precession speed. In Korea, a 900 MHz large magnetic resonance device (900 MHz NMR) set up in 2005 has been widely used in analyzing the structures of natural substances (Fig. 10.13).

In the case of hydrogen, the precession oscillation frequency for a magnetic field strength of 14,000 G is around 60 MHz. Coupling of the two electric fields occurs when the oscillation frequency during precessional movement equals the oscillation frequency of hydrogen's charge. When this happens, energy is transmitted from the

Fig. 10.13 900 MHz NMR
device (NMR; Silverstein
et al. 1981)

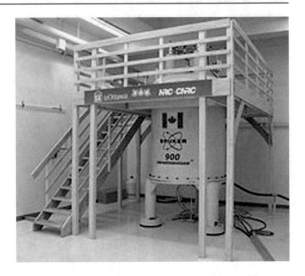

radiated light to the nucleus. The resulting change in the nucleus's spin is known as resonance. It is similar to the way a falling top gradually rights itself forcefully when spurred from outside, spinning upright once maximum external force is reached.

The follow information can be obtained from NMR:

- A substance's molecular structure can be ascertained. In natural substances, this is an efficient means of analyzing the number of hydrogen and carbon atoms within a molecule; the form in which they are present; branching and ring structures; multiple linkages; and isomers.
- Substances can be identified through comparative analysis with the spectra of previously known substances.
- Compound quantities can be determined.

(1)   1H NMR

In NMR, chemically different protons exhibit different absorption peaks. 1H NMR provides a way of determining the presence of different kinds of protons, where absorption peaks are exhibited in the range of 0–10 ppm. When they have the same positional relationship within the molecule, this is known as chemical equivalence: they have the same resonance frequencies and typically manifest through a single signal. Coupling occurs when they exhibit mutual influence with other nearby nuclei. In this case, signals appear in divided forms, including singlet, doublet, triplet, quartet, and multiplet peaks. The gap between these signals is known as the coupling constant J, which is represented in hertz (Hz). Typically, the number of separate hydrogen peaks equals $n + 1$, depending on the number of equivalent hydrogens right next to the hydrogen and attached to carbon atoms. The area for each

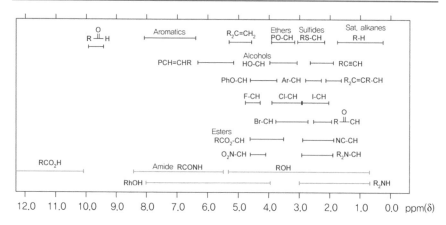

Fig. 10.14 Representative resonance ranges by molecular environment

signal is proportional to the number of hydrogen atoms exhibiting it, which represents the relative amount of hydrogen (Fig. 10.14) (Yadav 2013).

(2) ^{13}C NMR Spectrum and Structure

   1. ^{13}C NMR

      ^{13}C NMR absorption regions range roughly between 0 and 220 ppm. Chemical movement occurs over a broad range, resulting in little overlapping. Peaks are proportional to the amount of carbon in complex compounds, allowing for precise determination of the number of carbon atoms. Methene, methylene, methyl, and quarternary carbon can be distinguished, although it cannot be determined where these carbon atoms belong (Fig. 10.14).

   2. ^{13}C DEPT (Distortionless Enhancement by Polarization Transfer)

      This approach involves applying a third pulse to attached proteins within a molecule and altering the range of those pulses to 45°, 90°, and 135° to distinguish CH_3, CH_2, and CH. These are represented in the spectrum as positive and negative; quaternary carbon does not appear. The peak shapes and excellent quantitative assessment allow the production of entirely CH_3 or entirely CH_2 and CH spectra by adding or subtracting individual spectra; for instance, CH information can be obtained from DEPT 135°, while CH_3 can be obtained through comparison with DEPT 90° (Doddrell et al. 1982).

(3)   2D NMR

   1.   Correlation Spectroscopy (COSY)

      COSY allows for analysis between protons or between protons and carbon. A protein in a state of thermal equilibrium is shifted to the y-axis through 90 pulses. After time has passed, another 90 pulses are applied, and the z-axis component generated with the resulting information can be used to give a spin coupling spectrum. Through this approach, correlated peaks can be used to distinguish spin-spin coupling states for hydrogen atoms and between hydrogen and carbon (Fig. 10.14) (Keeler 2011).

2. Heteronuclear Multiple Quantum Correlation (HMQC)
   The HMQC method can be used to obtain information about carbon peaks and the hydrogen atoms bound directly to those carbon atoms.

3. Heteronuclear Multiple Bond Correlation (HMBC)
   C–H spin coupling is known as long-distance spin coupling when it passes through two or more bonds. Long-distance spin coupling can be used to obtain important structural analysis information not only for aromatic compounds with few hydrogen atoms and a high degree of nonsaturation, but also for compounds with quaternary carbon, including CO groups (Downard 2007).

D. Mass Spectrometry (MS)

In simple terms, the principle of mass spectrometry involves the ionization of molecules to isolate and measure them in terms of the ratio of mass to charge (m/z) to determine molecular weight. MS can be used to precisely determine molecular mass and composition formulas, while the fragmentation of molecules that occurs incidentally during ionization can be used to identify linked functional groups and part of the structure.

The horizontal axis in the mass spectrum represents the ratio of mass to ion charge (m/z), while the vertical axis is the relative ion strength percentage. The most important of the peaks produced is the molecular ion (also known as the mother ion, M^+). In many cases, the molecular ion peak represents the largest mass in the spectrum, although the molecular ion peak may be small or may not be observed at all, depending on the ionization method or molecular ion. Molecular ions of M^+-Na, M^+-H, and M^+-NH$_4$ have also been observed. Mass peaks smaller than the molecular weight that are observed as a result of molecular cleavage are referred to as fragment ions (daughter ions); the strongest of these peaks is called the base peak. Peaks are typically represented in terms of their relative size, where the base peak represents 100%.

10.3   Bioactive Substances from Marine Organisms

A bioactive substance is a substance that promotes or inhibits bodily functions over the course of an organism's life. They may be defined as those substances whose job is to correct any pathological states resulted from a deficiency or excess secretion of substances involved in coordinating functions within the body. As such, they are very important in allowing an organism to live a better, healthier life. New bioactive substances can be obtained from plants, animals, and other natural products, as well as through extraction and refinement from metabolites in microorganism, plant, and animal cell strains or through chemical synthesis.

The chief focus in natural substance bioactivity has been on antioxidation, anti-inflammation, anticancer, antitumor, cytotoxic, antibacterial, antifungal, and antiviral (e.g., HSV, HIV, cytomegalovirus) effects. Increasingly, however, attention is focusing on antimycotic activity and on bioactivity related to specific mechanisms, including the inhibition of particular enzymes or active substances, ion channel regulation, and signal transmission. More and more leading groups are shifting away from general activity toward mechanism-centered activity.

Marine natural products exhibit clear characteristics according to organism type. Over 90% of natural substances from corals and other coelenterates consist of terpenoids or the mixed biomaterials that form their framework, while oxylipins account for the remainder. Virtually no alkaloid content has been detected. In contrast, alkaloids are the main components of natural substances from sea squirts, while terpenoids are present only in trace amounts.

Most natural substances from mollusks are polypropionates, which are rarely found in other marine plants or animals. Echinoderms such as starfish and sea cucumbers are well known for their saponin content. While saponin is known to be present in large quantities in ginseng and some other land-based life forms, it is only rarely found in marine flora and fauna.

In contrast, substances from nearly every type of biosynthetic origin can be found in sponges, which occupy a central place in new marine substance research. The natural substance bioactivity that has been the driving force in pharmaceutical development has been focused chiefly on anticancer and antimicrobial effects. Indeed, over 40% of new substances reported have exhibited anticancer activity such as cytotoxicity and angiogenic inhibition. Antibacterial activity consists largely of Gram-positive and Gram-negative bacteria inhibition and intractable infectious diseases was infected with bacteria and fungus, tuberculosis, and malaria.

Research over the past 30 years has led to the discovery of new bioactive substances and substances with outstanding bioactive properties that have been isolated and refined from various forms of marine organisms, including sponges, large algae, coral, plankton, actinomycetes, fungi, and other marine microbes. Some of these have shown strong anticancer and anti-inflammatory properties and effects against cardiovascular disease. Several achievements have also been made in research on these components' pharmacology, medicinal effects, and toxicity (Hartmann et al. 2015).

Table 10.1 shows some of the approved marine organism-derived pharmaceuticals to date, as well as those undergoing clinical testing. Zinotide has already been approved as a new pain reliever and trabectedin (ecteinocidein 743), ω-3 unsaturated fatty acid (an EPA-DHA mixture), and eribulin mesylate (a halichondron derivative) as anticancer agents. In Japan, EPA is currently being used as a treatment for arteriosclerosis. While it is not listed in the table, the seaweed-derived polysaccharide carragelose has reportedly been approved as a cold medication in Austria. Currently in their third stage of clinic testing are the *Dolabella auricularia*-derived Brentuximab vedotin (SGN-35) and depsipeptides (aurilide-antibody

Table 10.1 Marine organism-derived pharmaceuticals that have been approved or are undergoing clinical testing (Malve 2016)

Clinical stage	Substance name	Product name	Source	Substance class	Target molecule(s)	Country	Disease/condition
Approved	Cytarabine (Ara-C)	Cytoasr-U	Sponge	Nucleoside	DNA polymerase	U.S.	Cancer
	Vidarabine (Ara-A)	Vira-A	Sponge	Nucleoside	DNA polymerase	U.S.	Viral diseases
	Ziconotide	Prialt	Cone shell	Peptide	N-type, calcium channel	U.S.	Pain
	Trabectedin	Yondelis	Colonial sea squirt	Alkaloid	DNA	Spain	Cancer
	ω-3 fatty acid ethylester	Lovaza	Fish	Fatty acid	Triglyceride synthase	U.S.	Hypertriglyceridemia
	Eribulin mesylate	Halaven	Sponge	Macrolide	Microtubules	U.S.	Cancer
Stage III	Brentuximab vedotin (SGN-35)		Mollusks	Antibody peptide complex	CD30, microtubules	U.S.	Cancer
	Plitidepsin	Aplidine	Colonial sea squirt	Depsipeptide	Rac1, JNK	Spain	Cancer
Stage 2	DMXBA (GTS-21)		Lineus sp.	Alkaloid	Nicotinic acetylcholine receptor	U.S.	Dementia
	Plinabulin (NPI-2358)		Fungi	Diketopiperazine	Microtubules, JNK stress proteins	U.S.	Cancer
	Elisidepsin	Irvalec	Mollusks	Depsipeptide	Cell membrane fluidity	Spain	Cancer
	PM00104	Zalypsis	Mollusks	Alkaloid	DNA	Spain	Cancer
	CDX-011		Mollusks	Antibody peptide complex	NMB, microtubules	U.S.	Cancer
Stage I	Zen2174		Cone shell	Peptide	Norepinephrine	Australia	Pain
	Marizomib (salinosporamide A)		Actinomycetes	β-lactone-γ-lactone	20S proteasome	U.S.	Cancer
	PM01183 (trabectidin analog)		Colonial sea squirt	Alkaloid	DNA	Spain	Cancer
	SGN-75		Mollusks	Antibody peptide complex	CD70, microtubules	U.S.	Cancer
	ASG-5ME		Mollusks	Antibody peptide complex	ASG-5, microtubules	U.S.	Cancer

(continued)

Table 10.1 (continued)

Clinical stage	Substance name	Product name	Source	Substance class	Target molecule(s)	Country	Disease/condition
				Antibody peptide complex			
	Hemiasterlin (E7974)		Sponges	Peptide	Microtubules	U.S.	Cancer
	Bryostatin 1		Sea moss, *Pectinatella magnifica*	Macrolide	PKC	U.S.	Cancer
	Pseudopterosin		Fan coral	Diterpene	Eicosanoid complex	U.S.	Lacerations

complexes), as well as the Caribbean colonial sea squirt-derived depsipeptide pli-tidespin (aplidine) (Li et al. 2009).

Among the substances in the second clinical stage is elisidepsin, a peptide isolated and refined from Hawaiian sea hares. Plinabulin is a diketopiperazine derivative developed as a secondary metabolite lead compound from filamentous bacteria. PM1183, which is in the first clinic stage, is a derivative of ecteinocidin 743, while hemiasterin is a peptide obtained from sponges. As this indicates, promising natural compounds have been discovered in marine organisms. In many cases, however, pharmaceutical development has ended up being abandoned due to very limited specimen quantities.

10.3.1   Substances with Antioxidant Properties

Most organisms on earth survive using the energy obtained from breathing and oxidizing oxygen in the air. Toxic, cell-damaging substances are unavoidably created as byproducts of this oxygen-requiring metabolic process. This is known as active oxygen. Active oxygen attacks organismal tissues and oxidizes and damages cells, and is also referred to as "free radicals." Large amounts of active oxygen forms such as peroxides (O_2^-) and hydrogen peroxide (H_2O_2) are also produced as part of the body's defensive process of eliminating pathogens and foreign matter. In some cases, their powerful bactericidal properties help to protect the body from pathogens.

Active oxygen causes damage to cells and their organelles and disrupts protein functioning by oxidizing amino acids in various proteins within the body. It is also damaging to nucleic acids and can cause mutations or cancer by altering or sepa-rating nucleic acid bases, truncating linkages, and oxidizing and degrading sugars.

As a first step, the production of active oxygen must be minimized. Smoking, which produces large amounts of active oxygen, must be avoided, and exposure to harmful environments (including air pollution, UV rays, and food additives) must be kept to a minimum. Stress should be released before it builds up too much, and moderate (not excessive) exercise is essential. The more food we consume, the greater the quantity of active oxygen created; as such, light eating is recommended. Consuming antioxidants is one way to prevent active oxygen from being produced within the body. Eating large amounts of fresh fruits and vegetables rich in vitamins and minerals is also recommended, and green or black tea is recommended instead of coffee.

As we grow older, more active oxygen is produced and the body's antioxidation capabilities decline. When this happens, it becomes impossible to remove active oxygen through fruit or vegetables alone. In such cases, one approach is to consume tablets of antioxidant-rich vitamin E (tocopherol), vitamin C, beta-carotene, sele-nium, melatonin, and/or propolis.

Antioxidation means the inhibition of oxidation. It is a concept that frequently comes up in explanations of cell aging and its prevention. Cell aging is essentially the same thing as cell oxidation. The oxygen we breathe into the body has

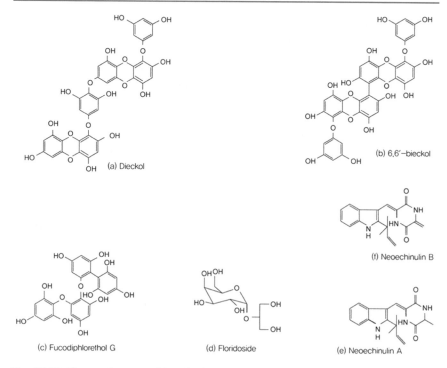

Fig. 10.15 Nature substances with anti-aging effects from marine organisms

beneficial effects, but active oxygen is created in the process. Active oxygen refers to oxygen existing in an unstable state, which has negative effects on the bodies of animals. Eliminating active oxygen is thus key to preventing cell oxidation, and thus cell aging. Typically, alkaloids and compounds with conjugated double bonds, phenyl structures, and –SH groups are known to exhibit antioxidation activity.

Powerful antioxidation bioactivity was observed in dieckol (A), 6-6′-bieckol (B), and fucodiphlorethol G (C) separated and refined from *Ecklonia cava*, an edible marine alga collected by the author off Jeju Island Li et al. 2009.

Antioxidation activity was also observed in floridoside (D) isolated from the red alga *Laurencia undulata*, the peptide Leu-Leu-Met-Leu-Asp-Asn-Asp-Leu-Pro-Pro from the skin of the cod *Gadus microcephalus*, and neoechinulin A (E) and neoechinulin B (F) obtained from the marine fungus *Microsporum* sp. (Fig. 10.15).

10.3.2 Substances with Anti-inflammatory Bioactivity Properties

Inflammation reactions play an important role in pathological conditions. The sequence and symptoms of inflammation include heat, redness, pain, swelling, and loss of function. During the inflammation reaction process, macrophages play the role of quickly deploying a defense mechanism against foreign substances. As macrophages are activated, they produce large amounts of inflammatory factors, including tumor necrosis factor-a (TNF-a), interleukin (IL), leukotrienes, and nitriooxide (NO). Inflammation occurs because of the propagation of white blood cells to prevent pathogens from proliferating further in infected regions where tissue has been damaged. A chief cause of inflammation is the accumulation of dead leukocytes and pathogens in the infected area.

Anti-inflammatory agents are substances or treatments that have the property of eliminating inflammation. More than half of anti-inflammatories are also pain relievers. Because non-opioid pain relievers (as opposed to opioid pain relievers such as opium) eliminate inflammation and reduce pain, many pain relievers also have an anti-inflammatory function.

A review of the relevant literature shows anti-inflammatory bioactivity effects from crassumolide (A) and durumolide (B), which have been isolated from the ocean-dwelling coral *Lobophytum crissum*. Between 2003 and 2010, the new compound fragilide (C) was isolated from the marine coral *Junceella* sp. Most of these exhibit distinctive structures and anti-inflammatory bioactivity.

Also exhibiting anti-inflammatory bioactivity are the lyngbyastatin derivative lyngbyastatin (D) from the marine actinomycete *Salinispora pacifica*, and the steroid glycoside linckoside (E) from the starfish *Linckia laevigata*. Together, they suggest possibilities for the prevention or treatment of inflammatory conditions.

Additionally, anti-inflammatory bioactivity has been seen in 1-(2-hydroxy-4-methoxyphenyl)ethanone (F) and 1-(5-bromo-2-hydroxy-4-methoxyphenyl) ethanone (G) from the sea horse *Hippocampus kuda Bleeler* (Fig. 10.16) (Kim et al. 2013).

10.3.3   Substances with Anticancer Bioactivity Properties

Cancer is the disease with highest incidence rate in South Korea. The incidence of cancer is expected to increase further in the future as a result of environmental issues, longer lifespans, and Westernized diets. As a global trend, the population with cancer is increasing by around 30 million each year, an estimated 20 million of whom are expected to die of the disease. Despite ongoing research into cancer, the diversification of metastatic and emergence mechanisms means that new anti-cancer agents still need to be developed with few side effects and the ability to overcome resistance, and new anticancer treatments continue to go on the market.

Cancer is thus an area where the failure to develop necessary medications points to strong market potential. Effective treatments need to be developed, and major international and domestic pharmaceutical companies are committed to developing and marketing such treatments.

Chemotherapeutic agents are a class of drugs that interfere with the metabolic pathways of cancer cells. Typically, they work directly on DNA, blocking its replication, transcription, or translation or disrupting the synthesis of nucleic acid precursors to inhibit cell division, thereby exhibiting anticancer activity (i.e.,

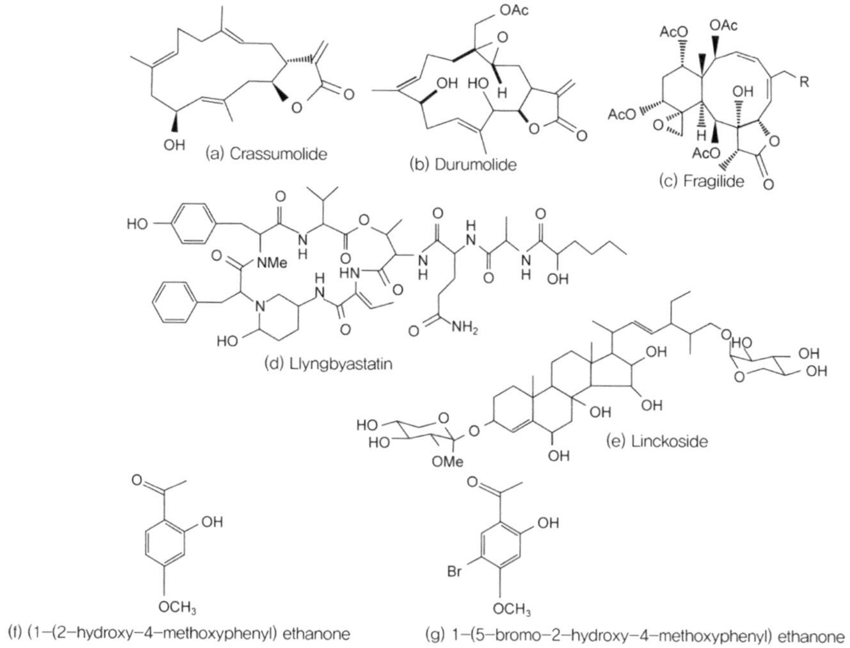

(a) Crassumolide

(b) Durumolide

(c) Fragilide

(d) Llyngbyastatin

(e) Linckoside

(f) (1–(2–hydroxy–4–methoxyphenyl) ethanone

(g) 1–(5–bromo–2–hydroxy–4–methoxyphenyl) ethanone

Fig. 10.16 Natural marine substances with anti-inflammatory effects

cytotoxicity toward cancer cells). Examples of chemotherapeutic agents include alkylating agents, metabolic antagonists, antibiotics, plant-derived alkaloids, and other natural substances and hormones. Examination of the relevant literature shows reports of natural anticancer agents isolated from marine organisms, including didemnin B (A), dolastatin (B), and cyanosafracin B (C). Perhaps the most notable example of pharmacological effects from marine organism-derived bioactive substances is antitumor activity. Examples of bioactive substances isolated and refined from marine animals include pateamine A (D), psammaplin A (E), haterumainide (F), neoamphimidine (G), and smenospongorine (H) (Fig. 10.17) (Shin 2013).

10.3.3.1   Substances with Antibiotic Bioactivity Properties

As a rule, antibiotics are used to treat and prevent bacterial infection. The category typically includes antibacterial and antifungal agents, which are classified according to similarities in antibiotic mechanism, scope, and structure. Natural antibiotics are chemicals produced by microorganisms such as bacteria and fungi, which allow them to inhibit the growth of or kill germs.

Around 100 species are currently used to treat infection. The first of the natural antibiotics to be discovered was benzylpenicillin, while other examples include streptomycin, chloramphenicol, tetracycline antibiotics, and macrolide antibiotics. Among the bioactive marine substances found to have strong antibiotic properties are flexibilide (A), sinulariolide (B), indolequinone (C), ascochytatin (D), isoaptamine (E), and batzelladine L (F) (Fig. 10.18).

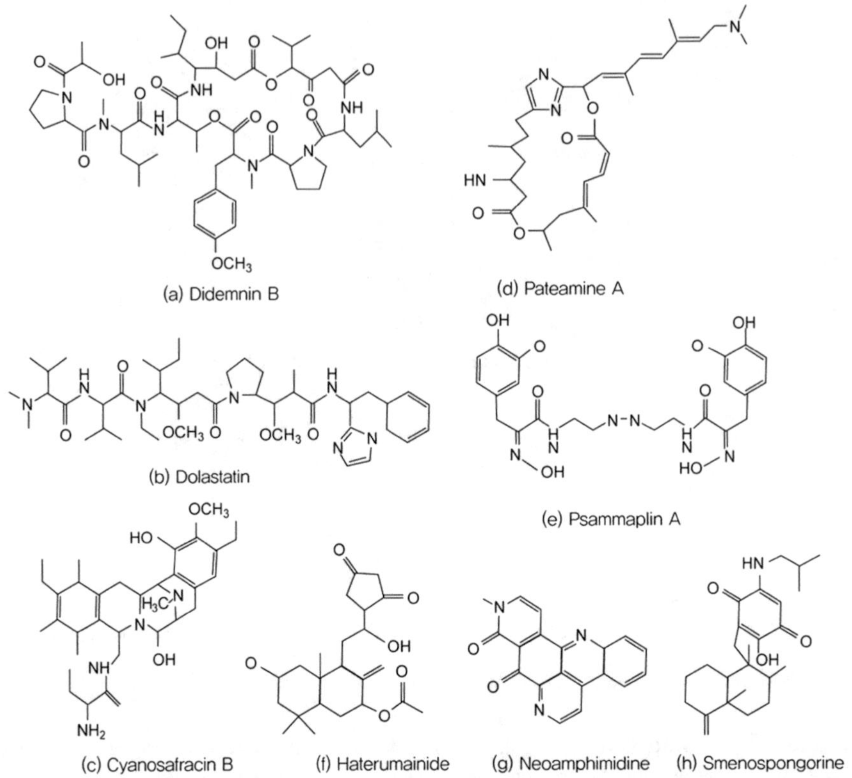

(a) Didemnin B

(b) Dolastatin

(c) Cyanosafracin B

(d) Pateamine A

(e) Psammaplin A

(f) Haterumainide   (g) Neoamphimidine   (h) Smenospongorine

Fig. 10.17 Natural marine substances with anticancer bioactivity properties

Also exhibiting antibiotic properties are eckol (G) and 8,8'-bieckol (H), which are phlorotannin compounds isolated from seaweed. Seaweed-derived bromophenol (A), callipeltin J (B), and holothurin B (C) have been found to exhibit powerful antifungal properties (Fig. 10.19).

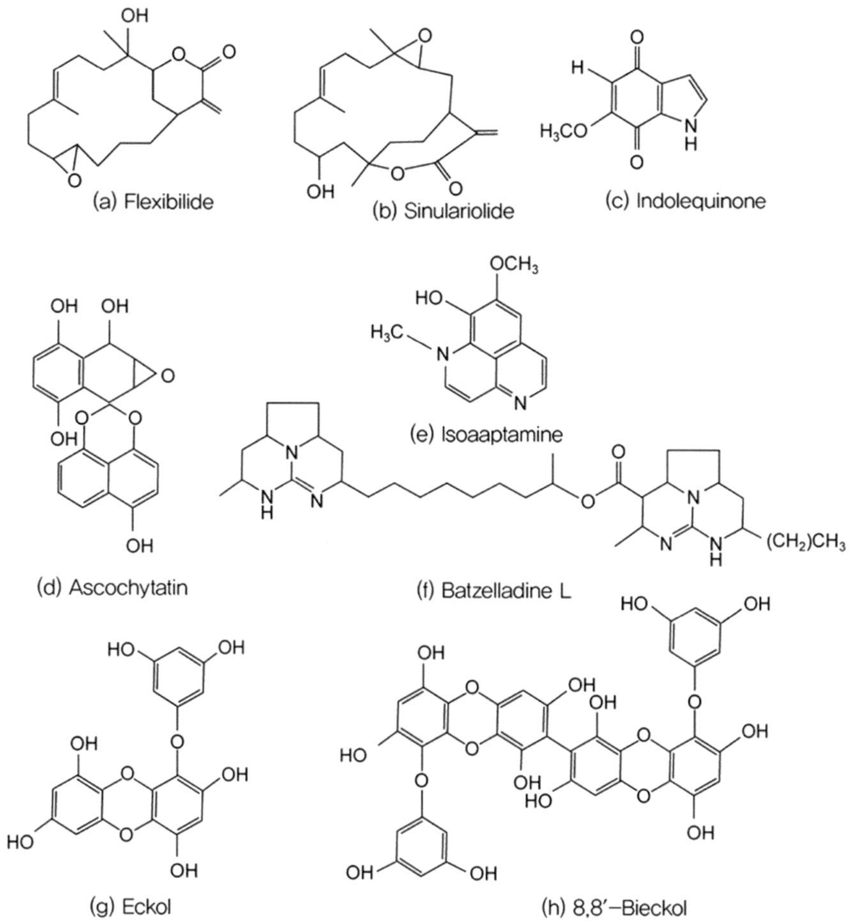

(a) Flexibilide

(b) Sinulariolide

(c) Indolequinone

(d) Ascochytatin

(e) Isoaaptamine

(f) Batzelladine L

(g) Eckol

(h) 8,8′–Bieckol

Fig. 10.18   Natural marine substances with antibiotic bioactivity properties

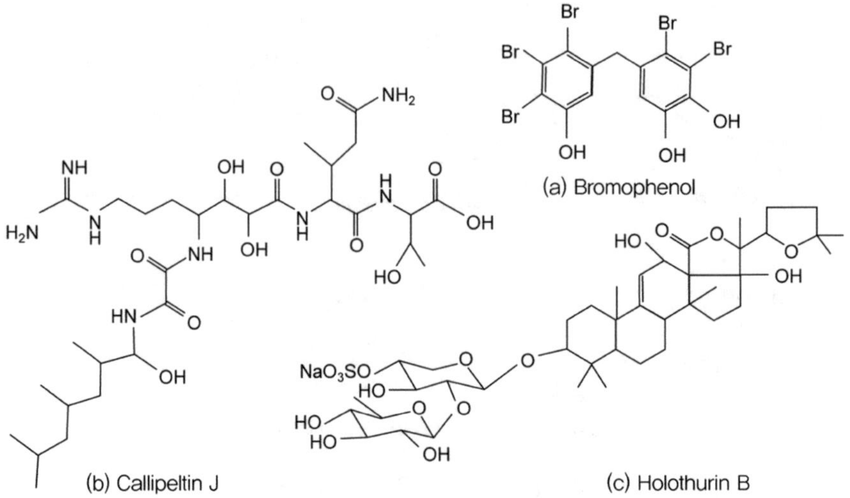

Fig. 10.19 Natural marine substances with antifungal bioactivity properties

10.4   Chapter Summary and Conclusion

Natural substances from land-based organisms have been an area of active research and application to date. As land-based organism resource quantities dwindle over time, attention has turned to natural marine substances as a new natural resource. Numerous constraints exist on research into natural marine substances, including difficulties with specimen collection owing to the particularities of the ocean environment. While the area was not a focus of active research in the past, the recent development of different techniques has made specimen collection relatively easy, while the development of analytical equipment has further boosted interest in natural marine substances.

Reports indicate more than 1000 types of natural compounds found each year in marine organisms, yet only a few natural marine substances are currently in use as pharmaceuticals. This stems from the fact that the natural substances present in marine organisms tend to exist in very small amounts, are for the most part highly toxic, and have very complex structures that make them difficult to synthesize chemically.

With so many new natural substances being isolated from marine organisms each year, it appears likely that some of them will have powerful and unique bioactivity properties that can be developed into game-changingly effective pharmaceuticals in the future.

References

Arnone, M. I., et al. (2015). Hemichordata. In: A. Wanninger (Ed.), *Evolutionary developmental biology of invertebrates 6.*

Barnes, J. (1992). *High performance liquid chromatography.* New York: Wiley.

Cavallari, D. C. (2015). Shells and bytes: Mollusks in the 16-bit era. *Journal of Geek Studies, 2*(1), 28–43.

Doddrell, D., Pegg, D., & Bendall, M. R. (1982). Distortionless enhancement of NMR signals by polarization transfer. *Journal of Magnetic Resonance (1969), 48*(2), 323–327.

Downard, K. (2007) *Mass spectrometry: A foundation course.* Royal Society of Chemistry.

Engelhardt, H. (2012). *High performance liquid chromatography.* Springer Science & Business Media.

Gerber, F., Krummen, M., Potgeter, H., Roth, A., Siffrin, C., & Spoendlin, C. (2004). Practical aspects of fast reversed-phase high-performance liquid chromatography using 3 μm particle packed columns and monolithic columns in pharmaceutical development and production working under current good manufacturing practice. *Journal of Chromatography A, 1036*(2), 127–133.

Hartmann, A., Becker, K., Karsten, U., Remias, D., & Ganzera, M. (2015). Analysis of mycosporine-like amino acids in selected algae and cyanobacteria by hydrophilic interaction liquid chromatography and a novel MAA from the red alga Catenella repens. *Marine drugs, 13*(10), 6291–6305.

Heathcock, C. H. (1996). As we head into the 21st century, is there still value in total synthesis of natural products as a research endeavor? In C. Chatgilialoglu & V. Snieckus (Eds.), *Chemical synthesis: Gnosis to prognosis* (pp. 223–243). Dordrecht: Springer Netherlands.

Hentschel, U., Piel, J., Degnan, S. M., & Taylor, M. W. (2012). Genomic insights into the marine sponge microbiome. *Nature Reviews Microbiology, 10*(9), 641.

Keeler, J. (2011). *Understanding NMR spectroscopy.* New York: Wiley.

Kim, S.-K. (2015). *Springer handbook of marine biotechnology.* Berlin: Springer.

Kim, S. K., Vo, T. S., & Ngo, D. H. (2013). Marine algae: Phamacological values and anti-inflammatory effects, In S. K. Kim (Ed.), *Marine pharmacognosy: Trends and applications* (pp. 273–280). New York, US: CRC Press.

Larkin, P. (2017). *Infrared and Raman spectroscopy: Principles and spectral interpretation.* Amsterdam: Elsevier.

Lathe, G., & Ruthven, C. (1956). The separation of substances and estimation of their relative molecular sizes by the use of columns of starch in water. *Biochemical Journal, 62*(4), 665.

Li, Y., Qian, Z.-J., Ryu, B., Lee, S.-H., Kim, M.-M., & Kim, S.-K. (2009). Chemical components and its antioxidant properties in vitro: an edible marine brown alga, *Ecklonia cava. Bioorganic & Medicinal Chemistry, 17*(5), 1963–1973.

Malve, H. (2016). Exploring the ocean for new drug developments: Marine pharmacology. *Journal of pharmacy & bioallied sciences, 8*(2), 83.

IUPAC. (1997). IUPAC. *Compendium of chemical terminology* (2nd ed.) (the "Gold Book"). Compiled by A. D. McNaught and A. Wilkinson. Oxford: Blackwell Scientific Publications (1997). XML on-line corrected version: http://goldbook.iupac.org (2006–) created by M. Nic, J. Jirat, B. Kosata; updates compiled by A. Jenkins. ISBN 0-9678550-9-8. https://doi.org/10.1351/goldbook.

Mercer, D. W. (1974). Separation of tissue and serum creatine kinase isoenzymes by ion-exchange column chromatography. *Clinical Chemistry, 20*(1), 36–40.

Shin, H. J. (2013): Anticancer compounds from marine microorganisms, In S. K. Kim (Ed.), *Marine pharmacognosy: Trends and applications* (pp. 409–419). New York, US: CRC Press.

Nicolaou, K., Vourloumis, D., Winssinger, N., & Baran, P. S. (2000). The art and science of total synthesis at the dawn of the twenty-first century. *Angewandte Chemie International Edition, 39*(1), 44–122.

NMR. https://en.wikipedia.org/wiki/Nuclear_magnetic_resonance_spectroscopy.

Paudler, W. W. (1987). Nuclear magnetic resonance: General concepts and applications.

Peterson, E. A., & Sober, H. A. (1956). Chromatography of proteins. I. Cellulose ion-exchange adsorbents. *Journal of the American Chemical Society, 78*(4), 751–755.

Rieman, W., III. (2013). Ion exchange. *Physical Methods in Chemical Analysis, 4,* 133.

Silverstein, R. M., Bassler, G. C., & Morrill, T. C. (1981). *Spectroscopic identification of organic compounds*. New York: Wiley.

Skoog, D. A., Holler, F. J., & Crouch, S. R. (2017). *Principles of instrumental analysis*. Cengage Learning.

Stoddard, J. M., Nguyen, L., Mata-Chavez, H., & Nguyen, K. (2007). TLC plates as a convenient platform for solvent-free reactions. *Chemical Communications* (12), 1240–1241.

Whelan, W. J. (2001). Partition chromatography revisited. *IUBMB Life, 51*(5), 329–330.

Yadav, L. D. S. (2013). *Organic spectroscopy*. Springer Science & Business Media.

Marine Microorganism Resources and Biotechnology

11

Contents

11.1 Introduction

The seas are a eutrophic environment with different characteristics from land, including high pressure, low temperatures, and salinity. Microorganisms survive even in oligotrophic environments further out in the ocean, where concentrations of organic and inorganic matter are low in comparison with coastal regions. Bacteria live at a rate of 10^7–10^8 cells/mL in the organic matter-rich coastal environment and in sediment on the ocean floor, and 10^4–10^6 cells/mL even in the oligotrophic high sea environment (Roszak and Colwell 1987).

Marine microorganisms exhibit great diversity, inhabiting the oceans not only vertically from surface to depths, but also horizontally, living even in the mud at the ocean floor. Researchers have been astonished to find microorganisms in highly acidic, high- or ultra-high-temperature, and high-pressure environments and around hydrothermal vents on the deep sea bottom. The discovery of these amazing organisms and their ability to survive in abundance has drawn new attention to the role of marine microorganisms in this. Occupying a pivotal role in the circulation of matter in the seas, marine microorganisms are involved both directly and indirectly in the breakdown of many kinds of organic matter, helping to sustain the oceanic environment (Munn 2011).

© Springer Nature Switzerland AG 2019
S.-K. Kim, *Essentials of Marine Biotechnology*,
https://doi.org/10.1007/978-3-030-20944-5_11

Recently, these organisms have drawn notice for their ability to clean up pollutants from land and degrade petroleum. Marine microorganisms are also directly involved in the fixing of carbon dioxide and the oxidation of hydrogen sulfide and methane, making them directly connected to our living environment.

Marine microorganisms may attach to plants and animals inhabiting the ocean, exist in symbiotic relationships within animal intestines or plant issues according to their benefit or convenience, or exist in parasitic or predatory relationships. The fact that marine microorganisms maintain such close interactive relationships with plants and animals means that various substances (including bioactive ones) are produced and shared among them. Examples of these bioactive substances include antibiotic and antiviral substances, toxic substances, enzymes and their inhibitors, hormones, and transmitters. Marine microorganisms are also necessarily implicated in issues that are currently the focus of global attention, including the control of pollution within the marine environment. This includes the prevention of damage to coastal fishery environments due to pollution or contamination, the prevention of red tides, degradation of petroleum and pesticides, and the control of pollutants in fish farms (Pandey and Purushothaman 2006).

Much of the research conducted on marine microorganisms has concerned *E. coli*, yeasts, and actinomycetes present in coastal regions due to pollution. In contrast, research on marine bacteria inhabiting specific environments is in its early stages. Microorganisms are present in low concentrations in the open seas; typically, the amount of bacteria cultured in agar medium ranges between 1/100 and 1/1000 when viewed under a microscope. Studying microorganisms that inhabit special environments requires the use of molecular biology metholods, as well as advanced analytical equipment to perform on-site structural analysis on microorganism community.

The physiological and ecological properties of the ultra-high-temperature bacteria (200–350 °C) found around deep-sea hydrothermal vents have been identified by various research teams around the world, and the special substances contained in them are currently under development for different purposes.

The fact that one of the known substances, the puffer fish poison tetrodotoxin, is produced by microorganisms from the genera *Vibrio* and *Alteromonas* suggests that new, as yet undiscovered functional substances could be developed from microorganisms inhabiting special environments.

Also found in marine microorganisms are unsaturated fatty acids produced by marine microalgae, including *r*-linolenic acid, EPA, fucothiamine, and β-carotene. This shows the important role that microorganisms play in the circulation of matter within the oceans, including the synthesis, degradation, and formation of organic materials.

Marine photosynthetic bacteria can help to purify the environment by breaking down organic materials. New bacteria have been developed to serve as basic feed in seedling production, serving as a basis for new strain development. The future may yet see the use of these marine microorganism-produced materials to develop potentially helpful substances for the treatment of incurable diseases and other issues (Kim and Dewapriya 2013).

11.2 Characteristics of Marine Microorganisms

Marine microorganism habitats possess a number of characteristics that distinguish them from land environments, including high salinity, high pressure, low temperature, and low nutrients. To adapt to these complex living environments, marine microorganisms have adopted halophilic, psychrophilic, barophilic, photic, and polymorphic characteristics (Arrigo 2004).

1. Halophilic: This is the most general characteristic of marine microorganisms: they require seawater as a basic environment to achieve optimal growth. These organisms typically exhibit optimal growth at a 30% saline concentration. Also obtainable from seaweed are trace elements and organic salts necessary for marine microorganism growth and metabolism, including potassium, magnesium, and calcium (Kokare et al. 2004).
2. Psychrophilic: Over 90% of ocean volume consists of environments with temperatures of 5 °C or less. Marine microorganisms must therefore be capable of growth at low temperatures. Marine microorganisms typically cannot grow at temperatures above 37 °C (D'Amico et al. 2006).
3. Barophilic: This refers to marine microorganisms' ability to inhabit high-pressure environments. It is typically found in the organisms that inhabit the bottom of ocean, where pressures exceed 380 atm. Some bacteria in the Pacific Ocean have been found inhabiting sea bottom areas with maximum pressures of around 1155 atm. Because of this resistance to high pressures, barophilic bacteria have been able to adapt to deep-sea environments (Horikoshi 1998).
4. Low-nutrient: Seawater does not contain many nutrients, which means that some marine bacteria must be cultured in low-nutrient media. When cultured in nutrient-rich media, marine bacteria die off quickly as they form their initial colony.
5. Photic: Bioluminescent bacteria represent one of the more interesting groups of bacteria inhabiting the seas, thanks to their ability to convert chemical energy to light energy and produce green or blue light. Attempts are currently being made to take advantage of this bioluminescence phenomenon by using photobacterium as indicators for investigating seawater pollution (Baharum et al. 2010).
6. Polymorphic: Having adapted to grow in complex marine environments, marine microorganisms exist in various forms. This phenomenon is observable under a microscope. In particular, polymorphism can be universally observed in Gram-negative coccus bacteria.

11.2.1 Psychrophilic Microorganisms

Marine environments are inhabited by many microorganisms that reflect the characteristics of those environments. Apart from the surface layers in tropical and subtropical regions, oceans typically possess low temperatures of around 2–3 °C and are inhabited by numerous psychrophilic bacteria.

Psychrophilic bacteria were first discovered in 1887 by Forster, who conducted an experiment by placing an ice cube at 0 °C inside a test tube. The bacteria were found growing alongside other bacteria that propagate at room temperature. In 1902, Schmidt-Nielsen used the term "psychrophile" to describe bacteria that required temperatures of 0 °C to live and were capable of growing at such temperatures. This name was suggested to be inappropriate, however, after experiments with these bacteria showed that some propagated at 0 °C while others had optimal growth temperatures of 20 °C or more; the name "cold-tolerant bacteria" was adopted for the latter. Psychrophilic bacteria as those capable of growth at temperatures of 5 °C or more.

Characteristics of psychrophilic bacteria include enzyme synthesis within the organism in low-temperature environments, types and compositions of lipids within the cell membrane, and the importance of other bioactive substances. Because of their particular environment (i.e., low-temperature and high-pressure), these bacteria are the focus of growing attention to their characteristic bioactive substances, which are not found in land-based organisms. Recently, research into bioactive substances with bioengineering methods has become more active.

As shown in Fig. 11.1, examination of the growth temperature for psychrophilic bacteria isolated from the waters near the South Pole showed maximum growth at a temperature of 4 °C and no growth at all at temperatures above 9 °C. For experiments with these psychrophilic bacteria, the entire process from sampling to isolation, culturing, and preservation must be conducted in low-temperature environments of around 5 °C. Equipment and media must also be cooled ahead of time (Imada 2009).

Most land-based bacteria, in contrast, do not grow at these low temperatures; their optimal growth temperature is around 37 °C, or roughly human body temperature. Enzymes in psychrophilic bacteria are typically more unstable with respect to heat than those in mesophilic bacteria. For example, the psychrophilic *Vibrio marinus* MP-1 bacterial strain possesses malic dehydrogenase within its cell

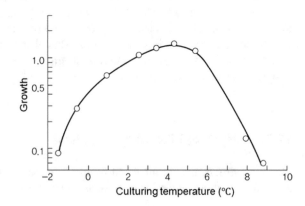

Fig. 11.1 Effects of temperature on the growth of psychrophilic bacteria isolated from sea waters near the South pole

| Temperature (°C) | Hours | | | |
	1.25	3.0	6.25	9.0
19.0	+[a]	+	+	+
21.0	+	+	+	+
23.0	+	+	+	+
25.0	+	+	+	+
27.0	+	+	−	−
28.8	+	+	−	−
30.8	−	−	−	−
32.7	−	−	−	−
34.8	−	−	−	−
36.9	−	−	−	−

Table 11.1 Changes in the survival rate of the marine bacterium *Vibrio marinus* (isolated from seawater collected at a depth of 1200 m) over time at various temperatures

[a]+: survived, −: died

body. This enzyme exhibits stable activity at a range of 0–15 °C, but activity decreases markedly at higher temperatures (Michener and Elliott 1964; Larkin and Stokes 1968; Inniss 1975; Gounot 1986; D'Amico et al. 2006; Margesin and Miteva 2011).

Table 11.1 shows results from a study of changes over time in the survival rates at various temperatures for marine bacteria isolated from seawater collected at a depth of 1200 m. These bacteria typically die within around six hours if placed in environments at temperatures of 28.8 °C; they die out completely within around one hour at temperatures of 33 °C or greater (Imada 2009).

It is chiefly the cell membrane and enzymes within the cell that are affected by low temperatures. The cell membrane filters nutrients and secretes or excretes substances from within the cell, while also functioning to maintain a certain level of homeostasis within the cell in response to changes in the external environment. The important components of this membrane are proteins and lipids; lipids have an especially crucial role in maintaining the membrane's functioning. Lipids in the cell membrane change phases with temperature (Fig. 11.2). The high levels of

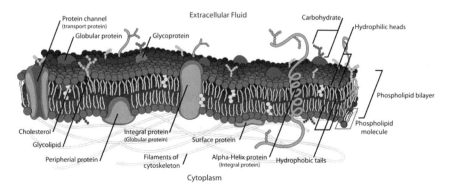

Fig. 11.2 Two-dimensional cell membrane structure (https://en.wikipedia.org/wiki/Fluid_mosaic_model#/media/File:Cell_membrane_detailed_diagram_en.svg)

unsaturated and free fatty acids that are typically reported in the cell membranes of psychrophilic bacteria are also related to the low-temperature environment.

11.2.2 Halophilic Microorganisms

Seawater has salt content of roughly 3.5%, and the microorganisms that inhabit the oceans are capable of growth even in the presence of high saline concentrations. Marine bacteria typically flourish at concentrations of 0.3–0.8 mol NaCl. Some bacteria require a high salt concentration to survive, such as those that live in the Dead Sea in the Middle East or the Great Salt Lake in North America, both of which have higher saline concentrations than seawater.

While marine and land bacteria may be distinguished by examining the relationship between salt concentration and growth, marine bacteria can also be identified simply through placement in distilled water. The cell walls of marine bacteria will rupture in the presence of pure water containing no salt, as they cannot regulate osmotic pressure.

When salt concentration within the cell is higher than outside, the imbalance typically leads to water gradually entering the cell from outside. The external cell membrane possesses mechanisms for allowing only what is needed within the cell and blocking unnecessary things from entering. Low molecular weight substances like water are not regulated by these mechanisms, and water permeates the cell through physical diffusion. As a result, the outward pressure exerted within the cell increases and the cell bursts.

The NaCl that marine bacteria require for growth is not needed for osmotic pressure regulation. Whereas a non-halophilic bacterium obtains energy through the movement of protons (H^+) in and out of the cell membrane, marine bacteria have developed mechanisms to harness energy from the movement not only of protons, but also of the sodium ions in NaCl (Margesin and Schinner 2001).

Because seawater possesses roughly 3.5% salt content, most marine bacteria cannot propagate at a 0% concentration, and propagate best at concentrations equivalent to those in seawater, as shown in Fig. 11.3. In contrast, land-bacteria typically exhibit maximum growth at a 0% salt concentration, and their growth is inhibited as the salt concentration rises.

Fish sauce is an example of a fermented food made through the industrial application of halophilic bacteria. Fish sauce is a nutrient-rich fermented condiment made by adding salt to hydrolyze the fish's flesh with protein-dissolving enzymes located inside the fish's viscera. The forms marketed in Korea, Japan, and Southeast Asia are made by traditional methods, but their commercial potential is diminished by the fact that the process takes a long period of 6–18 months. One approach that could be applied commercially to produce fish sauce quickly is the use of protein-dissolving enzymes extracted from media obtained through mass culturing of halophilic bacteria from seawater; adding these enzymes to fish such as sardines will result in flesh being broken down relatively easily (Imada 2009).

Fig. 11.3 Growth of marine bacteria at various salt concentrations

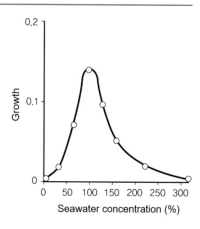

11.2.3 Pressure-Resistent and Barophilic Microorganisms

Seawater pressure increases by 1 atm for every additional 10 m of depth. This translates into pressure of 100 atm at deep-sea depths of 1000 m.

An open empty cola bottle will shatter from the pressure at a depth of around 4000 m. Because they are not filled with air like the cola bottle, microorganisms will not immediately rupture at such depths, but their propagation typically halts as pressure increases, resulting in death at high pressure levels. Why do microorganisms die when subjected to pressure? Most bacterial enzymes are vulnerable to pressure. Enzymes extracted from shallow-sea marine bacteria will stop functioning at pressures of around 100 atm. Yet some enzymes are capable of withstanding pressures of 1000 atm. This difference in resistance to pressure is explained by differences in the sizes and structures of enzyme protein molecules. In general, large enzymes and enzymes made up of several different protein molecules tend to be vulnerable to pressure. Changes in the cell membrane under pressure prevent them from taking in nutrients, causing the cell to die. The ability of certain bacteria to adapt to deep-sea environments and survive under high pressures is thus believed to be the result of special enzymes.

The deep ocean is an extreme environment with low temperatures and high pressure. Microorganisms that inhabit this environment are typically barophilic, although some are pressure-resistant as well. Barophilic bacteria are microorganisms that generally exhibit high metabolic activity and fast growth at pressures of 400 atm rather than 1 atm.

Enzymes are made up of proteins, and their activity will be diminished or halted under pressure due to changes in those proteins' three-dimensional structure. Some of the enzymes produced by microorganisms that inhabit depths of over 1000 m in

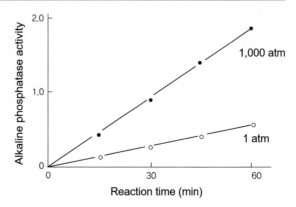

Fig. 11.4 Activity of alkaline phosphatase, an enzyme produced by marine bacteria separated from seawater at a depth of 1000 m

Table 11.2 Effects of pressure on the growth of actinomycetes isolated from sea water at depths of 5 and 1000 m

Pressure (atm)	Growth (μg/mL)	
	5 m	1000 m
1	3.5 ± 0.2 (100)[a]	3.0 ± 0.7 (100)
100	1.0 ± 0.7 (29)	2.2 ± 0.6 (73)
200	0.8 ± 0.5 (23)	1.7 ± 0.7 (57)
300	0.7 ± 0.7 (20)	1.0 ± 1.2 (33)

[a]Relative growth (%)

the ocean exhibit higher activity under higher pressure. While the mechanism of this resistance to pressure is not fully understood, it is believed to result from differences in the arrangement or three-dimensional structure of amino acids relative to ordinary enzymes.

As shown in Fig. 11.4, activity in the DNA-dephosphorylating enzyme known as alkaline phosphatase, which is produced by barophilic marine bacteria isolated from deep-sea water, was observed at 1 and 1000 atm, but activity was around three times higher at 1000 atm than at 1 atm.

Table 11.2 shows the results of an examination of pressure effects on growth in actinomycetes isolated from sediment at depths of 5 and 1000 m. In contrast with the 5 m actinomycetes, those separated at a depth of 1000 m showed little growth inhibition under pressure (Imada 2009).

11.2.4   Hyperthermophilic Microorganisms

In 1977, ecological researchers studied the area around a new heat source found emitting hot liquid during a sea bottom investigation by divers at a depth of 2600 m in the Pacific Ocean near the Equator. This marked the discovery of hyperthermophiles—curious organisms that were found inhabiting the environment. Hydrothermal vents on the ocean floor emit water temperatures of 200–380 ° C and a flow rate of 1–2 m/s. This water contains large amount of hydrogen sulfide (H_2S), hydrogen (H_2), methane (CH_4), ammonia (NH_3), sulfate ions (SO_4^{2-}), nitrogen dioxide (NO_2), iron ions (Fe^{2+}), and manganese ions (Mn^{2+}) (Fig. 11.5) (Stetter, Fiala et al. 1990; Holden et al. 1998).

Samples extracted near the vents isolated sulfur-oxidizing bacteria such as *Thiobacillus* sp., *Thiomicrospira* sp., and *Thiothrit* sp. The bacteria fixed CO_2 as an energy source to oxidize hydrogen sulfide (H_2S) and thiosulfate ions ($S_2O_3^{2-}$)

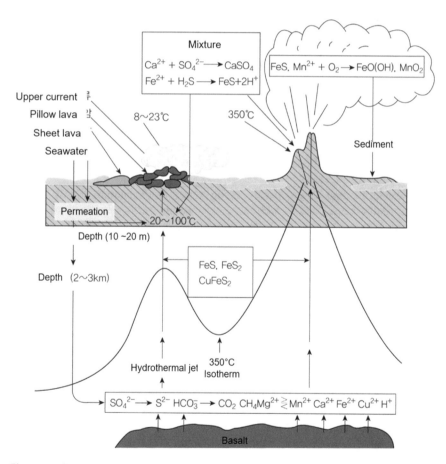

Fig. 11.5 Geochemical reaction process near ocean-floor hydrothermal vents

(Fig. 11.4). Also isolated were diagnostic bacteria, hydrogen-oxidizing bacteria, iron- and manganese-oxidizing bacteria, and methane-using bacteria. These special bacteria are expected to prove useful in the future. Tube worms, shellfish, shrimp, sea anemones, and other marine organisms were also found inhabiting the vicinity of hydrothermal vents.

11.3   Bioactive Substances from Marine Microorganisms

Among marine microorganisms, bacteria and fungi are very important resources in the hunt for future medicines and pesticides and the leading compounds for them. Table 11.3 shows the number of new compounds isolated from marine bacteria and fungi and the number of related reports. One important point concerning bioactive substances from marine microorganisms is the production of secondary metabolites from symbiosis or coexistence between microorganisms and other organisms (Kobayashi et al. 1992).

While compounds in marine-derived actinomycetes and filamentous fungi are often found to have similar structures to their land-based counterparts, the likelihood of discovering new compounds with special structures is higher than for land-based fungi. This is believed to be due to changes in the secondary metabolic system to adapt to a different living environment from land, including its salinity, pressure, and temperatures.

Research attempts are currently being made at developing highly marine environment-adapted bacteria and fungi, or those present in specific marine environments, into resources with greater usage value. Antimicrobial activity testing is relatively simple, and active explorations are under way on antimicrobial substances in marine bacteria and fungi. Tables 11.4 and 11.5 show lists of new

Table 11.3 Chronology of new compounds from marine bacteria and fungi and related reports

Year		~86	87	88	89	90	91	92	93	94	95
Bacteria	No. of reports	7	1	0	4	0	4	5	6	11	10
	No. of compounds	15	1	0	11	0	5	10	10	15	22
Fungi	No. of reports	1	1	1	4	1	4	2	2	6	7
	No. of compounds	1	1	1	12	4	6	5	3	19	18

Year		96	97	98	99	00	01	02	03	Total
Bacteria	No. of reports	7	14	10	10	10	12	6	16	133
	No. of compounds	12	31	16	25	22	26	15	38	274
Fungi	No. of reports	12	7	20	14	17	18	29	23	169
	No. of compounds	24	20	44	31	48	34	79	54	404

Table 11.4   New antibacterial, antifungal, and antiviral substances isolated from marine bacteria

Compound	Bacterium responsible	Source
(1) Antibacterial substances		
Albyssomicin B–D	*Verrucosispora* sp.	Sea mud
Andrimid, Noiramide A–C	*Pseudomonas fluorescens*	Sea squirt
Aplasmomycin A–C	*Streptomyces griseus*	Sea mud
Propylore	*Alcaligenes faecalis*	Mollusks
Bioxalomycins	*Streptomuces* sp.	Sea mud
Bogorol A	*Bacillus laterosporus*	Annelids
Bonactin	*Streptomyces* sp.	Sea mud
Chalcomycin B	*Streptomyces* sp.	Sea mud
Diazepinomicin	*Micromonospora* sp.	Sea squirt
2,4-Dibromo-6-chlorophenol	*Pseudoalteromonas luteoviolacea*	Seaweed
Himalomycin A·B	*Streptomyces* sp.	Sea mud
Istamycin A·B	*Streptomyces tenjimariensis*	Sea mud
Kahakamide A·B	*Nocardiopsis dassonicillei*	Sea mud
Korormicins	*Pseudoalteromonas* sp.	Seaweed
Loloatin A–D	*Bacillus sp.*	Annelids
Lornemide A·D	*Actinomycete*	Shells
Macrolactin G–M	*Bacillus* sp.	Seaweed
Maduralide	*Maduromycete*	Sea mud
Magnesidins	*Vibrio gazogenes*	Sea mud
Marinones	*Actinomycete*	Sea mud
Massetolide A–H	*Pseudomonas* sp.	Seaweed
Pentabromopseudilin	*Pseudomonas bromoutilis*	Sea grass
Quinolinol	*Pseudomonas* sp.	Seawater
Thiomarinol A–G	*Alteromonas rava*	Seawater
Trisindoline	*Vibrio* sp.	Sponges
Urachimycin A·B	*Streptomyces* sp.	Sponges
Wailupemycins	*Streptomyces* sp.	Sea mud
Diketopiperazine, Phenazine	*Bacillus cereus*	Sponges
α-Furan	*Pseudomonas* sp.	Sponges
Propylore	*Chromovacterium* sp.	Seawater
Diketopiperazine	*Pseudomonas aeruginosa*	Sponges
Phenazine	*Streptomyces* sp.	Sea mud
(2) Antifungal substances		
Basiliskamide A·B	*Bacillus laterosporus*	Annelids
Haliangicins	*Haliangium ochraceum*	Seaweed
Halolitoralin A–C	*Halobacillus litoralis*	Sea mud
		Sea mud
(3) Antiviral substances		
Caprolactin A·B	Gram-positive bacteria	Sea mud
Macrolactin A–F	Gram-positive bacteria	Sea mud

Table 11.5 New antibacterial, antifungal, and antiviral substances isolated from marine fungi

Compound	Bacterium responsible	Source
(1) Antibacterial substances		
Acetyl Sumiki's acid	*Cladosporium herbarum*	Sponge
Ascochital	*Kirschsteiniothelia maritima*	Wood
Aspergillitine	*Aspergillus versicolor*	Sponges
Auranticin A·B	*Preussia aurantiaca*	Sea mud
Exophilin A	*Exophiala pisciphila*	Sponges
Guisinol	*Emericella unguisu*	Jellyfish
Isocyclocitrinols	*Penicillium citrinum*	Sponges
Lunatin	*Curvularia lunata*	Sponges
Modiolide A·B	*Paraphaeosphaeria* sp.	Shells
Pestalone	*Pestalotia* sp.	Seaweed
Unguisin A·B	*Emericella unguis*	Jellyfish/mollusks
Varixanthone	*Emericella variecolor*	Sponges
Chloroasperlactones	*Aspergillus ostianus*	Sponges
(2) Antifungal substances		
Cladospolide D	*Cladosporium* sp.	Sponges
Dihydrocolletodiols	*Varicosporina ramulosa*	Seaweed
Fumiquinazoline H·I	*Acremonium* sp.	Sea squirts
Keisslone	*Keissleriella* sp.	Sea mud
Mactanamide	*Aspergillus* sp.	Seaweed
Microsphaeropsisin	*Microsphaeropsis* sp.	Sponges
Phomopsidin	*Phomopsis* sp.	Crabs
Stachybotrin A·B	*Stachybotrys* sp.	Wood
Xestodecalactone B	*Penicillium* cf. *montanense*	Sponges
Yanuthones	*Aspergillus niger*	Sea squirts
Zopfiellamide A·B	*Zopfiella latipes*	Sea mud
(3) Antiviral substances		
Halovir A–E	*Scytalidium* sp.	Sea grass
Sansalvamide A	*Fusarium* sp.	
(4) Antimalarial substances		
Aigialomycin D	*Aigialus parvus*	Crabs
Ascosalipyrrolidinone A	*Ascochyta salicorniae*	Seaweed
Drechslerine E–G	*Drechslera dematioidea*	Seaweed
(5) Antimicroalgal substances		
Bipolal	*Bipolaris* sp.	Leaves
Exumolide A·B	*Scytalidium* sp.	Plants
Halymecin A	*Fusarium* sp.	Seaweed

antibacterial, antifungal, and antiviral substances respectively isolated from marine bacteria and fungi.

An active search is also under way for antitumor substances in marine bacteria and fungi. As growth inhibition activity testing can be conducted relatively easily with cultured tumor cells, metabolites from marine microorganisms are also being actively screened. The number of new antitumor substances found to date in marine bacteria and algae is even greater than the number of antimicrobial substances. Tables 11.6 and 11.7 provide a list of new antitumor substances respectively isolated from marine bacteria and fungi.

Tables 11.8 and 11.9 list anti-inflammatory and other bioactive substances isolated from marine bacteria and fungi, respectively. In addition to anti-inflammatory substances, searches for enzyme inhibitors and antioxidants have also been reported (Namikoshi 2005).

Table 11.6 New antitumor substances isolated from marine bacteria

Compound	Bacterium responsible	Source
Abratubolactam C	*Streptomyes* sp.	Mollusks
Agrochelin	*Agrobacterium* sp.	Sea squirts
Altemicidin	*Streptomyces sioyaensis*	Sea mud
Alterramide A	*Alteromonas* sp.	Sponges
Aureoverticillactam	*Streptomyces aureoverticillatus*	Sea mud
Bisucaberin	*Alteromonas haloplanktis*	Sea mud
Chandrananimycin A–C	*Actinomadura* sp.	Sea mud
Cyclomarin A–C	*Spreptomyces* sp.	Sea mud
-δ-Indomycinone	*Spreptomyces* sp.	Sea mud
-γ-Indomycinone	*Spreptomyces* sp.	Sea mud
Halichoblelide	*Spreptomyces hygroscopicus*	Fish
Halichomycin	*Spreptomyces hygroscopicus*	Fish
Halobacillin	*Bacillus* sp.	Sea mud
Homocereulide	*Bacillus cereus*	Mollusks
Lagunapyrone A–C	*Actinomycete*	Sea mud
Lomaiviticin A·B	*Micromonospora lamaivitiensis*	Sea squirts
Neomarinones, Marinones	*Streptomyces* sp.	Sea mud
Octalctin A·B	*Pelagiobacter* sp.	Coral
Pelagiomicin A–C	*Salinospora* sp.	Seaweed
Salinosporamide A	*Micromonospora marina*	Sea mud
Thiocoraline	*Streptomyces* sp.	Coral
Caprolactones	*Micromonospora* sp.	Sea mud
Staurosporins	*Streptomyces* sp.	Sponges
Indoles		Invertebrates

Table 11.7 New antitumor substances isolated from marine fungi

Compound	Bacterium responsible	Source
Acetophthalidin	*Penicillium* sp.	Sea mud
Asperazine	*Aspergillus niger*	Sponges
Aspergillamide A·B	*Aspergillus* sp.	Sea mud
Aspergillicin A–E	*Aspergillus carneus*	Sea mud
Brocaenol A·B	*Penicillium brocae*	Sponges
Communesin A·B	*Penicillium* sp.	Seaweed
Communesin C–D	*Penicillium* sp.	Sponges
Cyclotryprostatin A–D	*Aspergillus fumigatus*	Sea mud
Dankasterone	*Gymnascella dankaliensis*	Sponges
Evariquinone	*Emericella variecolor*	Sponges
Fellutamide A·B	*Penicillium fellutanum*	Fish
Fumiquinazoline A–G	*Aspergillus fumigatus*	Fish
Fusaperazine A	*Fusarium chlamydosporum*	Sponges
Gymnastatin A–E	*Gymnascella dankaliensis*	Sponges
Gymnasterone A·B	*Gymnascella dankaliensis*	Sponges
Harzialactone B	*Trichoderma harzianum*	Sponges
Herbarin A·B	*Cladosporium herbarum*	Sponges
Insulicolide A	*Aspergillus insulicola*	Seaweed
Kasarin	*Hyphomycetes* sp.	Coral
Leptosin A–S	*Leptosphaeria* sp	Seaweed
Macrosphelide	*Periconia byssoides*	Sea urchins
N-Methylsansalvamide	*Fusarium* sp.	Seaweed
Penochalasin A–H	*Penicillium* sp.	Seaweed
Penostatin A–I	*Penicillium* sp.	Seaweed
Pericosine A·B	*Periconia byssoides*	Sea urchins
Pyrenocine E	*Penicillium waksmanii*	Seaweed
Sansalvamide A	*Fusarium* sp.	Sea grass
Scytalidamide A·B	*Scytalidium* sp.	Seaweed
Spirotryprostatin A·B	*Aspergillus fumigatus*	Sea mud
Trichodenone A–C	*Trichoderma harzianum*	Sponges
Trichodermamide B	*Trichoderma virens*	Seaweed
Tryprostatin A·B	*Aspergillus fumigatus*	Sea mud
Varitriol	*Emericella variecolor*	Sponges
Virescenoside M–U	*Acremonium striatisporum*	Sea cucumbers
Anserinones	*Penicillium corylophilum*	Sea mud
Phosphorohydrorazide thioate	*Lignincola laevis*	Sea grass
Trichothecenes	*Myrothecium verrucaria*	Sponges
Verticillins	*Penicillium* sp.	Seaweed

Table 11.8 New anti-inflammatory and other bioactive substances isolated from marine bacteria (Williams 2009)

Compound	Bacterium responsible	Source
(1) Anti-inflammatory substances		
Cyclomarin A–C	*Streptomyces* sp.	Sea mud
Lobophorin A·B	*Actinomycete*	Brown algae
Salinamide A–E	*Streptomyces* sp.	Crabs
(2) Enzyme inhibitors		
B-90063	*Blastobacter* sp.	Seawater
Flavocristamide A·B	*Flavobacterium* sp.	Shells
Pyrostatin A·B	*Streptomyces* sp.	Sea mud
(4) Other		
Aburatubolactam A	*Streptomyces* sp.	Mollusks
Anthranilamid	*Streptomyces* sp.	Sea mud
Komodoquinone A·B	*Streptomyces* sp.	Sea mud

Table 11.9 New anti-inflammatory and other bioactive substances isolated from marine fungi

Compound	Bacterium responsible	Source
(1) Anti-inflammatory substances		
Oxepinamide A	*Acremonium* sp.	Sea urchins
Phomactin A–G	*Phoma* sp.	Crabs
(2) Enzyme inhibitors		
Cathestatin C	*Microascus longirostris*	Sponges
Chlorogentisylquinone	*Phoma* sp.	Sea sand
Epolactaene	*Penicillium* sp.	Sea mud
Nafuredin	*Aspergillus niger*	Sponges
Phenochalasin A·B	*Phomopsis* sp.	Sponges
Roselipins	*Gliocladium roseum*	Seaweed
Sculezonone A·B	*Penicillium* sp.	Shells
Ulocladol	*Ulocladium botrytis*	Sponges
Xyloketal A	*Xylaria* sp.	Crabs
Betaenone	*Microsphaeropsis* sp.	Sponges
(3) Antioxidents		
Anomalin A	*Wardomyces anomalus*	Seaweed
Dihydroxyisoechinulin A	*Aspergillus* sp.	Seaweed
Epicoccone	*Epicoccum* sp.	Seaweed
Golmaenone	*Aspergillus* sp.	Seaweed
Parasitenone	*Aspergillus parasiticus*	Seaweed
Hydroquinone	*Acremonium* cf. *roseogriseum*	Sponges

(continued)

Table 11.9 (continued)

Compound	Bacterium responsible	Source
(4) Other		
Aspermytin A	*Aspergillus* sp.	Shells
Obionin A	*Leptosphaeria obiones*	Sea grass
Paecilospirone	*Paecilomyces* sp.	Seawater
Terreusinone	*Aspergillus terreus*	Seaweed

Methods studied in *E. coli*, *Bacillus subtilis*, and yeast can be applied in genetic recombination for marine microorganisms. For example, plasmids isolated from marine photosynthetic bacteria can be used as vectors for the cloning of fish growth hormone genes, enabling their mass production.

The placement of marine bacteria responsible for plant hormones on the surface of seaweed has been reported not only to influence growth and differentiation, but also to help recover heavy metals and produce ultrafine magnetite particles. Magnetotactic bacteria produce ultrafine magnetite particles covered in organic thin-film within their bodies. Other microorganisms produce energy in seawater, including methane-producing bacteria and hydrogen-producing photosynthetic bacteria. Reports show that these microorganisms can be fixed in highly polymerized gel for serial bioreactor production of methane and hydrogen (Tarman et al. 2013).

11.4   Development of Biocatalysts from Microorganisms

As biocatalysts, enzymes promote various biochemical reactions within cells, helping to sustain life activities. Enzymes exhibit strong activity under normal temperature and pressure and neutral pH conditions and are highly substrate-specific. These characteristics are strong advantageous in the use of enzymes as catalyzing elements in industrial bioprocesses. Since low-purity substrates can be used in enzyme reactions without need for any special pressurizing or heating equipment, costs can be saved in terms of reaction devices, operation energy, and reaction resources (Böttcher et al. 2004).

Long before they were ever identified, human beings had used enzymes (or enzyme-producing micoorganisms) for casting and the making of bread and cheese. The range and amount of uses has expanded markedly as biotechnology has advanced in recent years. An enzyme market established in the U.S. grew in scale from roughly US$400,000 in 1997 to US$100 million by 2011, and the uses of enzymes have expanded in research, pharmaceuticals, test reagents, food processing, starch processing, sugar manufacturing, detergents, fibers, paper making, brewing, and the livestock and dairy industries (Zhao et al. 2002; Heux et al. 2015).

Table 11.10 International enzyme classification system[a]

Class number	Class name	Type of reaction catalysed
1	Oxidoreductases	Transfer of electron (hydride ions or H atoms)
2	Transferases	Group transmission reactions
3	Hydrolases	Hydrolysis reactions (transfer of functional group to water)
4	Lyases	Addition of groups to double bonds or formation of double bonds by removal of groups
5	Isomerases	Transfer of group within molecules to yield isomeric forms
6	Ligases	Formation of C–C, C–S, C–O, and C–N bonds by condensation reactions coupled to cleavage of ATP or similar cofactor

[a]Most enzymes catalyze the transmission of electrons, atoms, or active groups. Accordingly, they are assigned names with a classification number based on transmission reaction type, active group donor type, or active group receptor type

Based on the types of reactions that they catalyze, enzymes are classified by the International Union of Biochemistry and Molecular Biology (UBMB) into ① oxidoreductases, ② transferases, ③ hydrolases, ④ lyases, ⑤ isomerases, and ⑥ ligases. As shown in Table 11.10, enzymes in each category are assigned Enzyme Commission Numbers by reaction form, consisting of "EC" followed by a four-digit number.

For example, α-amylase, which catalyzes starch hydrolysis, has the number EC 3.2.1.1. The initial "3" in the four-digit number indicates that the enzyme catalyzes a hydrolysis reaction. The "2" shows that it operates on glycosyl compounds, the first "1" indicates that it operates on O-glycosyk linkages, and the second "1" means that it cuts 1,4-α-D-glucan (starch) glycoside linkages through internal decomposition. Enzymes representing over 5000 types of reactions have been registered with Enzyme Commission Numbers in the database to date, and the number continues to grow each year (The Comprehensive Enzyme Information System: http://www.brenda-enzymes.org/).

Many different environments exist on earth, and the organisms that inhabit different environments have enzymes with differing characteristics due to environmental adaptations. Over the years, this has led to searches for enzymes with new characteristics in organisms living in different environments. Marine environments are likewise diverse, ranging from tropical to polar waters, shallow waters, deep-sea environments, and hydrothermal vents on the ocean floor. Amazing new enzymes have been sought in the microorganisms, seaweeds, invertebrates, and vertebrates living there. Many of the enzymes obtained from there have great industry value.

This section will introduce marine organism-derived enzymes that are found in deep-sea microorganisms, soil on the ocean floor, and the metagenomes of sponges. Most of the enzymes here are polysaccharide lyases. This is because growing recent interest in the production of functional oligasaccharides from seaweed polysaccharides and the saccharification of seaweed biomass has led to active research in related enzymes.

11.4.1   Enzymes from Deep-Sea Microorganisms

The oceans that cover around 70% of the earth's surface have an average depth of around 3800 m, such that most of the marine environment consists of deep-sea environments. At a depth of 3800 m, the oceans are a low-temperature (2–4 °C) and high-pressure (38 MPa) environment. Some localized environments, such as the areas around hydrothermal vents, also contain methane and hydrogen sulfide at temperatures of over 300 °C (Ojima 2011).

While these marine environments are thought of as strict habitat conditions for organisms, they are inhabited by a truly great number of arthropods, echinoderms, annelids, and microorganisms, indicating great potential for the discovery of enzymes with special, environmental adaption-related characteristics in organisms that inhabit deep-sea environments.

11.4.2   Heat-Resistance Agarase

One of the chief components of agar, agarose is a recalcitrant polysaccharide with a polymerized structure consisting of β-D-galactose and 3,6-anhydro-α-L-galactose. It is found in red algae such as Gelidiales, Gigartinales, and Ceramiales. Within agarose molecules, these constituent polysaccharides form agarobiose units (4-O-β-D-dalactopyranosyl-3,6-anhydro-L-galactopyranose) or neoagarobiose units (3-O-α-3,6-anhydro-L-galactopyranosyl-D-galactopyranose), where agarose polymers have structures linking these together (Pickett et al. 1981).

Agarose hydrolase is an enzyme that hydrolyzes the β-1-4 or α-1,6 linkage in agarose, producing agarooligosaccharides such as neoagarobiose or agarobiose. Agarose hydrolase that severs the beta-1,4 linkage to form neoagarooligosaccharides is called β-agarose hydrolase (EC 3.2.1.81), while the type that severs the α-1,3 linkage to form agarooligosaccharides is known as α-agarose hydrolase (EC 3.2.1.158).

Agarose hydrolase production is rare in land-based microorganisms, but relatively common in microorganisms that inhabit sea mud. The sea floor microorganisms that produce agarose hydrolase are believed to use agar from settling red algae detritus by breaking it down into agarose.

The neoagarooligosaccharides and agarooligosaccharides produced through enzymatic breakdown of agarose exhibit a range of bioactivity in human beings, including antitumor, antioxidation, immunostimulation, hydration, and whitening effects. For this reason, efforts are currently under way to find agarose hydrolases that can be used to efficiently produce these bioactive oligosaccharides.

Agarose possesses a structure in which agarobiose units—consisting of β-D-galactose (β-D-Gal) and (3,6) anhydro-α-L-galactose (α-L-Gal) connected by

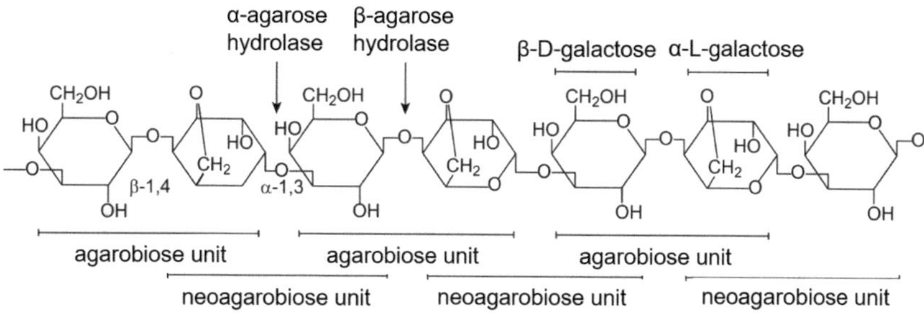

Fig. 11.6 Structure of agarose and active sites of where agarose hydrolases

β-1,4 linkages—are themselves connected by α-1,3 linkages. α-agarose hydrolases sever the α-1,3 linkage to produce agarooligosaccharides, while β-agarose hydrolases sever the β-1,4 linkage to produce neoagarooligosaccharides (Fig. 11.6) (Lahaye and Rochas 1991).

Recently, Japanese scientists succeeded in producing a recombinant version of the agarose hydrolase agar A, cloning genes for the β-agarose hydrolase from JAMB-A94 (a *Microbulbifer* species found in sea mud at depths of 2460 m) and using a bacillus expression system. The enzyme belongs to glycoside hydrolase family 16 (GHF16), with a molecular weight of approximately 46,000, an optimal pH of 7.0, and an optimal temperature of 55 °C. The chief product of agarose breakdown was neoagarotetraose. Agar A is a heat-resistant enzyme that does not diminish in activity when heated for 15 min at 60 °C. Its activity also did not diminish in the presence of 100 mM ethylene diamine tetraacetic acid (EDTA) and 30 mM sodium dodecylsulfate (SDS).

Because the enzyme exhibits a high level of agarose breakdown activity at high temperatures (40 °C) where agarose does not solidify, it may be used for DNA extraction through the heating and dissolving of gel following agarose electrophoresis. DNA extraction kits using the enzyme are already being marketed. The gene *aga A7* from Aga A7, a heat-resistant β-agarose hydrolase similar to aga A, can also be obtained from the A7 strain of the same genus (Ohta et al. 2004). The gene *aga O* from Aga O, a β-agarose hydrolase (GHF86) forming chiefly neoagarohexose from the breakdown of agarose, has similarly been obtained from bacteria in the same genus. As β-agarose hydrolases mainly produce tetrose sugars, Aga O is a rare example of the enzyme that mainly produces hexose sugars.

The β-agarose hydrolase gene Aga A11 has also been obtained from bacteria of the genus *Agarivorans* isolated from sea mud at a depth of 4152 m in Japanese trench. The main product of recombinant Aga A11 was neoagarobiose. Since neoagarobiose has whitening properties, this enzyme is useful in manufacturing disaccharides with whitening effects.

A gene for α-agarose hydrolase (Aga A33) has been obtained from JAMB-33 strain, a bacterium of the *Thalassomonas* genus isolated from sea mud from a depth of 230 m. Recombinant agarose A33 produced with a bacillus host resulted mainly in agaroterose through clipping of the α-1,3 linkage in agarose. Porphyran, an agaroheteropolysaccharide present in red algae, was previously known to have antioxidant properties, but the agarose hydrolase A33 amplified those properties by breaking it down. Moreover, because antioxidant activity is not amplified when porphyran is broken down with β-agarose hydrolase, this property is unique to α-agarose hydrolase.

As this shows, agarose hydrolases with various properties are being obtained from bacteria isolated from deep-sea waters or soil. These include many with diverse and exceptional properties, including high heat tolerance and the formation of specific reaction products, which may prove useful for industrial purposes in the future (Fu and Kim 2010).

11.4.3   Heat-Resistant Cellulase

Pyrococcus horikoshii, a hyperthermophilic obligate anaerobe from the Archaea domain that grows under anaerobic conditions at temperatures of 100 °C and greater, has been isolated from hydrothermal deposits off the coast of Japan. Analysis of this microorganism's genome by the National Institute of Technology and Evaluation resulted in cloning of a gene for EGPh, an endo-type cellulase with a molecular weight of 42,000 that belongs to GHF5. EGPh is produced in *E. coli*, but a differential scanning calorimeter (DSC) analysis showed a denaturing temperature of 96 °C. Its optimal temperature for activity was found to be near 100 °C, and it did not lose activity when heated up to 95 °C, indicating that it is a heat-resistant cellulose (Zhang et al. 2014; Kawarabayasi et al. 1998; Ando et al. 2002).

Additionally, EGPh has proven degradable not only with the soluble cellulase substrate carboxymethylcellulose (CMC), but also with avicel, an insoluble substrate that is difficult to degrade because of its high crystallinity. Protein engineering research was subsequently performed on EGPh's structure and functions, and important amino acid residues and local structures are being surmised for the enzyme's functions.

Cellulose saccharification with cellulase is applicable in various processes for the improvement of products through the breakdown of cellulose chains, including pulp bleaching and fiber cleaning. As a heat-resistant cellulase, EgPh is used as an enzyme for cotton fiber processing, which takes place at around 70 °C, and has proven effective at improving fiber flexibility and color in jeans. It may also be useful in the future saccharification of cellulose biomass and the production of cellooligosaccharides. The enzyme can be mass-produced through an expression system with *Brevibacillus brevis* as host.

11.4.4   Cold-Adapted Enzymes

For the most part, deep-sea waters are a low-temperature environment in the range of 2–4 °C. Even at these low temperatures, microorganisms are known to live at similar densities to those found in warm waters (10^5–10^6 cell/mL) (Zeng 2006).

In adapting to low-temperature environments, microorganisms have shown a number of physiological and biochemical changes. For this reason, they are believed to possess enzymes with various properties related to low temperature adaptations. Indeed, many psychrophilic or psychrotrophic bacteria inhabiting low-temperature environments possess cold-adapted enzymes that exhibit high levels of activity even at low temperatures (Siddiqui and Cavicchioli 2006; Marx et al. 2007).

Cold-adapted enzymes are generally known to exhibit higher levels of specific activity at temperatures of 5–15 °C than the enzymes in microorganisms that inhabit mesophilic regions. The specific activity of cold-adapted subtilisin at 10 °C, for example, is roughly five times greater than that of ordinary subtilisin, while the specific activity of cold-adapted amylase is around three times greater than that of ordinary amylase (D'Amico et al. 2003). The reason for these higher levels of specific activity stems from the fact that the rate of decrease in energy for reaction activity by cold-adapted enzymes is greater than is the case for regular enzymes (Smalås et al. 2000).

Additionally, cold-adapted enzymes exhibit lower heat stability than ordinary enzymes, which means that their optimum temperature is also lower than that of ordinary enzymes (10–20 °C lower for many enzymes). The use of cold-adapted enzymes presents several advantages over regular enzymes. Perhaps the greatest of these is that they exhibit high levels of activity even at low temperatures, which is beneficial when they are operating on a substrate that is unstable with respect to heat. Smaller quantities of enzymes can also be used than would be the case with ordinary enzymes. In addition, while cold-adapted enzymes are unstable with respect to heat, this actually presents an advantage in terms of being able to halt a reaction easily through a short period of heat treatment. Since all activity in the cold-adapted enzymes can be halted with brief heating, this allows for a decrease in thermal changes in reaction products and their degradation through residual activity.

To date, cold-adapted proteases, lipases, amylases, and cellulases have been developed as cleaning supplements. These enzymes can be used to obtain strong cleaning supplement effects even at low washing temperatures. Cold-adapted enzymes are used in foods, including protease to tenderize meat, β-galactosidase to break down lactose in milk, and pectinase for juice extraction and transparency. These forms of food processing must be performed at low temperatures. Cold-adapted enzymes are also suitable for uses in which enzymes must be completely deactivated with heating after a reaction, including the ligases and kinases used in gene manipulation experiments.

Cold-adapted enzymes are likely to be used in various low-temperature bioprocesses in the future. This is likely to entail the use of various source organisms from low-temperature environments such as deep-sea and polar waters.

11.4.5 Alginate Lyase Functioning Under Alkaline Conditions

Alkaline enzymes are enzymes with an optimum pH in the alkaline range above 9. Most of them are produced outside the body by alkaliphilic microorganisms. Many proteases and cellulases are known to be alkaline enzymes. Bacteria from the genus *Agarivorans* isolated from deep-sea mud have been found to produce the alginate lyase A1m, which functions under alkaline conditions. Based on protein electrophoresis (sodium dodecyl sulfate polyacrylamide gel, or SDS-PAGES), A1m is predicted to have a mass of 31 kDa. Its optimum pH is 10, and its activity was found to increase by 1.8 times with the addition of 0.2 M NaCl (Hattori et al. 1997).

Alginate lyases are found in bacteria, fungi, brown algae, mollusks, and viruses. Most are semi-alkaline enzymes with an optimal pH in the weak base range. A1m is thus regarded as an alginate lyase with optimum conditions at the highest end of the pH range. Because solubility of the alginate substrate increases sharply in the highly alkaline range, an alkaline alginate lyase exhibiting strong activity under these conditions can be viewed as well suited to alginate breakdown.

Other alginate lyases found in mollusks are useful for the manufacturing of various bioactive alginic acid oligosaccharides or the use of alginic acid in saccharification and biomass. It is expected that these will be put to many effective uses in the future with seaweed biomass.

A number of marine microorganisms contain alginate lyases, most notably bacteria from the *Alteromonas*, *Pseudoalteromonas*, and *Vibrio* genera. These are believed to play the role of final decomposer of the alginic acid containing in brown algae detritus.

11.4.6 Metagenomically Derived Esterase

Ninety-nine percent of microorganisms on the planet are regarded as "viable but nonculturable" (VBNC) bacteria due to the difficulty of isolating and culturing them. The term metagenome refers to VBNC bacterial DNA in its mixed state when directly extracted from its environment. This can be developed into a library (metagenomic library) allowing for metagenomics analysis through the reading of its base sequence. Next-generation DNA base sequence measurement equipment can be used to produce large volumes of base sequence information in a short time, while advancements in bioinformatics technology and high-throughput activity detection technology have facilitated the hunt for new enzyme genes.

Metagenomic analysis has been performed on a number of marine organisms to date, including VBNC bacteria present in seawater and sea mud and sponges and corals containing large amounts of symbiotic microorganisms. This has led to the discovery of numerous enzyme genes. This section will focus on the esterases (EC 3.1.1.1) obtained from the metagenomic libraries of sponges and sea mud (Chu et al. 2008; Steele et al. 2009).

A. Esterases from Sponge Metagenomes

Many antibacterials and bioactive substances with anticancer and other properties have bene found in sponges. Most of them are produced by the microorganisms living in a symbiotic relationship within the sponges. Since the majority of these symbiotes are VBNC bacteria, a metagenomics analysis of sponges is believed to be a useful avenue for identifying these substances' biosynthetic pathways. New enzyme genes may also be identified from the symbiotes through the metagenomics analysis of sponges (Okamura et al. 2010).

Recently, genes for EstHE1, an esterase with a molecular weight of around 25,000, were isolated from the metagenomic library for the Japanese sponge *Hytios erecta* (Okamura et al. 2010). A homology search showed that the enzyme belonged to the SGNH hydrolase superfamily, consisting of serine (Ser), glycine (Gly), asparagine (Asn), and histidine (His) as catalyzing amino acid residues. Recombinant EstHE1 expressed in *E. coli* hydrolyzed fatty acid ρ-nitrophenyl esters with two to four carbon atoms, but not fatty acid esters with a greater number of carbon atoms. From this, it was determined to be an esterase that breaks down short-chain fatty acid esters, rather than a lipase that breaks down fatty acid esters with 10 carbon atoms or more. In addition, it had an optimal temperature of 40 °C, and while it showed 50% or more of maximum activity at a range of 25–55 °C, activity dropped to 58% when cultured for 12 h at 40 °C. This indicates that is not a thermophilic enzyme, but a mesophilic enzyme that operates in a normal temperature range (Selvin et al. 2012).

At the same time, the enzyme tolerance to high concentrations of salt: activity dropped to 55% up to an NaCl concentration of 1.9 M, but slowly increased at higher concentrations, eventually reaching 62% at 3.8 M. This is thought to relate to the sponge's living environment, namely seawater containing approximately 0.6 M NaCl. This property of EstHE1 is something that had not been obtained in previous genetic engineering improvements of esterases, showing that metagenomics analysis is an effect means of obtaining genes for enzymes with new characteristics

B. Esterase from Deep-Sea Mud Metagenomes

Recently, a recombinant EstF enzyme was produced by cloning the esterase gene *estF* from a South China Sea deep-sea mud metagenomics library and using *E. coli* as an expression system (Fu et al. 2011). The enzyme showed its strongest decomposition performance with *p*-nitrophenyl butyrate (C4), but did not work on fatty acid esters with 10 or more carbon atoms. EstF is a cold-adapted enzyme that exhibits high levels of activity even at low temperatures. For example, it has shown roughly 20% of its maximum activity (50 °C) even at a reaction temperature of 0 ° C. This property shows the enzyme to a product of adaptation to the low-temperature deep-sea environment (Schmeisser et al. 2007).

EstF is also an alkaline enzyme that exhibits maximum activity at a pH of 9.0, which drops sharply at a pH below 7. Its catalytic site consists of 331 residues, and

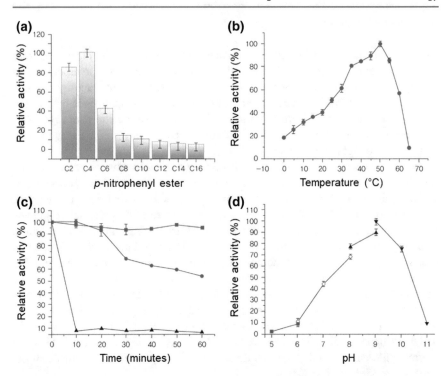

Fig. 11.7 Biochemical characteristics of recombinant EstF. **a** <u>Determination</u> of substrate specificity for refine recombinant EstF. C2: *p*-nitrophenyl acetic acid, C4: *p*-nitrophenyl butyric acid, C6: *p*-nitrophenyl caproic acid, C8: *p*-nitrophenyl octanoate, C10: *p*-nitrophenyl capric acid, C12: *p*-nitrophenyl lauric acid, C14: *p*-nitrophenyl myristic acid, C16: *p*-nitrophenyl palmitic acid; **b** optimal temperature for recombinant EstF (*p*-nitrophenyl butyric acid (C4) selected as substrate); **c** heat stability of recombinant EstF: residual specific activity for EstF esterase following culture at different temperatures [40 °C (filled square-filled square), 50 °C (filled circle-filled circle), 60 °C (filled triangle-filled triangle)]; **d** optimal pH for recombinant EstF (*p*-nitrophenyl butyric acid (C4) selected as substrate)

it is a high-novelty enzyme with a primary structure quite different from that of previously known bacterial esterases (Fig. 11.7).

Esterases are useful enzymes for industry, as with their selective decomposition of position isomers in acyl ester compounds. EstHE1 and EstF are regarded as leading examples in the search for enzymes from marine metagenomes. As more efficient screening techniques develop for different enzymes in the future, an even greater variety of enzyme genes may yet be discovered in marine-derived metagenomes (Gao et al. 2016).

11.5   Enzyme Inhibitors Produced by Marine Microorganisms

11.5.1   Protease Inhibitors

Various enzymes must perform roles within the body for the organism to survive. When those activities are too strong or too weak, the homeostasis that maintains the body in a certain state cannot be sustained.

Enzyme inhibitors are substances that bind with enzymes and block their activities. They play an important role in maintaining enzyme activity at normal levels.

Protease inhibitors are among the most widely studied enzyme inhibitors, with a broad range of applications in the treatment of various diseases caused by irregular protease activity.

These enzyme inhibitors are important not only in the analysis of enzyme reaction mechanisms, but also in the fields of medicine, pharmacology, and agricultural and fishery science. Accordingly, they have been the subject of a great deal of exploratory research to date, principally in land-based organisms.

The protease inhibitors isolated to date can be broadly divided into polymers such as proteins and low-molecular weight organic compounds. The former are found principally in plants and animals, while the latter have been derived from actinomycetes and other land-based micoorganisms, but rarely from marine microorganisms.

Recent years have seen growing attention to the isolation and characteristics of new enzyme inhibitors from marine microorganisms, with potential future applications in the fields of medicine, pharmacology, and agricultural and fishery science (Paterson et al. 2000; Imada 2009).

A. The Searching Process for Protease Inhibitor-Producing Bacteria

A marine microorganism's ability to produce metabolites such as antibiotics or enzyme inhibitors is known to vary drastically with culturing conditions. This requires changes in seawater concentrations, salt concentrations, pH, carbon and nitrogen sources, culturing time, and other conditions to find the optimal conditions for inhibitor production.

To locate protease inhibitor-producing bacteria, marine bacteria are inoculated into a seawater medium with added casein and cultured for several days at 27 °C while protease solution is sprinkled on the medium surface. When left for several hours, the casein proteins are broken down by the action of the protease, causing the medium to become transparent. In the area around the inhibitor-producing bacteria colony, the protease activity is diminished by the inhibitor that has been produced, leaving an opaque halo of undegraded casein. This provides an easy means of distinguishing bacteria that do and do not produce inhibitors (Fig. 11.7; the inhibitor-producing bacteria are at the center).

This method can be used to easily identify protease inhibitor-producing marine bacteria.

Table 11.11 Protease inhibitor properties

Property	Marinostatins [a]C-1, [b]C-2	Monastatin
Composition	Simple peptides	Glycoproteins
Molecular weight	1418, 1644	$\sim 20{,}000$
pH stability	4–8	2–12
Types of protease inhibited	Serine protease (not including trypsin)	Cystine protease (including protease for bacteria causing fish diseases)

[a]C-1: Phe-Ala-Thr-Met-Arg-Tyr-Pro-Ser-Asp-Ser-Asp-Glu
[b]C-2: Gln-Pro-Phe-Ala-Thr-Met-Arg-Tyr-Pro-Ser-Asp-Ser-Asp-Glu

From the protease inhibitor-producing strains obtained with this approach, those exhibiting high levels of inhibition activity and production stability can be selected for analysis of taxanomic character, and the chemical structures for the isolated and refined inhibitor substances can be identified.

As shown in Table 11.11, the B-10-31 strain of *Pseudoalteromonas sagamiensis*, a Gram-negative bacillus that uses a single polar flagellum to move, was found to produce two types of protease inhibitors: a glycoprotein with completely different characteristics known as monastatin, and a simple, low-molecular weight peptide inhibitor without any sugar known as marinostatin.

These inhibitors exhibited great differences in aspects ranging from molecular weight to pH stability and types of protease inhibited.

The lack of reported research to date of microorganisms such as these that produce two completely different types of inhibitors gives an inhibition of the diversity of physiological functioning in marine bacteria.

In some cases, protease in land-based animals ranging from higher organisms to microorganisms exhibit homology in amino acid sequences near the inhibition reaction site. From this, it can be determined that the inhibitors evolved from common ancestry, and that the inhibition reaction site has remained well preserved over the long evolutionary process (Table 11.12).

In contrast, marinostatin from marine bacteria does not exhibit the same kind of homology with the protease inhibitors found in land-based animals, suggesting that marine organisms have undergone a different evolutionary process from those animals.

B. Industrial Applications of Protease Inhibitors

This section introduces some of the known uses for protease inhibitors, including areas with potential industry applications.

1. **Medicinal applications**: Low-molecular weight inhibitors are currently being isolated and refined from various microorganism culture media to elucidate their role in regulating reactions and potentially apply them to medicinal treatments.

Table 11.12 Comparison of amino acid sequences near inhibition reaction sites for various protease inhibitors

Origin	Protease inhibitor	Amino acid sequence near reaction site p4 p3 p2 p1 p1′ p2′ p3 p4′
Microorganisms	Marinostatin	-Phe-Ala-Thr-Met[a]-Arg-Tyr-Pro-Ser-
	Plasminostreptin	-Ala-Cys-Thr-Lys[a]-Gln-Phe-Asp-Pro-
	S-SI	-Met-Cys-Pro-Met[a]-Val-Tyr-Asp-Pro-
Plants	Soybean	-Ala-Cys-Thr-Lys[a]-Ser-Asn-Pro-Pro-
	Lima bean	-Leu-Ser-Thr-Lys[a]-Ser-Ile-Pro-Pro-
	Fava bean	-Met-Cys-Thr-Arg[a]-Ser-Met-Pro-Gly
Animals	Leech	-Val-Cys-Thr-Lys[a]-Glu-Leu-His-Arg-
	Egg white	-Leu-Cys-Thr-Lys[a]-Asp-Phe-Ser-Phe-
	Pig semen	-Phe-Cys-Thr-Arg[a]-Gln-Met-Asn-Pro-
	Cow pancreatic juice	-Gly-Cys-Pro-Arg[a]-Ile-Thr-Asn-Pro-
	Pig pancreatic juice	-Gly-Cys-Pro-Lys[a]-Ile-Thr-Asn-Pro-
	Sheep pancreatic juice	-Gly-Cys-Pro-Arg[a]-Ile-Thr-Asn-Pro-
	Human pancreatic juice	-Gly-Cys-Thr-Lys[a]-Ile-Thr-Asn-Pro-

[a]Inhibition reaction site

These functions include reducing pain from burns and inhibiting blistering and inflammation of the skin. Immune system strengthening and inflammation prevention effects have also been observed (Haard 1992; Haddar et al. 2009; Rao et al. 2009).

2. **Agricultural applications**: While it does not involve marine bacteria, mushroom farming is an example of proteases being applied in agriculture.

   Artificial mushrooms such as *Pleurotus ostreatus* are known to produce acidic protease during farming. Excess production of this enzyme can result in autodigestion, preventing the mushroom from going further. By adding the protease inhibitor S-PI (isolated from land-based actinomycetes) that serves to inhibit the enzyme to the soil in which they are farmed, artificial mushrooms with profuse fruit bodies can be produced, as shown in Fig. 11.9.

3. Applications for the removal of false enzymes during isolation and purification: Isolation and refinement of protease from organism extract or microorganism culture medium can sometimes be confounded by autodigestion of enzyme proteins. One method that can be used in such cases is affinity chromatography, with the enzyme's inhibitor as a ligand.

The crude enzyme solution used at industrial scales sometimes contains proteases other than the target enzyme. These proteases can act upon the target enzyme and cause unforeseen damage. By adding inhibitors, it is possible to stop the protease from acting and achieve the target.

11.5.2   Amylase Inhibitors Produced by Marine Actinomycetes

A. The Search for Amylase Inhibitor-Producing Bacteria

Amylase inhibitors are relatively rare substances. Because they block amylase activity and specifically inhibit the digestion and absorption of starch, they have potential applications in antiobesity (diet) and other medications. Several inhibitors have been isolated and purified to date from land-based plants and actinomycetes.

Amylase inhibitors are known to fall into two major categories: plant-derived high-molecular weight protein substances and actinomycetes-derived low-molecular weight glycoside substances. To date, however, little research has reported the production of these substances from marine microorganisms.

In a study of amylase inhibitor productivity for around 5000 strains of marine microorganisms isolated from seawater, Imada found actinomycetes in sea-floor sediment that produced such an inhibitor. As seen in Fig. 11.10, when grown in an agar medium with starch, these actinomycetes inhibit the activity of amalyse added after culturing, resulting in a purple halo (<u>urea</u>-starch reaction) due to production of the inhibitor around the colony. As a result, producing and non-producing bacteria can be easily distinguished.

Figure 11.8 shows the average physical changes observed in around ten healthy male and female subjects who consumed 10 mg of the wheat-derived amylase inhibitor A1 with every meal for a month, with no constraints on eating or physical activity. Clear weight decreases were observed for both men and women with A1 construction, suggesting possible applications in antiobesity medications (Murao et al. 1977; Imada 2009).

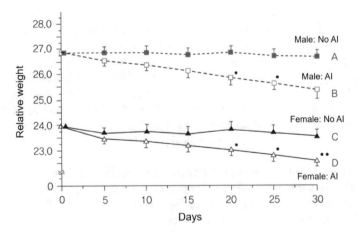

Fig. 11.8 Changes in body weight for human subjects consuming a wheat-derived amylase inhibitor (AI) with an exercise regimen

B. *N*-acetylglucoamidase Inhibitors

N-acetylglucoamidase is an enzyme that separates *N*-acetylglucosamine from the sugar chains in glycoproteins and glycolipids present on the cell surface. Rising levels of activity in the urine, however, indicate disorders in the kidney's renal tubules. Its activity is also known to increase in the blood of patients with diabetes, leukemia, and other forms of cancer.

Because they inhibit the activity of this enzyme, *N*-acetylglucoamidase inhibitors have potential treatment applications.

Recently, *N*-acetylglucoamidase inhibitor-producing strains have been sought through isolation of marine actinomycetes. The *N*-acetylglucoamidase inhibitors pyrostatin A and B have been isolated from these marine actinomycetes and their structures identified (Fig. 11.9).

11.5.3 Tyrosinase Inhibitors Produced by Marine Filamentous Bacteria

Tyrosinase inhibitors are metallic enzymes with copper at the center of their activity. They are a form of quinone-oxidizing enzyme found across a wide range of animals, plants, and microorganisms (Fig. 11.10).

These enzymes work by catalyzing oxidation of the essential amino acid tyrosine into dopa (3,4-dihydroxy phenylalanine) and dopa quinone. In dermatology, tyrosinase is involved in the early stages of melanin production, indicating its importance as an enzyme for melanin formation (Fig. 11.11).

Human skin is a membrane that covers the organism, consisting of an epidermis and dermis. Epidermal cells include keratinocytes and melanocytes. Following cell division, keratinocytes differentiate into four distinct layers: the stratum basale, the stratum spinosum, the granular layer, and the stratum corneum (Fig. 11.12) (Mayer and Hamann 2005).

The process of epidermal basal cells developing into their final granular layer (epidermal turnover) takes approximately four weeks.

At the same time, melanocytes do not move from the basal layer; the melanin that is formed is degraded by surrounding keratinocytes, which participate in

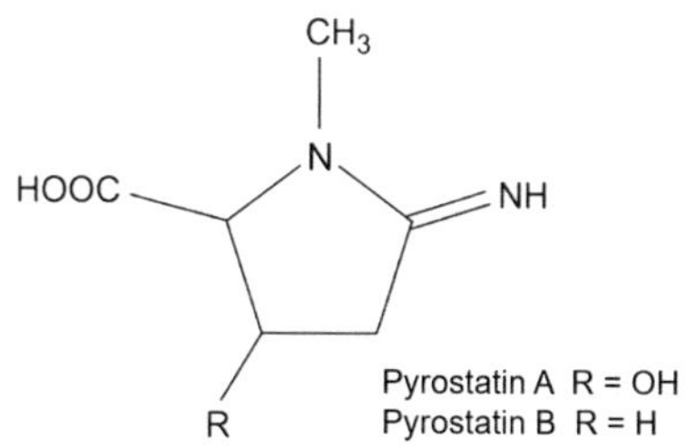

Fig. 11.9 Chemical structures of the *N*-acetylglucoamidase inhibitors pyrostatin A and B

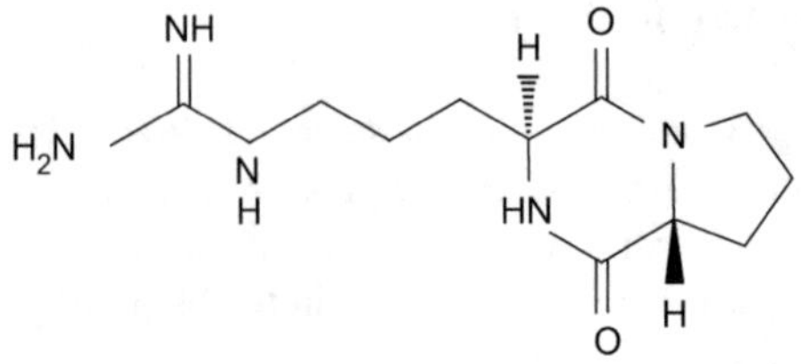

Fig. 11.10 Chemical structure of the tyrosinase inhibitor CI-4

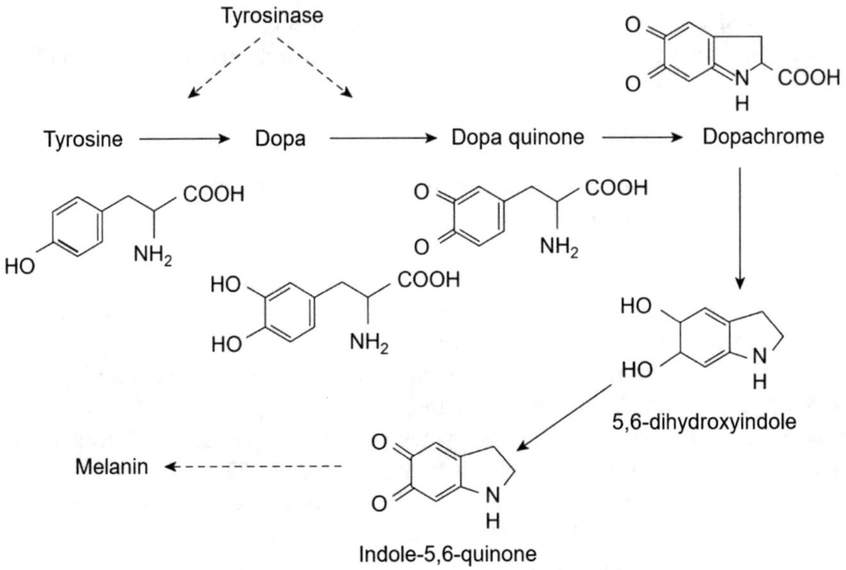

Fig. 11.11 Pathway of melanin synthesis from tyrosine

epidermal turnover and transmit them into the granular layer. Melanin therefore does not accumulate in the epidermis as long as a healthy epidermal turnover is sustained. In human skin, however, a form of pigmentation commonly known as "black spots" occurs with age. This pigmentation happens as excess melanin produced by melanocytes in the epidermis's basal layer remains within the epidermis due to a variety of factors.

Since they inhibit melanin formation, the tyrosinase inhibitors that inhibit tyrosinase activity are being applied in the field of whitening cosmetics. Recent years have seen a marked tendency toward the isolation and use of natural ingredient for whitening cosmetics, and there has been a demand for natural materials as sources for whitening agents.

Arbutin is an ingredient found in bearberries from the family Ericaceae, while 4-butylresorcinol is found in Siberian firs. Ellagic acid is a polyphenol found in abundance in red raspberries from the family Rosaceae. Ascorbic acid (vitamin C)

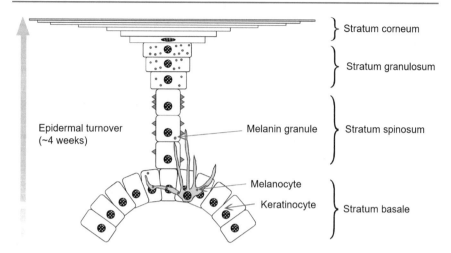

Fig. 11.12 Diagram of the human epidermis

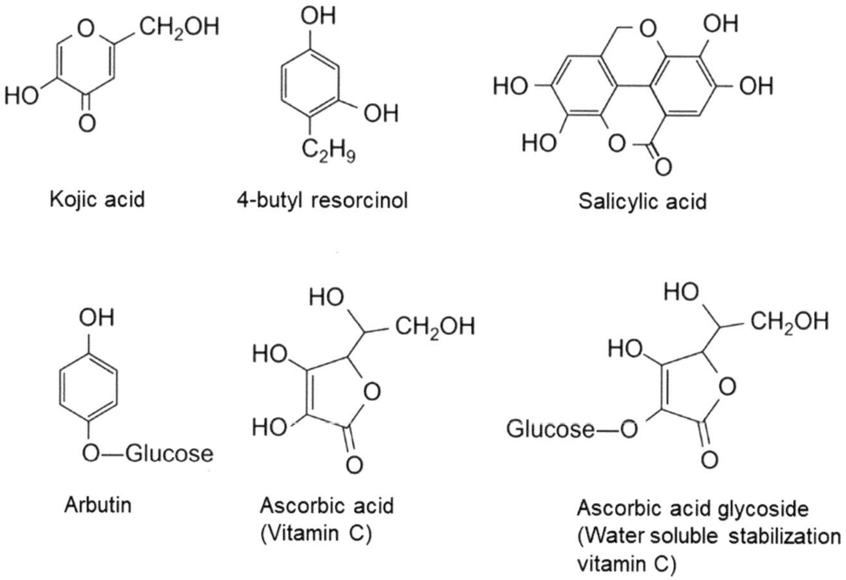

Kojic acid

4-butyl resorcinol

Salicylic acid

Arbutin

Ascorbic acid
(Vitamin C)

Ascorbic acid glycoside
(Water soluble stabilization
vitamin C)

Fig. 11.13 Notable effective ingredients in whitening cosmetics

is an antioxidant present in fruit; due to problems with decreased stability during the mixture process for addition to cosmetics, attempts have recently been made to develop various stabilization derivatives such as ascorbic acid glycosides (Fig. 11.13).

The best known of the microorganism-derived whitening agents is kojic acid, which is produced by land-based filamentous fungi from the *Aspergillus* and *Pencillum* genera. In addition to inhibiting tyrosinase activity, kojic acid also blocks melanin formation and is currently being developed as a new whitening agent. While it can be mixed with new whitening cosmetics, however, stability issues have resulted at one point in a ban on such mixtures. Although its effects are outstanding, it has been unable to show them.

This has led to a search for marine microorganisms to take the place of kojic acid in producing new tyrosinase inhibitors.

Imada isolated the H1-7 strain of the filamentous fungus *Trichoderma viride* from sea sediment at a depth of 100 m. The tyrosinase inhibitor 5-hydroxy-3-isocyano-5-vinylcyclopent-2-enone was isolated from a culture medium for this fungus.

Melanin formation suppression effects were examined for an Arbutin control group and culture supernatant for the H1-7 strain of *T. viride* using melanocytes from normal human subjects. As seen in Fig. 11.1, melanocytes cultures with the addition of culture supernatant to a final concentration of 10% were found to exhibit melanin formation suppression effects equal to or greater than those of Arbutin, suggesting potential applications as a new whitening agent.

Perhaps the future may yet usher in an age when new bioactive substances isolated from marine microorganisms can be used in the treatment of currently intractable disease (Imada 2009).

11.6 Chapter Summary and Conclusion

The oceans have a total volume of 1,369,000,000 km^3, accounting for over 98% of the hydrosphere. They are also very complex microbiologically, with many types of microorganisms inhibiting environments of differing pressure, salinity, and temperature. Having developed metabolic processes and physiological capabilities to adapt within extreme environments, marine microorganisms are very likely to contain metabolites that are quite different from those present in land-based microorganisms.

In terms of the physical and chemical properties of marine and land-based organisms, the marine environment consists of water and thus requires the production of large amounts of chemicals for smooth information exchange among organisms.

In addition, because the absorption of solar energy is confined to the marine environment's surface layer and protein-dominant biota possesses dominance over carbohydrates, the food chain is highly complex. The result is a very diverse mixture of epibionts and symbionts. Marine microorganisms can therefore be seen as an unexplored treasure trove of industrially valuable compounds.

Marine bacteria possess outstanding capabilities in producing high-molecular weight substances and enzymes. Additionally, some species are used as potential producers of enzymes such as deoxyribonuclease, lipase, alginate lyase, protease, agarase, cellulase, and esterase. As previously noted, numerous bioactive substances have been found in marine microorganisms, including those with antimicrobial, antifungal, antiviral, antitumor, anti-inflammatory, antioxidant, and enzyme-inhibitors. While difficulties with mass-producing source materials create problems in the use of marine plants and animals, marine microorganisms are beneficial from a usage standpoint in that they can be cultured.

References

Ando, S., Ishida, H., Kosugi, Y., & Ishikawa, K. (2002). Hyperthermostable Endoglucanase from Pyrococcus horikoshii. *Applied and environmental microbiology, 68*(1), 430–433

Arrigo, K. R. (2004). Marine microorganisms and global nutrient cycles. *Nature, 437*(7057), 349.

Baharum, S., Beng, E., & Mokhtar, M. (2010). Marine microorganisms: Potential application and challenges. *Journal of Biological Sciences, 10*(6), 555–564.

Böttcher, H., Soltmann, U., Mertig, M., & Pompe, W. (2004). Biocers: Ceramics with incorporated microorganisms for biocatalytic, biosorptive and functional materials development. *Journal of Materials Chemistry, 14*(14), 2176–2188.

Chu, X., He, H., Guo, C., & Sun, B. (2008). Identification of two novel esterases from a marine metagenomic library derived from South China Sea. *Applied Microbiology and Biotechnology, 80*(4), 615–625.

D'Amico, S., Gerday, C., & Feller, G. (2003). Temperature adaptation of proteins: engineering mesophilic-like activity and stability in a cold-adapted α-amylase. *Journal of molecular biology, 332*(5), 981–988

D'Amico, S., Collins, T., Marx, J. C., Feller, G., & Gerday, C. (2006). Psychrophilic microorganisms: challenges for life. *EMBO Reports, 7*(4), 385–389.

Fu, X. T., & Kim, S. M. (2010). Agarase: review of major sources, categories, purification method, enzyme characteristics and applications. *Marine Drugs, 8*(1), 200–218

Fu, C., Hu, Y., Xie, F., Guo, H., Ashforth, E. J., Polyak, S. W., Zhu, B., & Zhang, L. (2011). Molecular cloning and characterization of a new cold-active esterase from a deep-sea metagenomic library. *Applied microbiology and biotechnology, 90*(3), 961–970

Gao, W., Wu, K., Chen, L., Fan, H., Zhao, Z., Gao, B., et al. (2016). A novel esterase from a marine mud metagenomic library for biocatalytic synthesis of short-chain flavor esters. *Microbial Cell Factories, 15*(1), 41.

Giovannoni, S. J., Britschgi, T. B., Moyer, C. L., & Field, K. G. (1990). Genetic diversity in Sargasso Sea bacterioplankton. *Nature, 345*(6270), 60–63

Gounot, A.-M. (1986). Psychrophilic and psychrotrophic microorganisms. *Experientia, 42*(11–12), 1192–1197.

Haard, N. F. (1992). A review of proteotlytic enzymes from marine organisms and their application in the food industry. *Journal of Aquatic Food Product Technology, 1*(1), 17–35.

Haddar, A., Agrebi, R., Bougatef, A., Hmidet, N., Sellami-Kamoun, A., & Nasri, M. (2009). Two detergent stable alkaline serine-proteases from *Bacillus mojavensis* A21: Purification, characterization and potential application as a laundry detergent additive. *Bioresource Technology, 100*(13), 3366–3373.

Hattori, M., Ogino, A., Nakai, H., & Takahashi, K. (1997). Functional improvement of β-lactoglobulin by conjugating with alginate lyase-lysate. *Journal of Agricultural and Food Chemistry, 45*(3), 703–708.

Heux, S., Meynial-Salles, I., O'Donohue, M., & Dumon, C. (2015). White biotechnology: State of the art strategies for the development of biocatalysts for biorefining. *Biotechnology Advances, 33*(8), 1653–1670.

Holden, J. F., Summit, M., & Baross, J. A. (1998). Thermophilic and hyperthermophilic microorganisms in 3–30 °C hydrothermal fluids following a deep-sea volcanic eruption. *FEMS Microbiology Ecology, 25*(1), 33–41.

Horikoshi, K. (1998). Barophiles: Deep-sea microorganisms adapted to an extreme environment. *Current Opinion in Microbiology, 1*(3), 291–295.

Imada, C. (2009), *Utilization of Marine Microbiology,* (pp. 25–36). Tokyo, Japan: Seizando-Shoton Publishing Co.

Imada, C. (2009): Protease inhibitors produced by marine microorganisms. *Utilization of marine organisms* (pp. 50–68). Japan: Seizando-Shoten Publishing Co.

Inniss, W. E. (1975). Interaction of temperature and psychrophilic microorganisms. *Annual reviews in microbiology, 29*(1), 445–466.

Kawarabayasi, Y. (1998). Complete sequence and gene organization of the genome of a hyper-thermophilic archaebacterium, pyrococcus horikoshii OT3. *DNA Research, 5*(2), 55–76

Kim, S. K. & Dewapriya, P. (2013). Anticancer potentials of marine-derived fungal metabolites, In S. K. Kim (Ed.), *Marine microbiology* (pp. 237–243). Weinheim, Germany: Wiley

Kobayashi, J. I., Ishibashi, M., & Shigemori, H. (1992). Bioactive substances from marine microorganisms. *Journal of synthetic organic chemistry, 50*(9), 772–785. Japan.

Kokare, C., Mahadik, K., Kadam, S., & Chopade, B. (2004). Isolation, characterization and antimicrobial activity of marine halophilic *Actinopolyspora* species AH1 from the west coast of India. *Current Science, 86*(4), 593–597.

Lahaye, M., & Rochas, C. (1991). Chemical structure and physico-chemical properties of agar. In *International workshop on gelidium.* New York: Springer.

Larkin, J., & Stokes, J. (1968). Growth of psychrophilic microorganisms at subzero temperatures. *Canadian journal of microbiology, 14*(2), 97–101.

Margesin, R., & Miteva, V. (2011). Diversity and ecology of psychrophilic microorganisms. *Research in microbiology, 162*(3), 346–361.

Margesin, R., & Schinner, F. (2001). Potential of halotolerant and halophilic microorganisms for biotechnology. *Extremophiles, 5*(2), 73–83.

Marx, J-C., Collins, T., D'Amico, S., Feller, G., & Gerday, C.(2007) Cold-adapted enzymes from marine antarctic microorganisms. *Marine biotechnology, 9*(3), 293–304

Mayer, A. M., & Hamann, M. T. (2005). Marine pharmacology in 2001–2002: Marine compounds with anthelmintic, antibacterial, anticoagulant, antidiabetic, antifungal, anti-inflammatory, antimalarial, antiplatelet, antiprotozoal, antituberculosis, and antiviral activities; affecting the cardiovascular, immune and nervous systems and other miscellaneous mechanisms of action. *Comparative Biochemistry and Physiology Part C: Toxicology & Pharmacology, 140*(3–4), 265–286.

Michener, H. D., & Elliott, R. P. (1964). Minimum growth temperatures for food-poisoning, fecal-indicator, and psychrophilic microorganisms. *Advances in Food Research, 13,* 349–396. (Elsevier).

Munn, C. (2011). *Marine microbiology.* Garland Science.

Murao, S., Ohyama, K., & Ogura, S. (1977). Isolation of amylase inhibitor-producing microorganism. *Agricultural and biological chemistry, 41*(6), 919–924.

Namikoshi, M. (2005). Medical materials and reagents of research, In N. Fusetani (Ed.), *Biotechnological applications of marine natural substances,* (pp. 8–39). Tokyo, Japan: CMC Publishing Co.

Ohta, Y., Hatada, Y., Nogi, Y., Li, Z., Akita, M., Hidaka, Y., Goda, S., Ito, S., Horikoshi, K., & Miyazaki, M. (2004). Enzymatic properties and nucleotide and amino acid sequences of a thermostable β-agarase from a novel species of deep-sea Microbulbifer. *Applied microbiology and biotechnology, 64*(4), 505–514

Okamura, Y., Kimura, T., Yokouchi, H., Meneses-Osorio, M., Katoh, M., Matsunaga, T., et al. (2010). Isolation and characterization of a GDSL esterase from the metagenome of a marine sponge-associated bacteria. *Marine biotechnology, 12*(4), 395–402.

Okamura, Y., Kimura, T., Yokouchi, H., Meneses-Osorio, M., Katoh, M., Matsunaga, T., & Takeyama, H. (2010). Isolation and characterization of a GDSL esterase from the metagenome of a marine sponge-associated bacteria. *Marine biotechnology, 12*(4), 395–402

Pandey, P., & Purushothaman, C. (2006). Marine microbial biotechnology and aquaculture—An overview. *Microbial Biotechnology in Agriculture and Aquaculture, 2,* 319.

Paterson, D. L., Swindells, S., Mohr, J., Brester, M., Vergis, E. N., Squier, C., et al. (2000). Adherence to protease inhibitor therapy and outcomes in patients with HIV infection. *Annals of Internal Medicine, 133*(1), 21–30.

Pickett, C., Jeter, R., Morin, J., & Lu, A. (1981). Electroimmunochemical quantitation of cytochrome P-450, cytochrome P-448, and epoxide hydrolase in rat liver microsomes. *Journal of Biological Chemistry, 256*(16), 8815–8820.

Rao, C. S., Sathish, T., Ravichandra, P., & Prakasham, R. S. (2009). Characterization of thermo-and detergent stable serine protease from isolated *Bacillus circulans* and evaluation of eco-friendly applications. *Process Biochemistry, 44*(3), 262–268.

Roszak, D., & Colwell, R. (1987). Survival strategies of bacteria in the natural environment. *Microbiological Reviews, 51*(3), 365.

Schmeisser, C., Steele, H., & Streit, W. R. (2007). Metagenomics, biotechnology with non-culturable microbes. *Applied microbiology and biotechnology, 75*(5), 955–962.

Selvin, J., Kennedy, J., Lejon, D. P., Kiran, G. S., & Dobson, A. D. (2012). Isolation identification and biochemical characterization of a novel halo-tolerant lipase from the metagenome of the marine sponge *Haliclona simulans. Microbial Cell Factories, 11*(1), 72.

Siddiqui, K. S., & Cavicchioli, R. (2006). Cold-adapted enzymes. *Annual Review of Biochemistry, 75,* 403–433.

Smalås, A., Leiros, H., Os, V., & Willassen, N. (2000). Cold adapted enzymes. *Biotechnology Annual Review, 6,* 1–57.

Steele, H. L., Jaeger, K. E., Daniel, R., Streit, W. R. (2009). Advances in recovery of novel biocatalysts from metagenomes. *Journal of molecular microbiology and biotechnology, 16*(1-2), 25–37.

Stetter, K. O., Fiala, G., Huber, G., Huber, R., & Segerer, A. (1990). Hyperthermophilic microorganisms. *FEMS Microbiology Letters, 75*(2–3), 117–124.

Tarman, K., Lindequist U., & Mundt S. (2013). Metabolites of marine microorganisms and their pharmacological activities, In S. K. Kim (Ed.), *Marine microbiology: bioactive compounds and biotechnological applications,* (pp. 393–416). Weinheim, Germany: Wiley.

Ojima, T. (2011): Biocatalyst, In N. Fusetani, (Ed.), *Trends in marine biotechnology* (pp. 207–223). Tokyo, Japan: CMC Publishing Co.

Williams, P. G. (2009). Panning for chemical gold: Marine bacteria as a source of new therapeutics. *Trends in Biotechnology, 27*(1), 45–52.

Zeng, R., Xiong, P., & Wen, J. (2006). Characterization and gene cloning of a cold-active cellulase from a deep-sea psychrotrophic bacterium Pseudoalteromonas sp. DY3. *Extremophiles, 10*(1), 79–82

Zhang, J., Yue, L., Kong, Q., Liu, Z., Zhou, X., Zhang, C., et al. (2014). Sustainable, heat-resistant and flame-retardant cellulose-based composite separator for high-performance lithium ion battery. *Scientific Reports, 4,* 3935.

Zhao, H., Chockalingam, K., & Chen, Z. (2002). Directed evolution of enzymes and pathways for industrial biocatalysis. *Current Opinion in Biotechnology, 13*(2), 104–110.

Correction to: Essentials of Marine Biotechnology

Correction to:
Chapter 6 in: S.-K. Kim, *Essentials of Marine Biotechnology*,
https://doi.org/10.1007/978-3-030-20944-5_6

In the original version of the book, belated author corrections have been incorporated and Figure 6.14 has been updated. The current version of the book has been updated with the changes.

The updated versions of these chapters can be found at
https://doi.org/10.1007/978-3-030-20944-5_6

© Springer Nature Switzerland AG 2021
S.-K. Kim, *Essentials of Marine Biotechnology*,
https://doi.org/10.1007/978-3-030-20944-5_12

Glossary

Accessory pigment A colored molecule that absorbs light energy and transfers it to a reaction center of chlorophyll for use in photosynthesis

Acritarch The general term for unicellular fossils whose relationship are uncertain; many are regarded as resistant cyst stages of planktonic algae, often prasinophyceans or dinoflagellates.

Actinomycetes Group of gram-positive, sporeforming, prokaryotic microorganisms belonging to the bacteria, which grow as slender branched filaments (hyphae). They are found in soil, river muds and lake bottoms and include many species (e.g. *Streptomyces*) that produce antibiotics.

Active oxygen Strongly oxidizing oxygen that functions to maintain life by generating energy within the body to sustain the organism through respiration.

Adrenal cortical hormone Category of steroidal hormones secreted from the adrenal cortex.

Adrenocorticotropic hormone, adrenocorticotropin (ACTH) Polypeptide hormone synthesized by anterior pituitary and which acts on adrenal cortex, stimulating its growth and the synthesis and release of adrenocortical steroids. Also has effects on other tissues, stimulating lipid breakdown and release of fatty acids from fat cells.

Adventitious bud Bud that forms in a place where buds typically do not form, including on leaves and roots or between stem joints.

Adventive embryo Embryo that forms from plant somatic cells through the same morphological changes as in fertilization.

Agar A sulfated polygalactan extracted from walls of various red algae that is used as a gelling agent.

AIDS Disease of diminished immune functioning caused by infection with the human immunodeficiency virus (HIV).

Akinete A thick-walled spore that functions in asexual reproduction, frequently serving as a resistant stage that undergoes a period of dormancy.

Albino An animal in which pigment formation is absent or a plant lacking chromophores. This mutation results in a genetic lack of ability to synthesize pigments.

Algaenans Decay-resistant polymers of unbranched hydrocarbons present in the cell walls of some algae.

Algal bloom (bloom) Visible growth of planktonic algae, often associated with nutrient-enriched waters.

Alginic acids Polysaccharides (a mixture of mannuronic and guluronic acids) extracted from the walls of brown algae for industrial applications; may occur in the salt form (alginates or algin).

Aliasing Sampling problem resulting from the too-infrequent collection of samples.

Alkaloid Any of a group of nitrogen containing organic bases found in plants and which are toxic or physiologically active in vertebrates, such as caffeine, morphine, nicotine, strychnine.

Alkylating agents Highly reactive organic chemicals (e.g. mustard gas) that attach alkyl groups to bases in DNA, and to other molecules. They are potent mutagens as well as causing damage to tissues.

Allele Genes corresponding to allelomorphic characteristics that are located on corresponding sites on homologous chromosomes (i.e., corresponding gene loci). Multiple allelomorphs have groups of corresponding genes (multiple alleles). The relationships between alleles arising from mutation and regular (wild-type) genes may be categorized as follows: (1) amorphs, in which the mutant gene does not possess any of the normal gene's trait expression functions; (2) hypomorphs, in which the mutant gene's functioning is not qualitatively different from the normal gene's, but is quantitatively lesser; (3) antimorphs, in which the mutant gene functions in a direction opposite that of the normal gene; and (4) neomorphs, in which mutation results in functions entirely unrelated to those of the wild-type gene. The various relationships among these serve as the basis for determining complete or incomplete dominance in hetero individuals, as well as traits such as gene redundancy.

Allelism test A form of complementarity testing to examine whether mutants with similar phenotypes are the results of mutation at the same gene locus. Pairs of chromosomes with each of the mutant genes are inserted in the same cell to test whether the wild-type traits are expressed. If the test shows the two mutants exhibiting a wild-type phenotype due to complementarity, it can be determined that there is no allelism, and the two are mutations at different gene loci. If the same test shows no complementarity and a mutant type, it can be determined that allelism is present and the two are mutations at the same gene locus. This is a simple means of determining whether genetic loci are the same or different.

Allelism A state in which a gene involved in particular phenotype is an allele at a specific gene locus. In cases involving neighboring genes with very similar functions (multiple gene families), it is not possible to use genetic exchange to test whether a single gene locus has different gene types (alleles) or non-alleles at different gene loci. The allelism ratio is the percentage of alleles (identical gene loci) for two genes randomly chosen from among these alleles and non-alleles. The term "lethal allelism" refers to groups of genes that exhibit lethality due to phenotype.

Allelomorph Phenotype expressed as a result of an allele; in wild types, this refers to traits governed by mutant alleles. When two or more wild-type alleles are recognized, they are referred to as multiple alleles. Wild-type traits are typically dominant and mutant traits recessive (allelomorphic character).

Alloenzymes are variant forms of an enzyme that are coded by different alleles at the same locus.

Allophycocyanin A type of phycobiliprotein produced by cyanobacteria (except chlorophyll a and b-containing taxa), glaucophytes, red algae, and cryptomonads.

Alloxanthin A xanthophyll pigment that is characteristic of photosynthetic cryptomonads.

Alternation of generation A life history type in which there are two (or more, in some red algae) multicellular stages that can be distinguished by type of reproductive cell produced and sometimes also be morphological features.

Alveolata A eukaryotic super group that includes ciliates,

Amoeba A unicellular organism that generates motions via cellular projections

Amoeboid A type of cell organization in which a wall is absent and the protoplasm undergoes rapid shape changes

Amphiesma The covering of dinoflagellate cells, which, in addition to an overlying plasma membrane, consists of membranous alveoli that may contain little or no material or may contain cellulosic thecal plates of varying thickness.

Amplified restriction fragment length polymorphism (AFLP) Technique in which PCR is used to amplify only a selected part of the genome to analyze polymorphism in the length of the products.

Androspores Flagellate cells of oedogoniales that settle on body cells near oogonia and undergo a few divisions to produce a dwarf male filament, some cells of which produce and release flagellate sperm.

Aneuploid a. Having more or less than an exact multiple of the haploid number of chromosomes or haploid gene dosage; chromosomal abnormalities that disrupt relative gene dosage, such as deletions.

Aneuploidy State in which the number of chromosomes per cell in an individual or line is not a whole-number multiple of the basic number, but includes one or more fewer than the whole-number value, indicating an incomplete genome. Cells or organisms in this state are known as aneuploids. The term dysploid is also used when the basic number is not definitely known.

Angiogenesis factor General term for endogenous factors that induce the formation of new blood vessels. Found in tissues with high metabolism rates such as tumors and membranes, they are secreted from hypoxic macrophage cells on the edges or surfaces of wounds to induce regeneration of blood vessels as part of the healing process.

Angiogenesis n. Development of new blood vessels by sprouting from preexisting vessels.

Anisogamous reproduction Sexual reproduction involving gametes that are flagellate and structurally distinguishable.

Anisogamous A type of sexual reproduction in which gametes are flagellate and structurally distinguishable

Anisogamy A type of sexual reproduction characterized by two types of gametes that differ in size or behavior.

Ankylosing spondylitis Condition of unknown cause that causes joint stiffening and transformation with degenerative inflammatory symptoms in the vertebrae, shoulder joints, and knee joints.

Anoxygenic photosynthesis A type of photosynthesis that occurs in some prokaryotes in which water is not used as a reductant and oxygen is not released.

Anoxygenic Without production of oxygen; for example, use of hydrogen sulfide rather than water as a source of photosynthetic reductant by some cyanobacteria.

Antibiotic Any of a diverse group of organic compounds produced by microorganisms which selectively inhibit the growth of or kill other microorganisms. Many antibiotics are used therapeutically against bacterial and fungal infections in humans and animals.

Anticodon Group of three consecutive bases in tRNA complementary to a codon in mRNA.

Antifreeze protein Proteins secreted in the bodily fluids of fish and arthropods inhabiting polar regions, which function to lower those fluids' freezing point.

Antigen Any substance capable of binding specifically to an antibody or a T-cell receptor. An antigen may be unable to induce a specific immune response when administered on its own, but will do so if attached to a suitable carrier

Apical growth Growth that occurs by the division of one or more cells located at the tip (apex) of a multicellular body. Apicomplexans, and dinoflagellates, characterized by membrane sacs known as alveoli at the cell periphery.

Apicoplast A non-photosynthetic plastid occurring in the cells of most genera of apicomplexans, a group of parasitic protists that are classified in the Alveolata.

Aplonospore A nonflagellate spore that has the genetic potential to produce flagella under appropriate conditions; produced by subdivision of a parental cell.

Apogamy Form of parthenogenesis in which sporophytes are formed directly outside of egg cells in higher plants without fertilization.

Apolipoprotein Protein component of a lipoprotein especially of the lipoproteins that transport lipids in the blood. Apolipoprotein genes are diverse, resulting in differences in blood lipid levels. Genetic problems are known to result in congenital lipid metabolism irregularities.

Apospory A type of apomixis in which a diploid gamete is produced from the sporophyte without spore formation.

Apospory Form of parthenogenesis in the broad sense; observed in plants.

Aromatic compounds Organic compounds with a benzene ring in their molecule (benzene derivatives).

Asphaltene High-molecular weight substance in the non-melting component of asphalt petroleum ether that dissolves in the presence of benzene or carbon disulfide.

Association Phenomenon in which two or more of the same kind of molecule form two or more bonds due to intramolecular forces, creating a loose, regular aggregate.

Astaxanthin A red carotenoid pigment (3,3'-diketo-4,4'-dihydroxy-β-carotene) produced by some green algae that is commercially useful; also known as haematochrome.

Attaching repellent compounds Substances that deter sessile marine organisms from attaching to coral reefs, artificial structures, boat hulls, and fishing nets.

Aureochrome A blue-light photoreceptor present in diverse photosynthetic stramenopiles.

Autocolony A type of asexual reproductive colony that is a miniature of the adult colony; produced by single cells of the adult.

Autopolyploid Organism having more than two sets of homologous chromosomes; polyploidy in which chromosome sets are all derived from a single species.

Autoradiography Technique by which large molecules, cell components or body organs are radioactively labelled and their image recorded on photographic film, producing an autoradiograph or autoradiogram.

Autoreproduction Ability of a gene, virus, or nucleoprotein molecule to synthesize several molecules similar to itself from smaller molecules within the cell.

Autosome Any chromosome other than sex chromosomes. Sex chromosomes, which differ from autosomes in form, traits, and behavior, are also known as heterochromosomes or allosomes. Genetic phenomena governed by autosomal genes are known as autosomal inheritance, which is distinct from sex-linked inheritance.

Autospore A type of nonflagellate spore that lacks the genetic potential to produce flagella

Autotrophic microorganisms Microorganisms that are capable of growing by synthesizing all of their cellular components through CO_2 reduction; contrasted with heterotrophic organisms, which require organic compounds made by other organisms. This distinction is based on carbon assimilation and differs in definition from lithotrophy, a classification in terms of energy acquisition in which organic matter serves as an electron donor. Autotrophic microorganisms are classified into two types according to their means of energy acquisition. Photoautotrophs, which use light, include *Chlorella* and most algae, along with purple and green sulfur bacteria. Chemoautrophs obtain energy by oxidizing an electron donor through a chemical dark reaction; based on electron donor type, these include ammonium-oxidizing bacteria (e.g., *Nitrosomonas*), nitrous acid-oxidizing bacteria (*Nitrobacter*), sulfur-oxidizing bacteria, iron-oxidizing bacteria (*Thiobacillus*), and hydrogen bacteria (*Alcaligenes*).

Autotrophy Opposite of heterotrophy. Form of nutrition in which organisms synthesize the organic compounds they need to survive by absorbing carbon dioxide, water, and various inorganic salts without requiring organic matter as a source of nutrients. In photoautotrophs such as green plants, carbon assimilation is performed through photosynthesis. In chemoautotrophs such as chemosynthetic bacteria, it occurs through chemosynthesis. Some seaweed experience strong growth in the presence of organic matter (such as vitamins) as a nutrient source, while many require such a supply.

Auxiliary cell In the higher red algae, a cell into which the zygote nucleus or one of its mitotic progeny is deposited and that generates the carposporophyte generation by mitotic proliferation.

Auxospore A cell produced by diatoms that undergoes enlargement, compensating for the reduction in size that often occurs during population growth; commonly also the zygote of diatoms.

Avirulence gene Gene found in some bacterial plant pathogens that determines their ability to cause disease on a host plant containing a corresponding resistance gene. see gene-for-gene resistance.

Avirulent A strain of bacterium, virus or other potential pathogen that does not cause disease.

Bacterivores Organisms that consume bacteria as prey

Bacterivory The process by which cells ingest and digest bacterial prey

Baeocytes (endospores) Spores of cyanobacteria formed by internal division of vegetative cells (compare with exospores)

Base pair (bp) Single pair of complementary nucleotides from opposite strands of the DNA double helix. The number of base pairs is used as a measure of length of a double-stranded DNA.

Base ratio The ratio of the bases $(A+T)/(C+G)$ in DNA, which varies widely from species to species.

β-carotene An accessory carotenoid pigment lacking oxygen that occurs in major algal groups; the source of provitamin-A.

Binary fission The relatively simple process by which prokaryotic cells divide to form two equal-sized progeny cells.

Bioassay A procedure that uses organisms and their responses to estimate the effects of physical and chemical agents in the environment.

Biodiversity Convention Treaty concluded in June 1992 with the aim of preserving biodiversity.The need to protect biodiversity was first raised at the 14th meeting of the Governing Council for the United Nations Environment Programme (UNEP) in 1987. It was later determined that existing treaties were not adequate as a response, and the decision was made to enact a new convention. While the U.S., Japan, and other nations did sign, issues with detailed provisions in the convention resulting in being effectively postponed. At a meeting of signatory nations in November 1994, a discussion was held on the issues that had led to the postponement. Three of them had a particular impact in terms of bioengineering. One concerned the development of a protocol for biostability. Opinions submitted by developing and Northern European countries requested that a protocol be developed to determine the legal regulation of bio-related areas. While an opposition campaign was led by the U.S., Japan, Canada, Australia, and New Zealand, among others, the balance of support was generally in favor of a protocol. In addition to the possibility of advanced economies demanding new regulations if such a protocol is proposed, it may also serve as a non-tariff barrier on the export of future bio products. The second issue concerns intellectual property rights in connection with technology transfer. While the provisions mentioned the transferring of technology necessary for diversity preservation, they also included language about respecting intellectual property rights. European countries and the U.S. signed on after declaring their interpretation that technology transfers would not extend as far as disregarding intellectual property rights; other countries refused outright. The focus now is shifting to the question of what kind of technology transfer organization is to be created. The third issue concerns the rights of farmers. This includes the rights of farmers in developing countries who have contributed to the preservation and usage of plant gene resources. Developing countries are demanding an international fund be created to distribute the profits earned by advanced economies. The issue was initially addressed by the Food and Agriculture Organization (FAO), but was later taken up with the Biodiversity Convention. Advanced economies are very likely to face payment obligations.

Biofuel Gas such as methane or liquid fuel such as ethanol (ethyl alcohol) made from organic waste material, usually by microbial action.

Bioinformatics Discipline and methodology involving the use of computers to analyze biological data. This approach uses methods derived from statistics, linguistics, mathematics, chemistry, and biochemistry. It often involves structural data and data on nucleic acid and protein sequences; experimental findings from the data are also analyzed. Some data are taken from patient statistics and scientific papers. Bioinformatics research is focused on storage, searching, and analysis of data.

Biological assay The use of the amount of some substance that is essential or detrimental to organismal functions besides survival and growth as an indicator for said phenomena. It is often convenient to use substances that are effective in terms of growth or expression of functions even in very small amounts (such as vitamins and several other growth factors and hormones) as direct indicators for biological effects, rather than resorting to chemical means.

Biological half-life The number of elements and radioactive isotopes within an organism's tissues and organs decreases as a result of metabolism and excretion. The half-life is the time required for the initial amount of radioactive isotopes to decrease by half through these processes.

Biological pump The process by which some of the organic materials produced by marine phytoplankton are transported to deep ocean sediments, where they remain for thousands of years

Biological soil crust The community of organism including bacteria, cyanobacteria, eukaryotic algae, fungi, lichens, and mosses that covers much of the soil surface in arid and semiarid lands.

Biological species concept Distinction of species on the basis of breeding incompatibility.

Bioluminescence Production of light by living organisms. It is caused by an enzyme catalysed biochemical reaction in which an inactive precursor is converted into a light-emitting chemical.

Biomass Total weight, volume or energy equivalent of organisms in a given area; plant materials and animal wastes used as a source of fuel or other industrial products; (3) in biotechnology, the microbial matter in the system.

Bioreactor Device designed to produce substances by taking advantage of enzymes' ability to efficiently regulate highly specific reactions at constant temperature and pressure.

Bioremediation Biotechnology-based (i.e., microorganism-based) purification of contaminants in the environment.

Biphasic alternation of generations A sexual reproductive cycle thot involves two body forms that can be distinguished by chromosome level and type of reproductive cell produced.

Bipolar A type of centric diatom having elongate shape (i.e., having two poles).

Blades A type of algal body in which cells are arranged in flat sheets

Blastomere Any one of the cells formed by the first divisions of a fertilized egg.

Blunt end End of a double-stranded DNA molecule that lacks a single strand.

Blunt-end ligation Technique used in the construction of recombinant DNAs in which any two DNA molecules may be joined.

Bootstrap value An estimate of the validity of a branch in a phylogenetic tree that is determined by the number of times the branch appears after the data are repeatedly resampled.

Bottleneck effect This occurs in populations where the number of organs experiences a dramatic decline. An example is the increase in mutations of homozygotes due to inbreeding. A decrease in the size of the population results in a large genetic drift effect, lowering the amount of genetic polymorphism observed in that population.

Brackish water Water with a moderate saline concentration between those of seawater and freshwater.

Branched filaments A type of algal body in which branches emerge from a main filamentous axis.

Branches In algae, rows of cells that extend from a main filamentous axis.

Branching enzyme Enzyme that catalyses the transfer of a segment of 1,4-α-β-glucan chain to a primary hydroxyl group in a similar chain, forming a branchpoint in a polysaccharide chain.

Breeding Process of utilizing the genetic properties of crops or livestock to produce new species with potential benefits for farming or to improve existing varieties.

Brine shrimp bioassay The use of the amount of some substance that is essential or detrimental to organismal functions besides survival and growth as an indicator for said phenomena.

Broodstock Mature fish raised or kept for breeding purposes.

Bud In plants, structure from which shoot, leaf or flower develops; incipient outgrowth, as limb buds in animal embryo, from which limbs develop.

Budding Production of buds; method of asexual reproduction common in sponges, coelenterates and some other invertebrates, in which new individuals develop as outgrowths of the parent organism, and may eventually be set free; artificial vegetative propagation by insertion of a bud within the bark of another plant; cell from the parent cell; release of certain animal viruses from the host cell by their envelopment in a piece of plasma membrane which subsequently pinches off from the cell division by the outgrowth of a new cell.

Byssus Tuft of strong filaments secreted by the byssogenous gland of certain bivalve molluscs, by which they become attached to substrate; the stalk of certain fungi.

CAAT box Conserved sequence in the promoter region of some eukaryotic genes about 70–80 base pairs upstream from the start-point of transcription, and which is involved in control of initiation of transcription.

Caccolithophorids Unicellular members of the haptophyte algae that are characterized by a covering of small, ornate calcium carbonate scales.

Cachectin A protein also known as tumour necrosis factor. Its entirely separate role as cachectin involves inhibiting enzymes of lipid utilization, disturbing lipid metabolism and causing cachexia (wasting), as in patients with some cancers.

Cadherins One of the main families of cell adhesion molecules. They are cell surface proteins that can bind an identical cadherin molecule on another cell in a calcium dependent interaction, causing cells to bind strongly to each other. The family includes the E-cadherins, N-cadherins, P-cadherins and VE-cadherins.

Callus Plant tissue that has undergone a healing process following a wound or infection with a pathogen.

Canal raphe A tubelike structure extending longitudinally along the valves of some pennate diatoms that opens externally via the raphe slit.

Captive breeding program The capturing of gravedigger beetles in their natural state to for breeding and re-release into the wild.

Carbon concentrating mechanism (CCM) One of several types of processes that result in increasing the concentration of carbon dioxide in the vicinity of rubisco.

Carbon dioxide fixation the pathway incorporating carbon dioxide into carbohydrates which occurs in the stroma of chloroplasts; any reaction in which carbon dioxide is incorporated into organic compounds.

Carbon fixation Conversion of carbon dioxide into organic carbon as a result of the light-independent reactions of photosynthesis.

Carbon sequestration The burial of calcium carbonate or organic carbon such that these materials are not readily converted back to atmospheric carbon dioxide.

Carpogonium The nonmotile female gamete of red algae.

Carpospores In red algae, the spore released from a carposporangium; usually assumed to be diploid.

Carposporophyte (gonimocarp) A multicellular diploid phase in the life cycle of florideophycean red algae that generates carpospores via gonimoblast filaments; in all cases, attached to the female gametophytic body (i.e., not free living).

Carrageenans Mucilaginous sulfated polygalatans in the cell walls of red algae that are extracted for use as gelling agents in the food industry.

Carrying capacity The number of individual organisms that can be supported with available resources.

Cascade Series of enzymatic reactions in which the activated form of one enzyme catalyses the activation of the next, greatly amplifying the initial response.

Cassette model Description of the DNA rearrangements underlying switching of mating type in the yeast *Saccharomyces cerevisiae*, in which the mating-type locus, MAT, can be occupied by either of two genes (a or α) transposed from sites on either side of the locus.

Cation Positively charged ion which moves towards cathode, or negative pole, e.g. K^+,

cDNA library Collection (library) of DNA constituting the entirety of cDNA developed through cloning from cell types or specific organisms.

Cell cycle The period between the formation of a cell as one of the products of cell division and its own subsequent division. During this period all cells undergo replication of the DNA. In eukaryotic cells the cell cycle is divided into phases termed G1, S, G2 and M. G1 is the period immediately after mitosis and cell division, when the newly formed cell is in the diploid state and at which growth takes place. S is the phase of DNA synthesis, and is followed by G2, at which time the cell is in a tetraploid state and further cell growth may take place.

Mitosis (M) follows to restore the diploid state, accompanied by cell division. The interphase of the cell cycle comprises the G1, S and G2 phases.

Cell differentiation Process by which descendants of a shared mother cell obtain and maintain their structural and functional characteristics.

Cell fusion Fusing of two cells from different clones to form a single mixed cell. Cell fusion can be achieved relatively easily through use of dead Sendai virus.

Cell fusion The coming together of two cells to form one cell, not necessarily accompanied by fusion of the two cell nuclei.

Cell plate A planar array of vesicles containing cell wall material that assembles during early cytokinesis and gives rise by centrifugal extension to new cross-walls; present in certain green and brown algae and in land plants (embryophytes).

Cell sorting When groups of differentiated cells from multicellular organisms are artificially scattered and mixed together, different cells separate to form different agglomerations. This is known as cell sorting.

Cell-free system Any mixture of cell components reconstituted *in vitro* in which processes such as translation, transcription and DNA replication can be studied.

Central dogma Hypothesis on the interrelationships among the basic functions of DNA, RNA, and proteins. It holds that DNA serves as a model for its own replication and RNA, while RNA is a model for protein translation. The flow of genetic information thus proceeds from DNA to RNA to proteins.

Chambon's rule Splicing is the series of reactions through which unexpressed portions of RNA molecules created through gene transcription are removed and neighboring expressed sequences are linked together. In such cases, the 5' end of the unexpressed portion is typically a GU, while the 3' end is an AG. This is known as Chambon's rule.

Chimera Organism consisting of cells with two or more different genetic properties or tissues from different animal species.

Chi-square test Testing with the chi-square distribution is used to examine the appropriateness of theoretical and actual values calculated based on a specific hypothesis.

Chloroplast Green organelle found in the cytoplasm of the photosynthetic cells of plants and algae, and in which the reactions of photosynthesis take place. A chloroplast is bounded by a double membrane and contains a system of internal membranes (thylakoids) embedded in the matrix or stroma. The thylakoid membranes contain the green pigment chlorophyll and other pigments involved in light collection, electron-transport chains and ATP synthase, and are the site of the light reactions of photosynthesis in which ATP and NADPH are generated. The dark reactions of photosynthesis, in which carbohydrate is synthesized, take place in the stroma. A chloroplast also possesses a small DNA genome which specifies rRNAs, tRNAs and some chloroplast proteins.

Chromatic adaptation The ability of algae to modify the amounts or proportions of photosynthetic pigments in relation to changes in the light environment.

Chromatography Method of separating molecules within a mixture by repeatedly distributing molecules between mobile and stationary phases. The mobile phase is a liquid or gas, while the solid phase is a solid or liquid-covered solid. The molecular distribution between the two phases is participated in one or more of four basic processes: adsorption, gel filtration, ion exchange, and distribution. These processes result in differences in molecular movement speed according to the mobile phase's motion and the solid phase.

Chromonema Threadlike chromosomes or chromatids from the cell division stage that are the thinnest structures discernible with an optical microscope.

Chromosome set manipulation Technique for manipulating chromosomes in eukaryotic organisms. In plant breeding, this term refers to polyploid breeding. Chromosome from plants and animals are extracted and introduced into different cells to influence them and improve strains.

Chromosome walking Technique for mapping chromosomes from a collection of overlapping restriction fragments. Starting from a known DNA sequence, the overlapping sequences can be detected in other restriction fragments and a map of a particular area gradually built up.

Chromosome Located within the nuclei of eukaryotic cells, chromosomes consist of one or more large double-stranded chains of DNA molecules. Combining with RNA and histones, these DNA chains carry genes that store and transmit an organism's genetic information.

Chrondocyte Cell present in the lacunae of cartilage matrix that serves to synthesize and secrete cartilage matrix.

Cilia Motile hair-like outgrowth present on the surface of many eukaryotic cells, which makes whip-like beating movements. The synchronized beating of cilia propels free living unicells (e.g. protozoans) or, as with stationary cells (e.g. of nasal epithelium), produces a flow of material over the cell surface. A cilium is composed of a central core of microtubules (the axoneme) anchored to the cell by a basal body, the whole enclosed in plasma membrane. In multicellular organisms cilia chiefly occur on epithelial cells lining various internal passages; various other hair-like structures, esp. eyelash.

Ciliata Protozoans that use cilia to move by generating water flow.

Ciliate The informal name for a type of alveolate protist characterized by numerous cilia, presence of two types of nuclei (a larger macronucleus and a smaller micronucleus), and a cell mouth (cytostome).

Cladogram Tree-like diagram showing the evolutionary descent of any group of organisms or set of protein or nucleic acid sequences.

Classic food web An early model of the aquatic food web in which large (> 20 μm)phytoplankton are consumed by zooplankton, and zooplankton are consumed by fish in a simple linear food chain.

Cleavage Eggs are fertilized through the combination of a sperm and an oocyte, creating a multicellular structure. To become a multicellular organism, a fertilized egg must go through numerous processes of cell division. Cleavage refers to the cell division that occurs between fertilization and the gastrulation period during which tissue differentiation takes place in earnest. During cleavage, cells experience division without any growth in size terms; as a result, cells gradually decrease in size as cleavage proceeds.

Clinical trial Testing in which a new pharmaceutical, surgical technique, or medical equipment is applied directly to the body to examine its safety and efficacy.

Clone library A collection of bacterial colonies that in aggregate contains clones of the DNA present in an original sample.

Clone Group of genetically identical individuals or cells derived from a single cell by repeated asexual divisions; DNA clone; animal or plant derived from a single somatic cell or cell nucleus is termed a clone of the individual from which the cell or nucleus came; an apomict strain; to isolate a single cell or DNA molecule, and multiply it.

Cloning vector Specially modified plasmid or phage into which "foreign" genes can be inserted for introduction into bacterial or other cells for multiplication.

Cloning Refers to the entire process of obtaining a pure clone by extracting a DNA fragment from a DNA donor or reverse-transcribing RNA to produce DNA and linking it to a vector to introduce it by some means into a host and produce a body with recombinant DNA.

Closed mitosis (intranuclear mitosis) Mitosis that occurs within an intact nuclear envelope; present in many protists.

Cnida Cell organ found in coelenterates.

Coccoid The morphology of unicellular algae that have cell walls and are often, but not always, spherical in shape.

Codominance Case where alleles present in a heterozygous state produce a different phenotype from that produced by either allele in the homozygous state; case where each of a pair of heterozygous alleles is expressed equally and makes an equal contribution to the phenotype.

Codon Sequence of three neighboring nucleotides carrying the code for a specific amino acid in nucleic acid.

Coliphage Any of a number of viruses that proliferate by infecting *E. coli*. Viruses that infect bacteria are known as bacteriophages or simply "phages," so a coliphage is a phage that infects *E. coli*. Coliphages are the most widely studied of all phages and have had the greatest variety isolated. Most research on phage proliferation mechanisms and phage genetics uses coliphages.

Colloblast Cell on tentacles and pinnae of ctenophores, which carries small globules of adhesive material.

Colloid Gelatinous substance found in some diseased tissues; heterogeneous material composed of submicroscopic particles of one substance dispersed in another substance.

Colonization Invasion of a new habitat by a species; occupation of bare ground by seedlings.

Colony A group formed by microorganisms in a solid medium. Unable to move about freely in solid media, microorganisms accumulate in fixed positions, resulting in agglomerations of millions of organisms that are observable with the unaided eye.

Colony A type of protist body consisting of a group of cells held together by mucilage or cell wall material.

Community biomass Total weight per unit area of the organisms in a community.

Community well-defined assemblage of plants and/or animals, clearly distinguishable from other such assemblages.

Companion cell (Phloem) Single or double strand of DNA complementary to mRNA that can be synthesized from an mRNA model using reverse transcriptase. Like other DNA polymerases, reverse transcriptase requires a suitable primer. Typically, a poly(A) sequence is present at the 3' end of mRNA in eukaryotes, with oligo(dT) used as a primer for this reaction. In higher organisms, genetic replication typically uses cDNA.

Complementary DNA (cDNA) DNA synthesized *in vitro* on an RNA template by reverse transcriptase.

Complementation test Method of testing whether multiple mutations involved in a single phenotype belong to the same functional unit in a gene.

Conceptacle A cavity that contains the reproductive cells of some algae (particularly coralline red algae and fucalean brown algae).

Conchocelis A filamentous, sporophytic phase in the life cycle of bangialean red algae that produces conchospores in conchosporangia; commonly occurs in shells or other calcareous materials.

Conditional mutation Mutation that only produces an effect under certain conditions, as e.g. of temperature or nutritional status.

Conformation Three-dimensional arrangement of atoms in a structure.

Conjugated double bond Bond that occurs in alternating sequences of double and single bonds. The π electron in a double bond also binds weakly to a single bond, resulting in a bond with characteristics intermediate between the two.

Conjugation tube A connection between cells of mating filaments of zygnematalean algae, formed by dissolution of the ends of modified branches, which allows gametes to make contact.

Conjugation Mating of zygnematalean green algae involving nonflagellate gametes.

Connecting cell (connecting filament, ooblast) In florideophycean red algae, a cell (often long and filament-like) through which a zygote nucleus is transferred from the fertilized carpogonium to an auxiliary cell.

Connective tissue Supporting tissues of the animal body, including bone, cartilage, adipose tissue and the fibrous tissues supporting and connecting internal organs. It is derived from the embryonic mesoderm.

Consensus sequence The "ideal" form of a DNA sequence, in which the base present in a given position is the base most often found in comparisons of experimentally determined sequence.

Consistency index An estimate of the degree of homoplasy (parallel or convergent evolution) in a phylogeny.

Control experiment Form of experiment conducted simultaneously within another experiment to determine the effects on the experimental system from factors other than the subject and to eliminate and observe them. When identifying the effects, influences, and significance of some factor or manipulation for a subject, it is often necessary to conduct a control experiment that is identical for conditions other than the factor or manipulation in question, and to compare the results of the two experiments.

Convergence Coordinated movement of eyes when focusing on a near point.

Copepods Aquatic crustaceans that feed on larger phytoplankton, protozoa, and the juvenile stages of other aquatic crustaceans.

Coronal cells The 5 or 10 cells found at the tips of the tubular cells that form an investment around oogonia of charalean green algae.

Corresponding plant Set of closely related plants recognized for a particular region when comparing flora from two or more regions. Corresponding plants exist for every classification unit. At the genus level, for example, *Epimedium* and *Eleorchis* in Asia correspond to *Vancouveria* and *Alethusa* in North America. At the species level, the Chinese tulip tree corresponds to the North American tulip tree. Similar relationships exist for animals, as with the gorilla in Africa and the orangutan in Asia.

Cortex In a fleshy algal body, the layer of cells or tissues lying between the epidermis on the outside and the medulla on the inside.

Cosmid Type of cloning vector consisting of a bacterial plasmid into which the cos sequences of phage lambda have been inserted. This both allows the vector to grow as a plasmid in bacterial cells and enables subsequent purification of vector DNA by packaging into phage particles *in vitro*.

Coupling constant Parameter showing the strength of interactions among particles. When used for natural units, an interaction in which the coupling constant has no length dimension is a Type 1 interaction, while an interaction with a length dimension is a Type 2 interaction.

Crossbreeding One of the most basic method of breeding, which uses hybridization. Offspring may not only exhibit combinations of traits from both parents, but also traits superior to either. Many crops and livestock have been bred through this method, including rice, wheat, silkworms, and chickens. The approach is also used in the breeding of useful microorganisms.

Crossing-over Exchange of genetic material between homologous chromosomes at DNA recombination during meiosis. The structure formed during recombination is called a chiasma or crossover and is visible under the light microscope.

Crown gall Form of plant growth that occurs in many dicotyledons and gymnosperms and a small minority of monocotyledons due to infection by the soil bacterium *Agrobacterium tumefaciens*.

Crustacea, crustaceans Subphylum of arthropods, considered as a class in older classifications. They are mainly aquatic, gill-breathing animals, such as crabs, lobsters and shrimps. The body is divided into a head bearing five pairs of appendages (two pairs of pre-oral sensory feelers and three pairs of post-oral feeding appendages) and a trunk and abdomen bearing a variable number of often biramous appendages which serve as walking legs and gills.

Cryophilic algae Algae found in ice and snow environments.

Cryptogams Plants reproducing by spores, such as the mosses and ferns. The term has also been used for plants without flowers, or without true stems, roots or leaves.

Cryptomonads An informal name for members of the phylum Cryptophyta, most of which contain plastids derived from a red alga and which contain the distinctive xanthophyll pigment alloxanthin.

Cultural inheritance Transmission of particular traits and behaviours from generation to generation by learning rather than by genetic inheritance

Culture collection A reference collection of different species and strains of microorganisms or cultured cells.

Cutting Method of asexual reproduction in which part of a plant (branches, leaves, eyes, roots) is removed, uprooted, and germinated.

Cyanelles (cyanellae) A term used for the blue-green plastids of glaucophyte algae.

Cyanobacteria (chloroxybacteria, blue-green algae, cyanophytes) Photosynthetic bacteria that perform oxygenic photosynthesis.

Cyanobacteria Major group of photosynthetic Bacteria. Although they are prokaryotic, they have an oxygen-evolving type of photosynthesis resembling that of green plants and contain chlorophyll a and phycobilin pigments. Unicellular, filamentous and colonial types are found. They live in aquatic and terrestrial environments, either free-living or in symbiotic associations, as with fungi in lichens. Some species can fix atmospheric nitrogen. Some cyanobacteria produce toxins which can become a health hazard in conditions where cyanobacterial "algal blooms" appear. Known also as the blue-green algae from their previous classification in the plant kingdom as the Cyanophyta or Cyanophyceae.

Cyanophycean starch Polyglycan granules of glycogen that serve as the carbohydrate storage material of cyanobacterial cells.

Cypris Larva in the zoea stage in cirripedia from the Crustacea subphylum of arthropods. Through nauplius-cypris molting, the organism discards its triangular carapace and forms another carapace similar to a bivalve shell. While molting, it forms six pairs of thoracic appendages, which it uses to swim. The first antenna grows forward outside the cell. Like the legs, it transforms to assume locomotion or search capabilities, but immediately adheres to the bottom. The larva subsequent undergoes cypris-adult molting, rotating its body 180° to become a juvenile organism. The thoracic limbs become the adult's cirri.

Cystatin Substance that inhibits the functioning of protein-degrading enzymes in a cell.

Cytokine Family of small secretion proteins such as interleukins and interferons that bind to the cell membrane in sensitive cells and promote cell division or differentiation.

Cytology Discipline that involves the study of cells as constituents of organisms. Examines the morphological and functional composition of cells in connection with cell physiology, growth, differentiation, heredity, and evolution.

Cytotoxic T cells T cells that differentiate and proliferate due to antigen sensitization, functioning to damage antigen-specific cells.

Dalton (D, Da, dal) mass unit equal (by definition) to one-twelfth of the mass of a single atom of carbon-12, which is ca. 1.66×10^{-27} kg.

Dead zones Coastal ocean regions that are depleted of oxygen and thus life forms as the result of the decay of large populations of phytoplankton due to ocean eutrophication.

Deletion Mutation involving loss of part of a chromosome, or loss of a base or a stretch of bases in a DNA sequence.

Deletion Structural transformation in which part of a chromosome is lost. Known as a terminal deletion when part of the end of a chromosome is lost and an interstitial deletion when a central portion is lost.

Demospongia Class of sponges (Porifera, q.v.) which often have the body wall strengthened by a tangled mass of spongin fibres (e.g. in the bath sponge Spongia). They may have silica spicules, in the form of simple needles or a four-armed spicule whose points describe a tetrahedron, or may have no spicules. They are found on shores, and down to depths of more than 5000 m.

Denaturation Alteration in the structural properties of a macromolecule such as a protein or a nucleic acid, leading to loss of function, as a result of heating, change in pH, irradiation, etc. In most cases denaturation refers to the disruption of noncovalent bonding leading to loss of secondary structure (e.g. unfolding of a protein chain or separation of the two strands of a DNA double helix).

Deoxyribonuclease (DNase) Any of various enzymes that cleave DNA into shorter oligonucleotides or degrade it completely into its constituent deoxyribonucleotides.

Development engineering Development engineering, as opposed to genetic or cell engineering, involves the experimental manipulation of early embryos, chiefly in mammals. Goals of development engineering include firstly explicating various phenomena that occur in animal genesis and differentiation and developing disease model animals to study genetic conditions and growth disorders; and secondly serving a practical role in breeding of useful animals (livestock improvement) or using animals to produce organic substances or develop treatments of genetic conditions in humans. Examples of development engineering techniques include nuclear transplantation, chimera production, and the creation of animals with added genes.

Development In biology, the changes that occur as a multicellular organism develops from a single-celled zygote, from the first cleavage of the fertilized ovum until maturity.

Dialysis Separation of large molecules such as proteins from small molecules and ions by the inability of the larger molecules to pass through a semipermeable membrane.

Diazotrophy The ability to convert diatomic, gaseous N2, gas into ammonium ion, which can be used by cells to produce amino acids.

Dielectric constant Ratio of relative permittivity showing the relative amount of charge a substance is capable of storing when a magnetic field is supplied. Dielectric materials are regarded as insulators. When two conductors are separated by a dielectric material, they may exist in a state of electrical stress in the absence of a continuous energy supply from outside the system. The dielectric constant is calculated as $\varepsilon/\varepsilon^0$, where ε is the permittivity when a condenser is filled with isotropic material and ε^0 is the permittivity for a condenser in a vacuum. The vacuum here is considered a dielectric substance; for water, the dielectric constant is around 80.

Differentiation (cell) Process by which groups of cells or cell clones acquire specialized functional biochemical or morphological characteristics that were not present before.

Differentiation In a general sense, the increasing specialization of organization of the different parts of an embryo as a multicellular organism develops from the undifferentiated fertilized egg; of cells, the development of cells with specialized structure and function from unspecialized precursor cells.

Diffuse growth A type of growth that does not involve a localized point of cell division.

Dinoflagellate A protist that is usually single celled, characteristically possesses saclike alveoli beneath the cell membrane, swims by means of two distinctive flagella (or, if not motile, produces dinospores having such flagella), and may be photosynthetic or not.

Dinosporin A decay-resistant compound deposited at the surfaces of the cells of some dinoflagellates and their cyst stages that allows fossilization to occur.

Diploid Organisms whose cells (apart from the gametes) have two sets of chromosomes, and therefore two copies of the basic genetic complement of the species. Designated 2n. cf. haploid; a diploid organism or cell.

Disease resistance Property found in plants that do not readily develop diseases when attacked by pathogens.

Displacement loop (D-loop) Loop structure that appears when a complementary single strand of DNA links with part of a double strand of DNA. It may also be seen at the beginning of DNA replication when one side of a double strand becomes transposed with a short newly synthesized single strand of DNA.

Diversity index A measure of the biological diversity (generally the species diversity) within an environment. There are various types of diversity index, which are calculated in various ways from the number of species present and their relative abundance. Such indices can be used to detect ecological changes due e.g. to stress on an environment. cf. biotic index.

DNA fingerprinting Method of ascertaining individual identity, family relationships, etc., by means of DNA analysis. The DNA fingerprint consists of a pattern of DNA fragments obtained on analysis of certain highly variable repeated DNA sequences within the genome, whose number and arrangement are virtually unique to each individual. DNA fingerprints can be obtained from a tiny quantity of blood, semen or hair, and are widely used in forensic work and also in ecological studies.

DNA hybridization Technique for determining the similarity of two DNAs (or DNA and RNA) by reassociating single strands from each molecule and determining the extent of double-helix formation; general method involving reassociation of complementary DNA or RNA strands, used to identify and isolate particular DNA or RNA molecules from a mixture.

DNA library Collection of cloned DNAs.

DNA microarrays Attachment of DNA on highly arrayed solids in order to investigate ordinary gene expression.

DNA polymerase Enzyme that catalyzes DNA synthesis from deoxyribonucleotide triphosphate using a single or double strand of DNA as a model.

DNA probe Small fragment of radioactively or fluorescently marked single-stranded DNA with the same nucleotide sequence as an area of interest in human DNA.

DNA replication Series of processes whereby parental DNA is actually replicated and passed down to daughter cells during cell proliferation through the synthesis of complementary nucleotide chains using double-stranded DNA from the parent cells as a model (semiconservative replication).

Docosahexaenoic acid Polyunsaturated fatty acid from the omega-3 family with 22 carbon atoms and six double bonds; found chiefly in blue-backed fish.

Dominance frequency In ecology, proportion of samples in which a particular species is predominant.

Dominance Property possessed by some alleles of determining the phenotype when present in one copy in a cell. As one member of a heterozygous pair they mask the effects of the other allele (the recessive allele) to give a phenotype identical to that when the dominant allele is present as two copies. This phenomenon is known as complete dominance. Incomplete dominance is exhibited when the effects of the other allele are not completely masked; the extent to which a particular species predominates in a community and affects other species.

DOPA 3,4-dihydroxyphenylalanine, formed from tyrosine in the adrenal medulla, brain, and sympathetic nerve terminals by the enzyme tyrosine hydroxylase. It is a biosynthetic precursor of noradrenaline, adrenaline and dopamine. Also oxidized by dopa-oxidase to a melanin precursor, e.g. in the basal layers of skin. l-dopa is used in the treatment of Parkinson's disease.

Duchenne muscular dystrophy Muscular disease characterized by progressive atrophy of skeletal muscle and diminished muscular strength.

Economic trait Livestock genetic trait with productive aspects.

Ecosystem Community of different species interdependent on each other, together with their non-living environment, which is relatively self-contained in terms of energy flow, and is distinct from neighbouring communities. Different types of ecosystem are defined by the collection of organisms found within them, e.g. forest, soil, grassland. Continuous ecosystems covering very large areas, such as the northern coniferous forest or the steppe grassland, are known as biomes.

Ecotope A particular kind of habitat within a region; the total relationship of an organism with its environment, being the interaction of niche, habitat and population factors.

Ecotypes Genetic varieties of a species that may show morphological differences or adaptations to different environments.

Efficiency of plating (EOP) Ratio of the number of the number of plaques made when a bacteriophage or other specific host is used as an instruction bacterium to the number of plaques obtained when a standard bacterium is the instruction bacterium.

Eicosapentaenoic   acid   (EPA) Known   by   the   systematic   name 5,8,11,14,17-eicosapantaenoic acid, this is an unsaturated fatty acidic with 20 carbon atoms and five double bonds. It has a molecular formula of $C_{20}H_{30}O_2$ and a molecular weight of 302. It is abundant in fish, especially sardines and mackerels.

Electrophoresis Technique for separating molecules such as proteins or nucleic acid fragments on the basis of their net charge and mass, by their differential migration through paper, or through a polyacrylamide or agarose gel (gel electrophoresis) in an electric field

Electroporation Method of gene introduction in which DNA is inserted in a cell that is suspended in DNA solution when a direct current high-voltage pulse is passed through it. The electricity results in a hole being produced in the cell membrane; the DNA molecule is thought to enter the cell simultaneously through electrophoresis. This method can be applied with a wide variety of cells, including those of animals, plants, and microorganisms. It is widely used, since a relatively high rate of efficiency in gene introduction can be obtained when the appropriate conditions are selected.

Embden-Meyerhof pathway Metabolic pathway in which glucose is broken down into glucose-6-phosphate, fructose-1,6-biphosphate, glyceraldehyde-3-phosphate, phosphoenolpyruvate, and finally pyruvic acid. It consists of 10 stages of enzyme reaction is named after two (or three) of the people who contributed to explicating it: G. Embden, O. Meyerhof, J. K. Parnas,

O. H. Warburg, and C. F. Cori. It accounts for a major portion of the metabolic system during glycolysis and alcohol fermentation and is a chief pathway through which sugars are broken down by respiration. The pathway is also used by some bacteria in butyric acid and homolactic fermentation. Several enzymes in the Embden-Meyerhof pathway are distributed within the cell's membrane.

Embedding Method that involves creating a thin tissue fragment to serve as a sample for microscope study.

Embryo transfer Reproductive technology in which very early embryos produced by *in vitro* fertilization or artificial insemination are transferred into a surrogate mother for further development, used in cattle and sheep breeding to produce many more offspring from a prize female than she could produce naturally.

Embryo Multicellular organism from the first cell division of the initial fertilized egg to the ontogenesis process prior to independent survival.

Embryogenesis Process of the development of a new individual and formation of an embryo from a fertilized egg.

Endangered species Plant or animal species threatened with extinction.

Endobionts Organisms living within the body of other organisms (e.g., the dinoflagellates that live inside corals).

Endoliths Microorganisms such as bacteria, cyanobacteria, eukaryotic algae, and lichens that grow inside the pore spaces between mineral grains within porous translucent rocks.

Endoreduplication The process by which the nuclear DNA undergoes repeated rounds of replication without intervening mitotic separation, yielding DNA levels higher than the haploid or diploid state.

Endosymbiosis The condition in which one or more organisms live within the cells or body of a host without causing disease or other conspicuous harmful consequences.

Enhancer DNA region that enhances gene transcription efficiency by promoting the bonding of RNA synthase to promotor genes through protein interactions.

Envelope Certain viruses, layer of lipid and protein surrounding capsid, the lipid being derived from host cell membrane as the virus is discharged from the cell, the protein being virus-encoded. Such viruses are known as enveloped viruses; bacterial envelope.

Environmental genomics A set of procedures that employ phylogenetic information and molecular methods to explore the species diversity of natural microbial communities without first growing the organisms in culture.

Eosinophil Type of white blood cell classed as a granulocyte. It contains granules that stain with the red acidic dye eosin and which contain substances that can induce inflammation when the granules are secreted by exocytosis. It is involved in destruction of parasites and in inflammatory reactions.

Epibacteria Bacteria that are attached to surfaces, such as the outsides of microalgal cells or macroalgal bodies.

Epibody A layer of cells that forms the outer surface of a coralline red alga.

Epidermal growth factor Cytokine that stimulates the division of epidermal and other cells, and has been used to promote wound healing.

Epidermis Outer layer or layers of the skin, derived from embryonic ectoderm. In vertebrates a non-vascular stratified tissue, often keratinized,

Epilimnion The upper layer of a stratified water body whose waters are warmer in summer (and are typically more oxygen rich) than bottom waters (hypolimnion).

Epimer Three-dimensional isomer with a different arrangement among multiple chiral centers, as seen with the relationship between D-glucose and D-mannose.

Epipelagic Upper warm ocean waters.

Epipelic Living on the surfaces of mud or sand.

Epiphyte An organism that grows on the surfaces plants or algae.

***Escherichia coli* K-12** Line of colon bacilli developed from germs isolated from a recovering diphtheria patient at Stanford University. Appearing as *Escherichia coli K*-12 (λ) F+, it happens to be a lysogenic strain of the phage λ and was later learned to be a male strain with an F factor. It is the most widely used colon bacillus strain, serving as a study resource in molecular genetics and genetic biochemistry.

***Escherichia coli* plasmid** Chromosomal genetic factor in *E. coli*. The F plasmid, which was the first to be discovered, is around 100 kbp in size and functions to truncate multiple genes known as *tra* and different plasmids and chromosomal DNA in other cells (experimentally, this has included *Saccharomyces* yeast other intestinal bacteria besides *E. coli*). This process is similar to the truncation of T-DNA by *Agrobacterium*. Other examples include R plasmids (with resistance to multiple drugs) and colicin factors.

Etching Use of chemicals to corrode the surfaces of metals, ceramics, semiconductors, and other materials.

Etiology Research or theories concerning the factors responsible for disease and the means by which infection occurs, or the discipline concerned with the causes of disease.

Eubacteria The "true" bacteria, unicellular prokaryotic microorganisms possessing cell walls, with cells in the form of rods, cocci or spirilla, many species motile with cells bearing one or more flagella. They are distinguished from the archaebacteria by the possession of peptidoglycan cell walls and ester linked lipids. They include the Gracilicutes, and the Firmicutes.

Euendoliths Microorganisms that live under the surface of rocks and actively bore their way through the rock. Several desert lichens are known to be euendoliths, as are certain cyanobacteria found on coral reefs.

Euglenoids Unicellular protists that primarily occur as flagellates having a distinctive surface composed of proteinaceouspellicular strips and a characteristic storage known as paramylon granules; many but not all conduct photosynthesis.

Euglenozoa A protistsupergroup that includes euglenoidsand trypanosomes as well as some other protists.

Eukaryote Organism with a nuclear membrane and clearly defined nucleus, as well as organelles enclosed in a membrane, including endoplasmic reticula, a Golgi body, mitochondria, and chloroplasts. Mitosis occurs during cell division and introns are present in genes; distinct from prokaryotes in gene regulation and expression.

Euphotic zone The (upper) portion of the water column that receives enough light for photosynthesis to occur.

Eutrophic aquatic systems Waters that contain relative high levels of nutrients such as phosphate and/or combined nitrogen; typically exhibit high levels of primary productivity.

Eutrophication Phenomenon in which waters transform from a nutrient-poor to a nutrient-rich state.

Evolution The development of new types of living organisms from pre-existing types by the accumulation of genetic differences over long periods of time. It is studied by reference to the fossil record and to the anatomical, physiological and genetical differences between extant organisms. Present-day views on the process of evolution are based largely on the theory of evolution by natural selection formulated by Charles Darwin and Alfred Russel Wallace in the 19th century. Darwin's theory has undergone certain modifications to incorporate the principles of Mendelian genetics, unknown in his day, and the more recent discoveries of molecular biology, but still remains a basic framework of modern biology.

Evolutionarily stable strategy In evolutionary theory, a behaviour pattern or strategy which, if most of the population adopt it, cannot be bettered by any other strategy and will therefore tend to become established by natural selection. Using games theory the results of various different strategies (e.g. in contests between males) can be worked out and a theoretical ESS determined and compared with actual behaviour.

Exclusive Economic Zone Waters for which all sovereign rights are recognized within 200 nautical miles of coastal countries, including exploration, development, and preservation of marine resources, preservation of the marine environment, and scientific research activities.

Exon Block of DNA sequence encoding part of a polypeptide chain (or of tRNA or rRNA), which forms part of the coding sequence of a eukaryotic gene, and which is separated from the next exon by a noncoding region of DNA.

Exospores In cyanobacteria, spores that are cut off from one end of the parental cell. (compare with baeocytes)

Expressed sequence tag (EST) Data based on a cell-derived mRNA model that is developed only once into a base sequence from the end of cDNA synthesized with reverse transcriptase.

Extinction point Minimum level of illumination below which a plant is unable to survive in natural conditions.

Extracellular matrix (ECM) Materials generated by a cell that are secreted from, or produced on, the external surface; includes mucilage, cell walls, and loricas.

Facultative anaerobic bacteria Microorganism that is capable of metabolism by respiration or fermentation and can alternate between the two according to its environment (e.g., lactic acid bacteria, *Escherichia coli*).

Familial adenomatous polyposis (FAP) Autosomal dominant genetic disease in which hundreds to thousands of polyps form in the colon.

Fibronectin Glycoprotein of extracellular matrix, to which animal cells can bind by means of integrins in their plasma membranes. It is involved in interactions of animal cells with extracellular matrix.

Fidelity the degree of limitation of a species to a particular habitat; of DNA replication, transcription and translation, the probability of an error being made during the copying of DNA into DNA or RNA, or during the translation of RNA into protein.

Filter feeding A mode of food collection by herbivores that involves sieving large volumes of water for particles

Fingerprinting In biochemistry, a technique for detecting small differences in amino acid composition/sequence between different proteins by selective cleavage into small peptides which are then separated by electrophoresis in 1st dimension and chromatography in the 2nd dimension resulting in a pattern of peptide spots characteristic for each protein.

Fixation index In population genetics, a measure of genetic differentiation between subpopulations, being the proportionate reduction in average heterozygosity compared with the theoretical heterozygosity if the different subpopulations were a single randomly mating population.

Flagellar transformation A maturation process by which the younger flagella of a parental cell become the older flagella of a progeny cell.

Flagellates Diverse group of unicellular eukaryotic microorganisms, including photosynthetic and non-photosynthetic species, and classified in various schemes as protozoans, protists or algae. They are motile in the adult stage, swimming by means of flagella. They include both free living marine and freshwater species and some important commensals such as those living in the guts of ruminants, and human parasites such as trypanosomes.

Flagellates Unicellular or colonial protists whose cells bear one or more flagella.

Flagellum Long appendage that allows prokaryotes to move in water, or that allows fluids to move through an animal's various tissue surfaces.

Flavobacterium Saprophytic bacterium, chiefly involved in the decomposition of fish and shellfish.

Florideon starch In red algae, a branched a-1,4-linked glucose polymer with some α-1,6 linkages that occurs as granules within the cytoplasm.

Florideophytes An informal name for the members of the red algal class Florideophyceae; characterized by a triphasic reproductive cycle that includes a carposporophyte generation.

Flow cytometer An instrument that uses a laser and detectors to measure the optical properties of cells, such as pigment fluorescence.

Fluctuation Differences that arise among individuals of the same genotype due to changes in somatic cells stemming from incidental environmental influences during the genesis and growth processes.

Follicle cell Single- or multi-layered bubble-shaped epithelial cell found surrounding the exterior of animal tissues. Typical examples include in thyroid follicles and ovarian follicles. Ovarian follicles form a multiple layer of epithelium in mammalian ovaries and a single layer in insects to encircle oocytes. In sea urchins, the egg shell membrane surface is covered in markedly large follicle cells following egg discharge.

Food chain a sequence of organisms within an ecosystem in which each is the food of the next member in the chain. A chain starts with the primary producers, which are photosynthetic organisms (e.g. algae, plants, bacteria) or chemolithotrophic bacteria. These are eaten by herbivores (primary consumers) which are in turn eaten by carnivories (secondary consumers). Small carnivores may be eaten by larger carnivores.

Food quality The extent to which algae are able to provide essential nutrients when consumed as food.

Formulation Manipulation to ensure suitable shape and form to ensure ease of use and application and consistent efficacy when using pharmaceuticals on the human body.

Founder effect Example of genetic drift through which the frequency of a gene in a population varies according to chance or probability. A change in genetic frequencies between two populations that arises when a very small number of organisms break off from an original population to form a new one.

Founder principle Holds that when a small number of organisms become separated from their original population and become isolated founders in a new region, this serves as a factor in species differentiation.

Fucans Also known as fucoidins or ascophyllans, polymers of L-fucose and additional sugars that are sulfated

Fucosan polysaccharides composed of fucose units, found in vesicles (fucosan vesicles) in cells of brown algae where it may be a storage polysaccharides or a waste metabolic product.

Fucoxanthin A xanthophyll pigment that confers golden brown or brown pigmentation to diatoms, chrysophyceans, brown algae, and some other photosynthetic stramenopiles.

Fucoxanthin Brown xanthophyll carotenoid pigment found in brown algae, diatoms and golden-brown algae.

Fusion cell In red algae, (a) generally, a cell resulting from the coalescence of two or more non-gamete cells; (b) specifically, the cell produced by fusion of an auxiliary cell with one or more neighbors.

Gamete Representative cells with haploid genes, i.e., sperm and egg cells.

Gametic meiosis Meiosis occurring during the production of gametes

Gap site Portion of double-stranded DNA where one strand has come away, leaving only the other strand.

Gauss Electrical unit showing the intensity of magnetic induction (CGS unit showing magnetic flux density).

Gelling agent Substance with colloidal properties that exhibits viscosity when added to food and serves as a dispersion stabilizer, adhesion preservative, and coating material.

Gene bank Organization that systematically gathers and preserves genetic resources, including natural species, lines, varieties, wild-type strains, and genetic lines.

Gene expression Translation in the case of transcription and proteins to produce a final genetic product; genes are expressed when a biological product exists and is active.

Gene flow Propagation of genes. Occurs as genes continue to enter one genetic supply from another.

Gene Section of the chromosome that codes for a functional product (RNA or polypeptide resulting from its translation).

Genetic analysis Determination of the number of genes involved in a particular genetic trait, their location on the chromosomes, and their influences on phenotype, among other factors. In a broad sense, this involves physiological and biological analysis of gene function and structure, but it may also refer to analysis of genetic composition in Mendelian groups in a population genetics sense. Genetic analysis is founded on hybridization experiments, which requires the selection of marker genes with clear phenotypic characteristics.

Genetic character Properties determined by a particular gene or group of genes.

Genetic drift Phenomenon in which a gene becomes fixed or lost within a small population.

Genetic load Population genetics index showing the degree of natural selection at the genotype level; quantitatively expands on the concept of a deleterious gene that arises due to mutation becoming a burden on a population due to death or infertility among organisms. It is typical defined as the ratio (Wop-W)/Wop, where Wop is the fitness of the optimal phenotype and W is the average fitness of the population. While the definition of load is abstract, it is an important quantity when analyzing genetic changes in a natural population.

Genetic map Diagram showing the relative arrangement and positions of genes on chromosome molecules.

Genicula The uncalcified, flexible regions occurring as joints between calcified, non-flexible regions of the thalli of jointed coralline red algae and some ulvophycean green algae

Genome Project Collaborative international effort to determine the full base sequence of genes and decipher the totality of genetic information.

Genome Totality of chromosomes or genes contained in a gamete.

Genomics Field of molecular biology concerning the study of genome structure and the genetic base sequences that comprise it.

Girdle lamella A flat sheet composed of three thylakoids that extends just under the plastid envelope in some photosynthetic stramenopile algae.

Gland cell In red algae, a specialized cell that serves in secretion or storage

Glaucophytes An informal name for a group of eukaryotic algae capable of producing flagella, which have plastids whose envelope is composed of two membranes and which contain phycobilin pigments in phycobilisomes.

Globular protein Term for any functional protein with a spherical three-dimensional structure, including enzymes, hormones, and antibodies.

Gonidium In *Volvox* and related colonial green algae, an enlarged, nonmotile cell that can generate new (daughter) colonies

Gonimoblast In red algae, one or all of the filaments that bear carpospores (the filaments known in aggregate as the carposporophyte).

Growth hormone Polypeptide produced and secreted by acidophilic cells in the anterior pituitary gland that promote organismal growth.

Gynandrosporous Species of oedogonialean algae that produce androspores on the same filament that produces egg cells.

Gynogenesis Phenomenon in which fertilization occurs only in the female pronucleus for one reason or another, without fusion between male and female pronuclei.

Hair cells In green algae, cells that produce a long, hair like extension.

Haplo Used as a prefix before a symbol indicating a specific chromosome, this refers to organisms in which only one half of a chromosome pair is present in somatic cells.

Haploid Cell or organism with a chromosome number n equaling half the diploid number as a result of meiosis. This situation is also referred to as haploidy. In reference to nuclear phase, this means that only one-half of a chromosome pair is present. During alternation of nuclear phases, a haploid generation is produced when the number of chromosomes is reduced by half by meiosis. Haploid organisms regain diploid status through fertilization. In haploid individuals, only diploids are zygotes, and alternation of generations does not occur. In the parasexual life cycles of some bacteria, diploid cells transform into haploid cells, a phenomenon referred to as haploidization. Organisms are known as haplonts.

Haploid number Indicated with letter n, this refers to number of chromosomes per cell for the haploid generation. This includes the number of chromosomes for gametes. Haploids are cells with a single chromosome number or organisms made up of said cells; in these haploids, haploid generations are present, as well as instances in which the diploid generation's chromosome number is reduced by half as a result of parthenogenesis.

Haploid plant Plant in which the chromosome number has been halved in the life cycle through diploid meiosis. In the vegetative period in higher plants, diploid chromosomes (2n) are present in the diploid generation. The ability to easily acquire haploid (n) plants with half a chromosomal set would be useful in the following ways: (1) the ability to obtain pure lines in a short period of time by multiplying the chromosomes of haploid plants; (2) strong viability is pure lines through selection of fatal genes; (3) expression of dominant traits due to the lack of interaction among alleles in haploid plants, and thus greater ease in identifying genes; (4) potential applications in chromosome engineering; and (5) potential application of alloploid haploids in cell genetics. Haploid plants are small in size and highly infertile.

Haploidization Phenomenon that occurs in the parasexual life cycle of some species of bacteria, in which a diploid cell transforms into a haploid cell through the gradual loss of chromosomes without division.

Haploidy Condition in which the chromosome number is reduced by half. Two forms are used, respectively representing the number of chromosomes in gametes and the number of chromosomes for haploids as a form of polyploid. When indicating chromosome number as a series without respect to basic number, whole numbers and haploids are respectively represented as 2n and n; when basic values are used to represent chromosome numbers, the whole number values are shown as 2x, 3x, 4x, and so on, depending on chromosome number, while haploids are shown as x, 2x, 3x, 4x, and so on. Gametes and the number of chromosomes in their generation are described as haploid.

Haplotype Combination of alleles at a series of gene loci on the same chromosome when several genetic loci exist in dense series on a single chromosome. Can be used to distinguish genetic regions.

Hardy-Weinberg principle Principle of population genetics holding that the relative frequency of a genotype under conditions of random hybridization is a multiple of the frequency of alleles involved.

Heat shock protein Protein in which synthesis is induced when a cell, tissue, or organism is at a temperature 5–10 °C higher than its physiological temperature.

Helicase Enzyme needed to separate the forward double strand in a replication fork, as in prokaryotes.

Hepatocyte growth factor (HGF) Growth factor discovered, isolated, and refined for the earliest cultured liver cells. HGF is produced by various cells in the hepatic lobe system, with targets primarily including epithelial cells, neurons, endotheial cells, and some hepatic system cells. Can participate in promoting cell growth, exhibiting cell movement promotion and epithelium formation inducement activity. In development, it serves as a medium for epithelium/hepatic lobe interactions, participating in the formation of the liver, kidneys, lungs, and other internal organs, as well as placenta and the skeletal system.

Heteromorphic alternation of generations A type of sexual reproductive cycle involving two or more multicellular phases, at least two of which are morphologically distinct.

Heteromorphic Morphologically different; in algae, usually applied to distinctive gametophyte and sporophyte phases in sporic life cycles (alternation of generations).

Heteroploid Organism with one or more fewer chromosomes than a whole number multiple. Occurs as result of chromosomes failing to separate or becoming lost during cell division. Organisms are generally sterile. Used for genetic analysis and breeding for special traits.

Heterosis True heterosis occurs when a heterozygote shows stronger survival capabilities than a homozygote. The theory presumes that dominant genes have stronger survival capabilities than recessive ones, such that the first hybrid generation has superior survival capabilities than its parents because it receives different dominant genes from them.

High-performance liquid chromatography (HPLC) Form of chromatography in which improvements in the column's solid state filler and apparatus have enabled high-speed, high-performance separation, allowing for a separation speed around 100 to 1,000 times higher under high pressure.

Holocarpy In certain ulvophycean green seaweeds, conversion of all of a body's cytoplasm into reproductive cells, whose release results in the death of the parental alga.

Homologous chromosomes Chromosomes that bind in meiosis. In fully homologous chromosomes, the same number of identical genes or alleles array in the same sequence; in partially homologous chromosomes, only a portion is homologous.

Homology Property by which the organs of organisms that are being compared in developmental or evolutionary terms may be recognized as having descended from common ancestry. Examples include the pectoral fins in fish, the front legs in amphibians and mammals, and the wings of birds. Features are not necessarily similar in functional terms. May be used similar when comparing nucleic acid and protein structure in terms of molecular evolution. When two or more substances are deemed to have statistically similar structures regardless of function, they are thought to indicate common ancestry and described as homologous.

Homozygosity The possession of identical alleles when several alleles are present at one genetic locus; generally occurs between genetically similar gametes.

Homozygote Refers to a genetic state in which the same allele from among several for a single gene is passed down to offspring from its father and mother.

Host range Range of organisms using a parasitic organisms as a host, including bacteria, viruses, fungi, and other parasites.

Housekeeping gene Gene that codes for a protein essential to general cell functioning and is always expressed structurally. Includes genes that code for enzymes and compositional proteins in universal metabolic pathways, including the glycolytic system.

Huntington's chorea General psychoneural disease accurately recorded in 1872 by the U.S. physician G. Huntington. Autosomal dominant condition that emerges in adults and is clinically characterized by choreic movements, paralysis, and psychological disorders such as schizophrenia-like psychosis or dementia.

Hybridoma In the broad sense, this is a mixed form of cell with tumorlike properties created by artificial fusion of two kinds of cell. Generally refers to plasmocytomas and B cell hybridomas.

Hydrogen bond Unshared bond mediated by hydrogen that forms when a strongly electronegative atom Y (such as fluorine, chlorine, oxygen, or nitrogen) approaches a hydrogen atom possessing a shared bond with another strongly electronegative atom X (same as above), of the type X-H...Y. While the bond energy for the hydrogen bond is around 2–8 kcal, the combined effect of several hydrogen bonds contributes to stability.

Hypolithic algae Algae that grow underneath translucent rocks on the bottom surface where moisture collects.

Hypothalamus Central component of the autonomic nervous system located in the diencephalon of vertebrates. Accelerates, regulates, and integrates numerous physical functions, including secondary growth and development.

Immune globin Antiserum that includes specific antigens or antibodies for antigens.

Impregnation Transportation of a male sperm into a female's body.

In vitro Refers to experiments conducted in acellular systems. In some instances, this includes tissue culturing of cells from multicellular organisms under cell culturing conditions.

In vivo Refers to experiments performed in undamaged organism systems. These are conducted at the cellular level for microorganisms and at the whole-body level in animals.

Inbreeding Breeding between very closely related organisms, as in self-fertilization among plants or sibling mating among animals.

Inbreeding coefficient Value representing the extent of homozygosity among organisms as a result of inbreeding.

Inbreeding depression Phenomenon in which survival capabilities (including size, resistance, and fecundity) are generally diminished due to continued inbreeding over a long period of time.

Induced mutation Artificial inducement of a mutation. Methods include treatment with UV rays or chemicals.

Infectious pancreatic necrosis Acute viral infection that occurs primarily in juveniles of species such as the steelhead trout. Often occurs in juveniles weighing under 1g at the age of around eight weeks, with recorded two-week death rates in excess of 60%.

Infrared spectrophotometer Device in which a solid, liquid, or gaseous sample is separated with a 2.5–25 μm infrared spectroscope, with absorption calculated for each wavelength; the resulting absorption spectra are used for quantitative and qualitative analysis.

Infusion General term for liquids other than blood that are introduced into the blood vessels or under the skin for various purposes. Depending on purpose, they may use electrolyte, amino acid, or sugar solutions, but they are generally designed to have the same osmotic pressure as the blood.

Initiation codon AUG. Codes for the initial amino acid in a polypeptide chain. This initial amino acid is N-formylmethionine in prokaryotic cells and methionine in eukaryotic cells.

Insulin-like growth factor Growth factor with a similar structure to insulin consisting of polypeptides with a molecular weight of 7,500. Its reactions within serum are similar to those of insulin, but two structures have been identified that are not inhibited by insulin antibodies, with the names IGF-I and IGF-II. Both consist of four types of polypeptide chains (A–D), with the A and B chains roughly 45% identical in structure to insulin. In addition to mediating growth hormone activities in chondrocyte proliferation and protein biosynthesis, they exhibit similar physiological functions to insulin.

Integument Pellicle surrounding the ovule; one of the tissues forming the ovule.

Intercalary meristem In kelp brown algae, a region of cell division located between stipe and blade tissues.

Interferon IFN or IF for short. General term for proteins or glycoproteins with antiviral properties induced in animal cells by viruses, double-stranded RNA, or lectin. While they are known for virus inhibition effects not caused by antibodies, they are called interferons because the same effects are exhibited afterwards.

Interferon-γ Interferon produced in T lymphoctyes as a result of immune stimuli.

Intergeneric hybrid Hybrid produced from breeding between organisms of two different genera. While comparison with interspecific hybrids may suggest these hybrids would exhibit characteristic traits, hybrids are extremely difficult to produce and often exhibit growth difficulties and severe sterility.

Interleukin (IL) General term for active protein components produced by lymphocytes that influence their own differentiation, growth, or functioning or those of other lymphocytes. They form a complex network in which many liquid factors derived from immune system cells are involved in expression or regulation of immune response. These factors are known as lymphokines (derived from lymphocytes) or monokines (derived from monocytes or macrophages) and were previously defined purely in terms of their physiological activity. At the same time, the appearance of several different immune activities by the same molecules when purified, isolated, and refined is a major characteristic of these immune system factors. Researchers at an international workshop thus agreed to use the term IL for those factors clearly purified, isolated, and refined as molecules and the identify them sequentially. Over 10 IL molecules have been identified to date, most of them including gene and amino acid primary structure and specific receptors in addition to physiochemical properties.

Intervening sequence Base sequence portion that is not actually translated into mRNA genetic information; mostly found in the base sequences in eukaryote DNA.

Intraspecific Refers to phenomena that arise among members of a species, including competition and cooperation.

Intron Nucleotide sequence within a gene or its transcription product that is not included in the final RNA product of that gene.

Isomorphic alternation of generations A type of sexual cycle in which there are at least two multicellular stages that are morphologically similar.

Isoschizomer Refers to instances in which restriction enzymes isolated and refined from different bacteria have matching recognition sites. Known examples include *Rhodopseudomonas sphaeroides*-derived RSR I as an isoschizomer for *Escherichia coli* RY13-derived *Eco* RI, and *Gluconobacter industricas*-dervied Gin I as an isoschizomer for *Bacillus amyloliquefaciens* H-derived Bam HI.

Isozyme Refers to a sequence of enzymes within the same organism or cell that have protein molecules of different chemical structures but catalyze the same chemical reactions. In these cases, the enzymes have different genes and amino acid composition. Enzymes therefore have different isoelectric points and can be separated using differences in amount of movement through electrophoresis.

Jejunum Rear portion of the mammalian small intestine that connects the duodenum to the ileum.

Juvenile shell Young shell that has begun surviving benthically after a planktonic post-fertilization period.

Knowledge-based industry Industry that involves the use of knowledge to significantly improve products or services or to provide high value-added knowledge services.

Koilocyte Flattened cell that occasionally exhibits heterokaryosis, with holes appearing around the nucleus.

Label Radioactive atom introduced into a molecule to allow for easy observation of changes in the metabolism of substances.

Laminaran (laminarin) In brown algae and other photosynthetic stramenopiles, a soluble polysaccharide storage product composed of β-1,3-linked glucose *units* together with some branch-producing β-1,6-linkages.

Landlock type Phenomenon in which a fish that previously alternated between sea and river environments survives for generations in an inland location due to topographic or environmental changes. Often occurs among salmon, trout, and other fishes that enter rivers to lay eggs.

Larva Form of juvenile organism. Among fishes, larvae can be distinguished into yolk-sac larvae, pre-flexion larvae, flexion larvae, and post-flexion larvae.

Leukocyte Formed element in the blood that engages in phagocytic action toward germs or foreign substances, exhibiting amoeba-like wandering movement in eukaryotic cells and hematoceles.

Life cycle "Life history" refers to the characteristic process of a species from the organism's genesis to the appearance of second-generation offspring, while "life cycle" refers to the cycle of life from the organism's reproductive cells to those of the next generation. Processes indicated in the life cycle include alternation of generations, changes in nuclear phase, fertilization, and maturation division (meiosis).

Ligase Enzyme that forms a phosphodiester linkage between the 3' end of one DNA fragment and the 5' end of another while the fragments are forming base pairs with a model strand.

Light-dependent reactions Photosynthetic processes involve the use of light energy to split water and produce ATP and NADPH.

Light-harvesting complexes (LHCs) Aggregates of photosynthetic pigments, photoprotective pigments, and pro-teins that function to harvest light during photosynthesis.

Light-independent reactions Photosynthetic processes that use ATP and NADPH to transform carbon dioxide into organic compounds.

Linkage Phenomenon in which genes governing two or more different traits (non-alleles) are inherited together because they are present on the same chromosome.

Lipopolysaccharides (LPS) In general, polysaccharides having attached lipids; specifically, toxins produced by certain bacteria and bloom-forming cyanobacteria that can cause fever and inflammation in humans.

Lithic algae Algae that grow in some way in association with rocks or stony substrates such as concrete.

Litter size The number of offspring produced per birth. Animals such as cattle and horses were previously known as "single-birth animals" because they typically bear only one offspring at a time, while swine, dogs, cats, and other animals were known as "multiple-birth animals" because they bear multiple offspring at a time.

Localization Determination of the site or location of a synchronizer or lesion. Limits the formation of pairs or chiasmata in one part of the chromosome during the pachytene stage.

Lymphocyte Blood cell around 6 μm in diameter that is present in the blood, lymph, and lymphatic tissue. Cells are responsible for immune functions, recognizing specific antigens and initiating an immune response by the body or its fluids.

Lymphoid leukemia Form of leukemia that occurs in lymphocyte cells within the blood and marrow.

Macrogametes Relatively large reproductive cells consisting of female gametes produced from macrogamonts; believed to be female. Fertilization is completed and a zygote formed following fusion with a male gamete. (megagamete, oocarp↔microgamete)

Macrophage Large lymphocyte with phagocytic functions.

Magnetic field Force field operating between magnets, between electric currents, or between magnets and electric currents.

Magnetic resonance imaging (MRI) Computerized method in which a type of atomic nucleus is placed in a static field and electromagnetic wave energy is applied at a certain wavelength to generate resonance; the resulting energy emission signals are used to create a cross-sectional image. Whereas X-ray CT scans generally show morphological information, MRI is distinguished by the ability to detect and image chemical and functional changes in tissue.

Magnetization Process by which an object placed in a magnetic field becomes a magnet through magnetic induction.

Marek's disease virus Form of herpes virus that causes lymphoma in chickens.

Marine farm Facility for the artificial production, management, and cultivation of juvenile marine plants and animals or seedlings.

Marine snow Particulate aggregates of algal cells or their remains, fecal pellets, bacteria, and heterotrophic protists, held together by mucilage, that are important in the transformation and transport of organic carbon to deep-ocean sediments.

Mass doubling time When a logarithmic phase is maintained through three to five transplants following overnight culturing of a bacterium, the cells double over defined periods of time. The time needed for them to divide is known as the doubling time or generation time. These obviously vary with culturing conditions, and can be identified with the formula $G = t \log2/(\log b-\log a)$, where G is the time taken for one cell to divide into two, a is the initial number of bacteria, and b is the number of bacteria after t minutes. For example, the B/r strain of colon bacillus has a doubling time of 50 minutes, while the *Bacillus subtilis* W23 strain averages 55 minutes; other values include 9.8 minutes for marine *Vibrio* genera and 10 minutes for lactic acid bacteria present in the stomachs of cattle. Yeast has a doubling time of around 60 minutes; for the tuberculosis bacteria, it is six hours.

Mass spectrometer Mass analysis equipment used to separate the ions produced from specimen ionization through their mass/charge ratio.

Matrotrophy The provision of nutrients by cells of the parental generation to cells of the next generation that have been retained on the maternal body.

Medulla Cells or tissues occurring in the center of a fleshy multicellular algal body.

Meiosis Process by which diploid reproductive cells divide into haploid sex cells.

Melanoma Highly malignant tumor that forms in melanin-producing melanocytes and nevocytes (verrucous melanoma cells). Fast-growing and capable of metastasis throughout the body.

Meristem A cell or group of cells that is capable of repeated division and thus adds to the number of cells in a body.

Meroblastic cleavage Form of cleavage in which the blastomere boundaries are incomplete.

Mesoplankton A class of plankton consisting of organisms that are between 0.2 mm and 2 mm in diameter.

Messenger RNA (mRNA) RNA molecule complementary to a single strand of cellular DNA; functions to transport genetic information from the chromosomes to the ribosomes.

Metabolic antagonist Substance that functions antagonistically toward substances involve in an organism's normal metabolism.

Metabolite Any substance produced as a result of metabolism or metabolic processes within the body.

Metabolomics Discipline concerned with the general analysis of the production and transformation of metabolites as a result of protein functions within the body.

Metalimnion The layer of water lying beneath the upper epilimnion that is marked by a steep decline in temperature and increase in density of the water.

Metamerism Form of similarly sized partitions making up an animal's body.

Metaphyton Floating algae that have become detached from substrates where they were attached as periphyton. Metaphyton may make up a significant part of the species diversity of the phytoplankton.

Microbial food web The complex food web consisting of microorganisms less than 20 μm in longest linear dimension and including archaea, bacteria, cyanobacteria, small eukaryotic algae, and small protozoans.

Microbial loop The original term for the microbial food web that recognized the importance of secondary production by archaea and bacteria and their consumption by heterotrophic nanoflagellates and small ciliates to the larger aquatic food web.

Microinjection Method in which a micromanipulator is used to inject tiny amounts of a substance under a microscope.

Microplankton A class of plankton consisting of organisms that are between 20 μm and 200 μm in diameter.

Micropyle Small hole on an egg membrane through which a sperm can penetrate. While micropyles often serve as a passage for the sperm during fertilization, sperm may also pass through the membrane regardless of the micropyles. Sometimes serve as passageways for nutrients during the egg's growth phase.

Microsatellite DNA sequence typically consisting of two to three base pairs repeating roughly 15 to 40 times (indicated as (CA)n for C and A repeating n times). Commonly found in eukaryotes; in the case of (CA)n, around 5×10^4 and 5×10^5 repetitions are respectively found in humans and mice per haploid gene with the gene's regulatory region or introns. Recently, this sequence has entered wide use as a polymorph DNA marker for gene mapping. This is because polymorphs exhibit a high degree of variation with DNA strand length, allowing for fast and easy polymorph detection with PCR, and since dozens of replications are theoretically present per cM in the genome, it has become possible to draw the detailed maps (1cM maps) needed to label genes per cM. Microsatellites are also used for PCR selection of the yeast artificial chromosomes (YAC) and bacterial artificial chromosomes (BAC) that serve as materials for the physically mapping needed for positional cloning.

Microzooplankton Zooplankton in the size range from 20 μm to 200 μm. Microzooplankton are the main consumers of the microorganisms in the microbial food web.

Minisatellite DNA Form of repeated sequence present in chromosomes. The term refers to small-scale satellite DNA; satellites have long been known to exist as large repeating units in the centromere regions. Minisatellites are even smaller than microsatellites; as an example, a repeating CA sequence is referred to as microsatellite DNA. Microsatellites repeat in units ranging from roughly 10 to 50 base pairs, with total lengths ranging between 2 and 30 kbp. They are characterized by hypervariability, with several alleles effectively present.

Southern blotting using polymorphic minisatellite probes is known as DNA fingerprinting, which is used in identification of individuals and detection of family relationships.

Mixotrophy A form of nutrition in which both autotrophy and heterotrophy may be utilized, depending on the availability of resources. (*see also* phagotrophy)

Molecular barcodes DNA sequences accumulated in online databases that can be used to identify organisms in natural samples.

Molecular biology Discipline concerned with the systematic explication of the structure of organismal molecules and the biological phenomena resulting from their interactions.

Monochromatic radiation The visible light, ultraviolet rays, and X-rays emitted from typical light sources and X-ray generators contain mixtures of various wavelengths. With electromagnetic radiation energy, shorter wavelengths indicate higher energy, which results in changes in physiochemical reactions. Because clear wavelength dependency is present in some reactions, it is advisable to use only a specific wavelength when conducting experiments with electromagnetic radiation. Electromagnetic radiation from one specific wavelength is called monochromatic; visible light and UV rays may be obtained by using filters, prisms, or diffraction grating. Since X-rays cannot be rendered monochromatic with filters, special X-rays and gamma rays with specific wavelengths are used from the outset.

Monoclonal antibody Antibody that reacts to only one antigen determinant. Monoclonal antibodies have been obtained experimentally in test tubes from hybridomas made by fusing antibody-producing cells and myeloma cells.

Monogenea Class of flatworm in the phylum platyhelminthes. Possesses adhesive organisms at the front and back of its body; while it does not have cilia on its epidermis, it does possess tiny villi. It follows a life cycle with a single host and is an external parasite chiefly of fish, although it is found in rare instances as an internal parasite of amphibians and reptiles.

Monogenic Offspring of only one sex. *Drosophila affinis* possesses a line consisting only of males; even when mated with other lines, it produces only males and no females. The line is believed to possess a gene that kills off sperm possessing X chromosomes. In D. *bifasciata*, females that bear only female offspring have been reported.

Monohybrid Hybrid state between different (homo) parents for one pair of alleles only. The hybrid exhibits dominant or intermediate traits, depending on whether the dominance relationship between alleles is complete or incomplete. For the second hybrid generation, the ratios are 3:1 for complete dominance and 1:2:1 for incomplete dominance. = monogenic hybrid

Monophyletic Used to describe a group of organisms that have descended from a single common ancestor.

Morphogenesis Used when structural characteristic of internal tissue are differentiated or when explaining the origins of a particular form.

Morphological species concept The use of structural differences and similarities to distinguish species and classify them.

Multicellularity Composed of multiple adherent cells that have the capacity to communicate with each other and to specialize.

Mutant Organism possessing at least one mutation in its chromosomal DNA. Because mutations that do occur in the DNA cannot be detected unless expressed through changes in the organism's genetic traits, the term typically refers to organisms in which mutations have occurs or the phenotype has changed. Examples of mutated traits include colony forms, antigen structures, biochemical properties such as nutrient demands, drug resistance, and pathogenicity. Mutants can be obtained through natural mutations, but are also produced artificially through the inducement of mutation by various chemical or physical mutagens. They are used as gene labels in the genetic study of microorganisms: although the emergence of pathogenic strains and drug resistance in pathogenic bacteria have frequently resulted in serious problems for research and application. = variant

Mutation frequency The ratio of mutation for a specific trait within a population of cells or organisms (number of mutations/total population). For a cell population, the mutation frequency is determined by three variables. One is mutation rate, or how often mutations arise. Another is mutation emergence time: because mutations happen randomly, it cannot be predicted when or where they will occur. In the case of cell culturing, mutation frequency will increase when mutation occurs early during growth and decrease when it happens toward the end. The third variable is growth rate: mutation frequency changes depending on whether the mutation's growth rate is faster than that of the parent cells.

Mutation lag Phenomenon in which a mutation in a gene is only detected after several generations of cell division.

Mutation load One of the key factors in genetic load, this represents decreased fitness of a population due to a mutation as a value relative to maximum fitness.

Mutation rate Indicates the average frequency of a specific mutation (e.g., his→his+) per cell per division (generation), or the likelihood of mutation occurring of the course of a cell's division into two cells.

Mutation Permant change in genetic material that is passed down to offspring over generations. Mutations can be distinguished into gene mutations and chromosome mutations or irregularities.

Myeloma Malignant tumor occurring in the bone marrow due to a malignant mutation in antibody-producing Blymphocytes. Malignant cells enter the bone marrow and destroy tissues there.

Nanoplonkton A class of plankton consisting of organisms that are between 2 μm and 20 μm in diameter.

Natural killer cell Form of lymphocyte found in humans and other animals that recognizes and kills cancer cells.

Nauplius Term for the earliest juvenile form hatched from a crustacean egg (e.g., crab, shrimp, or barnacle).

Nemathocyst In certain dinoflagellates, a harpoon like ejectile structure that is morphologically distinct from the more common trichocyst.

New renewable energy Term referring to energy sources produced through changes to existing fossil fuels or to renewable energy, including sunlight, water, geothermal heat, and biomass.

Nitrogen fixation The process by which many cyanobacteria (and some non-photosynthetic bacteria) transform nitrogen gas into ammonia, fixed nitrogen.

Nitrogenase In cyanobacteria (and some other bacteria), the holoenzyme that performs nitrogen fixation conversion of diatomic N_2 gas into ammonium ion.

Node A site on an algal body from which branches arise.

Nomadism Form of collective lifestyle found in mammals. In contrast with a settlement-based lifestyle, it involves traveling from one place to another for daily survival.

Nuclear magnetic resonance spectrometer Spectrometer consisting of a magnetic section, an electromagnetic wave emission section, a sample probe wound in transmission and reception coils, and an amplifying recorder.

Nuclear-associated organelle (NAO) In red algae, a ringshaped structure that occurs at the spindle poles during cell division.

Nuclein Nucleoprotein degradation product intermediate between natural nucleoproteins and nucleic acid. Nucleic acid and base content varies according to nuclein type.

Nucleomorph In some algae having plastids of secondary origin (cryptomonads, chlorarachniophytes), a plastid based, double-membrane enclosed structure containing DNA arranged in small chromosomes and other features suggesting origin from a eukaryotic nucleus.

Null alleles Alleles that do not result in PCR products due to mutations in their primer section.

Okadaic acid A toxin produced by certain dinoflagellates that inhibits serine- and threonine-specific phosphates (which occur widely in eukaryotes); the cause of diarrhetic shellfish poisoning in humans.

Oligonucleotide probe A short piece of DNA that specifically binds the ribosomal RNA of particular species.

Oligosaccharide Sugar in which 2 to 10 monosaccharide residues are linked in straight or branching chains by glycoside bonds.

Oligotrophic aquatic systems Waters that are low in nutrients such as phosphate and combined nitrogen and consequently low in primary productivity and biomass but typically high in species diversity.

Oligotrophic Refers to a state lacking the necessary nutrients for the growth of aerobic photosynthetic organisms.

One gene-one enzyme hypothesis Hypothesis that single genes influence phenotype by participating in the formation of a single enzyme and governing its characteristics.

Oogamous Involving a larger nonmotile egg cell and a smaller motile sperm cell.

Oogomous reproduction Sexual reproduction involving egg and sperm.

Oogomy Sexual reproduction involving syngamy of a small flagellate male gamete and a larger, nonflagellate (or only transiently flagellate) female gamete.

Open reading frame (ORF) Portion of a DNA chain that is transformed into a codon defining a particular amino acid.

Organelle Structural unit within eukaryotic cells that performs various functions; akin to the various organs of the body.

Organotin compound Organic metal compound consisting of tin with butyl groups attached. Organotin compounds were previously used as PVC polymer stabilizers, plastic additives, industrial catalysts, pesticides, germicides, and wood preservatives. In the 1960s, they were found to be effective antifouling agents, and they are now widely used in anti-fouling paints for vessels and farm fishing gear.

Osmotic pressure Pressure that causes water or another solvent to from a solution with a low solute concentration to one with a high solute concentration through osmosis. Similar to the hydrostatic pressure that must be used to apply solvent to the more concentrated solution to stop the flow of water or solvent.

Osmotrophy A form of nutrition in which dissolved organic carbon is imported from the environment into cells.

Osteitis fibrosa cystica Condition that appears with hyperparathyroidism, in which bones undergo fibrous transformation that produces an inflammation reaction. Parathyroid hormones promote the breakdown of calcium from the bones and its absorption from the gastrointestinal tract, resulting ultimately in higher calcium concentrations in the blood. Due to calcium loss, bones develop holes and bend; cysts sometimes form as well.

Oxygenic photosynthesis Photosynthesis that generates oxygen by breaking water.

Palindrome Section of DNA in which a base sequence has an identical structure when read from left to right or from right to left.

Pallium A sheetlike extension of cytoplasm, also known as a feeding veil, that is produced by some phagotrophic dinoflagellates for the purpose of capturing and digesting prey.

Palmella A stage produced by algae genetically capable of producing flagella consisting of nonmotile cells embedded in an amorphous mucilaginous matrix.

Paraffin compound General term for paraffin wax or liquid paraffin made up of paraffin hydrocarbons or highly saturated hydrocarbons; characterized chiefly by weak reactivity and resistance to chemical medications. Present in heavy oil fractions and used for candles and electrical insulators.

Parasporangium In red algae, a sporangium that produces many spores; not equivalent to a tetrasporangium.

Paraxonemal rod A column of protein that stiffens the hair-covered flagella of euglenids and kinetoplastids.

Parthenogenesis Emergence of an organism from an egg without fertilization. May occur in the natural state or artificially as the result of experimental measures.

Parthenogenesis Production of a new individual from an unfertilized gamete.

Peduncle In dinoflagellates, a microtubule-containing tubular extension of cytoplasm that can be extended attach to prey cells or tissues and either extract their contents or engulf them.

Pelagic Living in open ocean waters rather than near shore coastal or inland waters.

Pellicle (a) In euglenoids, an often flexible, protein containing surface layer consisting of interlocking, helically oriented strips that occurs just inside the cell membrane; (b) in some dinoflagellates, a cellulosic a sporopollenin-containing layer beneath the theca that remains in place when the theca is shed, (c) the surface layers of some other algae.

Peptidoglycan A carbohydrate substance cross-linked peptides that forms much of the cyanobacterial cell wall

Perialgal vacuoles Membrane-bound vacuoles that protect enclosed endosymbionts from digestion within host cells.

Periaxial cells (pericentral cells) In uniaxial red algae cells that are cut off from the main axis that occur in whorl that surrounds the parental axial cell.

Periphyton Organisms that occur on the surfaces of plants, algae, and inorganic substrates in shallow benthic or nearshore littoral habitats.

Periplast In cryptomonads, the cell covering, consisting of intersecting plates lying beneath the cell membrane.

Peroxisome An organelle bound by a single membrane that occurs in the cytoplasm of embryophyte cells and those of some eukaryotic algae and that contains characteristic enzymes such as glycolate oxidase and catalase; this term is also sometimes used more generally as a synonym for microbodies-small single-membrane-bound structures that occur in some algae and contain catalase but not glycolate oxidase.

Phage Short for bacteriophage. Simplest organisms, consisting mostly of proteins and nucleic acid.

Phagocytosis Consumption by leukocytes and other macrophages to eliminate bacteria or other foreign matter.

Phagopod In dinoflagellates, a tubular structure lacking microtubules that can be formed on cells for the purpose of attaching to and extracting the contents of prey cells.

Phagotrophs Organisms thot use the process of endocytosis to engulf prokaryotes and/or small algae into cellular digestion structures known as food vacuoles.

Phagotrophy (phagocytosis) A form of nutrition in which particles such as cells are ingested by protists via invagination of the cell surface.

Phenetic Refers to actual expression of genetic traits, in contrast with genotype.

Phenotypic expression Expression and functionalization of traits (phenotypes) from genetically stored information (genotype) at the organismal level, including morphological and biochemical characteristics.

Phosphor dots Small fluorescent particles observed on a CRT screen used for imaging.

Photoautotroph An organism that obtains its organic nutrients by means of photosynthesis; obligate photoautotrophs are restricted to this form of nutrition.

Photosynthetic bacteria General term for prokaryotes that engage in photosynthetic activity to fix carbon with light energy, but differ from plants in not producing oxygen or using water as an electron donor.

Phototaxis Property of moving toward a light source in response to light stimuli. Examples include Pacific sauries, sardines, anchovies, mackerels, squids, and green laver spores.

Phragmoplast In post-mitotic cells of certain green algae and embryophytes, an array of microtubules arranged perpendicularly to the plane of division that, together with coated vesicles and endoplasmic reticulum, is involved with cytokinesis by centrifugal cell plate formation.

Phylogenetic tree A treelike diagram that represents a hypothesis regarding relationships of a group of organisms.

Physiologically active substance Substances in trace amounts with a large influence on an organism's functioning (physiology).

Phytoplankton Floating or swimming microscopic algae.

Picoplankton A class of plankton consisting of organisms such as bacteria and certain small eukaryotic algae that are between 0.2 μm and 2 μm in diameter.

Pili Threadlike structures located on the surfaces of at least some cyanobacteria (and many other Gram-negative bacteria) that are associated with twitching motility exhibited by certain cyanobacteria.

Placebo Substance with no pharmacological effect or slightly similar effects that is administered as a control when testing the effects of a clinic medication. Used to rule out possible psychological effects from medicinal administration.

Planing Cosmetic surgery technique in which an unsightly section of skin is ground down to minimize scar formation and promote epithelial regeneration.

Plankton General term for organisms that survive by floating in seawater, lakes, wetlands, and rivers. They either lack the ability to swim or have only minimal ability, leaving them unable to maintain their position when traveling against a current.

Plankton Microscopic organisms that are suspended or swim in the water column.

Plaque Round, transparent section that emerges over generations after a virus infects a location where cells have attached together to form a plate, killing or dissolving the immediately neighboring cells.

Plasmid Gene that is not part of the cell nucleus; examples include mitochondrial and chloroplast DNA.

Plurilocular sporangium In brown algae, a reproductive structure that is subdivided by cell walls into numerous small chambers (locules), each of which produces a single flagellate cell.

Polar body Small cell emitted during animal meiosis as an egg cell undergoes maturation division into primary and secondary egg cells.

Pollination Attachment of pollen to a stigma. Includes self-pollination, or the transmission of pollen from the same organism or with the same genes from stamen to stigma, and cross-pollination, or pollination with pollen from an organism with different genes.

Polyadenylic acid Adenylic acid sequence generally present at the 3' end of mRNA in eukaryotes.

Polygene Prefers to genes in a polygenic system involved in the expression of traits; while individual actions are very weak, multiple individuals act in concert and can be quantitatively measured.

Polymerase chain reaction (PCR) Method for the artificial amplification of genes. In PCR, DNA synthase is used to produce several clones of a DNA fragment all at once, allowing geometric amplification of genes at a particular site to be achieved in a short time with a very small amount of DNA.

Polynucleotide Linear nucleotide sequence connecting the 3' position on one nucleotide sugar to the 5' position and phosphate group on a neighboring nucleotide.

Polyploid Organism with two or more pairs of genomes.

Polyploidization Formation of a polyploid chromosome number.

Polyploidy Phenomenon in which the number of chromosomes in an organism is a multiple of the number in normal organisms.

Polysome State in which protein synthesis is performed by multiple ribosomes bonded to a single mRNA molecule.

Population genetics Discipline concerned with statistical analysis of genetic change among populations to examine them in connection with species evolution and methods of improving varieties.

Portable plastid hypothesis The idea that secondary red plastids could have originated more than once because they retain a higher proportion of the original primary endosymbiont's genome than do green plastids.

Positional cloning Method of cloning a gene by using information related to its position on the chromosome. When genes from higher plants and animals are being clones, standard gene markers and DNA polymorphs are first used (RFLP, VNTR, CA repeat), and a map of target genes within the vast genome DNA is drawn on the chromosome. Next, a detailed physical map of the area near that genetic locus is drawn. In cases of multiple mutants with irregularities in the target gene (patients with multiple conditions resulting from the same gene), the DNA structure near the target gene is examined for shared changes.
Identification of the target gene is made easier when chromosome irregularities are found in deletions or transpositions within the region including the target gene. When a DNA structural change is detected, the full genetic sequence must be compared individually with a known DNA marker to determine the target gene.

Practical salinity units (psu) The number of grams of salt contained in a liter of water. A liter of seawater with 35 grams of salt has 35 practical salinity units.

Precession Movement in which the axis of rotation for a rotating body gradually changes direction. Examples include the symmetrical movement of a top, the earth's revolution, and rotational movement in atomic nuclei and molecules. With phenomena resulting from precession, the term nutation is often used to refer to the regular portion and the term precession only for the irregular portion. In Larmor precession, nutation exclusively occurs.

Predators Organisms that attack and feed on other organisms, which are killed and consumed. For example, adult copepods are predators on protozoa and the juvenile stages of other aquatic crustaceans.

Primary endosymbiosis Incorporation of a free-living prokaryote into a host eukaryotic cell, with subsequent transformation into an organelle.

Primer Large molecule that initiates the synthesis of other large molecules in polymerization. Polynucleotide chain that serves as the starting point for the polynucleotide chain expansion in nucleic acid synthesis reactions. Possesses shared bonds with the reaction product.

Probe Marker fragment of nucleic acid with a complementary nucleotide sequence to the gene or gene sequence to be detected in hybrid formation experiments.

Prokaryote Organism that possesses no nuclear membrane within cells and lacks mitochrondria, chloroplasts, endoplasmic reticulum, and organelles surrounded by a Golgi body membrane. Photosynthesis and oxidative phosphorylation reactions occur in the cell membrane.

Promoter DNA region that specifically bonds with RNA synthase to initiate transcription. Basic size is around 60 base pairs in *E. coli* promoters. Different promoter types may include site with a transcription promotion factor, which differs from one promoter to the next.

Proteomics Discipline examining the interactions of all proteins within the cell.

Proteomics The study of types and relative amounts of proteins produced by a cell or other defined structure under defined conditions.

Protoplast Constituent of all cells, not including the cell wall. Refers to protoplasm surrounded by a cell membrane; in animals cells without cell walls, the cell itself is a protoplast.

Protozoa Terms for multicellular eukaryote observable only under a microscope.

Pseudopod Organelle involved in the locomotion of amoeboid cells; general term for temporarily formed protoplasm protuberances.

Pure line Line in which all genes are homo. A pure line has no genetic mutations; any mutations between organisms in a pure line is believed to be the result of environmental influences.

Pyrenoid A proteinaceous region in the plastids of many types of algae; known in some cases to contain rubisco and commonly associated with formation of storage compounds.

Quantitative character Quantitatively expressed characteristic (such as yield) found in organisms exhibiting continuous mutation.

Radial centric diatoms Fossil and modern diatoms having radially symmetrical valves of more or less circular outline.

Radial cleavage Form of cleavage found in animals in which blasts form radiating arrangements with respect to the egg axis. In cleavage, vertical cleavage (the meridional cleavage plane) is divided into 180°, 90°, and 45° types according to the partitioning of the mutual angle with the cleavage plane as it passes through the cleavage axis. In this case, the entire horizontal cleaage plane forms a perpendicular with the cleavage axis. Eggs exhibiting this can be found in sponges, coelenterates (cnidarians), echinoderms, protochordates, and amphibians.

Radioallergosorbent test (RAST) Blood testing method in which the presence of allergy-inducing factors is determined by the value of immunoglobulin E antibodies for a specific allergy antigen. In this approach, the specificity of a patient's IgE antibody is determined from the coefficient when an IgE antibody that generates a reaginic allergy is made to react with an allergen adsorbed to a solid phase, and a 125I-marked anti-IgE antibody is further made to react with the IgE antibody. Serological diagnostic method for determining allergens in a patient.

Rate-limiting factor Limiting factor that controls the rate of reaction for an entire system. Include a series of enzyme reactions; when the rate of the whole is determined by the slowness of a reaction phase due to a specific enzyme, that stage is considered the rate-limiting stage and the enzyme the rate-limiting factor. In organism reaction series, changes in the rate-limiting stage occur according to conditions, forming part of the regulation mechanism.

Recombination of genes Field of genetic engineering. In contrast with previous genetic recombination methods involving crossbreeding, it refers to the generation of activity by using micromanipulation to introduce genes from a different species or genus of organism (or artificially synthesized genes) into an organism.

Red algae The informal term for a lineage of algae lacking flagella and centrioles, whose plastids have an envelope consisting of two membranes and contain phycobilin accessory pigments.

Refractory carbon A form of organic carbon that is resistant to microbial, chemical, and physical degradation.

Repeated sequence Existence of an identical or very similar base sequence two or more times on a chromosome.

Replication origin Specific region of the chromosome where DNA replication is initiated. In eukaryotes and prokaryotes alike, chromosome replication occurs only once at a specific time in the cell cycle, and the time and frequency of replication initiation is strictly regulated by the replication origin and the interactions of the protein complexes operating on it.

Replication Self-copying of genetic material. In biosynthesis of genetic material, a single parent molecule serves as a model to create two offspring molecules with identical structure and function. This occurs as a result of semiconservative replication.

Resin Noncrystalline solid or semisolid made up of organic compounds and their derivatives. Divided into natural resins and synthetic resins (plastics), the latter of which are further subdivided into those produced at the time of petroleum refinement and those resulting from synthesis of pure monomers.

Respiratory syncytial virus (RSV) Member of the *Pneumovirus* genus in the family Paramyxoviridae. Containing a negative single strand of genetic material, it causes respiratory infection in children.

Restriction enzyme Endonuclease that recognizes double-stranded DNA and specific base sequences and truncates double-stranded DNA. Categorized into Types I, II, and III according to the factor demands for enzyme activity and the form of truncation. Found widely among bacteria, restriction enzymes come in a great variety of types, differing from species to species in enzyme types and sequences recognized.

Restriction fragment length polymorphism (RFLP) Each species has its own characteristic base sequence of chromosomal DNA. Close observation shows slight differences between individuals of the same species, which is known as DNA polymorphism. Due to differences in the lengths of fragments produced when these DNA polymorphs appear at restriction fragment recognition sites, they can easily be detected through Southern blotting; this is known as restriction fragment length polymorphism.

Retinoblastoma Malignant tumor that occurs in the retina; chiefly affects retinoblasts.

Reverse transcriptase Enzyme contained in the particles of various retroviruses, which synthesizes a complementary DNA nucleotide sequence using single-stranded RNA as a model.

Rhizaria A eukaryotic supergroup that includes the algae known as chlorarachniophytes as well as many types of non-photosynthetic protists.

Rotifers Aquatic animals characterized by a ring of cilia at their anterior ends. Rotifers mainly employ filter feeding to graze on algae and other microorganisms, but some practice raptorial feeding.

Sarcoma General term for malignant tumors that occur in non-epithial tissues (typically not including bone marrow lymph tissues), which serve as supporting tissues for organisms.

Scytonemin An indole-alkaloid that may color cyanobacterial sheaths yellow or brown and protects cyanobacterial cells from UV radiation.

Secondary endosymbiosis The incorporation of a photosynthetic eukaryote, whose plastid was derived from a prokaryote, into a eukaryotic host cell.

Segregative cell division In siphonocladalean green algae, cleavage of the multinucleate protoplast into spherical units that then expand, develop walls, and function as independent regions.

Selenious acid Chemical formula H_2SeO_3. Colorless, moisture-absorbing crystals forming hexagonal columns; toxic alkaloid agent that is more weakly acidic than selenic acid. Used as an oxidizing agent.

Self-fertilization Fertilization within sperm and egg that occurs within an animal possessing both male and female characteristics. Property of hermaphroditic plants that survive through self-fertilization and are incapable of producing offspring.

Semicell Portion of Desmidiaceae somatic cells, in which two semicells are separated by a constrained section and pupilla and joined by a narrow isthmus with a nucleus. Characteristics in genus and species classification include several protuberances and differentiated shapes on its surface.

Semiconservative replication Refers to double strand of DNA formed through replication, consisting of one strand from the synthesis parent cell and one newly synthesized strand.

Serratia marcescens Intestinal bacteria from the genus *Serratia*. Gram-negative, mesophilic aerobic bacterium that produces an insoluble red pigment in artificial medium and foods.

Sex chromosome In organisms that differentiate between male and female, these chromosomes exhibit sex-based differences in form and number for the same autosomes in males and females.

Sexual reproduction Form of reproduction in which two gametes combine or are fertilized into a zygote. Zygotes grow into new organisms or produce asexual reproduction cells.

Sheath A layer of polysaccharide mucilage on the body of an alga, particularly filamentous cyanobacteria.

Shuttle vector Complex plasmid produced to allow for replication in any part of two types of cells, even when their form of DNA replication differs.

Simian virus 40 (Simian vacuolating virus 40) SSV or SiSV for short. RNA virus isolated in 1971 from naturally occurring tumors in the woolly monkey. Fibrosarcomas form in monkeys within around one month of infection, and fibroblasts undergo transformation in culturing systems. Growth-defective virus that requires another virus to achieve growth. The virus genome consists of around 5,300 bases, with the roughly 1,000-base cancer-causing gene sis near the 3' end. The only example among primates of an RNA tumor virus with a cancer-causing gene.

Single nucleotide polymorphism (SNP) The human genome consists of some 3 billion base pairs. Individual comparison shows differences in 0.01% of the base sequence, which account for racial and other individual differences, as well as diseases.

Single-cell culture Aseptic culturing method used not only with unicellular organisms such as yeasts and bacteria, but also with single cells extracted from multicellular organisms.

Siphonaceous coenocytes Algal bodies characterized by relatively large, multinucleate cells lacking cross-walls except during the formation of reproductive structures.

smRNA RNA generally consisting of 200 or fewer bases. Term is typically used in connection with eukaryotes, although it is often used to exclude tRNA and rRNA. A portion is known to be involved in RNA editing.

Somatic cell division Biological phenomenon in which a single cell divides into two, increasing the number of cells.

Somatic cells In volvocalean green algae, body cells that are not capable of cell division.

Southern blotting Method in which radioactively labeled DNA or RNA probes are used to detect complementary base sequences in mixtures of DNA restriction enzyme fragments.

Specific growth rate Growth rate of a cell population by cell or unit cell quantity; calculated as $(1/x)(dx/dt)$. T = time, x = cell number or quantity.

Spectrophotometer Device that measures the intensity of light at different wavelengths. Consists of a component for splitting into monochromatic light and a component for quantitatively measuring the intensity of that light; mainly used to calculate reflectivity and transmission.

Spermatia The male gametes of red algae; no flagella are present.

Spindle fiber Thin, threadlike cell organelle that emerges from either end of the cell during division. Attaches to the chromosome's centromere to pull the chromosome toward either side.

Splicing Process occurring in the primary RNA transcription product of genes, which include introns containing no genetic information. The intron sections are removed, leaving only the exons that contain genetic information, which are linked together into mRNA for translation into a single polypeptide chain.

Spore In algae, a reproductive spore that may or may not be produced via meiosis; in land plants, cells generated by meiosis that become coated with sporopollenin.

Stabilizer Substance added when storing food to prevent chemical or physical transformation. Included as a way of maintaining product longevity, appearance, or effectiveness.

Standing crop Total number of surviving organisms in a specific population at a given time.

Sticky end DNA end section in which a few bases from the end of DNA molecules or fragments are linked into a single chain, forming a complementary base sequence to the chain formed by bases at the end of a molecule with a different base sequence. Also referred to as a complementary end or cohesive end.

Stratification Formation of a surface layer of warm water over deeper, cold water as a result of density differences that develop during warm-season heating.

Subaerial algae Algae that grow on the surfaces of substrates, where they are exposed to the atmosphere and sunlight.

Subunit Basic component in which proteins are joined by several basic unshared bonds to express some biological function.

Surfactant Substance that adsorbs to the surface in a watery solution and reduces its surface tension. Typically a substance that is amphiphilic at both ends, including an oleophilic and hydrophilic group on the same molecule.

Synapsis Phenomenon in which homologous chromosomes join during meiosis.

Termination codons Three codons that signal the end of a polypeptide chain in protein synthesis (UAA, UAG, and UGA).

Terpene Organic compound found in several plants, consisting of a multiple of five carbons ($5n$, where $n \geq 2$). In biosynthesis terms, a substance derived from a precursors consisting of n isoprenes or isopentanes.

Tetraploid Polyploid organism with four times the basic number of chromosomes.

Textile printing Method of applying beautiful color images to fabric and other textiles; a form of mass production of pattern dyes.

Thelytoky Form of parthenogenesis in insects in which only females are produced.

Thermotropic liquid crystal When a substance of unitary composition that typically exists in crystalline state at low or room temperature is slowly heated, it dissolves into a turbid, viscous liquid; when heated further, it becomes a transparent liquid. These may be classified into different types based on the molecular arrangement exhibited in the liquid crystal state with changes in temperature: nematic (threadlike form in which the major axis of a long molecule extends in a particular direction), smetic (form in which long molecules with their major axis extending in specific directions combine to form sheets, which are overlaid in layers), cholesteric (liquid crystal in which the molecules rotate in a spiral shape), and columns.

Thylakoid A flattened, saclike membranous structure in cyanobacterial cells and plastids of eukaryotic algae and plants.

Tissue culture Culturing of cells from a multicellular organism in liquid medium.

Toluidine blue Aminodimethyl aminotoluphenazthionium chloride; useful as a dye to detect substances that are basophilic and metachromatic.

Toxins Chemicals from biological sources that kill or disable cells or organisms.

Trabecula In some ulvophycean green algae, an extension of the cell wall into the cell lumen, which provides structural support.

Transcription Enzymatic reaction in which genetic information included on a single DNA strand determines a complementary mRNA base sequence.

Transduction The transfer of genes among prokaryotes via viruses.

Transferrin Beta-globulin with a molecular weight of roughly 75,000. Iron-transporting protein that bonds with iron (III) ions from two molecules absorbed in serum, using transferrin receptors to supply the cell with the iron needed for growth and hemoglobin production.

Transformation Phenomenon in which DNA taken from a cell or virus is introduced into another cell to alter its genetic properties.

Transgenic fish Fish that has acquired new functions or phenotypic characteristics through transplantation of foreign genes into its chromosomes.

Transit peptides Amino acid sequences at the ends of proteins that foster protein uptake into cell organelles.

Transition region (transition zone) In eukaryotic flagella, the zone between the flagellum and its basal body, at the point where the flagellum exits the cell; the specific structure of this region varies among major algal lineages.

Translation Process in which genetic information contained in an mRNA molecule determines the amino acid sequence in protein synthesis.

Transposon Transferred factor; carries label genes such as those for drug resistance, which can often be found in the chromosomes or plasmids of prokaryotes. Transposons are separated genetic units characterized by the ability to move from one cell to another, resulting in rearrangement of DNA in the recipient cell; this allows for insertion into various locations in plasmids and chromosomal DNA.

Trichogyne In red algae and some charophycean green algae, the elongated apical portion of the female gamete (carpogonium or oogonium) that is receptive to male gametes.

Trichothallic growth In some brown algae, active cell division occurring in an intercalary position, in a stack of short cells located at the base of filaments.

Triplet Genetic code consisting of three consecutive bases corresponding to an amino acid.

Triploid Cell or organism with three times the basic number of chromosomes. Containing three homologous chromosomes, it is represented as 3n and is a form of polyploid created through crossbreeding of a diploid and tetraploid. Sterility often occurs in triploids due to irregularities in chromosome separation during meiosis.

Trophic cascade A chain of effects that proceeds from one level in a food web to another. An increase in phytoplankton, for example, may cause an increase in grazing zooplankton and a further increase in planktivorous fish.

Type 1 diabetes mellitus (IDDM) Diabetes is a metabolic disease in which insulin is insufficiently secreted or does not perform its normal functions. It is characterized by high blood sugar as concentrations of glucose rise in the blood. This high blood sugar results in various symptoms; glucose is excreted in the urine. Diabetes exists in Types I and II; Type I is also known as juvenile diabetes and results from a failure to produce insulin at all.

Unialgal culture A culture that contains only one algal species, though other types of organisms, such as bacteria, may be present.

Unicell A protist body type consisting of a single cell.

Unilocular sporangium In brown algae, a sporangium in which all the spores are produced within a single compartment, usually by meiosis.

Variable number tandem repeats (VNTR) Form of genetic analysis inspired by the fact that DNA base sequences in the chromosome have defined structures of repetition. Takes advantage of the differences in the lengths of amplified DNA fragments resulting from differences in the number of repeated units in the VNTR and STR components of a particular section of DNA.

Vascularization Vascularization mechanisms are divided into the formation of plexuses through the fusion of hematopoeic cell groups created by angioblasts and the formation of tubes, and the formation of blood vessels as vascular endothelial cells bud from existing vessels to enter tissue as capillaries; the term is generally used to refer to the latter.

Vector Small DNA molecule with autonomous growth capabilities that is used in DNA recombination experiments to connect and amplify donor DNA fragments truncated by restriction enzymes.

Vegetative cells Non-reproductive cells.

Vegetative propagation Form of reproduction in which part of a parent frond breaks away to form a new body; sometimes used to refer to reproduction where no particular reproductive structure is formed.

Velum Flat wing-shaped organ found in veliger larvae of mollusks, with long, stiff cilia extending to the right and left from the back of the mouth.

Viscosity A measure of the stickiness or adhesive property of a fluid. Viscosity is measured as the mass in kilograms of a substance that will flow 1 meter in 1 second. Water is more viscous at lower temperatures than at higher temperatures.

Warburg-Dickens pathway Metabolic pathway that begins with glucose-6-phosphate molecules and involves complete oxidation into one molecule of glucose-6-phosphate, carbon dioxide, water, and phosphoric acid. This pathway is involved in carbon dioxide fixing in photosynthesis, creating reduction potential with the NADPH type for interconversion of monosaccharides.

Wild strain Normal form of a lineage that is found in the natural state.

Xanthophylls Oxygen-containing, yellow-pigmented carotenoids.

X-ray diffraction Pattern of light and darkness that occurs when X-rays are reflected from the molecules of a crystal; caused by refraction phenomena due to wave nature.

Zoospore A flagellate spore.

Zoospore Form of spore involved in asexual reproduction. Possesses flagella that allow it to achieve locomotion in water; in seaweeds and lower bacteria, zoospores result from asexual reproduction of zoosporangia.

Zygotic meiosis Meiosis occurring during zygote maturation or germination

Index

© Springer Nature Switzerland AG 2019
S.-K. Kim, *Essentials of Marine Biotechnology*,
https://doi.org/10.1007/978-3-030-20944-5